ELEMENTS OF GEOMETRY

ELEMENTS OF GEOMETRY

S BARNARD
J M CHILD

New Academic Science
An Imprint of
New Age International (UK) Ltd.
27 Old Gloucester Street, London, WC1N 3AX, UK
www.newacademicscience.co.uk • e-mail: info@newacademicscience.co.uk

27 Old Gloucester Street, London, WC1N 3AX, UK
www.newacademicscience.co.uk • e-mail: info@newacademicscience.co.uk

ISBN: 978 1 78183 033 8

British Library Cataloguing in Publication Data
A Catalogue record for this book is available from the British Library

Preface

This work is entirely distinct from *A New Geometry for Schools* by the same authors, and is the result of eleven years' experience in teaching on the lines suggested by the Mathematical Association.

The chief characteristics of the new work are as follows:

Euclid's methods have been reverted to in the treatment of parallels and tangents. The theorems of Euclid, Book II, are given in a substantive form, with proofs which are reasonably short.

Propositions are arranged in groups, those of each group dealing with a definite subject.

Constructions follow the theorems on which they depend, and are preceded by practical instructions on the use of instruments.

In the proofs of theorems, every reason is distinctly stated in words, no references being given.

A description of the forms of simple solids occurs early in the work; also an informal introduction to the properties of similar triangles, followed by definitions of the Trigonometrical Ratios and suitable problems thereon.

The examples, many of which are numerical, have been devised and arranged with great care. *Wherever a real difficulty occurs, a hint is given, so that the average pupil should be able to work through the exercises without much help from the teacher.*

Many teachers will follow the recommendations of the Board of Education and *will replace the formal treatment of angles at a point, parallels and congruent triangles by oral explanations.* But it was considered advisable to insert a formal treatment of these matters.

Without hesitation, Euclid's definition of a tangent to a circle has been adopted, as most suitable for the beginner. It has been found that very few pupils can reproduce a proof of the fundamental proposition, based on the "limit definition." If, however, the teacher prefers this method of proof, he will find it given as an alternative (see Section XXIX).

Unit 6 consists of miscellaneous propositions in elementary geometry.

The usual school course is covered by Units 1–5, with the easier sections of Unit 6.

Miscellaneous examples, arranged in groups for revision, occur at suitable intervals. Many recent examination papers are also given *in extenso*.

Many examples have been taken from recent papers set in examinations for the certificates of the College of Preceptors, Oxford and Cambridge Locals, London University Matriculation, Board of Education, Intermediate Board for Ireland, Scotch Leaving Certificates, and Oxford and Cambridge Entrance Scholarships. For permission to publish these questions, thanks are due to the Controller of His Majesty's Stationery Office, the Cambridge University Press, the Senate of the University of London, the Delegates of the Oxford Local Examinations, the Cambridge Local Examinations Syndicate, the College of Preceptors, and the Intermediate Board for Ireland.

The authors are glad to acknowledge their deep obligation to Mr. W. N. Wilson of Rugby School for many valuable hints and for permission to use certain constructions already published in his book on Geometrical Drawing. Thanks are also expressed gladly to Sir Richard Gregory and to Mr. A. T. Simmons for much valuable advice and for assistance in reading the proof sheets.

S. BARNARD
J.M. CHILD

September, 1914

Publisher's Note

The book has been thoroughly read by our subject experts and efforts have been made to present an error free edition. Though the overall format is not changed but the presentation and the rubric have been amplified. More clarity is incorporated wherever necessary and the figures are redrawn for better understanding. Suggestion for further improvement shall be received with appreciation.

Contents

UNIT 3
(143–204)

The Circle

UNIT 4
(205–248)

Areas of Rectangles

UNIT 5
(249–328)

Similar Figures

Abbreviations

∴	*for*	therefore.
=	,,	is equal to.
+	,,	together with.
∠s	,,	'angles' or 'the angles.'
m	,,	parallelogram.
sq.	,,	square.
‖	,,	is parallel to.
rt.	,,	right.
>	,,	is greater than.
<	,,	is less than.
Δs	,,	'triangles' or ' the triangles.'
rect.	,,	rectangle.
sqq.	,,	squares.
⊥	,,	is perpendicular to.

These are used very sparingly in the book work of Unit 1.

UNIT 1

The Straight Line

I. INTRODUCTORY

Geometry is the science of space and deals with the shapes, sizes and positions of things.

Euclid was a Greek mathematician of the third century B.C. who wrote a remarkable book called *The Elements*. For 2000 years, the first six books of this work have been used as an introduction to Geometry. Modern textbooks contain in a modified form the subject matter of Books I–IV, VI and XI of Euclid's work.

Axioms. All reasoning is based on certain elementary statements, the truth of which is admitted without discussion: such statements are called axioms.

Euclid gave a list of twelve axioms: of these, the following relate to magnitudes of all kinds*: axioms relating to geometrical magnitudes will be stated later.

1. Things which are equal to the same thing are equal to one another.
2. If equals are added to equals, the wholes are equal.
3. If equals are taken from equals, the remainders are equal.
4. Doubles of equal things are equal.
5. Halves of equal things are equal.
6. The whole is greater than its part.
7. If equals are added to unequals, the wholes are unequal.
8. If equals are taken from unequals, the remainders are unequal.

Surface, Line, Point. The space occupied by any object (say a brick) is limited by boundaries which separate it from surrounding space. These boundaries are called **surfaces**.

DEFINITION. A **surface** has length and breadth but no thickness.

Surfaces meet (or intersect) in *lines*. For example, a wall of a room meets the floor in a straight line.

DEF. A **line** has length but no breadth. Lines meet (or intersect) in *points*.

---

* The student should read these through in order to get a clear idea of what kinds of truths are assumed as axiomatic. It is quite unnecessary to commit them to memory.

DEF. A **point** has position but no magnitude.

In practice, the mark traced by a pencil-point on a sheet of paper is called a line. But it is not a line according to the definition: for, however thin it may be, it has some breadth.

Again, if we make a dot on the paper as a mark of position, the dot is not a geometrical point, for it has some magnitude.

Straight Lines. Lines are either *straight* or *curved.* Everyone knows what is meant by a *straight line*, but it is difficult to put the exact meaning into words. Euclid's definition was as follows:

DEF. A straight line is that which lies evenly between its extreme points.

To this he added the following:

AXIOM. Two straight lines cannot enclose a space.

POSTULATE* 1. Let it be granted that a straight line may be drawn from any one point to any other point.

POSTULATE 2. Let it be granted that a straight line may be produced to any length in a straight line.

Hence it follows that

(*i*) Two straight lines cannot intersect in more than one point. For, if they met in two points, they would enclose a space.

(*ii*) One and only one straight line can be drawn between two given points. When we draw this straight line, we are said to join the points.

In order to compare two straight lines, we may suppose one of them to be taken up and placed on the other. If the lines fit exactly, they are said to *coincide.*

DEF. *Equal straight lines* are those which can be made to coincide.

Equal straight lines are said to have the same length.

The distance between two points is the length of the straight line joining them.

If O is a point in a straight line AB, lengths measured from O towards B are said to be measured in the opposite sense to those measured from O towards A.

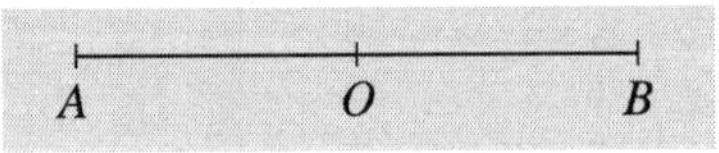

DEF. Any part of a straight line is called a **segment** of that straight line.

The Plane. In order to find out whether the surface of a board is what is called 'plane' or not, a carpenter applies a straight edge (the straight edge of his plane for instance) to the board, *in all directions*; *i.e.* he practically joins *any* two points in the surface with a straight line and sees whether this straight line is in contact with the board throughout its length.

---

*A postulate is a *demand.*

Def. A **plane** is a surface in which any two points being taken the straight line between them lies wholly in that surface.

A plane is often called a *plane surface.*

Angles. Def. Two straight lines, drawn from the same point, are said to contain an angle.

The straight lines are called the arms of the angle; their point of intersection is called the vertex of the angle.

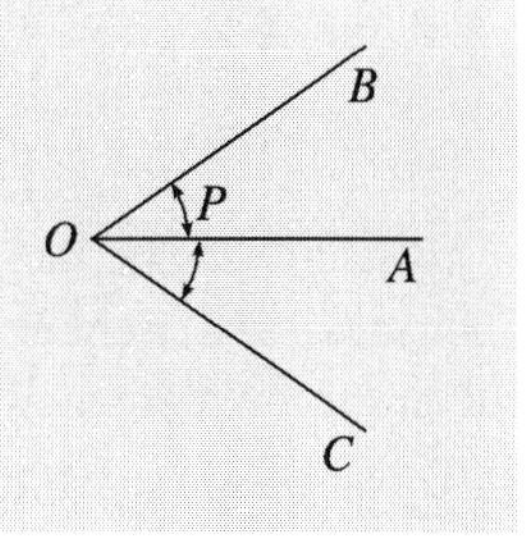

The angle contained by two straight lines *OA* and *OB* is called 'the angle *AOB*' or 'the angle *BOA*,' or, *if there is only one angle at the point*, simply 'the angle *O*.'

If a straight line *OP*, starting from the position *OA*, revolves about the point *O* as a hinge, until it reaches the position *OB*, it is said to turn through or describe the angle *AOB*.

In order to describe the angle *AOC* in the above figure, the revolving line would have to turn in the opposite direction or sense to that in which it turned to describe the angle *AOB*.

We therefore say that the angles *AOB*, *AOC* are described **in opposite senses**.

Def. Two angles are said to be **equal** when one of them can be placed so that its arms fall along the arms of the other.

Note that *the size of an angle does not depend on the lengths of its arms.*

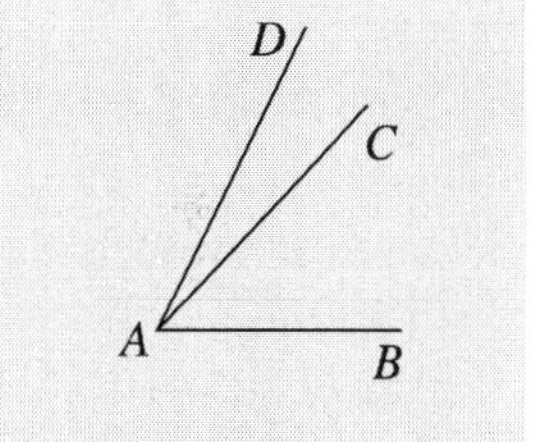

Def. Two angles (*BAC*, *CAD*), which are situated on opposite sides of a common arm and have a common vertex, are called **adjacent angles**.

The angle *BAD* is the sum of the angles *BAC*, *CAD*. The angle *CAD* is the difference of the angles *BAD*, *BAC*.

The bisector of an angle is the straight line which bisects it, *i.e.*, which divides it into two equal parts.

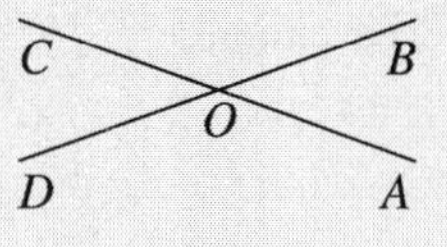

Def. If the arms of an angle *AOB* are produced through *O* to *C* and *D*, the angles *AOB*, *COD* are said to be **vertically opposite**: so also are the angles *BOC*, *DOA*.

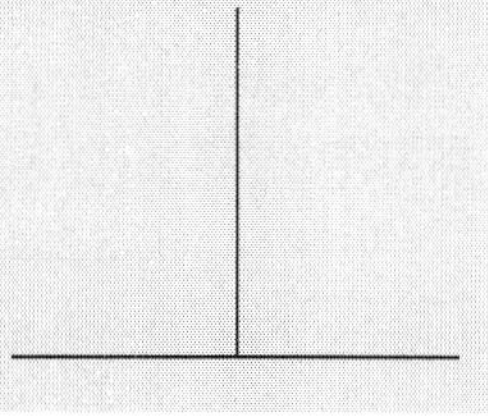

Def. When one straight line, standing on another straight line, makes the adjacent angles equal, each of these angles is called a right angle; and the straight line standing on the other is said to be **perpendicular** to it.

Axiom. All right angles are equal.

DEF. The straight line which bisects a given straight line, and is perpendicular to it, is called the **perpendicular*** bisector of the line.

The distance of a point from a straight line is the length of the perpendicular drawn from the point to the line.

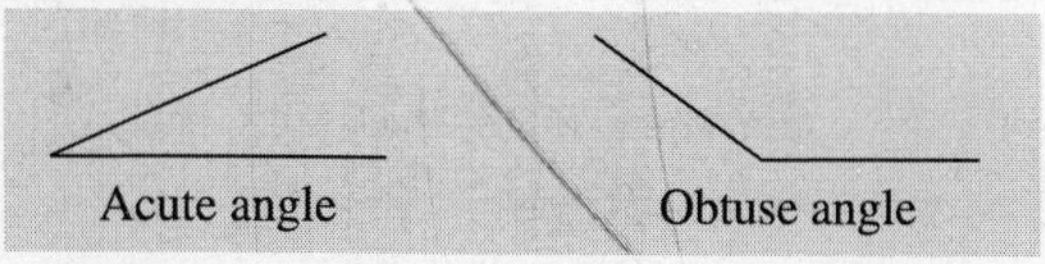

DEF. An **acute angle** is an angle which is less than a right angle.

DEF. An **obtuse angle** is an angle which is greater than a right angle.

DEE. Two angles are said to be **supplementary** when their sum is two right angles. In this case, each angle is called the supplement of the other.

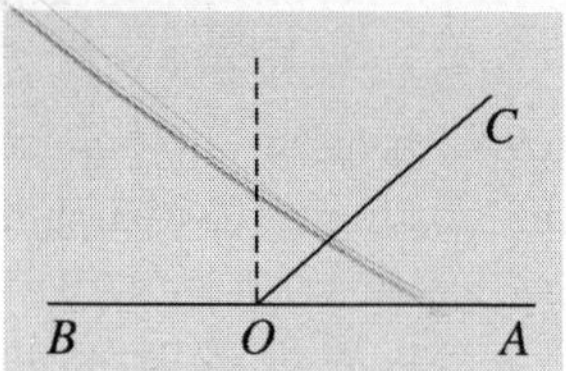

Thus, the angles *AOC*, *COB* are **supplementary**.

DEF. Two angles are said to be **complementary** when their sum is one right angle. In this case, each angle is called the complement of the other.

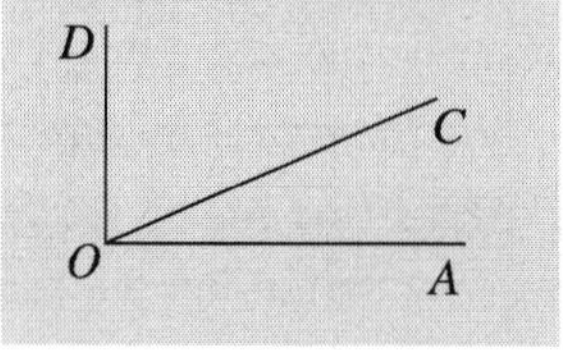

Thus, the angles *AOC*, *COD* are complementary.

The table of angular measurement is as follows:

$$1 \text{ right angle} = 90 \text{ degrees } (90°)$$

$$1 \text{ degree} = 60 \text{ minutes } (60')$$

$$1 \text{ minute} = 60 \text{ seconds } (60'').$$

Plane Geometry deals with the properties of lines and points which are in the same plane.

The discussion is divided into propositions which are either theorems or constructions.

A **theorem** is a proposition in which some statement has to be proved.

The statement itself is called the **enunciation**.

The enunciation of a theorem consists of two parts. The first part, called the **hypothesis**, states what is assumed. The second part, called the conclusion, states what has to be proved.

For example, a simple theorem in Algebra is as follows:

If $a = b$, then $na = nb$.

The hypothesis is $a = b$; the conclusion is $na = nb$.

If two theorems are such that the hypothesis of each is the conclusion of the other, then either of the theorems is said to be the converse of the other.

Thus, the converse of the above theorem is

If $na = nb$, *then* $a = b$.

*Sometimes also it is called the **right bisector**.

Note particularly that the converse of a theorem is not necessarily true.

Thus, consider the theorem *If* $a = b$, *then* $a^2 = b^2$.

The converse is *If* $a^2 = b^2$, *then* $a = b$, and this is not correct, for *if* $a^2 = b^2$ we can only conclude that *either* $a = b$ or $a = -b$.

When a theorem has been proved, it is sometimes found that other important theorems follow so easily that they hardly require formal proofs.

Such theorems are called corollaries.

> **NOTE:** In writing out Theorems, an accurate figure drawn with instruments is not expected. A neat freehand drawing is sufficient.

A **construction** is a proposition in which it is required to draw some particular figure.

It is convenient to arrange the theorems and constructions in separate groups.

In order to do this we make the following Postulates:

Let it be granted that

1. From the greater of two given straight lines, a part can be cut off equal to the less.
2. A straight line may be bisected (that is divided into two equal parts).
3. At any point in a given straight line, a perpendicular can be drawn to that straight line.
4. At any point in a given straight line, a straight line can be drawn, making with the given line an angle equal to a given angle.

Other postulates of this kind will be stated further on.

These postulates are sometimes called **hypothetical constructions**.

> **NOTE:** *It is suggested that, in a first course, the formal treatment of Theorems* 1, 2, 3, 4, 6, 7, 8, 9, 10, 14, 17, 18 *should be replaced by oral explanations, the facts stated in the enunciations being regarded as axiomatic.*

II. ANGLES AT A POINT

THEOREM 1 (Euclid I. 13)

If a straight line stands on another straight line the sum of the two adjacent angles is two right angles.

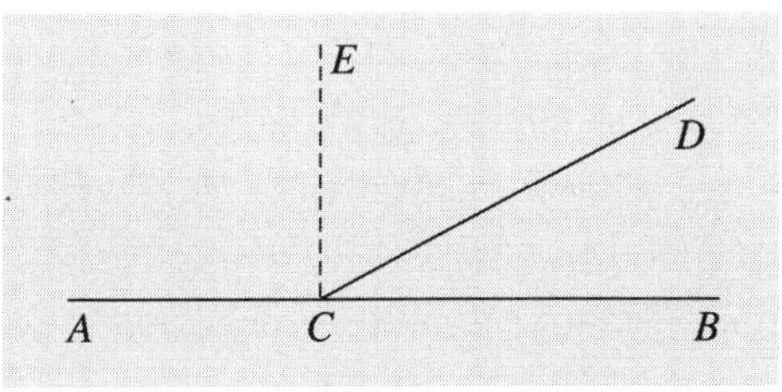

Let the straight line CD stand on the straight line ACB.

It is required to prove that

$$\angle ACD + \angle DCB = 2 \text{ right angles.}$$

Construction. Let CE be drawn perpendicular to AB.

Proof. $\angle ACD + \angle DCB = \angle ACE + \angle ECD + \angle DCB$.

Also $\angle ACE + \angle ECB = \angle ACE + \angle ECD + \angle DCB$.

$\therefore$ $\angle ACD + \angle DCB = \angle ACE + \angle ECB$.

But, by construction, $\angle$s ACE, ECB are right angles.

$\therefore$ $\angle ACD + \angle DCB = 2$ right angles.

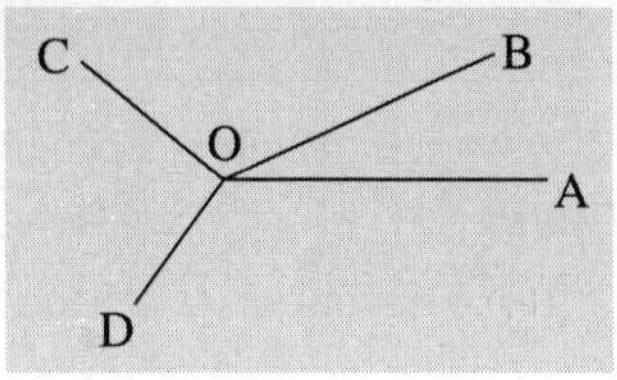

Corollary. If any number of straight lines are drawn from a given point, the sum of the consecutive angles so formed is four right angles.

Thus, in the figure,

$$\angle AOB + \angle BOC + \angle COD + \angle DOA = 4 \text{ right angles.}$$

Theorem 2* (Euclid I. 14)

If at a point in a straight line, two other straight lines, on opposite sides of it, make the adjacent angles together equal to two right angles, these two straight lines are in the same straight line.

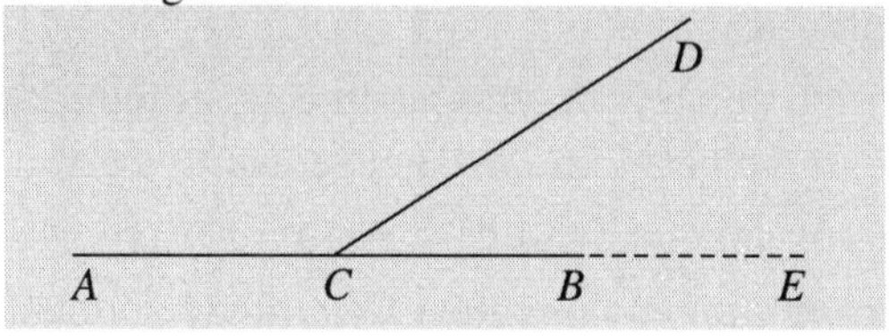

At the point C, in the straight line CD, let the straight lines CA, CB, on opposite sides of CD, make the adjacent angles ACD, DCB together equal to two right angles.

It is required to prove that CA, CB are in the same straight line.

Construction. Produce AC to some point E.

Proof. By construction, ACE is a straight line;

$\therefore$ $\angle ACD + \angle DCE = 2$ right angles.

But, it is given that

$$\angle ACD + \angle DCB = 2 \text{ right angles;}$$

$\therefore$ $\angle ACD + \angle DCB = \angle ACD + \angle DCE$.

From each of these equals, take the angle ACD:

$\therefore$ $\angle DCB = \angle DCE$;

$\therefore$ CB falls along CE.

But, by construction, CA and CE are the same straight line.

$\therefore$ CA and CB are in the same straight line.

*See 'Note' on page 5.

NOTE: In Theorem 1, it is given that *CA* and *CB* are in the same straight line, and it is proved that ∠*ACD* + ∠*DCB* = 2 rt. ∠s.

In Theorem 2, it is given that ∠*ACD* + ∠*DCB* = 2 rt. ∠s, and it is proved that *CA* and *CB* are in the same straight line.

These are therefore converse theorems (see p. 4).

THEOREM 3* (Euclid I. 15)

If two straight lines intersect, the vertically opposite angles are equal.

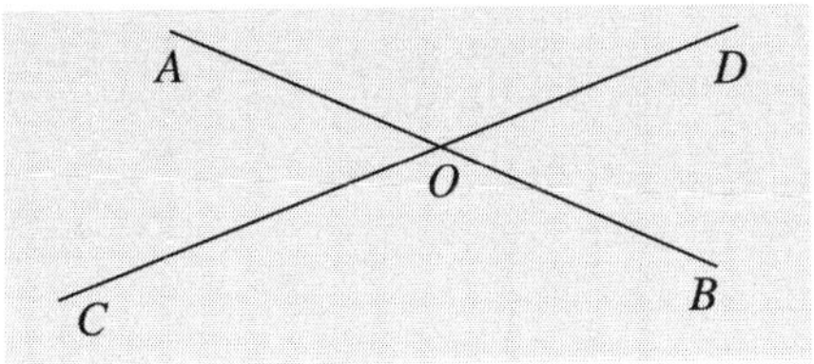

Let the straight lines *AB*, *CD* intersect at *O*. It is required to prove that

∠ *AOC* = ∠ *BOD* and ∠ *COB* = ∠ *DOA*.

Proof. Because *OC* stands on the straight line *AB*,

∴ ∠ *AOC* + ∠ *COB* = 2 right angles;

and because *OB* stands on the straight line *CD*,

∴ ∠ *BOD* + ∠ *COB* = 2 right angles;

∴ ∠ *AOC* + ∠ *COB* = ∠ *BOD* + ∠ *COB*.

From each of these equals take the angle *COB*;

∴ ∠ *AOC* = ∠ *BOD*.

Similarly it can be shown that

∠ *COB* = ∠ *DOA*.

EXERCISE I (on pages 1-7)

1. What is the angle in degrees between the hands of a watch at (*i*) 2 o'clock, (*ii*) 5 o'clock?
2. What angle does (*i*) the minute hand, (*ii*) the hour hand, turn through in 20 minutes?
3. What is the angle in degrees between the hands of a clock at (*i*) 12.15, (*ii*) 2.15 o'clock?
4. What are the supplements of 2/3rt. ∠, 40°, 120°, 98° 10′?
5. What are the complements of 1/4 rt. ∠, 40°, 35° 10′ 10″?
6. If in the figure of Theorem 1, the ∠ *BCD* = 32°, how large are the ∠s *ACD*, *ECD*?

*See 'NOTE' on page 5.

7. If in the figure of the corollary of Theorem 1, $\angle AOB = 30°$, $\angle BOC = 70°$, $\angle COD = 140°$, what is the $\angle DOA$?

8. If in the figure of Theorem 3, the $\angle AOC = 42°$, what are the other angles in the figure?

9. A straight line *AB* is bisected at *C*, and produced to *D*. Prove that $DA + DB = 2DC$.

10. A straight line *AB* is bisected at *C*, and any point *D* is taken in *CB*. Prove that $AD - DB = 2CD$.

11. In the figure of Theorem 3, prove that the bisectors of the angles *AOD*, *DOB* are at right angles.

12. In the figure of Theorem 3, prove that the bisector of the *AOD*, when produced, bisects the $\angle BOC$.

13. In the figure of Theorem 3, prove that the bisectors of the angles *AOC*, *BOD* are in the same straight line.

14. *A*, *B*, *C*, *D* are four points, and *AB*, *BC* subtend (*i.e.* are opposite to) supplementary angles at *D*: show that *A*, *D*, *C* are in the same straight line.

15. *OA*, *OB*, *OC*, *OD* are straight lines so drawn that $\angle AOB = \angle COD$ and $\angle BOC = \angle AOD$. Show that *AO*, *OC* and also *BO*, *OD* are in the same straight line.

16. *CAD* and *AB* are two straight lines and $\angle CAX = \angle BAD$; *B*, *X* being on opposite sides of *C*, *D*. Prove that *AB*, *AX* are in the same straight line.

17. *XOA*, *XOB* are angles on the same side of *OX*, and *OC* bisects the angle *AOB*. Prove that $\angle XOA + \angle XOB = 2\angle XOC$.

18. *AOX*, *XOB* are adjacent angles, of which *AOX* is the greater, and *OC* bisects the angle *AOB*. Prove that $\angle AOX - \angle XOB = 2\angle COX$.

III. TRIANGLES

Def. A **plane figure** is any part of a plane surface bounded by one or more lines, straight or curved.

Def. A **rectilineal figure** is one which is bounded by straight lines.

It will be assumed that any figure may be **duplicated** (*i.e.* copied exactly), or that it may be moved from any one position to any other position, and, if necessary, *turned over* or *folded.*

If a figure is taken up and placed on another figure in order to make a comparison, the first figure is said to be applied to the second. This process is called **superposition**.

Figures which occupy the same portion of space are said to coincide.

Def. Figures which can be made to coincide are called **congruent**. Congruent figures are said to be **equal in all respects**.

In two congruent figures, the sides and angles which coincide, when one figure is applied to the other, are said to correspond.

Def. The **area** of a plane figure is the amount of surface enclosed by its boundaries.

Def. **Equal figures** are those which have the same area.

Def. The **perimeter** of a plane figure is the sum of the lengths of its boundaries.

Def. A **triangle** is a plane figure bounded by three straight lines.

The straight lines *BC*, *CA*, *AB* which bound a triangle *ABC* are called its sides and the points *A*, *B*, *C* its angular points or vertices.

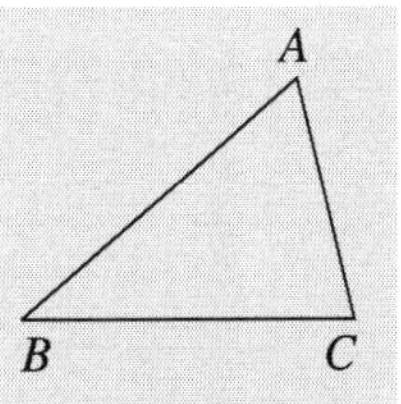

For distinction, one angular point is often called the vertex and the opposite side the base.

Def. An **isosceles triangle** is a triangle which has two equal sides.

The side which is unequal to the others is always called the **base**, and the angular point opposite to the base is called the **vertex**.

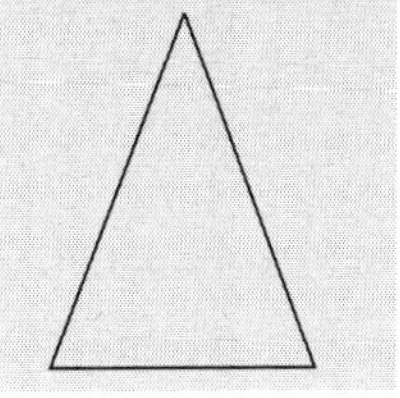

Def. An **equilateral triangle** is a triangle whose three sides are equal.

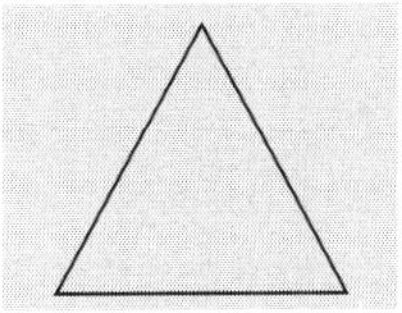

Def. A **right-angled triangle** is a triangle one of whose angles is a right angle.

Def. The side opposite the right angle in a right-angled triangle is called the **hypotenuse**.

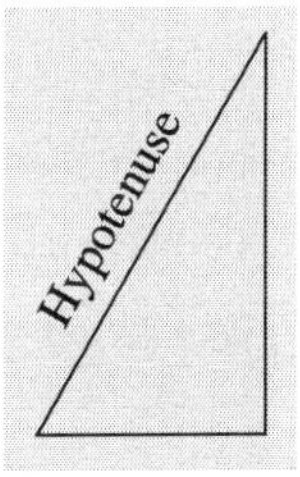

Def. An **obtuse-angled triangle** is a triangle which has one obtuse angle.

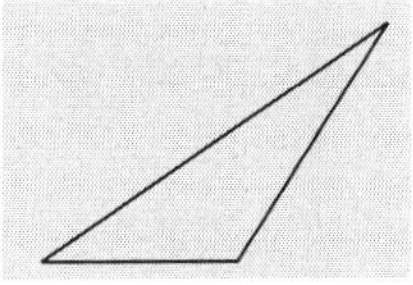

Def. An **acute-angled triangle** is a triangle which has three acute angles.

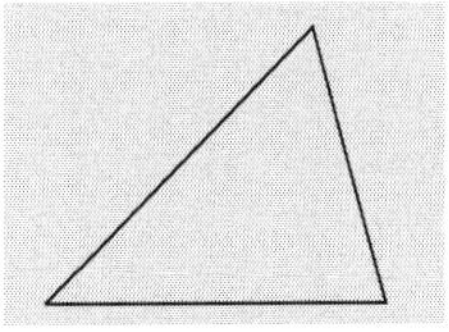

THEOREM 4* (Euclid I. 4)

Two triangles are congruent if two sides and the included angle of one triangle are respectively equal to two sides and the included angle of the other.

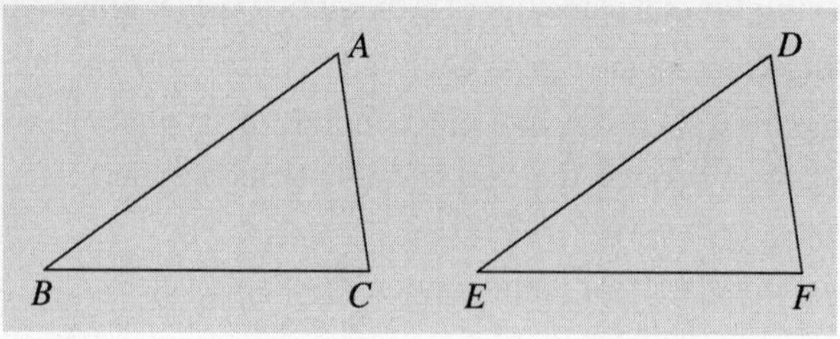

Let *ABC*, *DEF* be two triangles in which

$$AB = DE,$$
$$AC = DF$$

and the included $\angle A$ = The included $\angle D$.

It is required to prove that the triangles are congruent.

Proof. Apply the triangle *ABC* to the triangle *DEF* so that the point *A* falls on the point *D* and the straight line *AB* falls along the straight line *DE*.

Then because $AB = DE$ (*given*),

$\therefore$ The point *B* falls on the point *E*.

Also since *AB* falls along *DE* and it is given that

$$\angle A = \angle D,$$

$\therefore$ *AC* falls along *DF*; and since it is given that

$$AC = DF.$$

$\therefore$ The point *C* falls on the point *F*.

Hence *A* falls on *D*, *B* on *E* and *C* on *F*.

$\therefore$ The triangle *ABC* has been made to coincide with the triangle *DEF*.

$\therefore$ The triangles *ABC*, *DEF* are congruent.

Note on Theorem 4. In the triangles *ABC*, *DEF*, *it is given that* $AB = DE$, $AC = DF$, included $\angle A$ = included $\angle D$; *it is proved that* the triangles are congruent, and therefore (*i*) the triangles are equal in area, (*ii*) $BC = EF$, (*iii*) $\angle B = \angle E$, $\angle C = \angle F$.

Observe that the angles which are proved equal are opposite equal sides in the two triangles.

In congruent triangles, the sides and angles which coincide, when one triangle is applied to the other, are said to correspond.

In using Theorem 4, the work should be written out as in the following example:

---

*See 'NOTE' (italic) on page 5.

Example: *In the figure, if it is given that OA = OB and OC = OD, prove that CB = DA and ∠ ACB = ∠ BDA.*

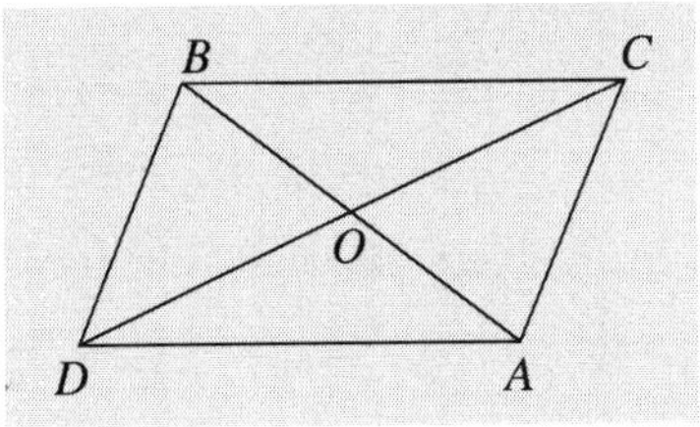

[It looks as if Δs *OCB*, *ODA* are congruent; we therefore begin thus:]

In the triangles *OCB*, *ODA*,

$$\begin{cases} OC = OD \text{ (given)}, \\ OB = OA \text{ (given)}, \\ \angle COB = \text{The vert. opp. } \angle DOA. \end{cases}$$

∴ The triangles are congruent,

∴ $$CB = DA;$$

and $$\angle OCB = \angle ODA.$$

[The rest is left as an exercise. Taking Δs *OAC*, *OBD*, prove that ∠ *OCA* = ∠ *ODB*, etc.]

Now observe the way in which the triangles are *named* and particularly the *order of the letters.* The angles *O*, *C*, *B* of Δ *OCB* correspond to the angles *O*, *D*, *A* of Δ *ODA*.

If we name the triangles in this way, we can write down the equal sides without even looking at the figure.

Next, everything belonging to Δ*OCB* is written on the left and everything belonging to Δ*ODA* is written on the right.

Theorem 5 (Euclid I. 16)

If one side of a triangle is produced, the exterior angle is greater than either of the interior opposite angles.

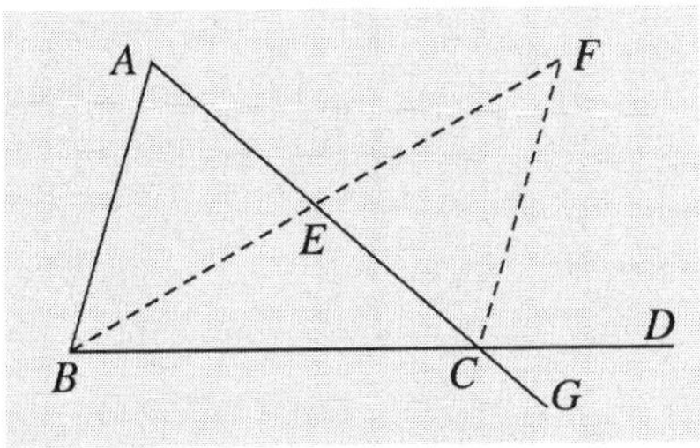

Let the side *BC* of the triangle *ABC* be produced to *D*. It is required to prove that the exterior ∠ *ACD* is greater than either of the interior opposite angles *BAC* or *ABC*.

Construction. Let AC be bisected at E. Join BE and produce it to F, making $EF = BE$. Join CF.

Proof. In the triangles AEB, CEF,

$$\begin{cases} AE = CE \ (const.), \\ EB = EF \ (const.), \\ \angle AEB = \text{The vert. opp. } \angle CEF. \end{cases}$$

∴ The triangles are congruent,

∴ $$\angle EAB = \angle ECF.$$

Now CF is within the $\angle ACD$;

∴ $$\angle ACD > \text{its part } \angle ECF.$$

∴ Also $$\angle ACD > \angle BAE.$$

Similarly, if BC is bisected and AC is produced to G, it can be shown that

$$\angle BCG > \angle ABC.$$

But $\angle BCG$ = the vert. opp. $\angle ACD$.

∴ Also $$\angle ACD > \angle ABC.$$

Hence $\angle ACD$ is greater than either of the $\angle$s BAC or ABC.

> **NOTE:** In the second part of the proof *it is not enough to say*, "Similarly it can be proved that $\angle ACD > \angle ABC$."

EXERCISE II (Theorems 4 and 5)

1. Finish the proof of the Example on p. 11.
2. A straight line AB is bisected at C, and through C a straight line CD is drawn perpendicular to AB. If P is any point in CD, prove that $PA = PB$.
3. Complete the enunciation of the following theorem, which is proved in Ex. 2: Any point in the perpendicular bisector of the straight line, joining two given points, is... .
4. In the triangle ABC, the sides AB, AC are bisected in D and E. Through D and E perpendiculars are drawn to AB, AC respectively, to meet in O. Prove that $OA = OB = OC$. (Take Δs OAD, OBD; then take Δs OAE, OCE.)
5. If two straight lines, AB and CD, bisect each other at right angles, and AD, DB, BC, CA are joined, prove that all the sides of the figure $ADBC$ are equal.
6. ABC is an isosceles triangle in which $AB = AC$. The bisector of the $\angle BAC$ meets BC in X. Prove that (*i*) the $\angle ABC$ = the $\angle ACB$; (*ii*) AX bisects BC; (*iii*) AX is perpendicular to BC.
7. Give (without reference to a figure) the enunciation of the theorem about an isosceles triangle proved in the last example.
8. Along the equal sides AB, AC of an isosceles triangle ABC, set of two equal lengths AX, AY. Prove that (*i*) the Δs AXC, AYB are congruent; (*ii*) the Δs BXC, CYB are congruent; (*iii*) the $\angle ABC$ = the $\angle ACB$.

9. Two isosceles triangles, whose vertical angles are equal, are placed so as to have their vertices coincident: prove that two of the lines joining their other angular points are equal.

10. Two quadrilaterals *ABCD*, *XYZW* have $AB = XY$, $BC = YZ$, $CD = ZW$, $\angle B = \angle Y$, $\angle C = \angle Z$; show that they are congruent. [Use the method of superposition.]

11. In the figure of Theorem 5, (*i*) if *A* and *F* are joined, show that the Δs *BEC*, *FEA* are congruent; (*ii*) show that the Δs *ABC*, *FBC* are equal in area.

12. Any two angles of a triangle are together less than two right angles. [Let *ABC* be the triangle. Produce *BC* to *D*. The $\angle ABC <$ the $\angle ACD$; to each add the $\angle ACB$.]

13. Every triangle has at least two acute angles.

14. *ABC* is a triangle and *O* is any point within it. Join *OB*, *OC*, and prove that the $\angle BOC >$ the $\angle BAC$. [Produce *BO* to cut *AC* in *X*, and consider the Δs *OXC*, *XAB*.]

IV. PARALLELS

Def. Straight lines, which are in the same plane and which do not meet, however far they are produced in either direction, are said to be **parallel**.

Playfair's Axiom. Two straight lines which intersect cannot both be parallel to the same straight line.

Postulate. Let it be granted that a straight line can be drawn, through a given point, parallel to a given straight line.

The distance between two parallel straight lines is the length of the perpendicular drawn from any point in one of the lines to the other.

Def. Any straight line drawn to cut two or more given straight lines is called a **transversal**.

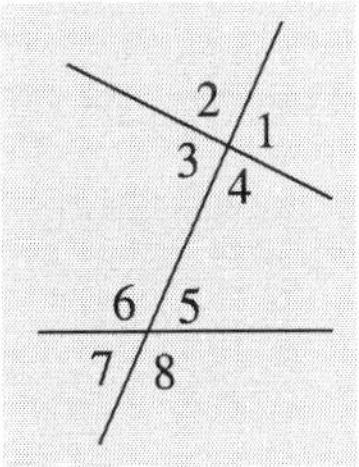

If a transversal makes with two straight lines angles marked as in the figure, then

1, 2, 7, 8 are called **exterior angles**;

3, 4, 5, 6 are called **interior angles**;

3, 5 are called **alternate** angles; so also are 4, 6;

1, 5 are called **corresponding** angles; so also are 2, 6; 8, 4; 7, 3.

The following definitions are required in the examples:

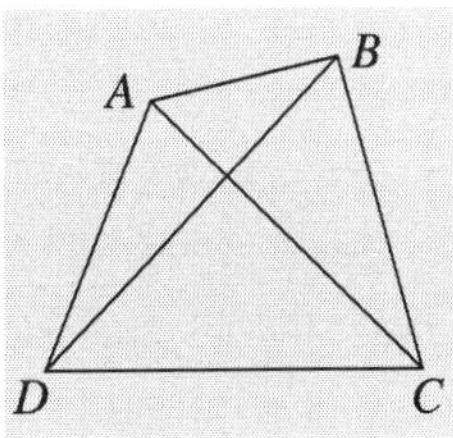

Def. A quadrilateral is a plane figure bounded by four straight lines.

The straight lines joining opposite angular points are called the **diagonals** of the quadrilateral.

Thus in the figure, *AC*, *BD* are the diagonals of the quadrilateral *ABCD*.

Species of Quadrilaterals

Def. A **trapezium** is a quadrilateral two sides of which are parallel.

Def. A **parallelogram** is a quadrilateral whose opposite sides are parallel.

Def. A **rectangle** is a parallelogram one of whose angles is a right angle.

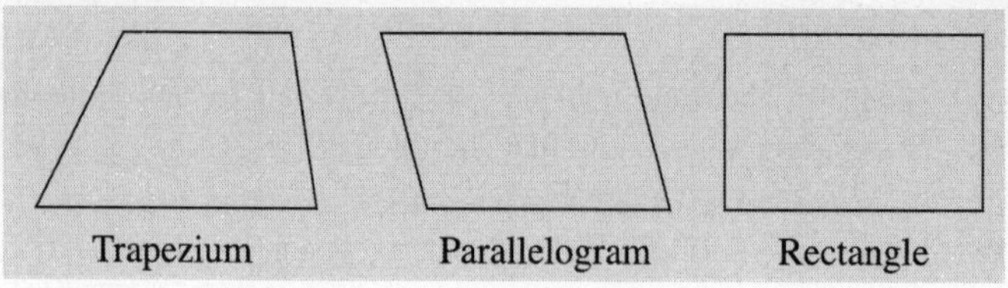

Def. A **square** is a rectangle, of which a pair of adjacent sides are equal.

Def. A **rhombus** is a quadrilateral with all its sides equal, but its angles are not right angles.

Note on the definitions. In defining anything, the definition should say just enough to describe the thing exactly. To say more than enough *spoils the definition.*

Starting with the definitions given above, and using the properties of parallel straight lines, *we can prove that*

The opposite sides of a parallelogram are equal.

All the angles of a rectangle or of a square are right angles.

A rhombus is a parallelogram.

But these facts must not be mentioned in the definitions.

Next, in proving properties of geometrical figures, the figure drawn should be *as general as possible.* It should have no special properties except those which are given.

Thus, in a question about a quadrilateral, draw an irregular figure and not a parallelogram, unless the given conditions result in its being a parallelogram.

Theorem 6* (Euclid I. 27)

If a straight line cuts two other straight lines so as to make the alternate angles equal, the two straight lines are parallel.

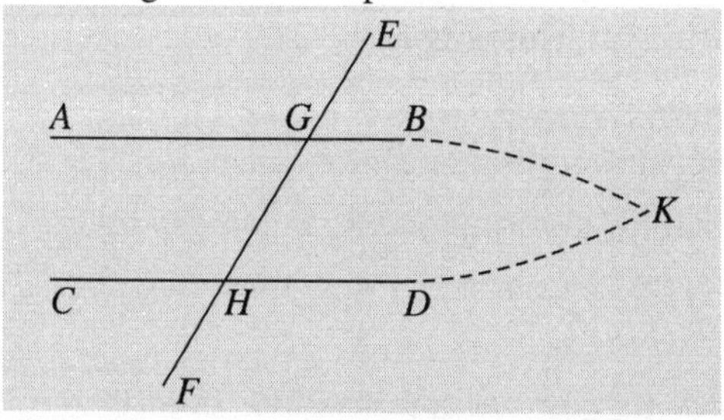

Let the straight line *EF* cut the straight lines *AB*, *CD* in *G* and *H*, so as to make the alternate angles *AGH*, *GHD* equal.

It is required to prove that *AB*, *CD* are parallel.

*See 'Note' (italic) on page 5.

Proof. If AB and CD are not parallel, they will meet if produced, either towards B, D or towards A, C.

If possible, let AB and CD meet at K, when produced towards B, D.

Then GHK is a triangle with the side KG produced to A.

∴ The exterior $\angle AGH$ is greater than the interior opposite $\angle GHD$.

But it is given that $\angle AGH = \angle GHD$.

∴ The $\angle AGH$ cannot be greater than $\angle GHD$.

∴ AB and CD cannot meet when produced towards B, D.

Similarly, it can be shown that they cannot meet when produced towards A, C.

∴ AB and CD are parallel.

THEOREM 7* (Part of Euclid I. 29)

This is the converse of Theorem 6.

If a straight line meets two parallel straight lines, it makes the alternate angles equal.

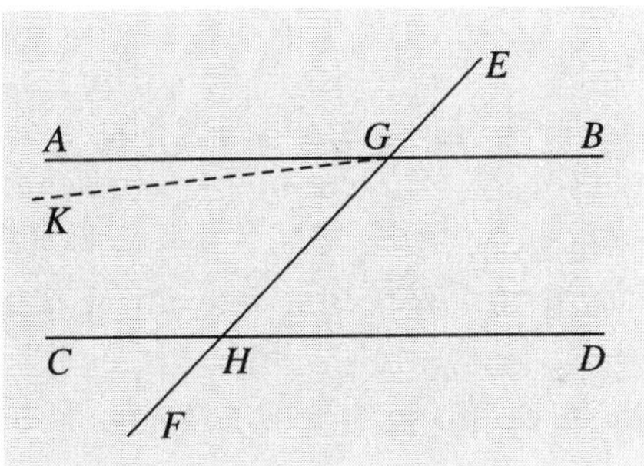

Let the straight lines AB, CD be parallel, and let the straight line EF meet them in G and H.

It is required to prove that the alternate angles AGH, GHD are equal.

Construction. If the $\angle AGH$ is not equal to the $\angle GHD$, let the straight line GK be drawn, making $\angle KGH$ equal and alternate to $\angle GHD$.

Proof. By construction, the $\angle KGH$ = the alternate $\angle GHD$;

∴ KG to parallel to CD. But it is given that AB is parallel to CD.

∴ AB and KG are both parallel to CD; and, by Playfair's axiom, this is impossible, for AB and KG intersect.

Hence it is wrong to suppose that the $\angle$s AGH, GHD are unequal;

∴ $$\angle AGH = \angle GHD.$$

> **NOTE:** The method of proof in Theorem 7 is called Reductio ad absurdum, and consists in showing that if the theorem is supposed to be untrue, an absurdity follows. This method is often used in proving the converse of a known theorem.

*See 'NOTE' (italic) on page 5.

THEOREM 8* (Euclid I. 28)

If a straight line cuts two other straight lines so as to make, (*i*) two corresponding angles equal; or (*ii*) the interior angles, on the same side of the line, supplementary, the two straight lines are parallel.

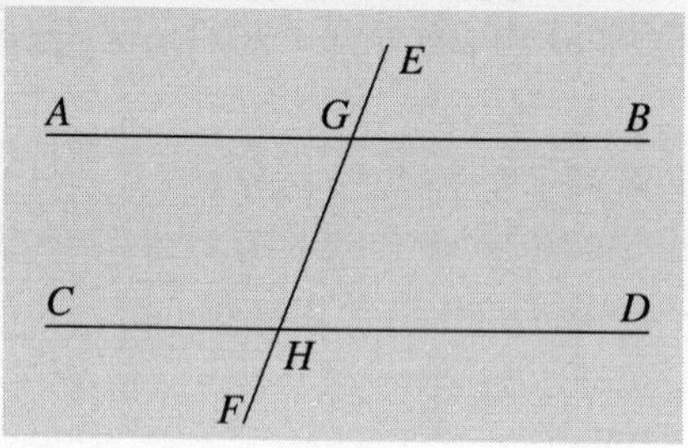

Let the straight line *EF* cut the straight lines *AB*, *CD* so as to make (*i*) the corresponding angles *EGB*, *GHD* equal.

It is required to prove that *AB* and *CD* are parallel.

Proof. It is given that $\angle EGB = \angle GHD$. But the $\angle EGB$ = The vert. opp. $\angle AGH$,

$$\therefore \quad \angle AGH = \angle GHD;$$

and these are alternate angles,

∴ *AB* and *CD* are parallel.

(*ii*) Let the interior ∠s *BGH*, *GHD* be supplementary. It is required to prove that *AB* and *CD* are parallel.

Proof. It is given that the $\angle GHD$ is the supplement of the $\angle BGH$. But since *AGB* is a straight line.

∴ The $\angle AGH$ is the supplement of the $\angle BGH$

$$\therefore \quad \angle AGH = \angle GHD;$$

and these are alternate angles,

∴ *AB* and *CD* are parallel.

THEOREM 9* (Part of Euclid I. 29)

This is the converse of Theorem 8.

If a straight line cuts two parallel straight lines it makes corresponding angles equal and also two interior angles on the same side of the line supplementary.

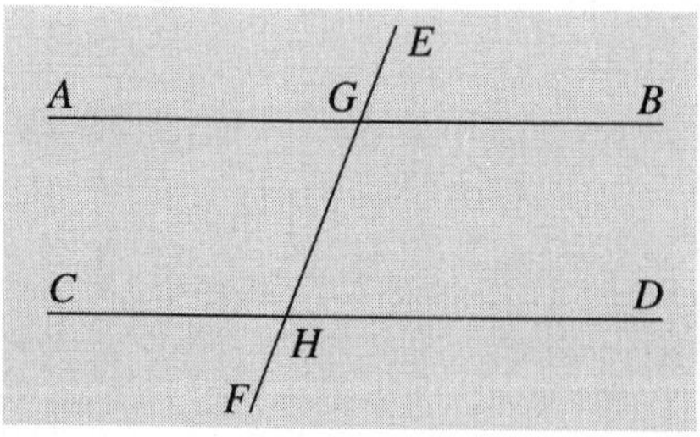

*See 'NOTE' (italic) on page 5.

Let the straight lines AB and CD be parallel, and let the straight line EF cut them in G and H.

(*i*) It is required to prove that the corresponding angles EGB, GHD are equal.

Proof. It is given that AB is parallel to CD;

∴ $\angle AGH =$ The alt. $\angle GHD$.

But $\angle AGH =$ The vert. opp. $\angle EGB$;

∴ $\angle EGB = \angle GHD$.

(*ii*) It is required to prove that the angles BGH, GHD are supplementary.

Proof. It is given that AB is parallel to CD,

∴ $\angle AGH =$ The alt. $\angle GHD$;

and because AGB is a straight line.

∴ $\angle AGH =$ The supplement of the $\angle BGH$,

∴ Also $\angle GHD =$ The supplement of the $\angle BGH$.

THEOREM 10* (Euclid I. 30)

Straight lines which are parallel to the same straight line are parallel to one another.

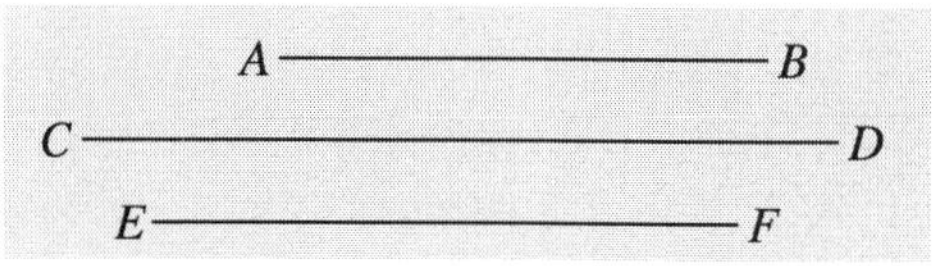

Let both of the straight lines AB and CD be parallel to the straight line EF.

It is required to prove that AB and CD are parallel.

Proof. By Playfair's axiom, two intersecting straight lines cannot be parallel to the same straight line; and since it is given that both AB and CD are parallel to EF.

∴ AB and CD cannot meet when produced.

∴ AB and CD are parallel.

EXERCISE III (Theorems 6–10)

1. Straight lines which are perpendicular to the same straight line are parallel.
2. Any straight line which is perpendicular to AB is also perpendicular to any parallel to AB.
3. If one angle of a parallelogram is a right angle, prove that all its angles are right angles.
4. Show that the opposite angles of a parallelogram are equal to one another. [Produce one of the sides.]

*See 'NOTE' (italic) on page 5.

5. The acute angle between any two intersecting straight lines is equal to the acute angle between any parallels to these lines.
6. Draw any triangle *ABC* and produce the side *BC* to *D*. Through *C* draw *CE* parallel to *BA*, and show that the angle *ACD* is equal to the sum of the angles *A* and *B*. Hence show that the sum of the angles of a triangle is two right angles.
7. If, in the quadrilateral *ABCD*, the sides *AB*, *DC* are equal and parallel, prove that the sides *BC*, *AD* are also equal and parallel.
8. If the diagonals of a quadrilateral bisect one another, prove that the figure is a parallelogram.

V. ANGLES OF RECTILINEAL FIGURES

Def. A **rectilineal figure** is a figure which is bounded by straight lines.

If all the sides of the figure are equal, it is said to be equilateral; if all its angles are equal, it is equiangular.

A **convex** rectilineal figure is one which has no angle greater than two right angles.

If a fly walks round a rectilineal figure, and each side is produced in the direction in which he walks along it, the sides are said to be produced in order.

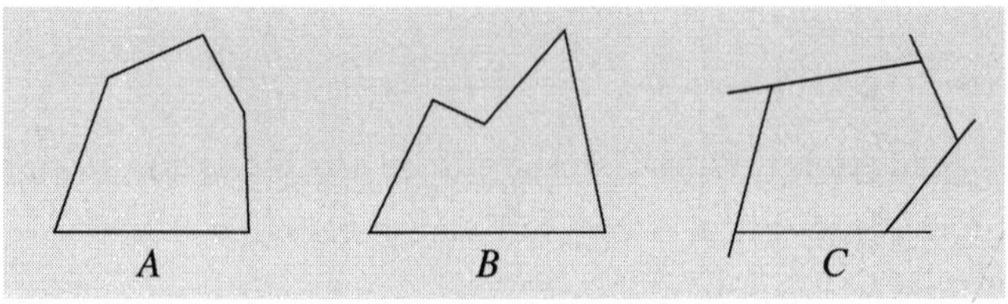

Thus, the figure *A* is convex, *B* is not convex, *C* has its sides produced in order.

Def. A **polygon** is a plane figure bounded by more than four straight lines.

A polygon is said to be **regular** when all its sides are equal and all its angles are equal.

A polygon with 5 sides is called a pentagon.
„ „ 6 „ „ hexagon.
„ „ 7 „ „ heptagon.
„ „ 8 „ „ octagon.
„ „ 10 „ „ decagon,

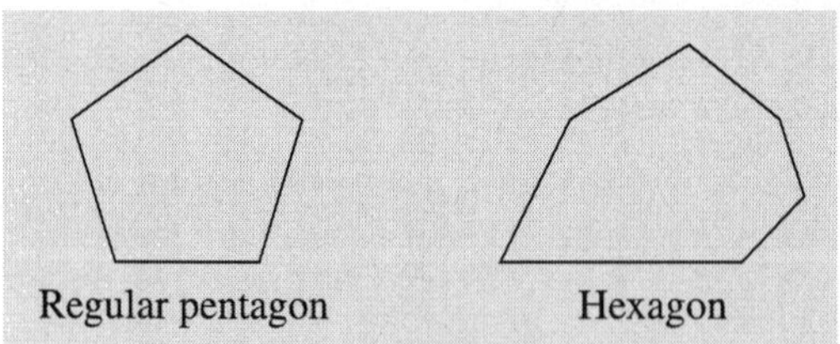

THEOREM 11 (Euclid I. 32)

If one side of a triangle is produced, the exterior angle is equal to the sum of the two interior opposite angles; also, the sum of the three angles of a triangle is two right angles.

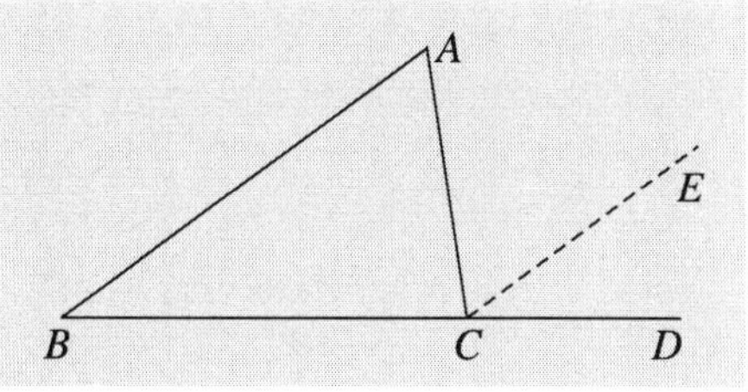

Let *ABC* be a triangle, and let *BC* be produced to *D*.

(*i*) It is required to prove that the exterior angle *ACD* is equal to the sum of the two interior opposite angles *A* and *B*.

Construction. Through *C* let *CE* be drawn parallel to *BA*.

Proof. By construction, *BA* and *CE* are parallel and *AC* meets them;

∴ $\angle ACE$ = The alternate $\angle A$.

Again, *BA* and *CE* are parallel and *BD* cuts them;

∴ $\angle ECD$ = The corresponding $\angle B$,

∴ The whole $\angle ACD = \angle A + \angle B$.

(*ii*) It is required to prove that the sum of the angles of the triangle *ABC* is two right angles.

Proof. It has been proved that

$$\angle ACD = \angle A + \angle B.$$

To each of these equals add $\angle ACB$;

∴ $\angle ACD + \angle ACB = \angle A + \angle B + \angle ACB.$

And since *BCD* is a straight line,

∴ $\angle ACD + \angle ACB$ = 2 right angles,

∴ $\angle A + \angle B + \angle ACB$ = 2 right angles.

COR. 1. The sum of any two angles of a triangle is less than two right angles.

COR. 2. If two angles of one triangle are respectively equal to two angles of another triangle, the third angles are equal and the triangles are equiangular.

> **NOTE:** Equiangular triangles are not necessarily congruent.

THEOREM 12 (Euclid I. 32, Cor. 1)

All the interior angles of a convex rectilineal figure, together with four right angles, are equal to twice as many right angles as the figure has sides.

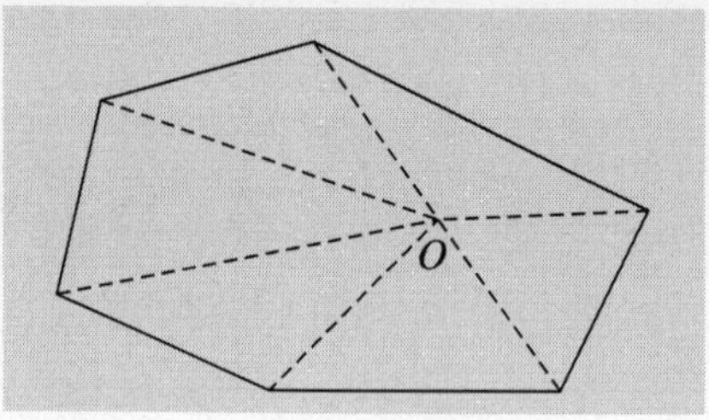

Let n be the number of sides of the figure.

It is required to prove that the angles of the figure + 4 rt. $\angle$s = $2n$ rt. $\angle$s.

Construction. Take any point O within the figure and join O to each vertex.

Proof. The figure has been divided into n triangles, and the three angles of each triangle together = 2 rt. $\angle$s.

∴ All the angles of all the triangles together = $2n$ rt. $\angle$s.

But all the angles of all the triangles make up the angles of the figure and the angles at O; also the angles at O together = 4 rt. $\angle$s;

∴ The angles of the figure + 4 rt. $\angle$s = $2n$ rt. $\angle$s.

THEOREM 13 (Euclid I. 32, Cor. 2)

If all the sides of a convex rectilineal figure are produced in order, the sum of the exterior angles is four right angles.

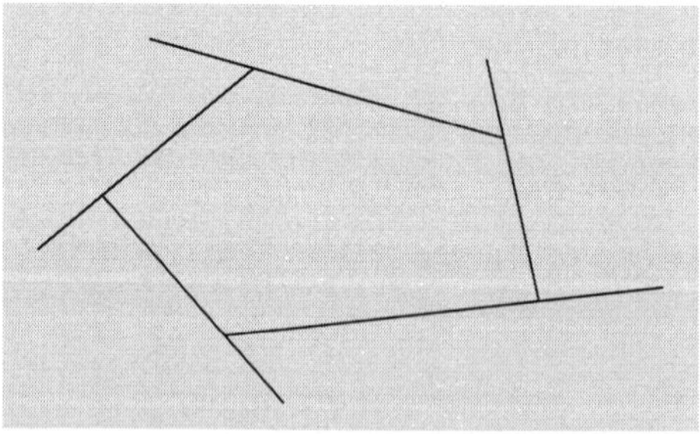

Proof. Let n be the number of sides of the figure;

∴ n is also the number of vertices.

Now at each vertex, the interior $\angle$ + the exterior $\angle$ = 2 rt. $\angle$s.

∴ All the interior $\angle$s + All the exterior $\angle$s = $2n$ rt. $\angle$s.

But, in a convex rectilineal figure, it has been shown that all the interior $\angle$s + 4 rt. $\angle$s = $2n$ rt. $\angle$s;

∴ All the int. $\angle$s + All the ext. $\angle$s = All the int. $\angle$s + 4 rt. $\angle$s.

∴ All the ext. $\angle$s = 4 rt. $\angle$s.

EXAMPLE: *Find, in degrees, an interior angle of a regular pentagon.*

The sum of the ext. ∠s = 90° × 4 = 360°

and the figure has 5 angles.

∴ An ext. ∠ = $\frac{1}{5}$ of 360° = 72°,

∴ An int. ∠ = 180° – 72° = 108°.

EXERCISE IV (Theorems 11–13)

Numerical

1. In a right-angled triangle, one angle = 50°; find the third angle.
2. In the triangle *ABC*, ∠ *A*= 100°, ∠ *B* = 20°; what is ∠ *C*?
3. In the triangle ABC, ∠ *A* = 20°, the exterior angle at *B* = 45°; what is ∠ *C?*
4. In a quadrilateral *ABCD*, ∠ *A* = 90°, ∠ *B* = 90°, ∠ *C* = 60°; what is ∠ *D*?
5. In a quadrilateral *ABCD*, ∠ *A* = 60°, ∠ *B* = 120°, ∠ *D* = 120°; what is ∠ *C?* What kind of quadrilateral is *ABCD*?
6. How many angular points has an *n*-sided rectilineal figure? How many triangles are formed by joining one of its vertices to all the remaining vertices? Express in degrees (*i*) an exterior, (*ii*) an interior angle of a regular pentagon, hexagon, octagon.
7. In a certain quadrilateral, the sum of two interior adjacent angles is 200 degrees. Find the obtuse angle contained by the bisectors of the two remaining interior angles.
8. A five-sided figure has four equal angles, and the fifth angle is half one of the other four; express the fifth angle as a fraction of a right angle.
9. A hexagon *ABCDEF* has the angles at *A*, *C*, *E* all equal, and each double any one of the angles at *B*, *D*, *F*, which are also all equal; determine the angle *A* as a fraction of a right angle.
10. If each angle of a rectilineal figure is equal to seven-eighths of two right angles, find the number of sides.

Theoretical

11. If one angle of a triangle is equal to the sum of the other two, prove that the triangle is right angled.
12. If a straight line meets two parallel straight lines, the bisectors of two interior angles on the same side of the line are at right angles.
13. Prove that the bisectors of the angles of a parallelogram form a rectangle. [Use Ex. 12.]
14. Prove that the sum of the angles of a quadrilateral is four right angles by drawing a diagonal and using Theorem 11.
15. In the quadrilateral *ABCD*, if ∠ *A* = ∠ *C* and ∠ *B* = ∠ *D*, prove that the figure is a parallelogram. [Use Ex. 14 to show that ∠ *A* + ∠ *B* = 2 rt. ∠s.]

16. Prove Theorem 12 by joining one vertex of the polygon to all the remaining vertices, thus dividing it into $(n - 2)$ triangles, where n is the number of sides of the polygon.

17. In the triangle ABC, the bisectors of the angles B and C are drawn, to meet in X. Prove that $\angle BXC = 90° + 1/2 \angle A$.

18. In the triangle ABC, the sides AB, AC are produced, and the bisectors of the exterior angles at B and C are drawn to meet in Y. Prove that $\angle BYC = 90° - 1/2 \angle A$.

19. In the triangle ABC, perpendiculars AD and BE are drawn to BC and CA respectively. If AD and BE meet in O, show that $\angle AOE = \angle C$. [Apply Ex. 14 to the quadrilateral $CDOE$.]

VI. TRIANGLES (*Continued*)

THEOREM 14* (Euclid I. 26)

Two triangles are congruent if two angles and a side of one triangle are respectively equal to two angles and the corresponding side of the other.

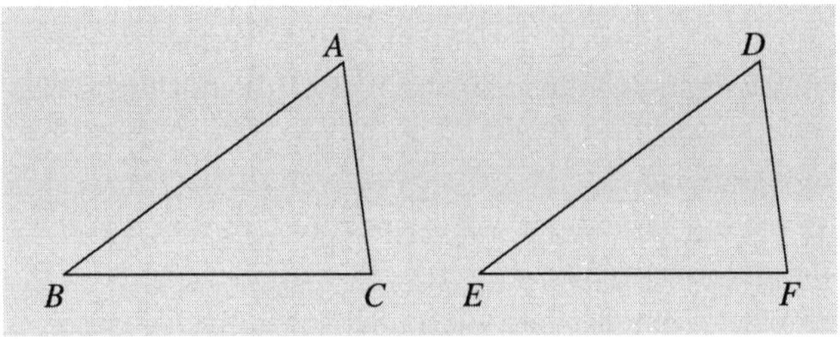

Let ABC, DEF be two triangles in which

$$\angle B = \angle E \text{ and } \angle C = \angle F.$$

Since the $\angle$s of each triangle are together equal to 2 rt. $\angle$s.

$\therefore$ The remaining $\angle A$ = The remaining $\angle D$.

Also let it be given that $BC = EF$.

It is required to prove that the triangles are congruent.

Proof. Apply the triangle ABC to the triangle DEF so that B falls on E and BC falls along EF.

Then, because $BC = EF$ (*given*),

$\therefore$ C falls on F.

Again, because BC falls along EF, and $\angle B = \angle E$ (*given*),

$\therefore$ BA falls along ED.

---

*See 'NOTE' (italic) on page 5.

Also, because *CB* falls along *FE*, and $\angle C = \angle F$ (*given*),

∴ *CA* falls along *FD*;

∴ *A*, which is the point of intersection of *BA* and *CA*, falls on *D*, which is the point of intersection of *ED* and *FD*.

Thus *A* falls on *D*, *B* on *E* and *C* on *F*.

∴ The triangle *ABC* has been made to coincide with the triangle *DEF*.

∴ The triangles are congruent.

EXERCISE V (Theorem 14)

1. If two isosceles triangles have equal bases and equal vertical angles, prove that they are congruent.
2. Prove that the opposite sides of a parallelogram are equal. Hence show that all the sides of a square are equal.
3. The diagonals of a parallelogram bisect one another.
4. *AB*, the perpendicular distance between two parallel straight lines, is bisected in *C*. Any straight line through *C* meets the parallels in *P*, *Q*. Show that $CP = CQ$.
5. Any point on the bisector of an angle is equidistant from the arms of the angle.
6. In any triangle *ABC*, *OB*, *OC* are the bisectors of the angles *B* and *C*, and *OX*, *OY*, *OZ* are drawn perpendicular to *BC*, *CA*, *AB* respectively. Prove that $OX = OY = OZ$.
7. In any triangle *ABC*, *AB* and *AC* are produced, and the bisectors of the exterior angles at *B* and *C* meet in *O*. Prove that *O* is equidistant from the three sides of the triangle.
8. *AB*, *CD* are two straight lines meeting in *O*, and *EF* is another straight line. Show that, in general, two points can be found in *EF* which are equidistant from *AB* and *CD*. When is there only one such point? [Consider the bisectors of ∠s *AOC*, *COB*, using Ex. 5.]
9. *ABCD* is a quadrilateral, and the diagonal *AC* bisects the angles at *A* and *C*; show that *AC* is at right angles to the other diagonal *BD*. [Let *AC*, *BD* meet in *O*. Prove Δs *ABC*, *ADC* congruent: then prove Δs *AOB*, *AOD* congruent.]
10. *ABCD* is a quadrilateral in which the diagonal *AC* bisects the diagonal *BD*. Prove that *B* and *D* are equidistant from *AC*. [Let *BE*, *DF* be perpendicular to *AC*, and let *AC*, *BD* meet in *O*. Consider Δs *OBE*, *ODF*.]
11. *ABCD* is a trapezium with *BC* parallel to *AD*. *AB* is bisected at *O*, and *XY* is drawn through *O*, parallel to *DC*, to meet *AD*, *CB* (produced if necessary) in *X* and *Y*. Prove that (*i*) Δs *AOX*, *BOY* are; congruent; (*ii*) the figures *ABCD*, *XYCD* are equal in area.

12. If two quadrilaterals *ABCD*, *EFGH* have the angles *A*, *B*, *C*, *D* equal to *E*, *F*, *G*, *H* respectively, and *AB*, *DC* equal to *EF*, *HG* respectively, and if *AD*, *BC* meet when produced; then the quadrilaterals are congruent. [Let *AB*, *BC* meet in *L*, and let *EH*, *FG* meet in *M*. Consider Δs *ABL*, *EFM* and then Δs *CDL*, *GHM*.]

THEOREM 15 (Euclid I. 5)

If two sides of a triangle are equal, the angles opposite to these sides are equal.

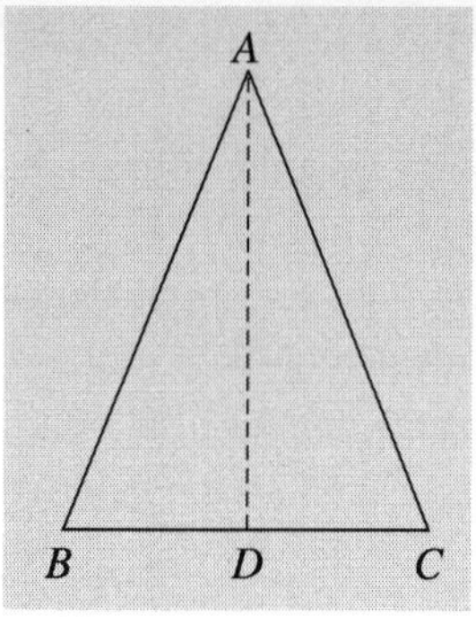

Let *ABC* be a triangle in which

$$AB = AC.$$

It is required to prove that

$$\angle B = \angle C.$$

Construction. Let the angle *BAC* be bisected by a straight line meeting *BC* in *D*.

Proof. In the triangles *ABD*, *ACD*,

$$\left\{\begin{array}{l} AB = AC \text{ (given)}, \\ AD \text{ is common to both triangles} \\ \text{and } \angle BAD = \angle CAD \text{ (construction)}. \end{array}\right.$$

∴ The triangles are congruent

∴ $\angle B = \angle C.$

COR. 1. The bisector of the vertical angle of an isosceles triangle

(*i*) bisects the base;

(*ii*) is perpendicular to the base.

For it has been shown that the triangles *ABD*, *ACD* are congruent;

∴ $BD = CD$

and $\angle ADB = \angle ADC,$

and these are adjacent angles;

∴ *AD* is perpendicular to *BC*.

COR. 2. An equilateral triangle is also equiangular.

THEOREM 16 (Euclid I. 6)

***This is the converse of Theorem* 15.**

If two angles of a triangle are equal, the sides which subtend* these angles are equal.

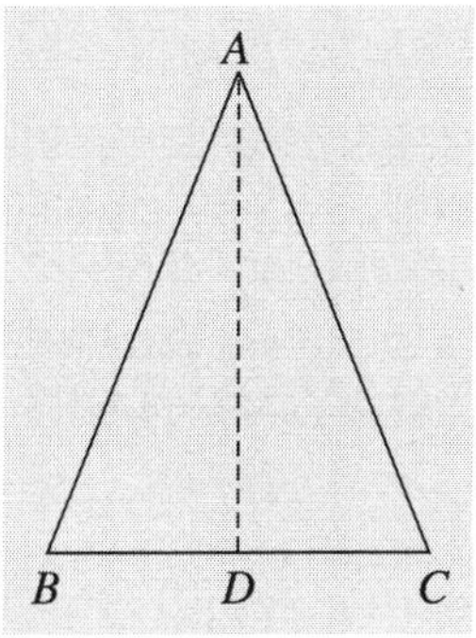

Let ABC be a triangle in which, $\angle B = \angle C$.

It is required to prove that $AB = AC$.

Construction. Let the angle BAC be bisected by a straight line meeting BC in D.

Proof. In the triangles ABD, ACD,

$$\left\{\begin{array}{l} \angle B = \angle C \text{ (given)}, \\ \angle BAD = \angle CAD \text{ (construction)} \\ \text{and } AD \text{ is common to both triangles.} \end{array}\right.$$

∴ The triangles are congruent

∴ The sides opposite to equal angles in each triangle are equal

∴ $$AB = AC.$$

COR. An equiangular triangle is also equilateral.

The following definitions are inserted for use in the Exercises:

DEF. A **circle** is a plane figure bounded by one line called the circumference, and is such that all straight lines drawn from a certain point within the figure to the circumference are equal.

These straight lines are called radii, and the point is called the **centre of the circle**.

Postulate. Let it be granted, that a circle may be described with any point as centre, and with radius equal to any given straight line.

DEF. A **diameter** of a circle is a straight line drawn through the centre and terminated both ways by the circumference.

DEF. A **semi-circle** is the figure bounded by a diameter of a circle and part of the circumference cut off by the diameter.

DEF. A **chord** of a circle is a straight line joining any two points on the circumference.

*'*To subtend' means 'to be opposite to.'*

Def. An **arc** of a circle is any part of the circumference.

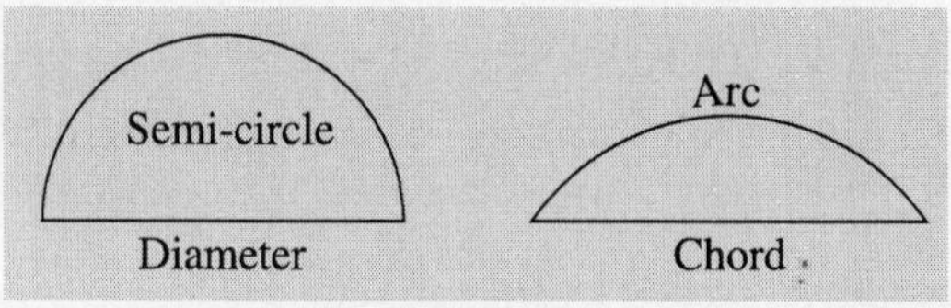

Def. A **segment** of a circle is the figure bounded by any straight line and one of the arcs into which it divides the circumference.

Def. A **sector** of a circle is the figure bounded by two radii and the arc intercepted between them.

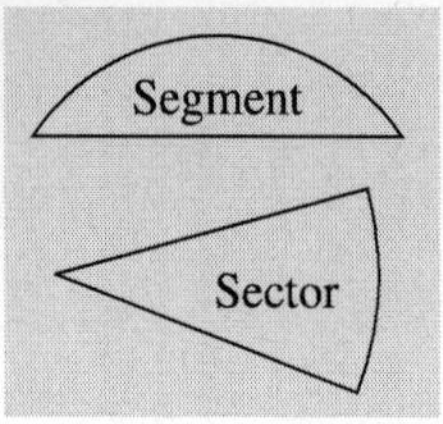

EXERCISE VI (Theorems 15, 16)

1. The vertical angle of an isosceles triangle is 90°; what are the base angles?
2. One of the base angles of an isosceles triangle is 70°; what is the vertical angle?
3. How many degrees are there in an angle of an equilateral triangle?
4. If the equal sides of an isosceles triangle are produced, the angles on the other side of the base are equal.
5. Points X, Y are taken in the base BC of an isosceles triangle ABC, so that $BX = CY$; show that AX, AY are equal and equally inclined to BC.
6. ACB, ADB are isosceles triangles on the same base AB. Show that $\angle CAD = \angle CBD$.
7. Equilateral triangles BAD, CAE are described on the sides AB, AC of an equilateral triangle ABC, externally to the triangle; show that DA is in the same straight line as AE.
8. On the sides BC, CA, AB of an equilateral triangle, the three equilateral triangles BCD, CAE, ABF are drawn, externally to the triangle; prove that D, E, F are the corners of an equilateral triangle.
9. Equilateral triangles ABH, ACK are described on AB, AC, sides of a triangle ABC, externally to the triangle. Show that $CH = BK$.
10. $ABCD$ is a parallelogram in which the diagonal AC bisects the angle at A. Prove that $ABCD$ is a rhombus.
11. If P is any point on the circumference of a semi-circle whose diameter is AB, then the angle APB is a right angle. [Let O be the centre. Join OP. Apply Theorem 15 to Δs OAP, OBP, and show that $\angle APB = \angle A + \angle B$.]
12. Show how to find points D and E on the sides AB and AC respectively of a triangle ABC, such that DE is parallel to BC, and equal to BD. [Observe that if $DB = DE$, then $\angle DBE = \angle DEB$.]
13. A circle is described on the side AB of the triangle ABC as diameter; XY is that diameter of the circle which is parallel to BC. Prove that BX, BY are the bisectors of the interior and exterior angles of the triangle at B.

14. *ABCDE* is a regular pentagon and *OA*, *OB* are the bisectors of the angles *A* and *B*. Prove that *OC*, *OD*, *OE* bisect the angles *C*, *D*, *E*. Hence show that *O* is equidistant from *A*, *B*, *C*, *D*, *E*. [Use Δs *BAO*, *BCO* to show ∠ *OCB* = ∠ *OAB*. Hence show that *OC* bisects ∠ *BCD*.]

THEOREM 17* (Euclid I, 8)

Two triangles are congruent if the three sides of one triangle are respectively equal to the three sides of the other.

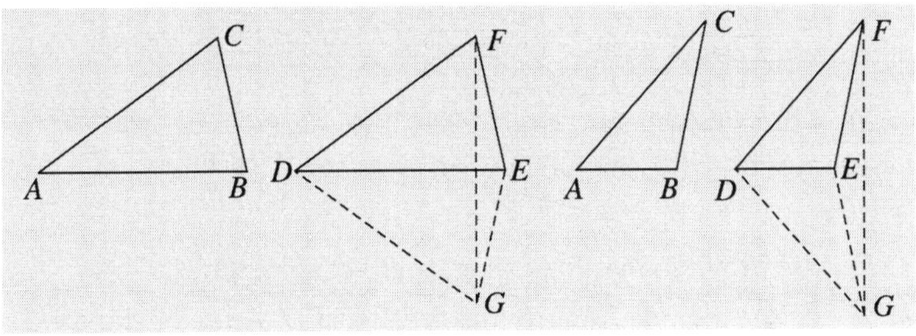

Let *ABC*, *DEF* be two triangles in which

$$AB = DE,\ AC = DF,\ BC = EF.$$

It is required to prove that the triangles are congruent.

Proof. Place the triangle *ABC* so that *A* falls on *D*, *AB* falls along *DE*, and *C* falls at some point *G*, on the opposite side of *DE* to *F*. Join *FG*.

Then, because $AB = DE$ (*given*),

∴ *B* falls on *E*.

Again, it is given that $AC = DF$,

and, by supposition, $AC = DG$;

∴ $DF = DG$,

and these are sides of the triangle *DFG*;

∴ $\angle DFG = \angle DGF$.

Similarly, it can be shown that

$$\angle EFG = \angle EGF.$$

∴ The whole or remainder ∠ *DFE* = The whole or remainder ∠ *DGE*.

But $\angle ACB = \angle DGE$

∴ $\angle ACB = \angle DFE$.

Hence, in the triangles *ABC*, *DEF*,

$CA = FD$ (*given*)
$CB = FE$ (*given*),
the incl. ∠ *ACB* = The incl. ∠ *DFE* (*proved*);

∴ The triangles are congruent.

*See ‘NOTE’ (italic) on page 5.

THEOREM 18*

Two right-angled triangles are congruent if the hypotenuse and a side of one are respectively equal to the hypotenuse and a side of the other.

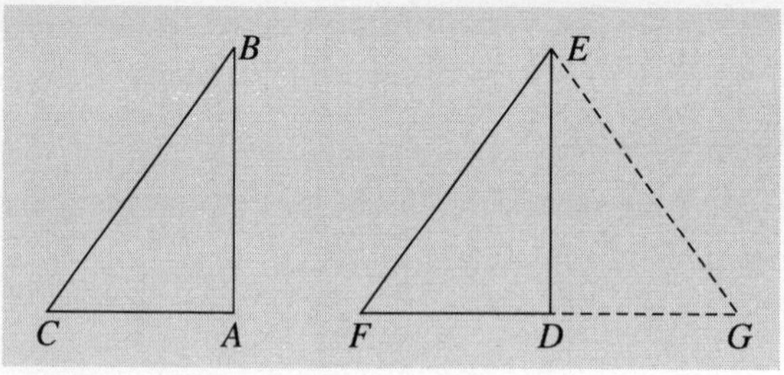

Let ABC, DEF be two right-angled triangles, in which
the hypotenuse BC = The hypotenuse EF
and $AB = DE$.

It is required to prove that the triangles are congruent.

Proof. Place the triangle ABC so that A falls on D, AB falls along DE, and C falls at some point G, on the opposite side of DE to F.

Then, because $AB = DE$ (*given*),

$\therefore$ B falls on E.

Also, it is given that $\angle$s BAC, EDF are right angles,

and $\angle BAC = \angle EDG$;

$\therefore$ The adjacent $\angle$s EDG, EDF are right angles

$\therefore$ DF and DG are in the same straight line.

Again, $BC = EF$ (*given*)

and, by supposition, $BC = EG$.

$\therefore$ EFG is a triangle in which $EF = EG$;

$\therefore$ $\angle EFD = \angle EGD$.

But $\angle BCA = \angle EGD$;

$\therefore$ $\angle BCA = \angle EFD$.

Hence, in the triangles ABC, DEF,

$$\begin{cases} \angle BAC = \angle EDF, \\ \angle BCA = \angle EFD, \\ AB = DE. \end{cases}$$

$\therefore$ The triangles are congruent.

Note on the Congruence of Triangles. It has been shown that two triangles are congruent in the following cases:

(*i*) If two sides and the included angle of the one are respectively equal to two sides and the included angle of the other. (*Theorem* 4)

*See 'NOTE' (italic) on page 5.

(*ii*) If two angles and a side of one are respectively equal to two angles and the corresponding side of the other. (*Theorem* 14)

(*iii*) If the three sides of one are respectively equal to the three sides of the other. (*Theorem* 17)

(*iv*) If two sides of one are respectively equal to two sides of the other and the angle opposite a pair of equal sides in each is a right angle. (*Theorem* 18)

The three sides and the three angles are sometimes called the parts of the triangle.

To prove two triangles congruent, we must have three parts of one respectively equal to three parts of the other, and these parts must be chosen in one of the ways just described.

Two triangles are **not necessarily** congruent in the following cases:

(*i*) If the three angles of one are respectively equal to the three angles of the other.

For example, in the figure *LM* is drawn parallel to *BC* and the triangles *ABC*, *ALM* are equiangular but are not congruent.

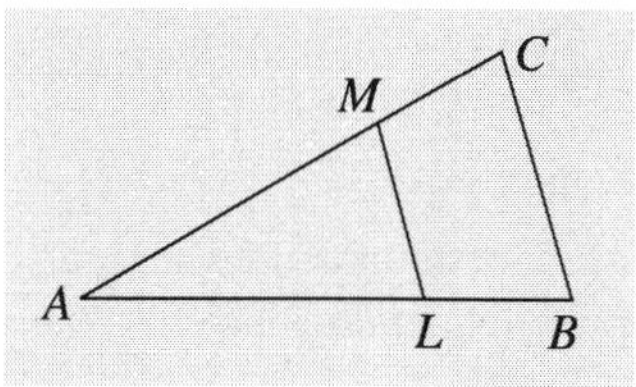

(*ii*) If two sides and an angle of one triangle are respectively equal to two sides and an angle of the other, the equal angles being opposite equal sides in each triangle. Thus, in the triangles *ABC*, *ABX*, *AB* is common, *BX* = *BC*, and the angle *A*, opposite the equal sides *BX*, *BC*, is common, but the triangles are not congruent.

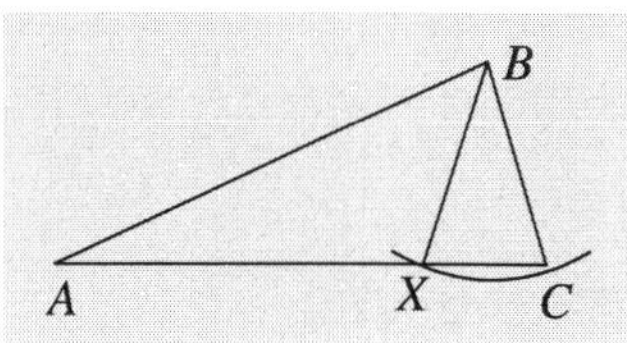

EXERCISE VII (Theorems 17, 18)

1. In the quadrilateral *ABCD*, if *AB* = *CD* and *BC* = *DA*, prove that *ABCD* is a parallelogram.

2. Prove that a rhombus is a parallelogram.

3. The diagonals of a rhombus bisect the angles of the figure and bisect each other at right angles.

4. A chord *AB* of a circle is bisected at *N*. If *O* is the centre, show that *ON* is perpendicular to *AB*.

5. A straight line cuts a circle in *A*, *B* and a concentric circle in *C*, *D*. Show that *AC*, *DB* are equal and that they subtend equal angles at the common centre.

6. *ACB*, *ADB* are two triangles, on the same base *AB* and on the same side of it, such that *AC* = *BD* and *AD* = *BC*. If *AD*, *BC* meet in *O*, prove that the triangles *OAC*, *OBD* are congruent.

7. *A* and *B* are the centres of two circles which cut at *C* and *D*. Prove that *AB* bisects the common chord *CD* and is perpendicular to it. [Let *AB*, *CD* meet in N. Consider Δs *ABC*, *ABD* and then Δs *ANC*, *AND*.]

8. If *AB* is a chord of a circle whose centre is *O*, prove that the perpendicular from *O* to *AB* bisects *AB*.

9. *AB* and *CD* are two equal chords of a circle whose centre is *O*, and *OX*, *OY* are the perpendiculars from *O* to *AB*, *CD* respectively. If *AB* = *CD*, prove that *OX* = *OY*. [Use Ex. 8 to show that *AX* = *CY*, then take Δs *OAX*, *OCY*.]

10. *AB* and *CD* are two chords of a circle whose centre is *C* and *OX*, *OY* are the perpendiculars from *O* to *AB*, *CD* respectively. If *OX* = *OY*, prove that *AB* = *CD*. [Take Δs *OAX*, *OCY* to show *AX* = *CY*, then use Ex. 8.]

11. In any triangle *ABC*, the bisectors of the angles *B* and *C* meet in *O*. Show that *AO* bisects the angle *A*. [Use Exercise V. 5]

12. In any triangle *ABC*, the straight lines bisecting the exterior angles at *B*, *C* formed by producing *AB*, *AC* meet in *O*. Show that *AO* bisects the angle *BAC*. [use Exercise V. 7]

13. In the triangle *ABC*, *AB* is bisected at *X*, *AC* is bisected at *Y*. The perpendiculars through *X* and *Y* to *AB* and *AC* respectively meet in *O*. Prove that *OA* = *OB* = *OC*. Hence show that the perpendicular from *O* to *BC* bisects *BC*.

14. *ABCDE* is a regular pentagon and *OA*, *OB* are the bisectors of the angles *A* and *B*. Prove that *O* is equidistant from the sides of the pentagon. [Use Exercise VI, 14]

VII. THE USE OF INSTRUMENTS

The following instruments are required:

(*i*) A hard pencil.

(*ii*) A pair of compasses with a hard pencil point.

(*iii*) Two set squares, one with angles 60°, 30°, 90°; the other with angles 45°, 45°, 90°.

(*iv*) A ruler with one edge graduated in inches and tenths of an inch, the other in centimetres and millimetres.

(*v*) A pair of dividers.

(*vi*) A protractor.

Your pencils and compass leads should be finished off with glass paper to a fine chisel edge.

Representation of Points

A point should be represented by the intersection of two short lines.

To ensure accuracy in practical work, all lines should be as thin and uniform as possible.

When ruling the straight line joining two points, first place the pencil-point at one of the points, bring the straight-edge up to it and, using the pencil as a pivot, swing the straight edge into position so that the other point is just clear, and rule the line as evenly and continuously as possible.

In drawing a circle of given radius, open your compasses to a slightly greater distance than that required and apply them to the scale, press the legs together until the points are at the right distance and the 'spring' is removed from the legs. Then draw the circle, taking care to hold the compasses by the head only.

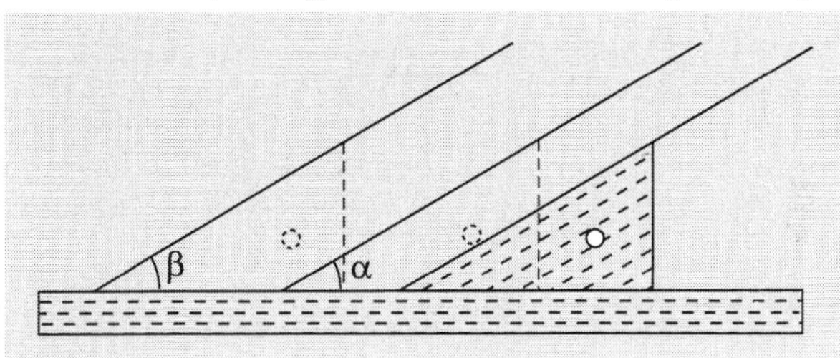

Set-squares are used for drawing parallels, perpendiculars and for making angles of 30, 45 and 60 degrees.

It must be borne in mind that the angles of set-squares are never to be relied on quite close up to the vertex.

Take a ruler and place a set-square with one of its edges along the square edge of the ruler. Hold the ruler firmly and slide the set-square along it into various positions. Draw straight lines using another edge of the set-square as a ruler.

It will be seen that the angles α and β which any two of the lines make with the edge of the ruler are equal, for each is one of the angles of the set-square. Also these angles are corresponding angles, and therefore the lines are parallel.

Through a given point (*C*), draw a straight line parallel to a given straight line (*AB*), using set-squares.

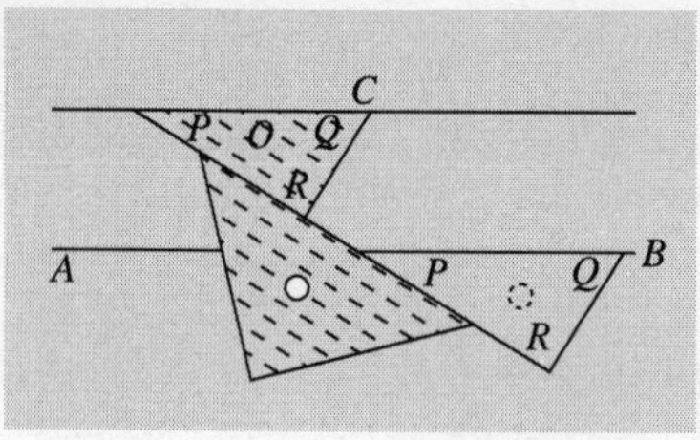

Place a set-square, *PQR*, with its hypotenuse *PQ* along *AB*.

Place another set-square with its hypotenuse in contact with *PR*.

Hold the second set-square firmly, and slide the first into the position *P'Q'R'*, so that *P'Q'* is just clear of *C*.

Draw a line through *C*, using the edge *P'Q'* as a ruler.

This line is the required parallel.

Draw a perpendicular to a given straight line (*HK*), through a given point (*O*), using set-squares.

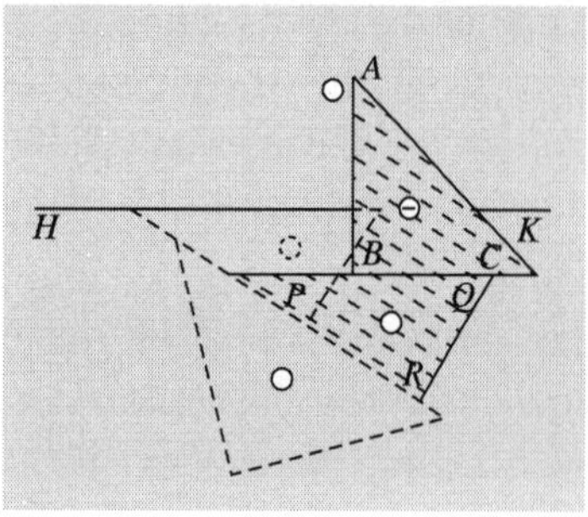

Place a set-square, *PQR*, with its hypotenuse *PQ* parallel to *HK*.

Slide a second set-square, *ABC*, along *QP*, with *BC*, one of the sides containing the right angle, in contact with *PQ*, until *AB* is just clear of *O*.

Using the edge *AB* as a ruler, draw a line through *O*.

This line is the required perpendicular.

Measurement of Length

Any suitable length may be selected as a unit. The unit of length employed in Practical Geometry is either an inch or a centimetre.

A scale, in which the upper numbered divisions are inches and the lower divisions are centimetres, is given below.

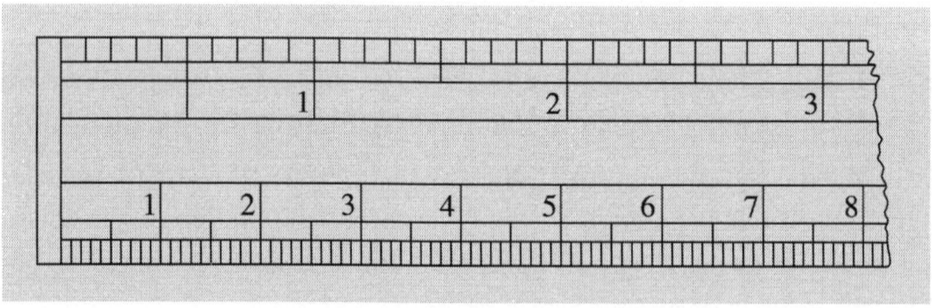

The abbreviation 2′ 7″ is often used to denote 2 feet 7 inches.

Remember that 1 centimetre (cm) = 10 millimetres (mm).

In measuring a line, be careful to tilt the scale, so as to bring the graduations as near to the line as possible. A scale with a thin bevelled edge is convenient.

EXERCISE VIII

Practical

In expressing the result of a measurement, the unit employed should always be stated.

1. What is the length of one of the small lower divisions of the scale on page 32?
2. By referring to the scale on page 32 or to your ruler, state as nearly as you can the number of millimetres in 3 inches. Hence deduce the number of millimetres in 1 inch.
3. In the last question, your final result ought to be more accurate than if you had directly estimated the number of millimetres in 1 inch. Why is this?
4. By referring to the scale on page 32, state as nearly as you can the number of inches and tenths of an inch in 7 centimetres, giving the result decimally.
5. Using the result obtained in the last question, express 1 centimetre as the decimal of an inch.
6. Measure the lengths *AB*, *BC*, *CD* in the figure below, in inches and tenths of an inch, giving the results decimally.

Find the length of *AD* by addition, and state your results as follows:

$$AB = \ldots \text{ in},$$
$$BC = \ldots \text{ in},$$
$$CD = \ldots \text{ in},$$
$$AD = AB + BC + CD = \ldots \text{ in}.$$

Verify the result by measuring *AD*.

7. Measure the distances *PQ* and *XY*, giving the results correct to the nearest millimetre.

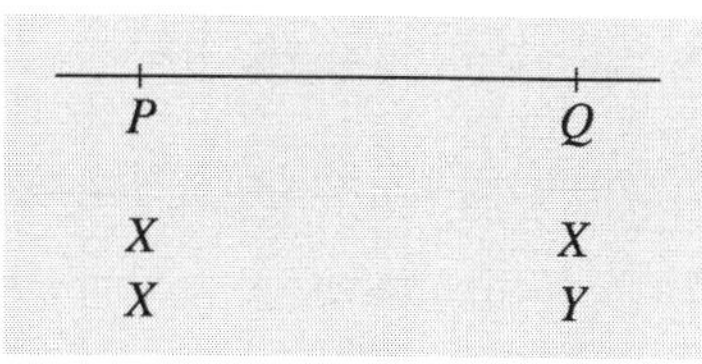

State as follows:

$$PQ = \ldots \text{ mm.},$$
$$XY = \ldots \text{ mm.}$$

It commonly happens that the length of a line, or the distance between two points, is not an exact number of tenths of an inch or an exact number of millimetres.

With practice, always taking care to bring the graduated edge of the scale into close contact with the paper, it is possible to estimate a length correctly to the nearest hundredth of an inch or to the nearest tenth of a millimetre.

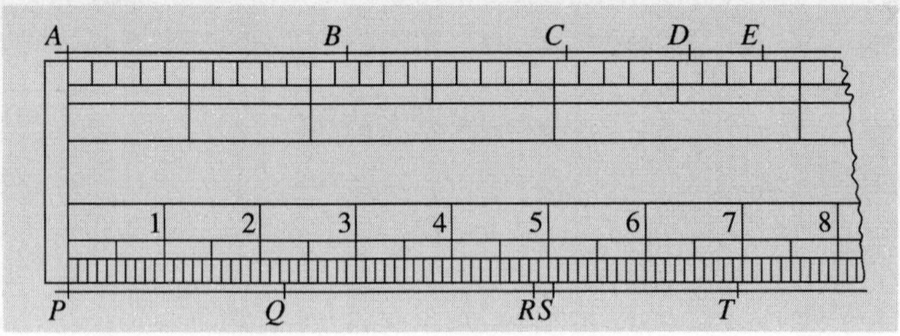

Thus, in the above figure,

$$AB = 1.14 \text{ in}, \qquad AC = 2.07 \text{ in},$$
$$AD = 2.59 \text{ in}, \qquad AE = 2.82 \text{ in},$$
$$PQ = 2.25 \text{ cm}.$$

Estimate the lengths of *PR*, *PS*, *PT*.

When a line has to be measured *accurately* to the nearest hundredth of an inch, a diagonal scale must be used (see p. 72).

The **Dividers** are used for comparing the lengths of straight lines and for measuring the distance between points, the distances being transferred to the scale.

In applying the dividers to the scale, hold them nearly flat, so that the points do not injure the scale.

Open the dividers to a slightly greater distance than that required, and press the legs together until the points are at the right distance and the 'spring' is removed from the legs.

In transferring the distance, hold the dividers by the head only, or you may alter the distance of the points.

Do not make holes, through the paper.

EXERCISE IX

Examples on Measurement and the Use of Set Squares

1. By applying a scale to the figure below, measure the lengths of *AB*, *BC*, *CD*, *DE*, *EF* correct to the nearest hundredth of an inch, and verify by measuring *AF*.

2. By applying a scale to the figure below, measure the lengths of *AB* and *AC*, giving the results in millimetres and tenths of a millimetre. Deduce the

length of BC and verify by measurement.

State your results as follows:

$$AC = \text{....}\ \text{mm.}$$
$$AB = \text{....}\ \text{mm.}$$
$$BC = AC - AB = \text{....}\ \text{mm.}$$

3. Repeat Exs. 1 and 2, using your dividers.
4. Draw a circle of radius 1.3 in. Find points A, B, C on the circumference such that AB = 2.1 in, AC = 2.3 in. Measure BC.
5. Draw a parallelogram $ABCD$ in which AB = 2 in, AD = 1.6 in. $\angle A = 60°$. Draw DE perpendicular to AB. Measure DE and AC.
6. Draw a triangle ABC in which AB = 2.5 in, $\angle A = 30°$, $\angle B = 45°$. Draw perpendiculars AX, BY, CZ to BC, CA, AB respectively. Measure CZ.
7. Draw a straight line AB. Draw AC perpendicular to AB. Along AC set off AD = 1 in. Through D draw DE parallel to AB. *You have now drawn a pair of parallels 1 inch apart.* Take any point O; through O draw a straight line making an angle of 60° with AB and DE, cutting them in X and Y. Measure XY.
8. Draw a pair of parallels 1.8 in. apart. Draw another pair of parallels 2 in. apart, making an angle of 60° with the first pair. Measure the sides of the parallelogram formed by these lines.
9. Draw a circle of radius 1.3 in. In it place a chord of length 2.4 in. Draw the perpendicular from the centre of the circle to the chord. Measure this perpendicular.

Measurement of Angles

Angles are measured with the protractor, which may be semi-circular or rectangular.

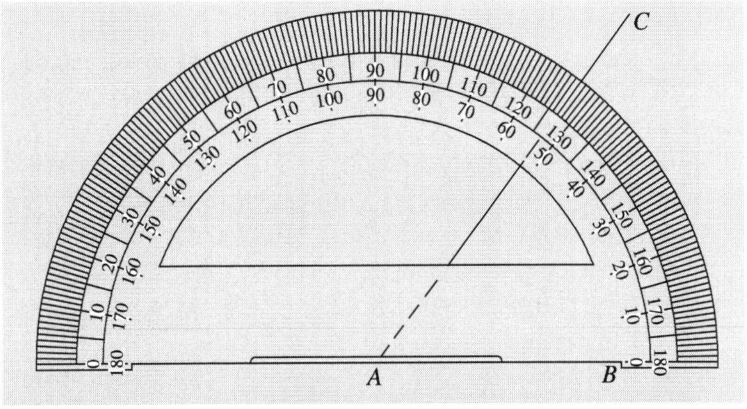

In order to measure an angle BAC with the semi-circular protractor, place the central notch at the vertex A of the angle and the straight edge of the protractor along one of the arms *AB*. Now read off the number of degrees in the angle by observing where *AC* cuts the edge of the protractor.

The best way to use a rectangular protractor for setting out an angle of a given number of degrees is indicated in the following example:

At the point P in the straight line AB, make an angle of 30°.

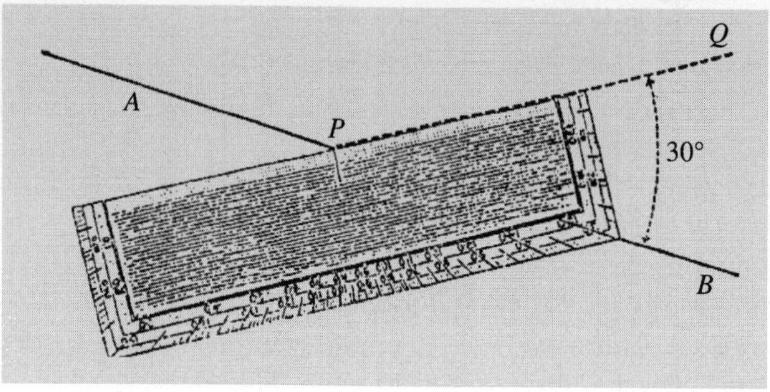

Place the protractor with the central notch at *P* and the division corresponding to 30° on *PB*. Now use the edge through *P* as a ruler and draw the second arm *PQ* of the angle.

EXERCISE X

The Use of Protractors

1. Why has the protractor two sets of numbers round it? How do you know which set to use?
2. Make an angle *BAC* of 37° and produce *CA* through *A* to *D*. Measure ∠ *BAD*. How can you verify your result?
3. At a point *A* in a straight line *AB*, make two angles *BAC*, *BAD* of 50° and 130°, respectively, on opposite sides of *AB*. If your figure is correctly drawn, what relation should *AD* have to *AC*?
4. From a point *O* draw straight lines *OA*, *OB*, *OC*, *OD* so that ∠ *AOB* = 45°, ∠ *BOC* = 50°, ∠ *COD* = 123°. Measure ∠ *DOA*, and verify your result by calculation.
5. Draw a straight line *AB* of length 2 in. and find a point *C* such that *CA* and *CB* are each 3 in. Measures ∠s *A* and *B* of the triangle *ABC*.
6. Construct a triangle *ABC* in which *BC* = 2 in, ∠ *B* = ∠ *C* = 50°. Measure *AB* and *AC*.
7. Draw any quadrilateral, and find the sum of its angles by measurement. Give the enunciation of a theorem which enables you to state what your answer should be.
8. Draw any pentagon, with its sides produced in order. Find by measurement the sum of the exterior angles. Give the enunciation of a theorem which enables you to state the correct result.

Points of the Compass

The directions, North, East, South, West, at any point of the earth's surface, are called the **Cardinal Points** of the **Compass**. Between each pair of cardinal points there are seven other directions called *points*, so that the angle between two adjacent points is an eighth part of a right angle.

A mariner's compass card is shown in the figure.

Note the method of abbreviated lettering. Thus, N.N.W. (nor'-nor'-west) denotes the point bisecting the angle between the points N. and N.W.

In nautical language *a point* is frequently used to denote an angle, and means in that case an eighth part of a right angle.

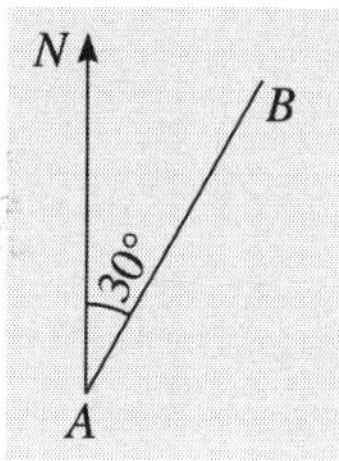

The **Bearing** of an object B as seen from a station A is the angle which the direction of the straight line drawn from A to B makes with North; thus, in the figure, the bearing of B, as seen from A, is 30° East of North. The abbreviation N. by 30° *E.* is often used to denote 30° East of North.

EXERCISE XI

The Points of the Compass

1. How many points of the compass are there?
2. The wind shifts through one point of the compass. What angle in degrees does the vane of a weather-cock turn through?
3. How many degrees are there in a point of the compass? Make a simple diagram showing the *thirty-two* points of the compass, representing each direction by a straight line drawn from a common centre, and name each point by the abbreviated lettering used on the compass card.
4. A ship sailing East turns South; name all the points of the compass it turns through in order.

5. Name the points of the compass which are (*i*) $22\frac{1^\circ}{2}$ South of West, (*ii*) $11\frac{1^\circ}{4}$ West of South, (*iii*) $33\frac{3^\circ}{4}$ East of North.

6. What angle in degrees does each of the following directions make with North : W, NE, NNE, SW?

7. You are marching W and turn NW; what angle do you turn through? You march WSW and turn S by E; what angle do you turn through?

Diagrams Drawn to Scale

By means of a scale, a plan of a large object, such as a field, can readily be drawn.

In order to do this, the unit of the scale is taken to represent some (much greater) length connected with the object.

If a length of 6 in. on the plan represents an actual length of 100 yd, this fact is expressed shortly thus:

$$6 \text{ in.} = 100 \text{ yd},$$

in which statement the sign of equality is an abbreviation for the word "represents."

EXAMPLE: *A ladder, 25 feet long, rests with one end on level ground, 7 feet from a vertical wall, its other end resting against the wall. Find, by means of a diagram, the height of the top of the ladder. (Take 1 in. to represent 20 ft.)*

20 ft. is represented by 1 in.

$\therefore$ 25 ft. .... $\frac{25}{20}$ = 1.25 in.

7 ft. .... $\frac{7}{20}$ = 0.35 in.

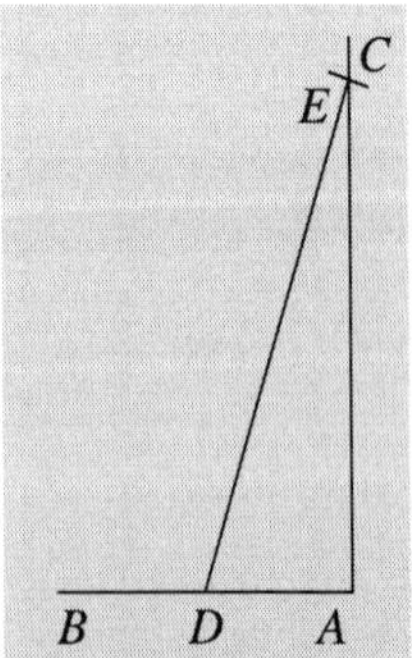

Scale 2 in 2.30 ft.

Draw two straight lines *AB*, *AC* at right angles. Along *AB* set off *AD* = 0.35 in. With centre *D* and radius 1.25 in. draw an arc cutting *AC* in *E*.

By measurement, we find that *AE* = 1.2 in.

Now 1 in. represents 20 ft;

$\therefore$ 1.2 in. .... 20×1.2 ft

$\therefore$ Height of top of ladder = 20×1.2 ft = 24 ft.

NOTE: The scale must always be indicated on the diagram.
If the scale to be used is not given in the question, then a suitable one must be chosen.
In deciding what scale to employ, remember that the larger the diagram, the more accurate is the result likely to be. The diagram should therefore be as large as your paper will allow.

EXERCISE XII

Diagrams to Scale

1. If 4 in. = 1 mile, what is the distance between two points on the plan whose actual distance is 2000 yd.?
2. Draw a straight line to represent a flag pole 35 ft high, scale 1/8 in. = 1 yd. What is the length of your line?
3. The distance from Rugby to Northampton by road is 21 miles. What is the length of the road as shown on a map, 6 in. to the mile?
4. I walk 60 yd. East, then 33 yd. North, then 4 yd. West; find by drawing a diagram (1 in. = 10 yd.) my distance from the starting point.
5. I stand by the hedge of a round garden, the diameter of which is 90 ft, and find that I can just reach the centre of the garden with water from a hose pipe. Draw a diagram, scale 1 in. = 22.5 ft, showing how much of the garden can be watered from this point. If A and B are the extreme points on the edge of the garden which can be reached, measure AB.
6. AB is a ladder, with its foot A on the ground, and B against a vertical wall, C being the foot of the wall. Draw a plan (scale 1 in.= 10 feet) in the following cases:
   (*i*) AB = 65 ft, BC = 63 ft; find AC.
   (*ii*) AB = 41 ft, AC = 9 ft; find BC.
7. The range of a gun is 11.7 miles, and its distance from a straight road is 10.8 miles. Find by a diagram (1 in. = 2 miles) what length of road is dangerous.
8. A gun is 8 miles from a straight road, and I find that it can shell a length of 12 miles of road. What is the gun's range?
9. A, B, C, D are points on the edge of a circular pond, of diameter 85 yd. Given that the diagonal AC = 85 yd., BD = 68 yd., AB = 77 yd., draw a plan (scale 1 in. = 20 yd.), and find the lengths of BC, CD, DA.
10. A man starts from a point A and walks 6 miles East, then 6 miles North, then 3 miles West, then three miles South, finally reaching a joint B. Draw a plan of the route, taking 1 cm to represent 1 mile.

    Find the distance and bearing of B from A.
11. A man starts from a point A and walks 3 miles West and then 3 miles in the direction 30° East of North, finally reaching a point B. Draw a plan, scale 1 in. = 1 mile, and find the distance and bearing of B from A.
12. Two vessels, A and B, start simultaneously from the same point, A in the direction 17° East of North and B in the direction 17° South of East. If each vessel steams 10 miles an hour, how far are they apart at the end of an hour, and in what direction is B as seen from A?

13. Three places, A, B and C, are situated at the corners of an equilateral triangle, and C is to the North of the line AB. If the bearing of B, as seen from A, is 50° East of North, find the bearings of C from A and B.

14. A and B are two stations 9.6 miles apart, and the bearing of B, as seen from A, is 20° East of South. C is a place to the East of the line AB, and distant 10 miles from A and 2.8 miles from B. Draw a plan showing the relative positions of the places, scale 1 cm. = 1 mile, and find the bearings of C from A and B.

15. B and C are two places 5 miles distant from A, and, as viewed from A, they are in the directions 36° West of South and 18° South of West respectively. Find the distance and bearing of B from C.

16. Shooting through a loop-hole which faces North, I can only cover 10° horizontally with my rifle. A man, running West, crosses my line of fire at the rate of 10 yd. a second. Find, how long he is under fire, if the range when he is nearest is 500 yd.

We determine whether a line is horizontal by means of a spirit level. The surface of water at rest is a horizontal plane.

The **vertical** is the direction of a plumb line.

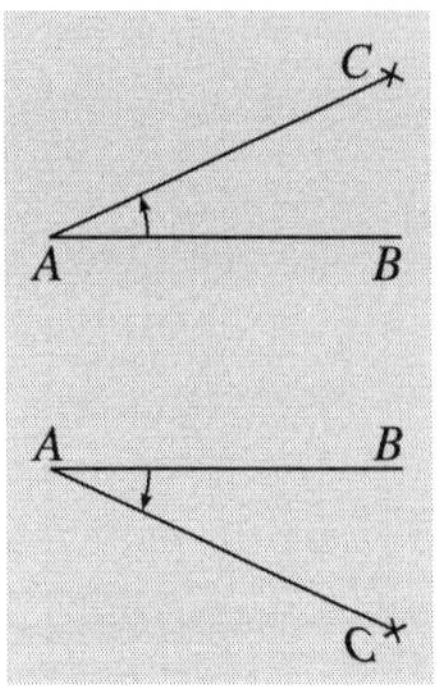

Any plane which contains a vertical line is called a **vertical plane**.

If, looking upwards from a point A, I see an object at C, and AB is the horizontal line in the same vertical plane as AC, the angle BAC is called the (angular) **altitude** or **elevation** of C.

If, again, I look downwards from A, the angle BAC is called the (angular) **depression** of C.

EXERCISE XIII(a)

Diagrams to Scale (*Continued*)

1. The angular elevation of the top of a chimney, from a point in the same horizontal plane as its base and 200 ft distant from the base, is 20°, find the height of the top.

2. I find from the Nautical Almanac that at noon, on a certain day, the sun's altitude is 40°. Find (by diagram) the height of a chimney that casts a shadow 100 ft long.

3. I am on a cliff 200 ft above the sea, and find that the angular depression of a boat is 30°. What is the boat's distance from me?

4. A road rises 1 in 10, the 10 being measured *along the road*; find, by diagram, the angle which the road makes with the horizontal.

5. A straight road whose inclination to the horizon is 10 degrees joins two places which are *marked on the map* as distant 5 miles apart. Find how high one place is above the other. Give the answer in feet.
6. A boat is 3000 yds. from a cliff which is 3000 ft high; what is the angular elevation of the top of the cliff as seen from the boat?
7. The angular elevation of the top of a mountain from a point in the plain distant 5 miles from the top, as shown by a map, is 15°. Find the height of the mountain above the plain in feet.
8. A railway cutting, 40 feet deep, is 118 ft wide at the top and 18 ft wide at the bottom. Find the angle of slope of the sides, these being equally inclined to the horizon.

VIII. ELEMENTARY CONSTRUCTIONS

***For exercises on these constructions, see p.* 50**

In purely geometrical constructions the ruler is used only for joining points and producing straight lines.

The compasses may be used

(*i*) to draw a circle with a given centre and radius;

(*ii*) from the greater of two given straight lines, to cut off a part equal to the less.

The ruler, and compasses may not be used for any other purpose.

The scale and protractor can be used only in numerical applications.

In examinations, questions are often set in which the use of set squares is forbidden. In such questions, parallels and perpendiculars must be drawn with the ruler and compasses only.

In writing out constructions, an accurate figure must be drawn, and all lines required for the construction must be shown clearly: a hard pencil with a very fine point, or a chisel-edge, should be used.

Any additional lines, which may be necessary for supplying a proof, should be dotted.

It is frequently demanded that the figure, which is the object of the construction, should be 'lined in.' This should be done with a pencil giving a *blacker*, though not necessarily a *wider*, line.

Where a point is determined by the intersection of two circles, it is only necessary to show intersecting arcs about one-tenth of an inch in length.

CONSTRUCTION 1

From a given point in a given straight line, draw a straight line making with the given line an angle equal to a given angle.

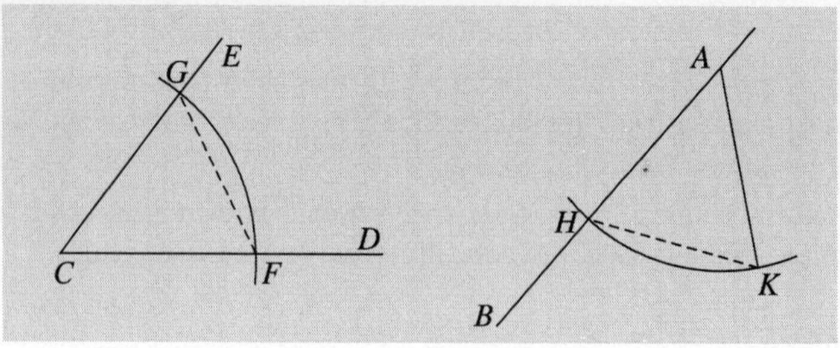

Let A be the given point, AB the given straight line and DCE the given angle.

Construction. With centre C and any radius, draw an arc of a circle, cutting CD in F and CE in G.

With centre A and the same radius, draw an arc of a circle HK, cutting AB in H.

With centre H and radius equal the chord FG, draw an arc of a circle to cut the arc HK in K.

Join AK.

Then AK is the required line.

Proof. Join GF, HK.

In the triangles AHK, CFG, by construction.

$$\begin{cases} AH = CF, \\ AK = CG, \\ HK = FG. \end{cases}$$

∴ The triangles are congruent;

$$\therefore \qquad \angle HAK = \angle FCG;$$

∴ AK is the required line.

CONSTRUCTION 2

Bisect a Given Angle

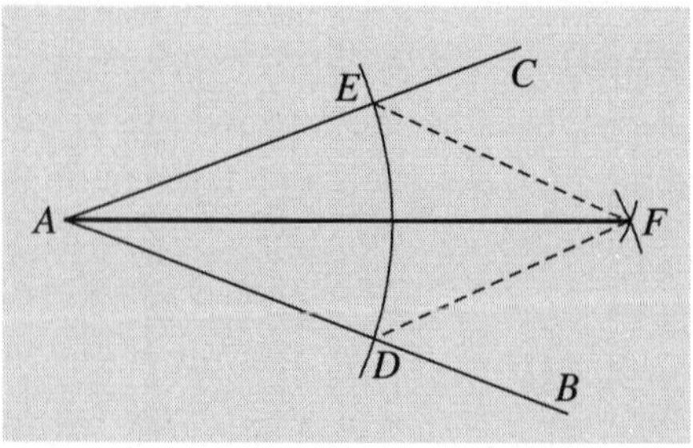

Let BAC be the angle which is to be bisected.

Construction. With centre A and any radius, draw an arc of a circle, cutting AB in D and AC in E.

Draw arcs of circles with centres D and E and any sufficient radius (the same for both circles), to cut in F.

Join AF.

Then AF bisects the angle BAC.

Proof. Join DF and EF.

By construction, DF and EF are radii of equal circles.

$$\therefore \quad DF = EF.$$

Hence, in the triangles ADF, AEF,

$$\begin{cases} AD = AE \text{ (construction)}, \\ DF = EF \text{ (proved)}, \\ AF \text{ is common.} \end{cases}$$

$\therefore$ The triangles are congruent

$$\therefore \quad \angle DAF = \angle EAF;$$

$\therefore$ AF bisects the angle BAC.

NOTE: The circles drawn with centres D and E will in general meet in two points. Theoretically it does not matter which of these points is joined to A. In practice the one which is further from A should be chosen, for the line joining two very near points cannot be drawn accurately.

CONSTRUCTION 3

Bisect a Given Finite Straight Line

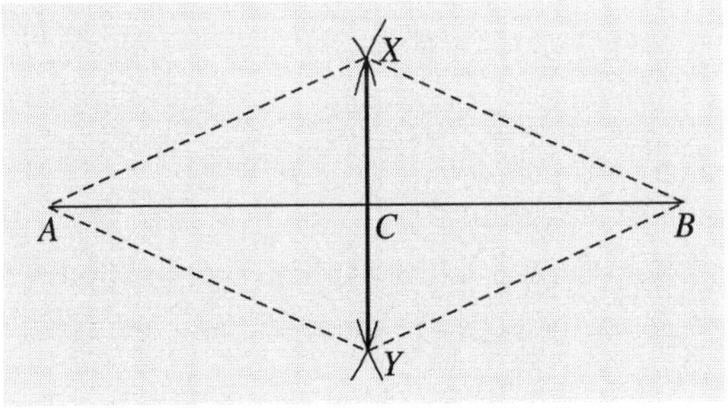

Let AB be the straight line which is to be bisected.

Construction. Draw arcs of circles with centres A and B and any sufficient radius (the same for both circles), to cut in X and Y.

Join XY, cutting AB in C.

Then AB is bisected at C.

Proof. Join AX, BX, AY, BY.

By construction, AX, BX, AY, BY are radii of equal circles;

$$AX = BX \text{ and } AY = BY.$$

Hence, in the triangles AXY, BXY,

$$\begin{cases} AX = BX, \\ AY = BY, \\ XY \text{ is common.} \end{cases}$$

∴ The triangles are congruent;

$$\therefore \quad \angle AXY = \angle BXY.$$

Again, in the triangles AXC, BXC,

$$\begin{cases} AX = BX, \\ XC \text{ is common,} \\ \text{incl. } \angle AXY = \text{incl. } \angle BXY \text{ (proved).} \end{cases}$$

∴ The triangles are congruent

$$\therefore \quad AC = BC.$$

NOTE: The angles at C are right angles, hence this construction gives the perpendicular bisector of AB.

CONSTRUCTION 4

Draw a straight line perpendicular to a given straight line from a given point within it.

First Method

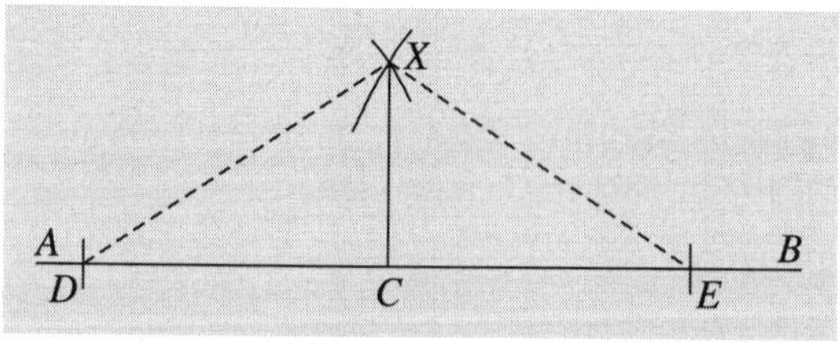

Let AB be the given straight line and C the given point in it. It is required to draw a perpendicular to AB from C.

Construction. Along CA and CB, set off two equal lengths, CD and CE.

Draw arcs of circles with centres D and E and any radius greater than CD (the same for both circles) to cut in X.

Join CX.

Then CX is perpendicular to AB.

Proof. Join DX, EX.

By construction, DX and EX are radii of equal circles;

$$\therefore \quad DX = EX;$$

Hence, in the triangles *CDX*, *CEX*,

$$\begin{cases} CD = CE \ (construction), \\ CX \text{ is common}, \\ DX = EX \ (proved). \end{cases}$$

∴ The triangles are congruent;

$$\therefore \qquad \angle DCX = \angle ECX;$$

∴ *CX* is perpendicular to *AB*.

Second Method*

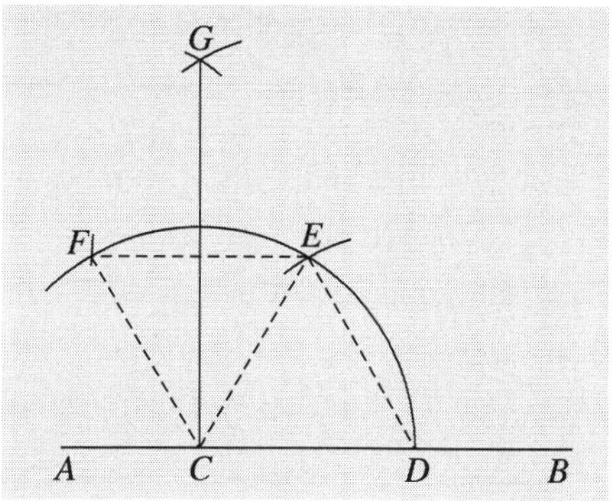

Construction. With centre *C* and any radius, draw an arc of a circle *DEF* cutting *CB* in *D*.

With centre *D* and the same radius as before, draw an arc cutting the arc *DEF* in *E*.

With centre *E* and the same radius as before, draw an arc cutting the arc *DEF* in *F*.

Draw arcs, with centres *E* and *F*, and the same radius as before, to cut in *G*.

Join *CG*.

Then *CG* is perpendicular to *AB*.

Proof. Join *CE*, *CF*, *DE*, *EF*.

By construction, *CD*, *CE*, *CF*, *DE*, *EF* are radii of equal circles.

∴ All these lines are equal.

∴ The triangles *CDE*, *CEF* are equilateral;

$$\therefore \qquad \angle DCE = \angle ECF = 60°.$$

But, by construction, *CG* bisects the angle *ECF*;

$$\therefore \qquad \angle ECG = 30°;$$

$$\therefore \qquad \angle BCG = DCE + \angle ECG = 90°.$$

∴ *CG* is perpendicular to *AB*.

* This method is especially convenient when constructing a square.

CONSTRUCTION 5

Draw a straight line perpendicular to a given straight line of unlimited length, from a given point outside it.

First Method

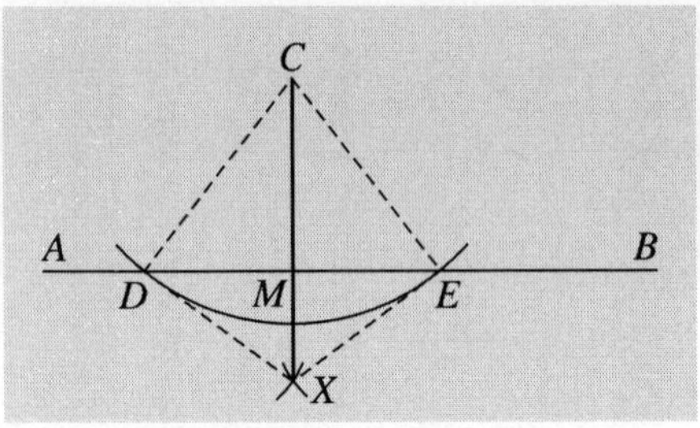

Let AB be the given straight line and C the given point.

It is required to draw a perpendicular from C to AB.

Construction. With centre C and any sufficient radius, draw an arc of a circle to cut AB in D and E.

Draw arcs with centres D and E and any sufficient radius (the same for both) to cut in X. Join CX.

Then CX is perpendicular to AB.

Proof. Let CX meet AB at M.

Join CD, CE, XD, XE.

By construction, DX and EX are radii of equal circles.

$\therefore$ $$DX = EX.$$

Hence, in the triangles CDX, CEX,

$$\begin{cases} CD = CE \text{ (construction)}, \\ CX \text{ is common}, \\ DX = EX \text{ (proved)}. \end{cases}$$

$\therefore$ The triangles are congruent

$\therefore$ $$\angle DCX = \angle ECX.$$

Again, in the triangles CDM, CEM,

$$\begin{cases} CD = CE, \\ CM \text{ is common}, \\ \angle DCM = \angle ECM \text{ (proved)}. \end{cases}$$

$\therefore$ The triangles are congruent;

$\therefore$ $$\angle CMD = \angle CME;$$

$\therefore$ CX is perpendicular to AB.

> **NOTE:** If the perpendicular from C to AB would obviously cut AB near one end, the last construction would be inconvenient.
> For the proof of a method to suit this case we require the following definition and theorem.

Def. The angle contained by the straight lines which join any point on the circumference of a semi-circle to the ends of the diameter is called the **angle in a semi-circle**.

Theorem. The angle in a semi-circle is a right angle.

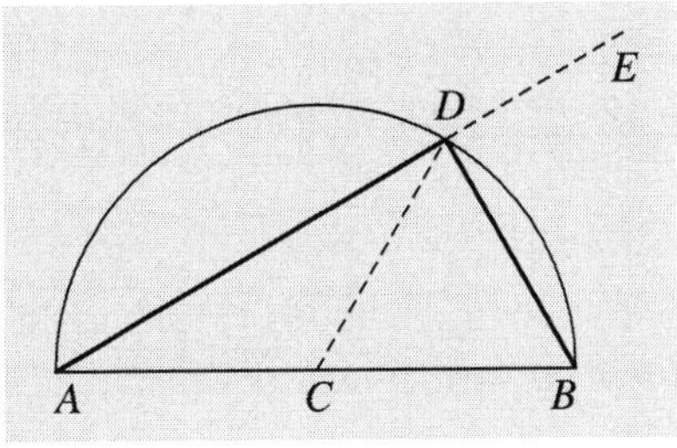

Let D be any point on the circumference of a semi-circle, of which AB is the diameter. Join AD, DB.

It is required to prove that the angle ADB is a right angle.

Construction. Let C be the centre of the semi-circle. Join CD. Produce AD to E.

Proof. Because $CD = CA$, $\therefore \angle CDA = \angle CAD$,

and because $CD = CB$, $\therefore \angle CDB = \angle CBD$;

$\therefore$ $\angle ADB = \angle CAD + \angle CBD$.

Again, the side AD of the triangle ADB is produced to E;

$\therefore$ ext. $\angle EDB = \angle CAD + \angle CBD$,

$\therefore$ $\angle ADB = \angle EDB$,

$\therefore$ $\angle ADB$ is a right angle.

Construction 6

Draw a straight line perpendicular to a given straight line (AB) of unlimited length from a given point (C) outside it.

Second Method

This method suits the case in which the perpendicular from C obviously cuts AB near an end, which is close to the edge of the paper.

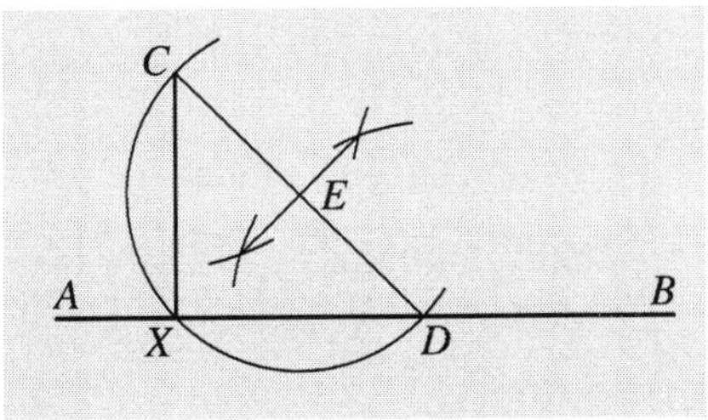

Construction. Through *C* draw any straight line, cutting *AB* at *D*.

Bisect *CD* at *E*.

With centre *E* and radius *ED*, draw an arc of a circle, cutting *AB* again at *X*.

Join *CX*.

Then *CX* is perpendicular to *AB*.

Proof. The angle *CXD* is an angle in the semi-circle *CXD*;

$\therefore \angle CXD$ is a right angle;

$\therefore$ *CX* is perpendicular to *AB*.

The next construction provides a method for drawing a circle to pass through any three given points *A*, *B*, *C*, which are not all in one straight line.

The circle through *A*, *B*, *C* is said to circumscribe the triangle *ABC*, and is called the **circumcircle of the triangle**. Its centre is called the **circumcentre** or the **triangle**.

CONSTRUCTION 7

Draw a circle to pass through three given points which are not in the same straight line.

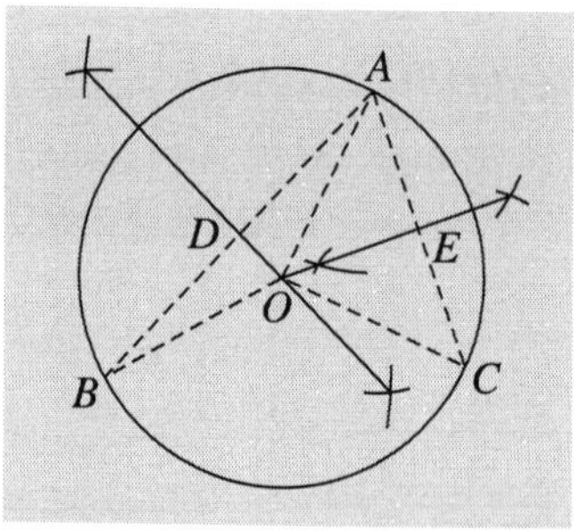

Let *A*, *B*, *C* be the three given points through which it is required to draw a circle.

Construction. Draw the perpendicular bisectors of *AB* and *AC*, meeting in *O*.

With centre *O* and radius *OA*, draw a circle. This circle is the required circle.

Proof. Join *AB*, *AC*, *AO*, *BO*, *CO*.

Let the perpendicular bisectors of *AB*, *AC* meet *AB*, *AC* respectively in *D* and *E*.

By construction, *AD* = *DB*, and the angle at *D* is a right angle;

Hence, in the triangles *ADO*, *BDO*,

$$\begin{cases} AD = DB, \\ DO \text{ is common}, \\ \angle ADO = \angle BDO. \end{cases}$$

$\therefore$ The triangles are congruent,

$$\therefore \qquad OA = OB.$$

Similarly, it can be shown that the triangles *AEO*, *CEO* are congruent.

$\therefore \qquad OA = OC;$

$\therefore$ *OA*, *OB*, *OC* are equal.

Hence the circle described with centre *O* and radius *OA* passes through *B* and *C*.

Construction 8

Draw a parallel to a given straight line through a given point.

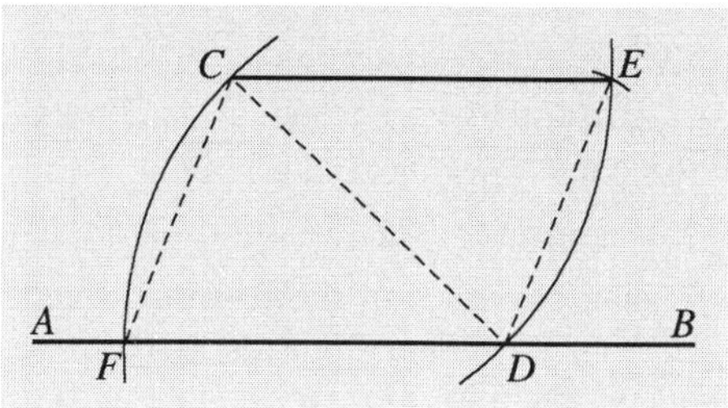

Let *AB* be the given straight line and *C* the given point.

It is required to draw a parallel to *AB* through *C*.

Construction. With centre *C* and any sufficient radius, draw an arc of a circle *DE*, cutting *AB* in *D*.

With centre *D* and the same radius, draw an arc *CF* to cut *AB* in *F*.

With centre *D* and radius equal to the chord *CF*, draw an arc to cut the arc *DE* at *E*, *E* being on the same side of *AB* as *C*.

Join *CE*.

Then *CE* is parallel to *AB*.

Proof. Join *CF*, *CD*, *DE*.

By construction, *CE*, *CD* and *DF* are radii of equal circles;

$\therefore \qquad CE = CD = DF.$

Hence, in the triangles *CDE*, *DCF*,

$$\begin{cases} CE = DF \ (\textit{proved}), \\ CD \text{ is common}, \\ DE = CF \ (\textit{construction}). \end{cases}$$

$\therefore$ The triangles are congruent,

$\therefore \qquad \angle ECD = \angle FDC;$

and these are alternate angles,

$\therefore$ *CE* is parallel to *AB*.

EXERCISE XIII(b) (Constructions 1–8)

In exercises where the figure is given as a hint, the student should complete the construction and proof.

1. On a given straight line, construct an equilateral triangle.
2. Using ruler and compasses only, make angles of 60°, 30°, 45°, 75°.
3. Draw any triangle *ABC*. Bisect the angles *A*, *B*, *C*. Verify that the bisectors meet in a point.
4. Draw any triangle *ABC*. Draw the perpendicular bisectors of *BC*, *CA*, *AB*. Verify that these lines meet in a point.
5. Draw any triangle *ABC*. Draw the perpendiculars from *A*, *B*, *C* to the opposite sides. Verify that these lines meet in a point.
6. Explain how to complete a circle of which an arc is given.
7. Prove the following construction for drawing a perpendicular from *A* to a straight line *AB*. Take any point *C* outside *AB*. With centre *C* and radius *CA*, draw a circle to cut *AB* at *D*. Join *DC*, and produce it to cut the circle at *E*. Join *EA*. Then *EA* is perpendicular to *AB*.
8. Prove the following construction for drawing a perpendicular to a straight line *AB* from a point *C* outside it. With centre *A* and radius *AC*, draw an arc of a circle. With centre *B* and radius *BC*, draw an arc meeting the former arc in *D*. Join *CD*. Then *CD* is perpendicular to *AB*.
9. Prove the following construction for bisecting a given angle *AOB*: "With centre *O* draw two arcs of circles cutting *OA* in *P*, *Q* and *OB*, in *R*, *S* respectively. Join *QR*, *PS*, cutting in *X*. Join *OX*. Then *OX* bisects the angle *AOB*."
10. Given two pieces of string of the same length and four pegs, show how to mark out a corner of a tennis ground. [See Const. 4, Second Method.]
11. Draw a circle to pass through two given points *A* and *B*, and have its centre on a given straight line *XY*. When does the construction fail?

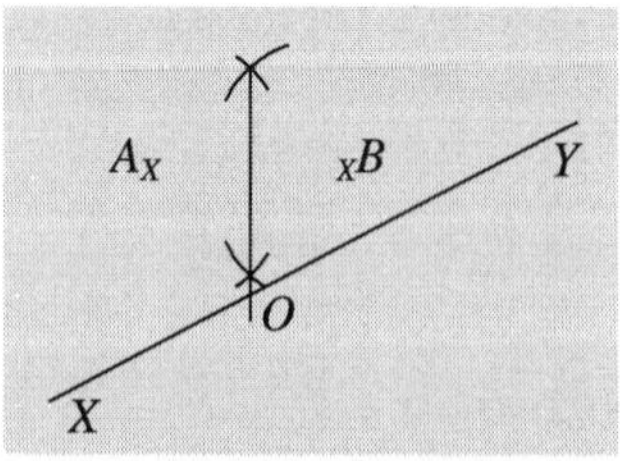

[In the figure, *O* is the centre of the required circle.]

12. Draw a circle to pass through two given points *A*, *B*, and have its centre on a given circle.

When is a solution possible? How many solutions are there?

[In the figure, *P* and *Q* are both centres of circles which satisfy the conditions.]

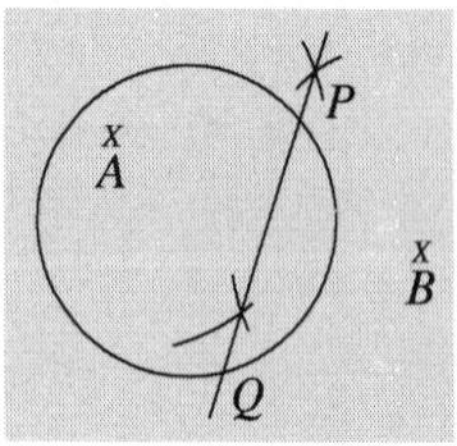

13. Draw a circle of given radius to pass through two given points. How many solutions are there? When is no solution possible?

14. Draw a circle with a given centre to pass through the ends of a diameter of a given circle.

15. *ABC* is a given triangle; *OX* and *OY* are two given straight lines meeting at *O*. Show how to draw a triangle equal in all respects to *ABC*, having a side along *OX* and an angular point in *OY*.

16. Draw a circle with centre *O*. Take a point *A* on the circumference. In the circle place chords *AB*, *BC*, *CD*, *DE*, *EF*, each equal to the radius. Join *FA*. Prove that $AF = AB$, and that all the angles of the figure *ABCDEF* are equal.

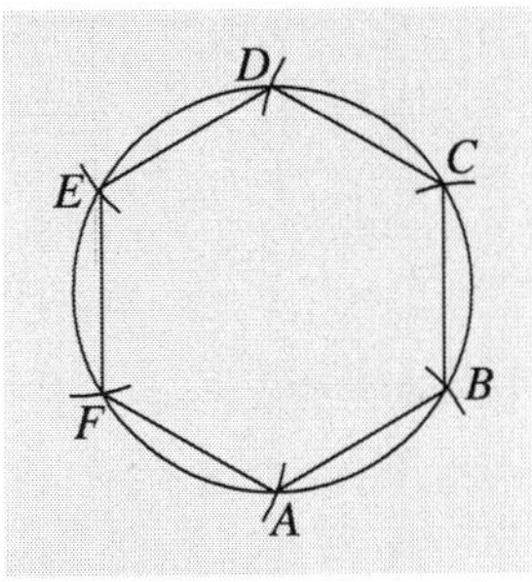

17. Explain how to draw a regular hexagon on a given straight line *AB*. [With centres *A*, *B*, and radius equal to *AB*, draw arcs meeting at *O*. With centre *O* and the same radius, draw a circle. Proceed as in Ex. 16.]

18. On a given straight line *AB* as base, draw an equilateral triangle *ABC*. Bisect the angles *A* and *B* by straight lines meeting in *O*. Draw *OD*, *OE* parallel to *CA*, *CB* respectively, and let them cut *AB* in *D*, *E*. Prove that *D*, *E* are the points of trisection of *AB*; *i.e.* $AD = DE = EB$.

IX. INEQUALITIES CONNECTED WITH THE TRIANGLE

THEOREM 19 (Euclid I. 18)

If two sides of a triangle are unequal, the greater side has the greater angle opposite to it.

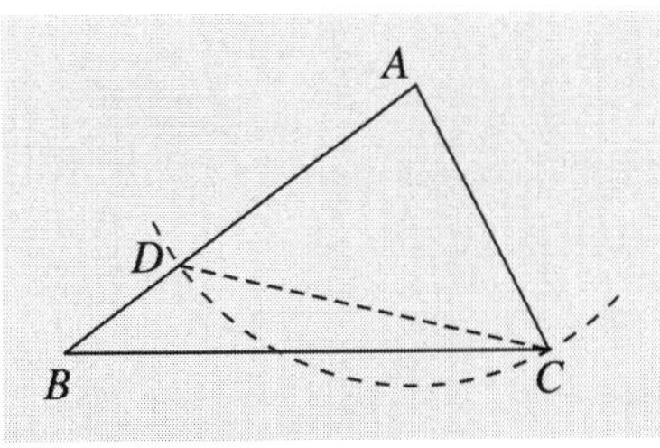

Let *ABC* be a triangle in which $AB > AC$.

It is required to prove that $\angle C > \angle B$.

Construction. From AB cut off AD equal to AC. Join CD.

Proof. In the triangle ADC, by construction,

$$AD = AC$$

$\therefore$ $\angle ADC = \angle ACD.$

Again, the side BD, of the triangle BDC is produced to A.

$\therefore$ The exterior $\angle ADC >$ the interior opposite $\angle DBC$,

that is, $\angle ADC > \angle ABC$;

$\therefore$ Also $\angle ACD > \angle ABC$.

But $\angle ACB > \angle ACD$,

$\therefore$ $\angle ACB > \angle ABC$.

THEOREM 20 (Euclid I. 19)

***This is the Converse of Theorem* 19.**

If two angles of a triangle are unequal, the greater angle has the greater side opposite to it.

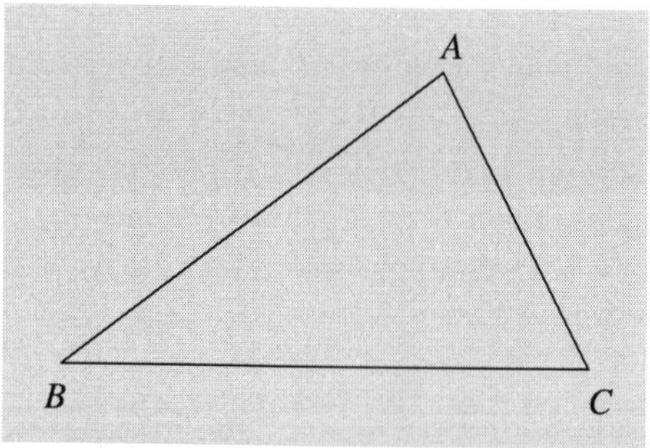

Let ABC be a triangle in which $\angle C > \angle B$.

It is required to prove that $AB > AC$.

Proof. If AB is not greater than AC, it must be either equal to AC or less than AC.

Now, if AB were equal to AC,

$\angle B$ would be equal to $\angle C$;

but this is impossible, for it is given that $\angle C > \angle B$.

Again, if AB were less than AC,

$\angle C$ would be less than $\angle B$;

but this also is impossible, since $\angle C > \angle B$.

Hence AB is neither equal to, nor less than, AC;

$\therefore$ $AB > AC$.

Theorem 21

Of all straight lines that can be drawn to a given straight line from a given external point, the perpendicular is the least.

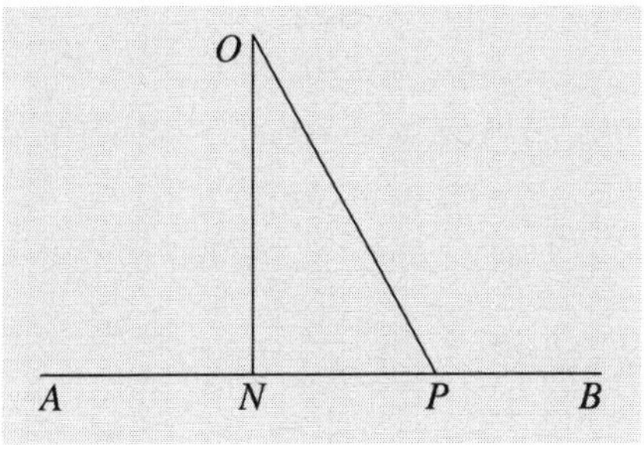

Let *AB* be the given straight line and *O* the given point. Let *ON* be the perpendicular from *O* to *AB*, and let *OP* be any other straight line drawn from *C* to *AB*.

It is required to prove that $ON < OP$.

Proof. The side *PN* of the triangle *OPN* is produced to *A*.

∴ The exterior $\angle ONA >$ the interior opposite $\angle OPN$;

and because *ON* is perpendicular to *AB*,

∴ $\angle ONA = \angle ONP$;

∴ $\angle ONP > \angle OPN$.

Now, in the triangle *ONP*, the greater angle has the greater side opposite to it.

∴ $OP > ON$;

that is, $ON < OP$.

Theorem 22 (Euclid I. 20)

Any two sides of a triangle are together greater than the third side.

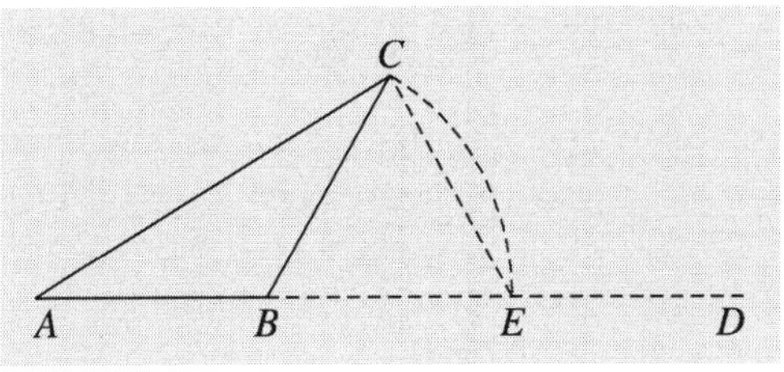

Let *ABC* be a triangle. Choose any pair of sides; for example, *AB*, *BC*.

It is required to prove that

$$AB + BC > AC.$$

Construction. Produce *AB* to *D*.

From *BD* cut off *BE* equal to *BC*.

Join CE.

Proof. In the triangle CBE, by construction,

$$BE = BC;$$

$\therefore$ $\angle BEC = \angle BCE.$

But $\angle ACE > \angle BCE;$

$\therefore$ $\angle ACE > \angle BEC;$

that is, $\angle ACE > \angle AEC.$

Now, in the triangle ACE, the greater angle has the greater side opposite to it;

$\therefore$ $AE > AC.$

But $AE = AB + BE$

and $BE = BC$ (*construction*),

$\therefore$ $AE = AB + BC;$

$\therefore$ $AB + BC > AC.$

EXERCISE XIV (Theorems 19–22)

1. In the figure of Theorem 19, show that

   (*i*) $\angle ADC = \frac{1}{2}(C + B),$

   (*ii*) $\angle BCD = \frac{1}{2}(C - B).$

   [Observe that $\angle ADC = \angle B + \angle BCD$ and $\angle ACD = \angle C - \angle BCD$.]
2. The angle opposite the greatest side of a triangle is greater than 60°.
3. The bisector of the angle A of the triangle ABC meets BC in X. If $AB > AC$, prove that AXC is an acute angle. [Prove that $\angle AXB > \angle AXC$.]
4. The bisector of the angle A of the triangle ABC meets BC in X. Show that $AB > BX$ and $AC > CX$.
5. The side BC of the triangle ABC is bisected in X. Show that (*i*) if $AX > BX$, the angle BAC is acute; (*ii*) if $AX < BX$, the angle BAC is obtuse.
6. In a right-angled triangle, the straight line joining the middle point of the hypotenuse to the opposite vertex is equal to half the hypotenuse. [Use Ex. 5.]
7. If the sides AB, BC, CD, DA of a quadrilateral are in descending order of magnitude, then the angle CDA is greater than the angle CBA,
8. If AN is perpendicular to a straight line BNC, show that AB is greater than, equal to or less than AC, according as BN is greater, equal to or less than NC. [Prove the theorem of equality first. Then if $BN > NC$, take a point D in BN such that $DN = NC$, and prove that $BA > AD$.]

9. If X is a point in the side BC of a triangle ABC, prove that either AB or AC is greater than AX. [Use Ex. 8.]

10. BC is the greatest side of the triangle ABC, and D, E are any points in BC, CA respectively. Prove that $BC > DE$. [Join BE. Show that $BC > BE$ and $BC > CE$. Use Ex. 9 to show that either BE or CE is greater than DE.]

11. BC is the greatest side of the triangle ABC, and E, F are any points in CA, AB respectively. Prove that $BC > EF$. [Use Δs BEC, BEF to show $BC > BE > EF$.]

12. No straight line can be drawn within a triangle greater than the greatest side. [Use Exs. 10, 11]

13. A is the greatest angle of the triangle ABC. Show that it is not possible to make a triangle with sides equal to AB, AC, $2BC$ respectively.

14. If x, y, z are the distances from the vertices of any point within a triangle whose sides are a, b, c, then

$$x + y + z > \frac{1}{2}(a + b + c).$$

[Let ABC be the triangle and O the point within it. Apply Theorem 22 to Δs OBC, OCA, OAB.]

15. The sum of the diagonals of a quadrilateral is greater than the semi-perimeter. [Let $ABCD$ be the quadrilateral, and let AC, BD meet in O. Apply Theorem 22 to Δs OAB, OBC, OCD, ODA.]

16. If x, y, z, w are the distances from the vertices of any point within a quadrilateral, whose sides are a, b, c, d, then

$$x + y + z + w > \frac{1}{2}(a + b + c + d).$$

[Method of Ex. 14.]

17. The sum of the diagonals of a quadrilateral is less than the perimeter, but greater than half the perimeter. [Let $ABCD$ be the quadrilateral. For the first part, apply Theorem 22 to the triangles ABC, ADC, BAD, BCD; for the second part, let the diagonals intersect in O, and apply Theorem 22 to the Δs AOB, BOC, COD, DOA.]

18. Find the point the sum of whose distances from the four angular points of a quadrilateral is least.

19. ABC is an equilateral triangle, and D is a point within it. Show that it is possible to construct a triangle with its sides equal to DA, DB, DC respectively.

20. ABC is a triangle, and M is the middle point of BC; prove that $AB + AC > 2AM$. [Produce AM to N, so that $MN = AM$. Join BN. Prove $BN = AC$.]

21. AB is a given straight line; C, D are two given points on the same side of it; find a point P in AB such that, if Q, is any other point in AB, then $CP + PD < CQ + QD$. [Draw CN perpendicular to AB and produce it to X*, so that $NX = CN$. Join DX, cutting AB in P.]

---

* X is called the image of C in AB. See Example 3, p. 84.

THEOREM 23 (Euclid I. 21)

If from the ends of a side of a triangle two straight lines are drawn to a point within the triangle, these lines are together less than the other two sides of the triangle, but they contain a greater angle.

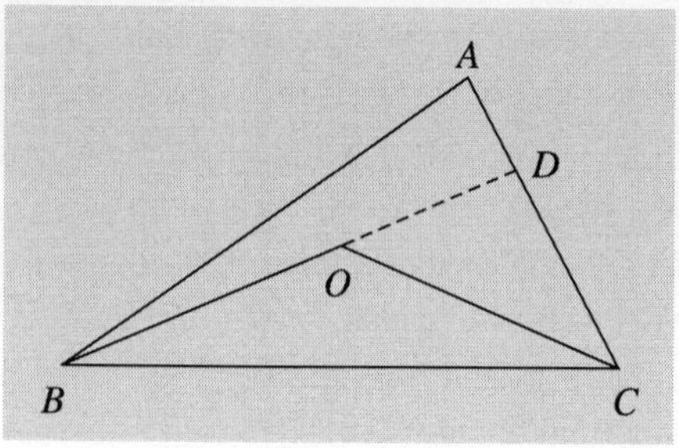

Let ABC be a triangle, and let straight lines be drawn from B, C to any point O within the triangle,

(*i*) It is required to prove that

$$BO + OC < BA + AC.$$

Construction. Produce BO to cut AC at D.

Proof. In the triangle BAD,

$$BA + AD > BD.$$

To each of these unequals add DC;

$$\therefore \quad AB + AC > BD + DC.$$

Again, in the triangle ODC,

$$OD + DC > OC.$$

To each of these unequals add BO;

$$\therefore \quad BD + DC > BO + OC.$$

But it has been shown that

$$BA + AC > BD + DC;$$

$$\therefore \quad BA + AC > BO + OC.$$

(*ii*) It is required to show that $\angle BOC > \angle A$.

The side DO of the triangle DOC is produced to B.

$\therefore$ The exterior $\angle BOC >$ the interior opposite $\angle ODC$.

Also, the side AD of the triangle BAD is produced to C;

$\therefore$ The exterior $\angle ODC >$ the interior opposite $\angle BAD$.

Hence also $\angle BOC > \angle BAD$.

THEOREM 24 (Euclid I. 24)

If two triangles have two sides of the one respectively equal to two sides of the other and the included angles unequal, then the base of the triangle with the greater included angle is greater than the base of the other.

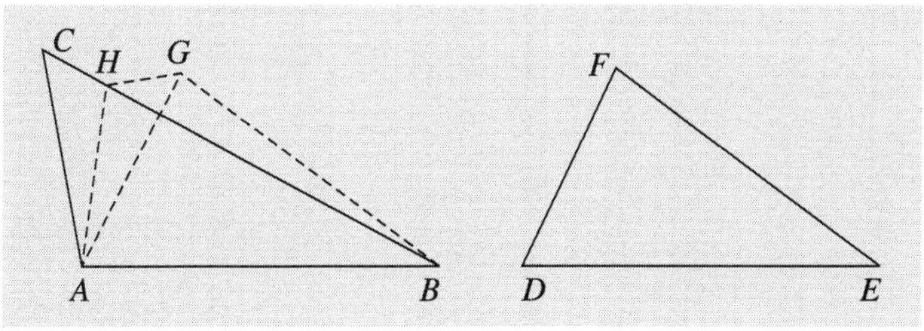

Let ABC, DEF be two triangles in which

$$AB = DE \text{ and } AC = DF$$

and the included $\angle BAC >$ the included $\angle EDF$.

It is required to prove that $BC > EF$.

Proof. Place the triangle DEF so that D falls on A, DE falls along AB and F falls at some point G, on the same side of AB as C.

It is given that $\angle BAC > \angle BAG$;

$\therefore$ AG is within the angle BAC.

Let AH be drawn to bisect the angle CAG and to meet BC in H.

Join GH.

Then, in the triangles ACH, AGH,

$$\begin{cases} AC = AG \text{ (given)}, \\ AH \text{ is common}, \\ \angle CAH = \angle GAH \text{ (construction)}. \end{cases}$$

$\therefore$ The triangles are congruent,

$\therefore$ $HC = HG$.

Now $BC = BH + HC$;

$\therefore$ $BC = BH + HG$.

Also any two sides of the triangle BHG are together greater than the third;

$\therefore$ $BH + HG > BG$;

$\therefore$ $BC > BG$;

that is, $BC > EF$.

THEOREM 25 (Euclid I. 25)

This is the converse of Theorem 24

If two triangles have two sides of the one respectively equal to two sides of the other and their bases unequal, then the angle contained by the sides of the triangle with the greater base is greater than the corresponding angle of the other.

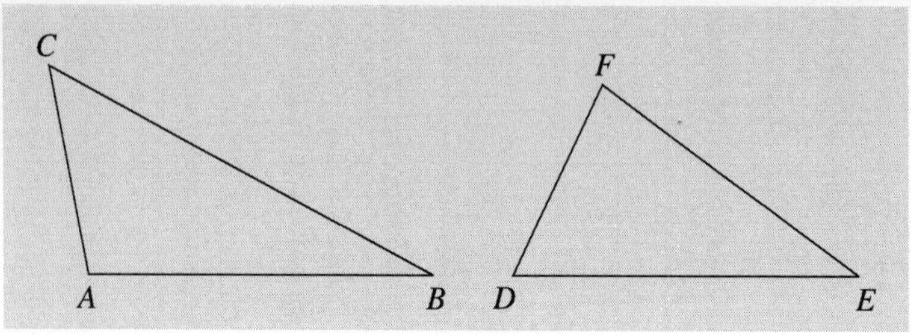

Let *ABC*, *DEF* be two triangles in which

$$AB = DE,$$
$$AC = DF$$

and $$BC > EF.$$

It is required to prove that $\angle A > \angle D$.

Proof. If $\angle A$ is not greater than $\angle D$, it must be either equal to or less than $\angle D$.

If $\angle A$ were equal to $\angle D$,

BC would be equal to *EF*;

but this is impossible, for it is given that $BC > EF$.

Again, if $\angle A$ were less than $\angle D$,

BC would be less than *EF*;

but this also is impossible, for $BC > EF$.

Hence $\angle A$ is neither equal to nor less than $\angle D$;

$\therefore$ $$\angle A > \angle D.$$

EXERCISE XV (Theorems 23–25)

1. *C*, *C′* are the centres of equal circles in which *AB*, *A′B′* are chords. Prove that if $AB > A'B'$, then $\angle ACB > \angle A'C'B'$.
2. State and prove the converse of the theorem in Ex. 1.
3. In the triangle *ABC*, *AB* is bisected in *X* and $AC > CB$. Prove that the angle *AXC* is obtuse, and that if *D* is any point in *CX*, then $AD > BD$.
4. If, within the triangle *ABC*, a point *O* can be found such that *OA* is equal to *AB*, show that *BC* is greater than *CO*.

5. If x, y, z are the distances from the vertices of any point within a triangle whose sides are a, b, c; then $a + b + c > x + y + z$.

6. AB is the diameter of a given circle, passing through a point C within the circle, and AC contains the centre O. Prove the following:

(*i*) CA is the greatest straight line that can be drawn from C to the circumference;

(*ii*) CB is the least straight line that can be drawn from C to the circumference;

(*iii*) If CX, CY are any other straight lines from C to the circumference, and if $\angle XOC > \angle YOC$, then $CX > CY$; and conversely.

7. C is any point outside a circle, whose diameter AB when produced passes through C, and AC contains the centre O. Prove the following:

(*i*) CA is the greatest of all straight lines which can be drawn from C to the circumference;

(*ii*) CB is the least of all such lines;

(*iii*) If CX, CY are any other straight lines from C to the circumference, and if $\angle XOC > \angle YOC$, then $CX > CY$; and conversely.

8. If, in the sides AB, AC produced of a triangle ABC, in which AC is greater than AB, points D, E be taken such that BD, CE are equal, then BE is greater than CD. [Apply Theorem 24 to Δs BCE, CBD.]

X. CONSTRUCTION OF A TRIANGLE WITH GIVEN PARTS

For exercises on these constructions, see p. 62

It has been shown that two triangles are congruent (and therefore of the same size and shape) if three parts of one triangle are respectively equal to three parts of the other, these parts being chosen as described on p. 29.

If, then, three properly chosen parts of a triangle are given, the size and shape of the triangle are determined. Various cases which arise are discussed later.

It is usual to denote the angles of the triangle ABC by the letters A, B, C, and the lengths of the sides opposite these angles by a, b, c respectively.

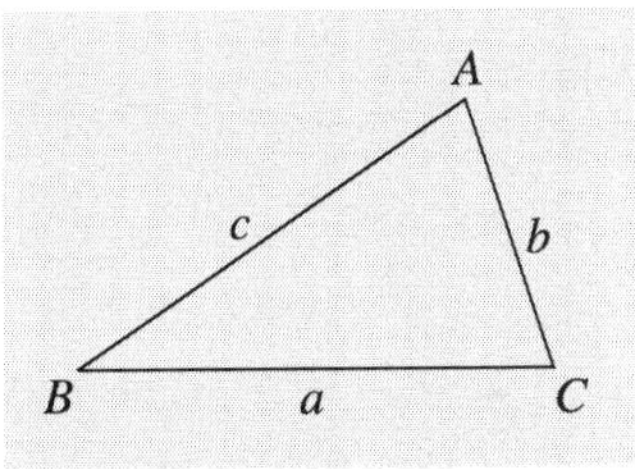

CONSTRUCTION 9

Describe a triangle with its sides respectively equal to three given straight lines, any two of which are together greater than the third.

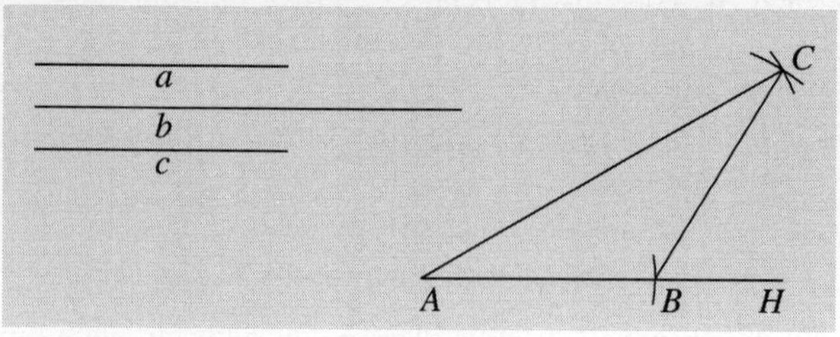

Let a, b, c be the three given straight lines.

Construction. Draw a straight line AH.

Along AH set off AB equal to c.

With centre A and radius equal to b, draw an arc of a circle.

With centre B and radius equal to a, draw an arc of a circle, cutting the former arc at C. Join AC, BC.

Then ABC is the required triangle.

CONSTRUCTION 10

Describe a triangle, given a side and two angles.

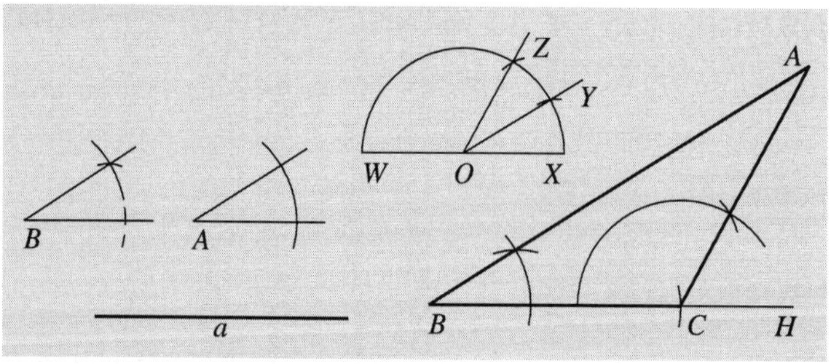

(i) Given two angles A, B and the side a opposite one of the given angles.

Construction. Draw a straight line BH.

Along BH set off BC equal to a.

Draw a straight line WOX.

Make the adjacent angles XOY, YOZ equal to A, B respectively.

Since the angles XOY, YOZ, ZOW are together equal to two right angles,

$\therefore \angle ZOW$ is equal to the third angle C of the required triangle.

At the points B and C in the line BH, draw straight lines, making with BC angles equal to B and C respectively.

Let these straight lines meet at A.

Then ABC is the required triangle.

(*ii*) Given two angles B and C and the side a adjacent to these angles.

Construction. This is obvious from the construction above.

CONSTRUCTION 11

Describe a triangle, given two sides and the angle opposite one of them.

Let a, b be the given sides and A the given angle.

Construction. Make an angle HAK equal to A.

Along AH set off AC equal to b.

With centre C and radius equal to a, draw a circle.

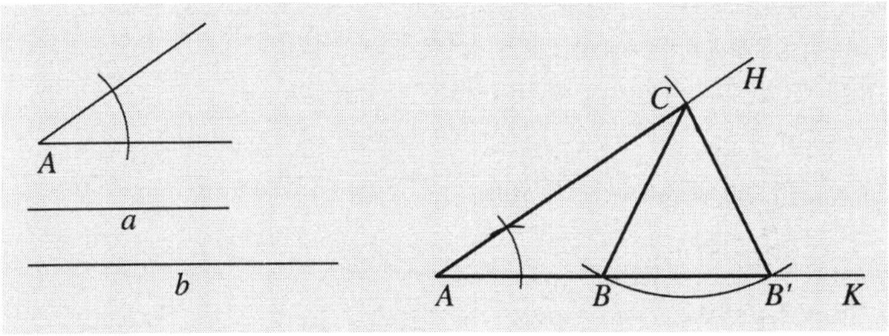

Several cases now arise:

(*i*) If, as in the figure, the circle cuts AK, produced to any length towards K, in two points B and B', there are two solutions of the problem, namely the triangles ABC, $AB'C$.

(*ii*) If the circle cuts AK in B and KA produced through A in K', there is only one solution, namely the triangle ABC.

(*iii*) If the circle meets AK, produced to any length, in only one point B, there is only one solution, namely the triangle ABC.

(*iv*) If the circle does not meet AK, produced to any length, no triangle exists with the given parts.

NOTE: Case (*i*) is called the **ambiguous case**. For two triangles can be drawn, different in size and shape, each having the given parts.

The student should draw figures to illustrate cases (*ii*)–(*iv*). He will then see that if p is the length of the perpendicular from C to AK,

Case (*i*) arises when $b > a > p$.

Case (*ii*) .... $a > b$.

Case (*iii*) .... $a = p$.

Case (*iv*) .... $a < p$.

Further, it will be seen that the ambiguous case cannot arise unless the given angle A is acute and $b > a$.

Hence, two triangles which have the above parts are not necessarily congruent unless the given angle is either (1) right; or (2) obtuse; or (3) if acute, is opposite the greater of the two given sides. In these three cases the triangles are congruent. The case in which the angle A is a right angle is so important that it will be considered separately.

CONSTRUCTION 12

Describe a right-angled triangle, given the hypotenuse and one side.

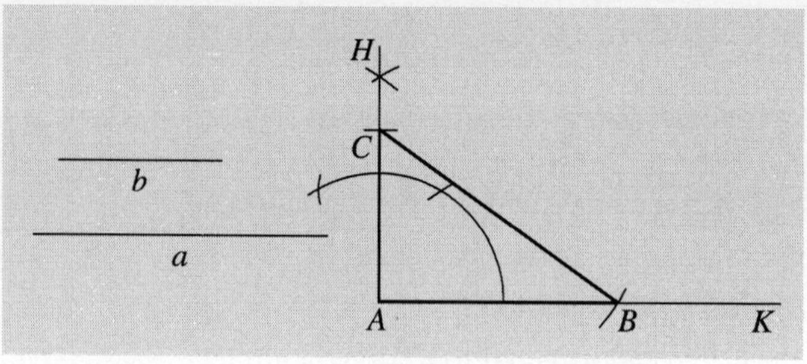

Let a be the given hypotenuse and b the given side. Draw two straight lines AH, AK at right angles. Along AH set off AC equal to b. With centre C and radius equal to a, draw an arc of a circle to cut AK in B.

Join BC.

Then ABC is the required triangle.

EXERCISE XVI (Constructions 9–12)

1. Draw figures to illustrate the failure of Construction 9 (as given on p. 60) in cases where any two of the given lines are not together greater than the third. Take lengths as shown below:

   (*i*) $a + b = c$. (*ii*) $b + c = a$. (*iii*) $a + b < c$. (*iv*) $b + c < a$.

   In Examples 2–12, *construct triangles to the given measurements and, in each case, measure the remaining sides and angles.*

2. $BC = 10$ cm, $CA = 8$ cm, $AB = 6$ cm.
3. $BC = 7$ cm, $CA = 10$ cm, $AB = 5$ cm.
4. $\angle BAC = 42°$, $AB = 2.55$ in, $AC = 3.19$ in.
5. $\angle ABC = 135°$, $BA = 4.02$ in, $BC = 2.88$ in.
6. $BC = 3.56$ in, $\angle ABC = 90°$, $\angle ACB = 53°$.
7. $AC = 8.4$ cm, $\angle CAB = 60°$, $\angle ACB = 36°$.
8. $AB = 6.2$ cm, $\angle ACB = 34°$, $\angle BAC = 34°$.
9. $a = 1.5$ in, $A = 27°$, $C = 19°$.
10. $b = 1.4$ in, $c = 2.8$ in, $B = 30°$.
11. $b = 7$ cm, $c = 3$ cm, $B = 45°$.
12. $b = 7$ cm, $c = 3$ cm, $B = 135°$.
13. Construct a right-angled isosceles triangle whose greatest side is $1\frac{1}{2}$ in. Measure the equal sides.

14. Construct a right-angled triangle in which the hypotenuse is 3 in. and one side 2 in. Measure the smallest angle.

In Examples 15, 16 *explain why it is impossible to construct triangles with the given parts. Illustrate each case with a figure.*

15. $a = 2$ cm, $b = 3$ cm, $c = 6$ cm.

16. $a = 6$ cm, $b = 2$ cm, $B = 30°$.

17. Construct an equilateral triangle, the length of the line joining the vertex to the middle point of the base being 1.2 in. Measure a side.

For Examples on the Construction of Quadrilaterals, see Exercises XVIII (*a*) *and* (*b*), *pp.* 67-68.

XI. PARALLELOGRAMS

THEOREM 26 (Euclid I. 33)

The straight lines which join the ends of two equal and parallel straight lines towards the same parts are themselves equal and parallel.

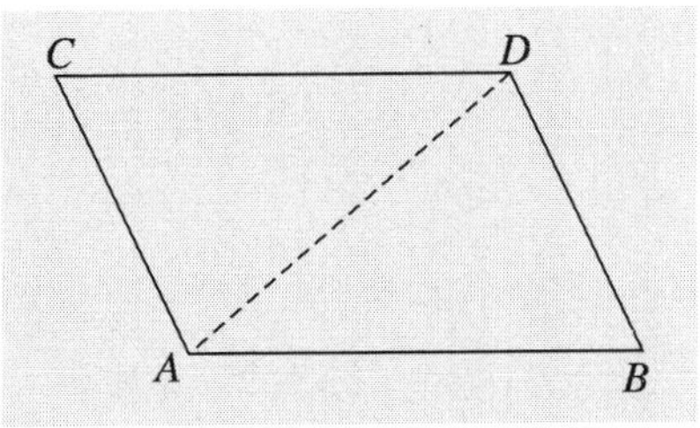

Let AB and CD be equal and parallel straight lines, and let them be joined towards the same parts by the straight lines AC and BD.

It is required to prove that AC and BD are equal and parallel.

Construction. Join AD.

Proof. Because AB is parallel to CD and AD meets them,

$\therefore$ $\angle ADC =$ The alternate $\angle DAB$.

Hence, in the triangles ADC, DAB,

$$\begin{cases} DC = AB \ (given), \\ AD \text{ is common}, \\ \text{incl. } \angle ADC = \text{Incl. } \angle DAB. \end{cases}$$

$\therefore$ The triangles are congruent;

$\therefore$ $AC = DB$,

and $\angle DAC = \angle ADB$.

Now these are alternate angles,

$\therefore$ AC is parallel to DB.

Hence AC and DB are equal and parallel.

THEOREM 27 (Euclid I. 34)

The opposite sides and angles of a parallelogram are equal, and each diagonal bisects the parallelogram.

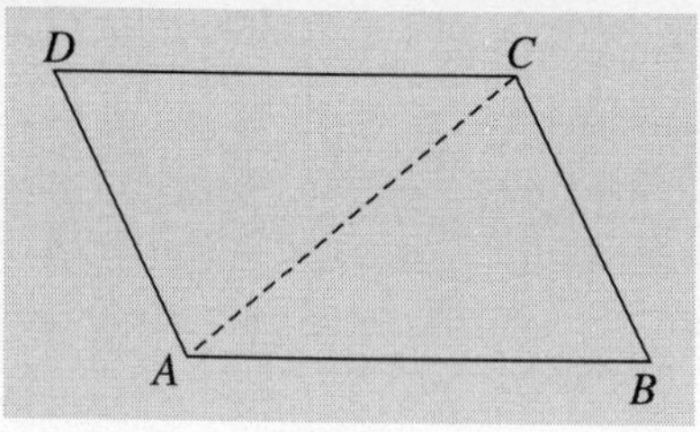

Let *ABCD* be a parallelogram.

It is required to prove that its opposite sides and angles are equal, and that each diagonal bisects the parallelogram.

Construction. Join *AC*.

Proof. By the definition of a parallelogram, *AB* is parallel to *CD* and *BC* is parallel to *DA*.

Because *AC* meets the parallels *AB*, *CD*,

∴ $\angle BAC =$ The alternate $\angle DCA$;

and because *AC* meets the parallels *BC*, *DA*.

∴ $\angle BCA =$ The alternate $\angle DAC$.

Hence, in the triangles *ACB*, *CAD*,

$$\begin{cases} \angle BAC = \angle DCA \text{ (proved)}, \\ \angle BCA = \angle DAC \text{ (proved)}, \\ AC \text{ is common.} \end{cases}$$

∴ The triangles are congruent;

∴ $AB = CD$,

$BC = DA$,

and $\angle ABC = \angle CDA$;

and the triangles *ACB*, *CAD* are equal in area, that is *AC* bisects the parallelogram.

Similarly, by joining *BD*, it can be shown that

$$\angle BAD = \angle BCD,$$

and that *BD* bisects the parallelogram.

THEOREM 28

The diagonals of a parallelogram bisect one another.

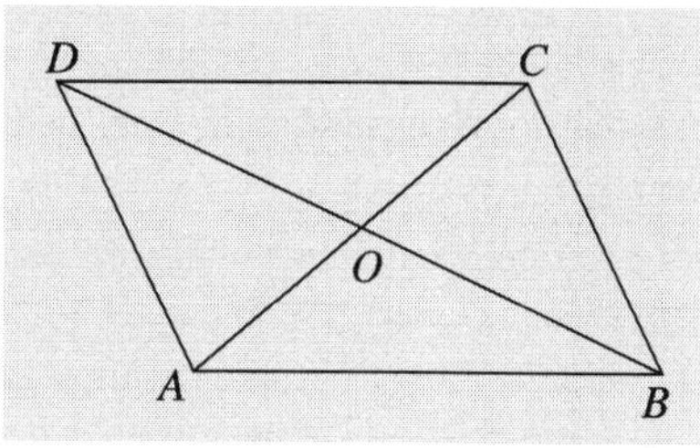

Let *ABCD* be a parallelogram, and let the diagonals *AC*, *BD* meet at *O*.

It is required to prove that $AO = CO$ and $BO = DO$.

Proof. Because *ABCD* is a parallelogram,

∴ *AB* is parallel to *CD* (*Def.*)

and $AB = CD$ (*by the last theorem*).

Again, because *AC* meets the parallels *AB*, *CD*,

∴ $\angle BAO =$ The alternate $\angle DCO$.

Also, because *BD* meets the parallels *AB*, *CD*,

∴ $\angle ABO =$ The alternate $\angle CDO$.

Hence, in the triangles *ABO*, *CDO*,

$$\begin{cases} \angle BAO = \angle DCO \text{ (proved)}, \\ \angle ABO = \angle CDO \text{ (proved)}, \\ AB = CD. \end{cases}$$

∴ The triangles are congruent.

∴ The sides opposite to the equal angles in each triangle are equal;

∴ $AO = CO$ and $BO = DO$.

> **NOTE:** *A quadrilateral is a parallelogram*
>
> (*i*) *if one pair of opposite sides are equal and parallel: or,*
>
> (*ii*) *if both pairs of opposite sides are equal: or,*
>
> (*iii*) *if both pairs of opposite angles are equal; or,*
>
> (*iv*) *if the diagonals bisect one another.*

(*i*) Follows from Theorem 26; (*ii*), (*iii*) and (*iv*) have already been given as exercise See Exercise VII, 1; Exercise IV, 15; Exercise III, 8.

Exercise XVII (Theorems 26-28)

Mostly Constructions

1. The distance between a pair of parallel straight lines is everywhere the same.
2. What is the condition that the diagonals of a parallelogram should bisect the angles? (See Exercise VI, 10)

3. If the diagonals of a parallelogram are equal, the parallelogram is a rectangle.
4. Draw a straight line *AB*. Construct a rectangle of which *AB* is one diagonal, the other diagonal making 30° with *AB*.
5. Given two parallel straight lines *AB*, *CD* and a point *O*. Through *O* draw a straight line *XY*, equal to a given straight line, with its extremities *X*, *Y* on *AB*, *CD* respectively.
6. Construct a parallelogram whose diagonals and one side are equal to three given straight lines. What condition must be fulfilled by the given lines that it may be possible to construct the parallelogram?
7. Draw a straight line *AB*. Construct a rhombus of which *AB* is one diagonal, the other diagonal being equal to a given straight line *CD*.
8. Draw a triangle *ABC*. Construct a rhombus *AXYZ*, the point *X* being somewhere in *AB*, *Y* in *BC* and *Z* in *CA*.
9. *OX*, *OY* are given straight lines and *P* is any point within the angle *XOY*. Draw a parallelogram with its diagonals intersecting at *P* and with adjacent sides along *OX* and *OY*.
10. Through a given point *P* between two given straight lines draw a straight line terminated by the given lines and bisected at *P*.
11. Given two sides of a triangle in position but not in magnitude, and the middle point of the base, construct the triangle.
12. Construct a parallelogram, given the lengths of two adjacent sides and the perpendicular distance of a pair of opposite sides.
13. Construct a parallelogram, given one angle and the perpendicular distances between pairs of opposite sides.
14. Construct a trapezium whose parallel sides are equal to two given straight lines b and $d (b > d)$, and whose other sides are equal to two other given straight lines a and c. When is no such construction possible? [Make a triangle whose sides are a, c, $b - d$.]
15. Place a straight line, equal and parallel to a given straight line *AB*, so as to have its extremities on two given straight lines *OX*, *OY*.

    [Begin thus,—Take any point *C* in *OX*. Draw *CD* equal and parallel to *AB*. Draw *DE* parallel to *XO*, etc.]
16. *X'OX*, *Y'OY* are given straight lines meeting at *O*, and *A*, *B* are given points within the angle *XOY*. Find a point *P* in *XOX'* and a point *Q* in *YOY'*, such that *AP* and *BQ* may be equal and parallel. [Let *AB* or *BA* produced cut *OX* in *A'*; on the line *AB*, produced if necessary, take a point *B'* on the same side of *A'* as *B* is of *A*, making *B'A'* = *BA* : draw *B'Q* parallel to *OX* to cut *OY* or *OY* produced in *Q*, draw *QP* parallel to *BA*.]
17. Upon the same base *AB*, and upon opposite sides of it, the parallelograms *ABCD* and *ABEF* are described so that the side *AD* of the first is equal to the diagonal *AE* of the second, and the diagonal *AC* of the first is equal to the side *AF* of the second. Prove that
    (*i*) *C*, *A*, *F* are in the same straight line.
    (*ii*) *D*, *A*, *E* are in the same straight line.

18. If E is a point in the diagonal AC of a parallelogram $ABCD$, such that EB equals ED, show that E is the middle point of AC. There is an exceptional case; state it. [Let AC, BD meet in O. Suppose that E does not coincide with O. Take Δs BOE, DOE.]

19. $ABCD$ is a quadrilateral; the four parallelograms $BCDP$, $CDAQ$, $DABR$, $ABCS$ are completed. Show that AP, BQ, CR, DS are equal and parallel.

20. If the sum of the distances of any angular point of a quadrilateral from the other three is the same for all four, the figure is a rectangle. [Prove that the opposite sides are equal and that the diagonals are equal.]

21. $ABCD$ is a parallelogram and O any point. The parallelograms $OAEB$, $OBFC$, $OCGD$, $ODHA$ are completed. Show that $EFGH$ is a parallelogram, whose area is double that of $ABCD$.

EXERCISE XVIII(a)

Construction of Quadrilaterals

1. Draw a straight line AB of length 7 cm, and on it construct a quadrilateral, having BC = 4 cm, CD = 3 cm, DA = 5 cm. How many such quadrilaterals can be drawn on AB?

2. In Ex. 1, it will be found that we can draw as many quadrilaterals as we choose, different in size and shape, with their sides of the given lengths. In other words, the given conditions are not sufficient to determine the quadrilateral.

Now suppose a fifth condition given, *e.g.* the length of the diagonal AC; suppose AC = 6 cm. You will now find that only one such quadrilateral can be drawn on one side of AB. Construct it and measure the diagonal BD.

3. *Construct the quadrilateral ABCD, having given the lengths of the sides and the inclination* α *of a pair of opposite sides AB and CD.*

Suppose that $ABCD$ is the required quadrilateral. Complete the parallelogram $ADCE$. Join BE.

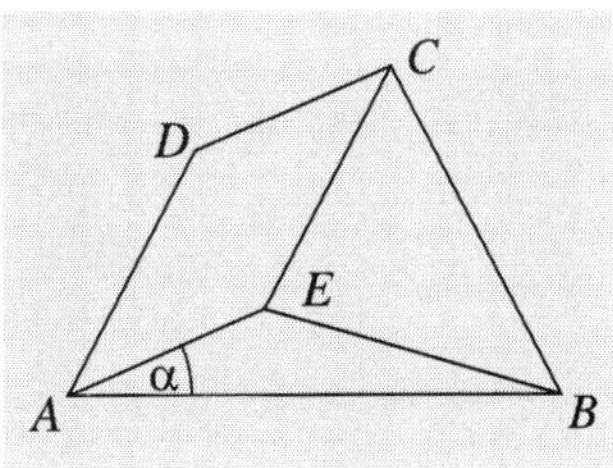

Consider ΔBAE; the lengths of AB and AE are given (for $AE = CD$), and $\angle BAE$ is equal to α, the inclination of AB and CD.

Hence the triangle BAE can be constructed. We can also construct Δ BCE, for the lengths of its sides are given.

4. Construct the quadrilateral $ACBD$, having given the lengths of a pair of opposite sides BC, AD, the lengths of the diagonals AB, CD, and the angle α between the diagonals. [This is really the same problem as that in Ex. 3, and the explanation applies, word for word, from "Complete" onwards.]

5. Construct the quadrilateral $ABDC$, having given the lengths of the diagonals BC, AD, the lengths of a pair of opposite sides AB, CD, and the inclination of these sides. [This again is the same as Ex. 3.]

EXERCISE XVIII(b)

Numerical Examples

In Examples 1–8 construct quadrilaterals to the given measurements; the order of the letters in each case is *A, B, C, D*. In each case measure the lengths of the parts into which the diagonals divide one another.

1. $BC = 2.5$ in, $CD = 2$ in, $DA = 1.3$ in, $BD = 2.4$ in, $AC = 2.3$ in.
2. $AB = CD = 1.3$ in, $BC = AD = 0.8$ in, $BD = 1.8$ in.
3. $AB = CD = 1.3$ in, $BC = 0.8$ in, $AD = 1.5$ in, and BC is parallel to AD.
4. $AB = CD = 1.3$ in, $BC = 0.8$ in, $AD = 1.5$ in, and the angle between BC and AD $= 25°$.
5. $AB = 2.8$ cm, $BD = 3.4$ cm, $CD = 2.6$ cm, $CA = 4.1$ cm, and the diagonals AC, BD are at right angles.
6. $AD = 3.4$ cm, $DB = 6.0$ cm, $BC = 6.2$ cm, $CA = 5.8$ cm, and the angle between the diagonals AC, BD is $80°$.
7. $AB = 0.75$ in, $BD = 2.84$ in, $DC = 1.12$ in, $CA = 2.57$ in, and the sides AB and CD are at right angles.
8. $AB = 0.75$ in, $BD = 2.84$ in, $DC = 2.57$ in, $CA = 1.12$ in, and the sides BA, CD are parallel.

XII. DIVISION OF A LINE INTO EQUAL PARTS

THEOREM 29

The straight line joining the middle points of two sides of a triangle is parallel to the third side and is equal to half the third side.

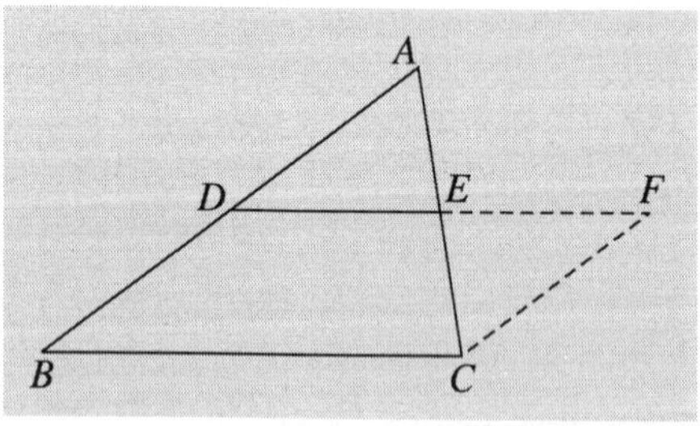

Let ABC be a triangle, and let AB be bisected at D and AC at E.

It is required to prove that DE is parallel to BC and equal to half BC.

Construction. Through C draw a straight line parallel to BA to meet DE produced at F.

Proof. Because AC meets the parallels AD, CF,

$\therefore$ $\angle DAE =$ The alternate $\angle FCE$.

Hence, in the triangles *DAE*, *FCE*,

$$\begin{cases} \angle DAE = \angle FCE \text{ (proved)}, \\ \angle DEA = \text{The vertically opposite } \angle FEC, \\ AE = CE \text{ (given)}. \end{cases}$$

∴ The triangles are congruent.

∴ The sides opposite to equal angles in each triangle are equal;

∴ $DE = FE$ and $DA = FC$.

But it is given that $DA = DB$;

∴ $DB = FC$;

also *DB* is parallel to *FC* (*construction*);

∴ *DB* and *FC* are equal and parallel;

∴ *DF* and *BC* are also equal and parallel.

Thus *DE* is parallel to *BC*.

Again, it has been shown that $DE = EF$ and $DF = BC$;

∴ $DE = \frac{1}{2} BC.$

THEOREM 30

If through the middle point of one side of a triangle a straight line is drawn parallel to another side, this straight line bisects the third side.

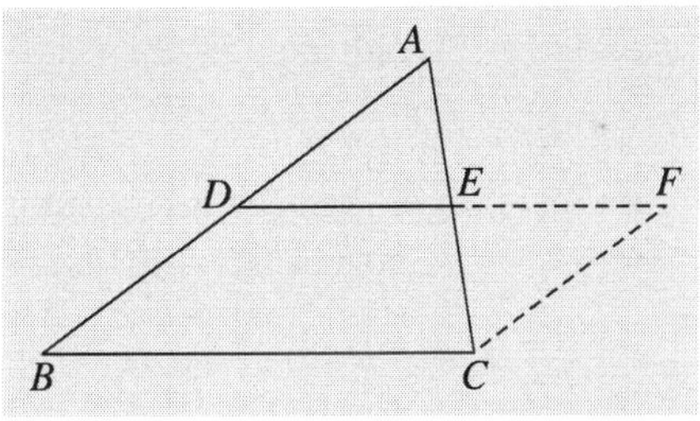

Let *ABC* be a triangle, and let *AB* be bisected at *D*. Through *D* let a straight line be drawn, parallel to *BC*, to meet *AC* at *E*. It is required to prove that $AE = EC$.

Construction. Through *C* draw a straight line parallel to *BA* to meet *DE* produced at *F*.

Proof. It is given that *DF* is parallel to *BC*, and by construction, *CF* is parallel to *BD*;

∴ *BCFD* is a parallelogram;

∴ $BD = CF$.

Now, it is given that $BD = AD$;

∴ $AD = CF$.

Again, because *AC* meets the parallels *AD*, *CF*,

∴ $\angle DAE$ = The alternate $\angle FCE$.

Hence, in the triangles *DAE*, *FCE*,

$$\begin{cases} \angle DAE = \angle FCE \text{ (proved)}, \\ \angle DEA = \text{The vertically opposite } \angle FEC, \\ AD = CF \text{ (proved)}; \end{cases}$$

∴ The triangles are congruent.

∴ The sides opposite to equal angles in the triangles are equal;

∴ $AE = CE$.

Theorem 31

If there are three or more parallel straight lines, and the intercepts made by them on any straight line that cuts them are equal, then the corresponding intercepts on any other straight line that cuts them are also equal.

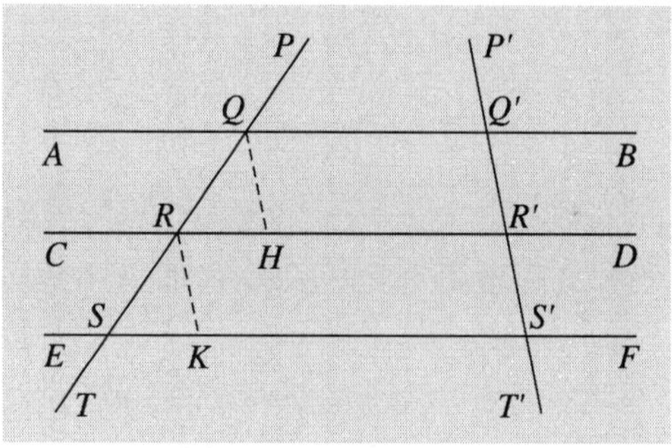

Let *AB*, *CD*, *EF* be parallel straight lines, and let a straight line *PT* cut them in *Q*, *R*, *S* respectively and make *QR* equal to *RS*. Let any other straight line *P'T'* cut them in *Q'*, *R'*, *S'*.

It is required to prove that $Q'R' = R'S'$.

Construction. Draw *QH*, *RK* parallel to *P'T'*, to meet *CD*, *EF* respectively in *H* and *K*.

Proof. Because *PT* meets the parallels *CD*, *EF*,

∴ $\angle QRH$ = The corresponding $\angle RSK$.

Also, because *QH*, *RK* are both parallel to *P'T'*,

∴ *QH* is parallel to *RK*;

and because *PT* meets the parallels *QH*, *RK*,

∴ $\angle RQH$ = The corresponding $\angle SRK$;

Hence, in the triangles *QRH*, *RSK*,

$$\angle QRH = \angle RSK \text{ (proved)},$$
$$\angle RQH = \angle SRK \text{ (proved)},$$
$$QR = RS \text{ (given)}.$$

∴ The triangles are congruent.

∴ The sides opposite equal angles in the triangles are equal;

∴ $QH = RK$.

Now the figures $QQ'R'H$, $RR'S'K$ are parallelograms;

∴ Their opposite sides are equal,

∴ $QH = Q'R'$ and $RK = R'S'$;

and it has been proved that $QH = RK$,

∴ $Q'R' = R'S'$.

CONSTRUCTION 13

Divide a straight line into a given number of equal parts (say five).

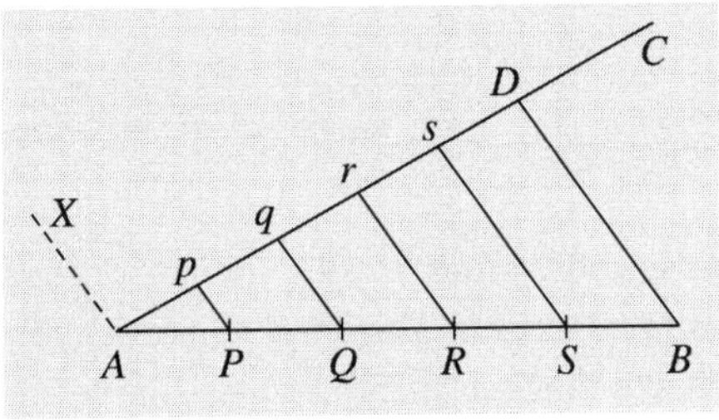

Let AB be the given straight line. It is required to divide AB into five equal parts.

Construction. Through A draw any straight line AC, making any convenient angle with AB.

From A, along AC, set off five equal lengths, Ap, pq, qr, rs, sD.

Join DB.

Through s, r, q, p draw straight lines parallel to DB, to cut AB in S, R, Q, P respectively.

Then P, Q, R, S are the required points of division.

Proof. Through A draw AX parallel to BD.

By construction, XA, pP, qQ, rR, sS are parallel to DB:

∴ XA, pP, qQ, rR, sS, DB are parallel to one another.

But, by construction, these parallels make equal intercepts on AC.

∴ They also make equal intercepts on AB;

∴ AB is divided into five equal parts at P, Q, R, S.

NOTE: 1. The foregoing construction can be used to cut off a given fractional part from a given straight line. Thus

$$AQ = \frac{2}{5}AB.$$

2. If through p, q, r, s parallels are drawn to AB, they will divide DB into five equal parts.

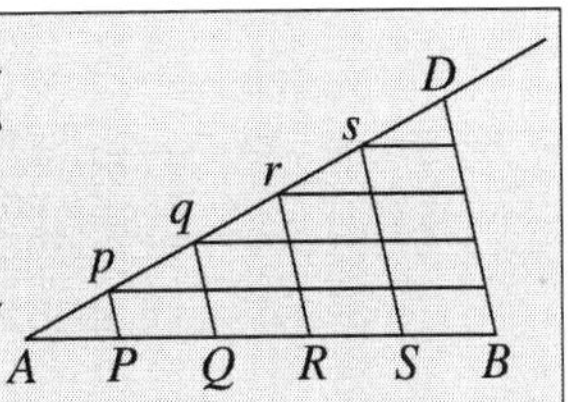

Hence, it appears that

$$pP = \frac{1}{5}DB,\ qQ = \frac{2}{5}DB,\ rR = \frac{3}{5}DB,\ sS = \frac{4}{5}DB.$$

The Diagonal Scale

The figure below is a diagonal scale showing inches, tenths and hundredths of an inch.

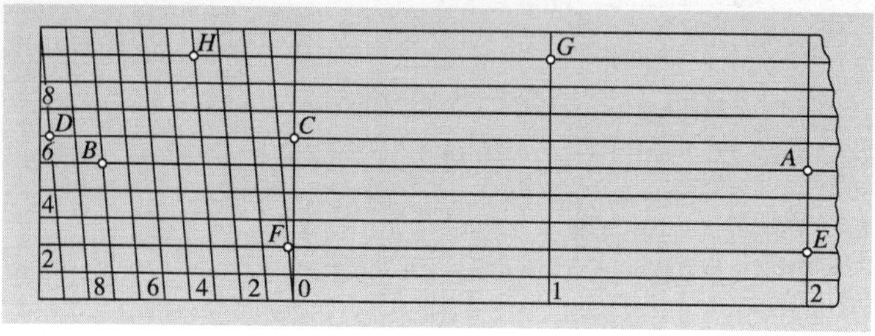

The large divisions of the bottom line are inches: of these, the one on the left is subdivided into tenths of an inch. The diagonal scale is formed by drawing ten equidistant lines parallel to the bottom line, the top line being subdivided similarly to the bottom, and joining up diagonally, so that the mark '3' (say) of the bottom line is joined to the mark '4' of the top line. In practice, the construction is more conveniently made by joining division 9 of the bottom line to division 10 of the top line, and drawing parallels through the other points of division of the bottom line.

On proceeding upwards along the diagonal starting from 'o,' as we pass each horizontal line, the horizontal distance from the vertical through 'o' increases by $\frac{1}{10}$ of a subdivision, *i.e.* by $\frac{1}{100}$ of an inch. Hence this scale reads to inches (main divisions), tenths (subdivisions of bottom line), and hundredths (given by the number opposite the horizontal line along which the measurement is made).

1. Set off a certain distance from the diagonal scale, say 2.75 inches. [Place one point of the dividers at the point of intersection of the vertical 2, with the horizontal 5, and the other at the point of intersection of the diagonal marked 7, with horizontal 5. These are the points marked *A* and *B* in the diagram of the diagonal scale on the page preceding. The distance between the points is now 2.75 inches, and can be transferred anywhere. Observe that in using the diagonal scale, the two points of the dividers must be *on the same horizontal line.*]
2. Read off the lengths from *C* to *D*, *E* to *F*, *G* to *H*, points marked on the diagram of the diagonal scale on the preceding page.

3. By using the diagonal scale, find the number of inches in (*i*) 3 centimetres, (*ii*) 4 centimetres, (*iii*) a decimetre, correct to two places of decimals.

EXERCISE XIX (Theorems 29–31)

1. Any point *X* is taken in the side *BC* of the triangle *ABC*. Show that *AX* is bisected by the straight line joining the middle points of *AB*, *AC*.

2. Construct a triangle having given the middle points of its sides.

3. The straight line *AB* is bisected at *C*, and perpendiculars *AL*, *BM*, *CN* are drawn to any straight line *OX*. Prove that

(*i*) If *A*, *B* are on the same side of *OX*, then *CN* is half the sum of *AL* and *BM*.

(*ii*) If *A*, *B* are on opposite sides of *OX*, then *CN* is half the difference between *AL* and *BM*.

4. The straight line joining the middle-points of the diagonals of a trapezium is parallel to the parallel sides. [Let *ABCD* be the trapezium, with *AB* parallel to *CD*. Through *E*, the middle point of *AD*, draw *EF* parallel to *AB*, and show that *EF* bisects *AC* and *BD*.]

5. In any quadrilateral, the middle points of the sides are the vertices of a parallelogram. [Draw the diagonals]

6. *ABCD* is a quadrilateral; *E*, *F* are the middle points of *AB*, *CD*; *X*, *Y* are the middle points of the diagonals *AC*, *BD*. Prove that the figure *EXFY* is a parallelogram. [Use Δs *BAC*, *BDC* to prove *EX* and *FY* parallel to *BC*.]

7. The straight lines joining the middle points of opposite sides of a quadrilateral, and the middle points of the diagonals, meet in a point, which is the middle point of all three lines. [Use Exs. 5 and 6.]

8. *ABC* is a triangle and *O* is any point; *OA*, *OB*, *OC* are bisected at *P*, *Q*, *R*. Show that the triangles *PQR*, *ABC* are equiangular, and that each side of *PQR* is half the corresponding side of *ABC*.

9. *X* is any point on the circumference of a given circle, with centre *C* and radius *r*. *O* is any given point, and *D*, *P* are the middle points of *OC*, *OX* respectively. Prove that *P* lies on a circle, with centre *D* and radius $\frac{1}{2}r$.

10. Prove the following construction for trisecting a straight line *AB* in *G* and *H*: On *AB* as diagonal, construct a parallelogram *ACBD*; bisect *AC*, *BD* in *E* and *F* respectively; join *DE*, *FC*, cutting *AB* in *G* and *H*.

11. Draw any triangle *ABC*. Bisect *AB* at *X*. Join *CX* and bisect *CX* at *Y*. Join *BY* and produce it to cut *AC* at *Z*. Prove that *AZ* = 2 *ZC*. [Draw *XH* parallel to *BZ* to meet *AC* at *H*.]

12. If D is the middle point of the hypotenuse BC of a right-angled triangle ABC, prove that $AD = \frac{1}{2}BC$. [Bisect AB at E. Join ED. Prove Δs ADE, BDE congruent.]

EXERCISE XX (Construction 13)

1. From a given straight line, cut off a length equal to three-sevenths of the line.

2. Draw a triangle ABC. Find points P, Q in AB, AC respectively, such that $AP = \frac{3}{5}AB$, $AQ = \frac{3}{5}AC$. Prove that PQ is parallel to BC and equal to $\frac{3}{5}BC$.

3. $ABCD$ is a quadrilateral, and O any point inside or outside $ABCD$. Join OA, OB, OC, OD. Along OA take $OP = \frac{2}{3}OA$. Draw PQ, QR, RS, parallel to AB, BC, CD (as in the figure), to cut OB, OC, OD respectively in Q, R, S. Join SP. Prove that

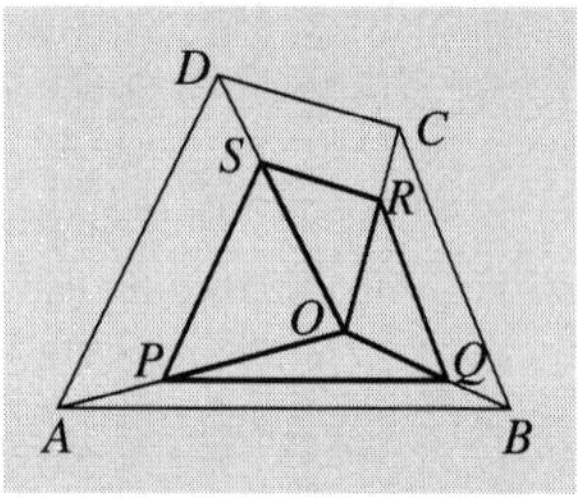

(*i*) $OQ = \frac{2}{3}OB$, $OR = \frac{2}{3}OC$, $OS = \frac{2}{3}OD$.

(*ii*) SP is parallel to DA.

(*iii*) Any side of $PQRS$ is two-thirds of the parallel side of $ABCD$.

(*iv*) The figures $PQRS$, $ABCD$ are equiangular.

4. Given a map 3 inches to the mile, explain how to construct the corresponding map 2 inches to the mile. [Let A, B, C, D be any four points in the given map. Take any point O and make the construction of Ex. 3. Then P, Q, R, S are the corresponding points in the required map.]

XIII. LOCI

Def. If a point moves so as to satisfy some given geometrical condition, the line which it describes is called the **locus of the point**.

When it is said that *the point moves so as to satisfy a geometrical condition*, the meaning is that the point has to move in some definite manner.

For example, if a point moves so as to be always 2 inches distant from a fixed point A, the locus of the point is a circle with centre A and radius equal to 2 inches.

In proving that the locus of a point which satisfies a given geometrical condition is a certain line, two things must be shown:

(*i*) that every point which satisfies the given condition lies on the line;

(*ii*) that every point on the line satisfies the given condition.

Very often, when the first of these things has been proved, the second is fairly evident.

In writing out examples, it is usual to write out only the first part of the proof; that is, to show that the point lies on the line.

In the Theorems 32 and 33, very important loci are discussed; and it is advisable to give both parts of the proof.

Consider now the problem of finding a point which satisfies two geometrical conditions.

EXAMPLE: *Given a straight line AB and a point O. Find a point distant 1.5 cm. from O and 1 cm from AB.*

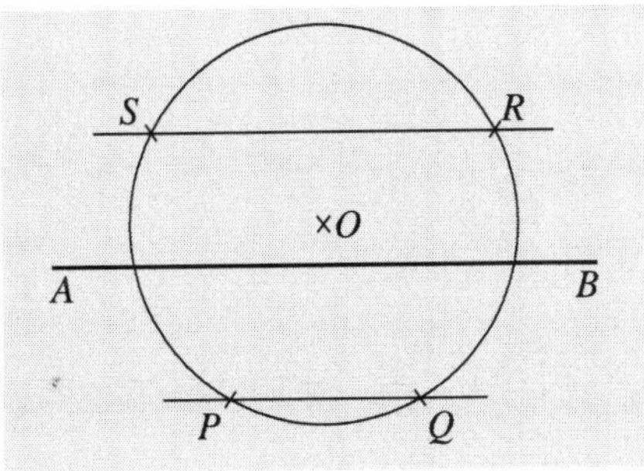

The locus of a point which moves so as to be always 1.5 cm distant from O is a circle with centre O and radius 1.5 cm.

The locus of a point which is 1 cm distant from AB is obviously a pair of straight lines parallel to AB, the distance between AB and each line being 1 cm.

Draw these two loci. Any point in which the circle cuts either of the parallels is a solution. In the above figure the points P, Q, R, S are solutions.

THEOREM 32

The locus of a point, which is equidistant from two given points, is the perpendicular bisector of the straight line joining the given points.

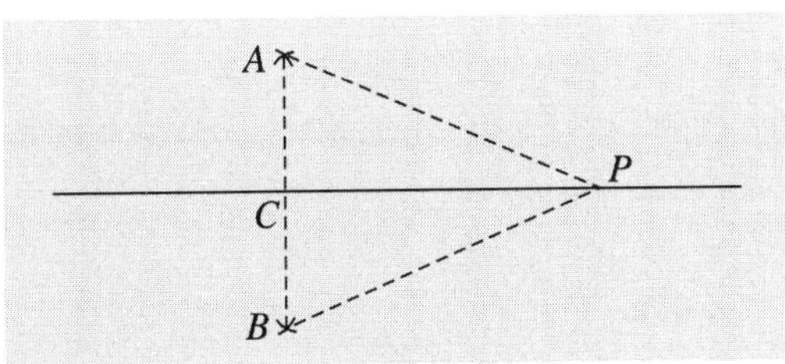

Let A, B be the given points

First, let P be any position of a point which moves so that $PA = PB$.

It is required to prove that P is on the perpendicular bisector of AB.

Proof. Let C be the middle point of AB. Join PC.

Then, in the triangles APC, BPC,

$$\begin{cases} AC = BC \text{ (construction)}, \\ CP \text{ is common}, \\ AP = BP \text{ (given)}. \end{cases}$$

∴ The angles are congruent;

$$\therefore \qquad \angle ACP = \angle BCP;$$

∴ CP is perpendicular to AB;

∴ P lies in the perpendicular bisector of AB. *Secondly*, let P be any point on the straight line which bisects AB at right angles in C.

It is required to prove that $PA = PB$.

Join PA, PB.

Then, in the triangles APC, BPC,

$$\begin{cases} AC = BC \text{ (given)}, \\ CP \text{ is common}, \\ \angle ACP = \angle BCP \text{ (given)}; \end{cases}$$

$$\therefore \qquad PA = PB.$$

Hence, the required locus is the perpendicular bisector of AB.

Theorem 33

The locus of a point, which is equidistant from two given intersecting straight lines, is the pair of bisectors of the angles between the lines.

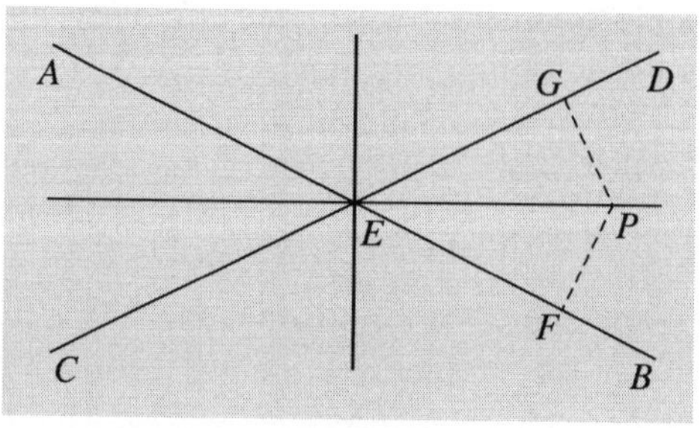

Let AEB, CED be the given straight lines.

(*i*) Let P be any position of a point which moves so that the perpendiculars PF, PG, drawn from P to the given lines, are equal.

It is required to prove that $\angle PEF = \angle PEG$.

Proof. In the right-angled triangles *PEF*, *PEG*,

$$\begin{cases} PE \text{ is the common hypotenuse,} \\ PF = PG \ (given). \end{cases}$$

∴ The triangles are congruent;

∴ $\angle PEF = \angle PEG.$

(*ii*) Let *P* be any point on the bisector of one of the angles between *AB* and *CD*.

It is required to prove that the perpendiculars *PF*, *PG*, drawn from *P* to the given lines, are equal.

Proof. In the triangles *PEF*, *PEG*,

$$\begin{cases} \angle PEF = \angle PEG \ (given), \\ \angle PFE = \angle PGE \ (given), \\ PE \text{ is common.} \end{cases}$$

∴ The triangles are congruent;

∴ $PF = PG.$

Hence, the required locus is the pair of bisectors of the angles between the lines.

EXERCISE XXI(a)

Loci

1. Draw a straight line *AB* and take a point *O* distant 1.5 in. from it. Find points which are

(*i*) 0.5 in. from *AB* and 2.5 in. from *O*;

(*ii*) 0.5 in. from *AB* and 2 in. from *O*;

(*iii*) 1 in. from *AB* and 1 in. from *O*.

2. Find the point on a given straight line *AB* which is equidistant from two given points *C*, *D*. When is this impossible?

3. Find a point which is equidistant from two given intersecting straight lines and at a given distance from a fixed point.

Show that there may be four possible solutions.

4. Draw an equilateral triangle *ABC* of 3 in. side. Find the points which are distant 2 in. from the middle point of *AB* and equidistant from *AB* and *AC* or from these sides produced through *A*.

Show that there are four solutions.

5. Find a point which is equidistant from three given points *A*, *B*, *C*. When is there no solution?

[Draw the locus of points equidistant from *A*, *B* and the locus of points equidistant from *A*, *C*.]

6. Find a point which is equidistant from the three sides of a triangle *ABC* (supposed to be produced to any length).

Show that there are always four solutions.

[Draw the pair of bisectors of the angles at *A* and the pair of bisectors of the angles at *B*.]

7. *A* is a given point and *BC* a given straight line. If the point *P* moves along *BC*, find the locus of the middle point of *AP*.

8. *ABC*, *DBC* are two triangles on the same base and on the same side of it. If *AB* = *BD*, prove that *AC* and *CD* are unequal.

9. Take a set square. Call it *ABC*, *A* being the right angle. Take two points *P*, *Q* on your paper. Slide the set square so that the edges *AB*, *AC* always pass through *P*, *Q* respectively. Mark a number of positions of *A*. Prove that the locus of *A* is a circle. [Use Ex. XIV, 6]

10. Draw two straight lines *OX*, *OY* at right angles. Take a straight edge *AB* and find its middle point *C*. (A short ruler or a strip of paper about 4 in. long will do.) Slide the straight edge so' that *A* moves along *OX* and *B* along *OY*. Mark a number of positions of *C*. Prove that the locus of *C* is a circle. [Use Ex. XIV, 6.]

11. Find a point *X* on the circle through three given points *A*, *B*, *C* such that *AX* is bisected by *BC*. [Let *P* be any point in *BC*. Join *AP* and produce it to *Q*, such that *PQ* = *AP*. Consider the locus of *Q*.]

EXERCISE XXI(b)

Special cases of two important Theorems

1. *A is a given point and P is any point on a given straight line BC. Draw a straight line AQ, equal to AP, and making a constant angle α with AP. Prove that the locus of Q is a straight line, and explain how to draw it.*

[Draw *AM* perp. to *BC*. Draw *AN* making $\angle\,\alpha$ with *AM*, the angles *MAN*, *PAQ* being measured in the same sense. Make *AN* = *AM*. Join *QN*. Prove $\angle$ *ANQ* a right angle. Hence the required locus is the straight line through *N* perpendicular to *AN*.]

2. *A* is a given point and *BC* a given straight line. In *BC* take any point *X* and draw a square *AXYZ* on *AX*. Prove that the locus of *Z* consists of two straight lines perpendicular to *BC*.

[Draw *AN* perpendicular to *BC*. On *AN* draw squares *ANHK*, *ANH'K'*. The locus is the pair of lines *HK*, *H'K'*. See Ex. 1.]

3. *A* is a given point and *OX*, *OY* are given straight lines, Draw a square *ABCD* with *B* on *OX* and *D* on *OY*. Show that there are, in general, two such squares.

[By Ex. 2, construct the locus of the vertex *D* of a square *ABCD*, with its vertex *A* at the given point, the vertex *B* being at any point in *OX*. The points at which the locus cuts *OY* will be the required positions of *D*.]

4. *ABC* is an equilateral triangle with the vertex *A* at a given point and the vertex *B* at any point in a given straight line. Construct the locus of *C*.

[Draw *AM* perpendicular to the given line. Draw the equilateral triangles *AMH*, *AMK*. Through *H*, *K* draw straight lines perpendicular to *AH*, *AK*. By Ex. 1, the locus consists of these lines.]

5. *A* is a given point and *OX*, *OY* are given straight lines. Draw an equilateral triangle with one vertex at *A*, another on *OX* and the third on *OY*. Show that there are, in general, two solutions.

6. *A is a given point, and P is any point on a given circle whose centre is C. Draw a straight line AQ, equal to AP, making a constant angle* α *with AP. Prove that the locus of Q is a circle, and explain how to draw it.*

[Draw *AD* equal to *AC*, making $\angle$ *CAD* = α, the angles *CAD*, *PAQ* being measured in the same sense. Join *DQ*. Prove Δs *ACP*, *ADQ* congruent; $\therefore$ *DQ* = *CP*; $\therefore$ the locus of *Q* is a circle with centre *D* and radius equal to *CP*.]

7. *ABCD* is a square with the vertex *A* at a given point and the vertex *B* at any point on a given circle whose centre is *C*. Construct the locus of *D*.

[Draw *HAK* perpendicular to *AC*. Make *AH* = *AK* = *AC*. With *H*, *K* as centres, draw circles with radii equal to that of the given circle. The locus consists of these circles.]

8. *ABC* is an equilateral triangle with the vertex *A* at a given point, and *B* at any point on a given circle. Construct the locus of *C*.

XIV. MISCELLANEOUS THEOREMS AND CONSTRUCTIONS

THEOREM 34

The perpendicular bisectors of the three sides of a triangle meet in a point.

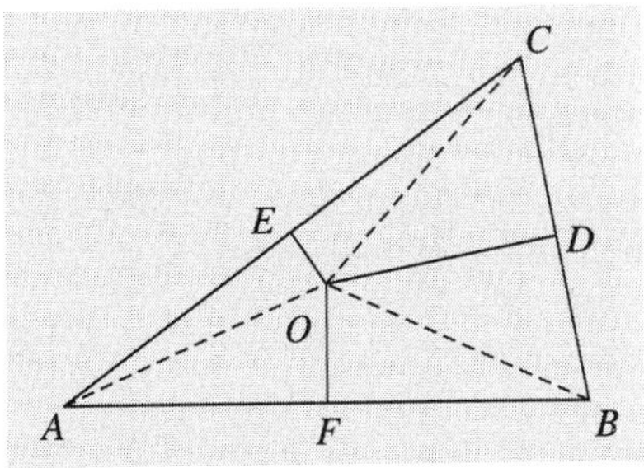

Let *ABC* be a triangle, and let *EO*, *FO* be the perpendicular bisectors of *CA*, *AB*, meeting in *O*.

Let *D* be the middle point of *BC*.

Join *OD*.

It is required to prove that *OD* is perpendicular to *BC*.

Construction. Join *AO*, *BO*, *CO*.

Proof. In the triangles AFO, BFO,

$$\begin{cases} AF = BF \text{ (given)}, \\ FO \text{ is common}, \\ \angle AFO = BFO \text{ (given)}. \end{cases}$$

∴ The triangles are congruent;

∴ $$AO = BO.$$

Similarly, from the triangles AEO, CEO, it can be shown that

$$AO = CO;$$

∴ $$BO = CO.$$

Hence, in the triangles BDO, CDO,

$$\begin{cases} BD = CD \text{ (by supposition)}, \\ OD \text{ is common}, \\ BO = CO \text{ (proved)}. \end{cases}$$

∴ The triangles are congruent;

∴ $$\angle BDO = \angle CDO;$$

∴ OD is perpendicular to BC.

THEOREM 35

The bisectors of the three angles of a triangle meet in a point.

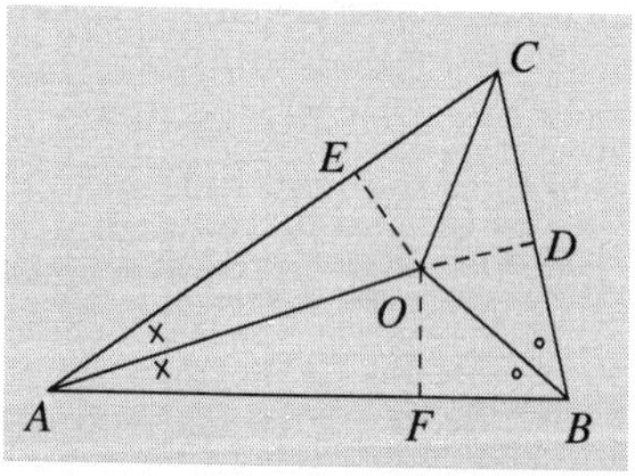

Let ABC be a triangle, and let AO, BO be the bisectors of the angles BAC and ABC, meeting in O.

Join OC.

It is required to prove that OC bisects the angle ACB.

Construction. Draw OD, OE, OF perpendicular to BC, CA, AB respectively.

Proof. In the triangles OAE, OAF,

$$\begin{cases} \angle OAE = \angle OAF \text{ (given)}, \\ \angle OEA = \angle OFA \text{ (construction)}, \\ OA \text{ is common}. \end{cases}$$

∴ The triangles are congruent;

∴ $$OE = OF.$$

Similarly, from the triangles *OBD*, *OBF* it can be shown that

$$OD = OF;$$

$\therefore$ $$OD = OE.$$

Hence, in the right-angled triangles *ODC*, *OEC*,

{ the hypotenuse *OC* is common,
and $OD = OE$ (*proved*).

$\therefore$ the triangles are congruent;

$\therefore$ $$\angle OCD = \angle OCE;$$

$\therefore$ *OC* bisects the angle *ACB*.

DEF. Three or more points are said to be **collinear** when they are in the same straight line.

DEF. Three or more straight lines are said to be **concurrent** when they intersect in the same point.

DEF. The straight line joining a vertex of a triangle to the middle point of the opposite side is called a **median**.

THEOREM 36

The three medians of a triangle meet in a point.

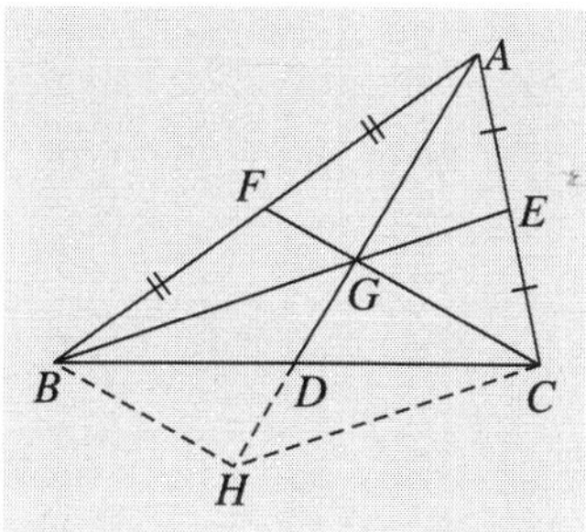

Let *ABC* be a triangle, and let *E*, *F* be the middle points of *AC*, *AB*. Let the straight lines *BE*, *CF* meet at *G*.

Join *AG*, and produce it to cut *BC* it *D*.

It is required to prove that $BD = DC$.

Construction. Through *B* draw a straight line parallel to *FC* to meet *AD* produced at *H*. Join *CH*.

Proof. In the triangle *ABH*,

$$AF = FB \text{ (given)},$$

and *FG* is parallel to *BH* (*construction*);

$\therefore$ $$AG = GH.$$

Again, in the triangle *AHC*,

$$AG = GH \text{ (proved)},$$

and $$AE = EC \text{ (given)};$$

∴ *GE* is parallel to *HC*.

Hence, the opposite sides of the figure *BGCH* are parallel;

∴ *BGCH* is a parallelogram.

Now, the diagonals of a parallelogram bisect one another;

∴ $$BD = DC.$$

Cor. In the above figure, *DG* is one-third of *DA*.

For *BGCH* is a parallelogram (*proved*);

∴ $$DG = DH;\ \therefore\ HG = 2DG.$$

But $$HG = GA \text{ (proved)};$$

∴ $$GA = 2DG;$$

∴ $$DA = 3DG.$$

Def. The point of concurrence of the medians of a triangle is called the **centroid** of the triangle.

From the corollary, it is seen that *the centroid is a third of the way up the median DA, measured from D*.

Theorem 37

The perpendiculars, drawn from the vertices of a triangle to the opposite sides, meet in a point.

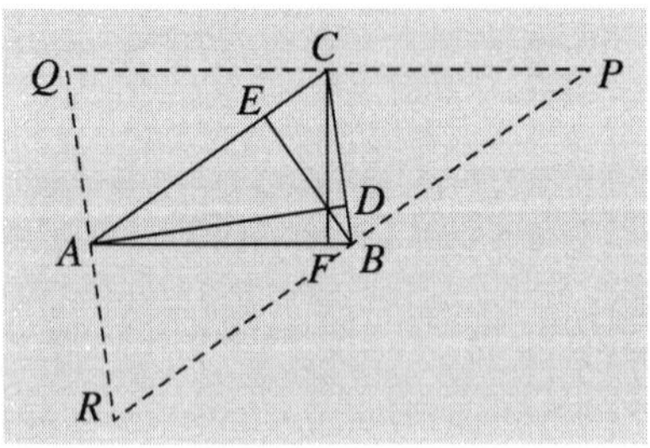

Let *ABC* be a triangle, and let *AD*, *BE*, *CF* be the perpendiculars from *A*, *B*, *C* to *BC*, *CA*, *AB* respectively.

It is required to prove that *AD*, *BE*, *CF* meet in a point.

Construction. Through *A*, *B*, *C* draw straight lines *QAR*, *RBP*, *PCQ*, parallel to *BC*, *CA*, *AB* respectively, forming the triangle *PQR*.

Proof. By construction, *PBAC* is a parallelogram;

∴ $$CP = AB;$$

and by construction *QABC* is a parallelogram,

$\therefore \qquad QC = AB,$

$\therefore \qquad QC = CP,$

Again, *AB* is parallel to *PQ*,

and *CF* is perpendicular to *AB*;

$\therefore$ *CF* is perpendicular to *PQ*,

$\therefore$ *CF* is the perpendicular bisector of *PQ*.

Similarly, *AD*, *BE* are the perpendicular bisectors of *QR*, *RP*.

But the perpendicular bisectors of the sides of the triangle *PQR* meet in a point;

$\therefore$ *AD*, *BE*, *CF* meet in a point.

DEF. The point of concurrence of the perpendiculars, drawn from the vertices of a triangle to the opposite sides, is called the **orthocentre** of the triangle.

Method of search for the solution of a geometrical problem.

EXAMPLE 1. *A, B are two points on the same side of a straight line CD. Find a point P in CD such that the angles APC, BPD are equal.*

We search for the solution by supposing that the required point *P* has been found, and then studying the properties of the figure.

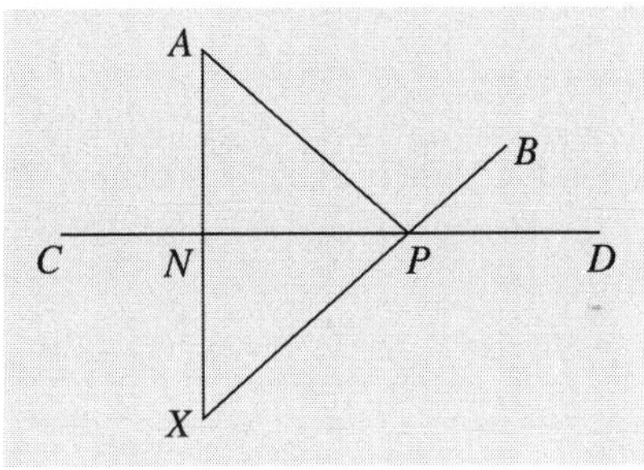

Suppose then that *P* is a point in *CD* such that $\angle APC = \angle BPD$.

If a perpendicular is drawn from *A* to *CD*, cutting *CD* at *N* and *BP* produced in *X*, we should have

$$\angle APC = \angle BPD = \text{The vert. opp. } \angle CPX.$$

Hence, we should have in Δs *ANP*, *XNP*,

$$\begin{cases} \angle ANP = \angle XNP, \\ \angle APN = \angle XPN, \\ NP \text{ common.} \end{cases}$$

$\therefore$ The triangles would be congruent, and *NX* would be equal to *AN*.*

The proper construction is now evident.

Construction. Draw *AN* perpendicular to *CD*, and produce it to *X* so that *NX* = *AN*.

* The work up to this point is called the **analysis** *(i.e.* the picking to pieces) of the problem. What follows is the **synthesis** *(i.e.* the putting together).

Join BX, cutting CD at P.

Then P is the required point.

Proof. Join AP.

In the triangles ANP, XNP,

$$\begin{cases} AN = XN \text{ (construction)}, \\ NP \text{ is common}, \\ \angle ANP = \angle XNP \text{ (construction)}. \end{cases}$$

$\therefore$ The triangles are congruent,

$\therefore$ $\angle APN = \angle XPN$.

But $\angle XPN =$ The vert, opp, $\angle BPD$;

$\therefore$ $\angle APC = \angle BPD$.

EXAMPLE 2. *A, B are two points on the same side of a straight line CD, and P is any point in CD. Show that PA + PB is least when the angles APC, BPD are equal.*

As in Ex. 1, find the point P in CD such that $\angle APC = \angle BPD$.

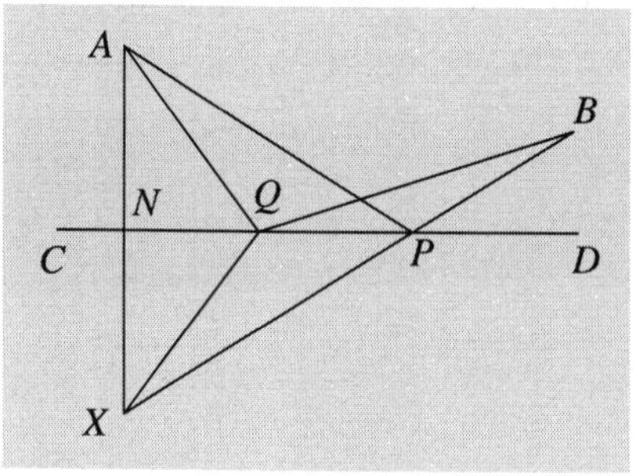

Take any other point Q in CD.

It has to be proved that $QA + QB > PA + PB$.

Outline of proof. Take Δs ANQ, XNQ, and show that $QA = QX$;

$\therefore$ $QA + QB = QX + QB$.

Next show that $PA + PB = XB$, and observe that $QX + QB > XB$.

Applications of Exs. 1 and 2.

When a ray of light is reflected at the surface of a mirror, it leaves the mirror at the same angle as that at which it strikes the mirror. The course of a billiard ball after striking a perfectly elastic cushion is determined by the same experimental law.

EXAMPLE 3. *An eye situated at B sees a bright point A by reflection at a plane mirror CD. Determine the path of the ray from A to B.*

The path of the ray is along AP, PB in the figure of Ex. 1. To the eye, the bright point A seems to be at X: the point X is therefore called the image of A.

EXAMPLE 4. *In Ex. 3, prove that the actual course of the ray from A to B is the shortest possible course (if A is seen by reflection).* This follows from Ex. 3.

A number of rays proceeding from a point is called a pencil of rays.

EXAMPLE 5. *In Ex. 3. the eye is treated as a point. Now consider the eye as having size, and draw the pencil of rays which, proceeding from A, fills the eye.*

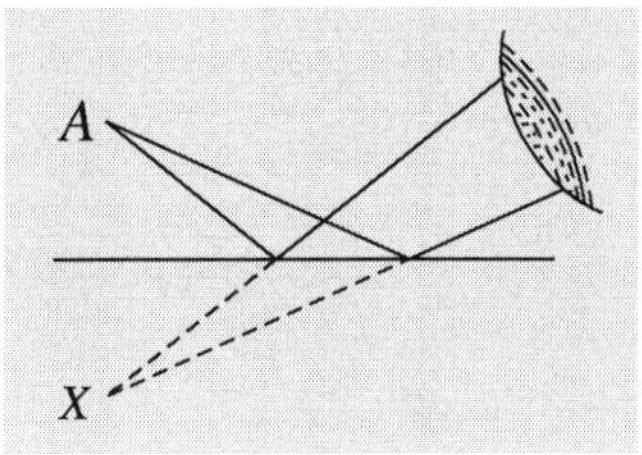

EXAMPLE 6. *ABCD is a billiard table. P, Q are two balls placed anywhere on the tables; P is struck so that, after hitting the cushions AD, DC, it hits Q. Construct the path of P.*

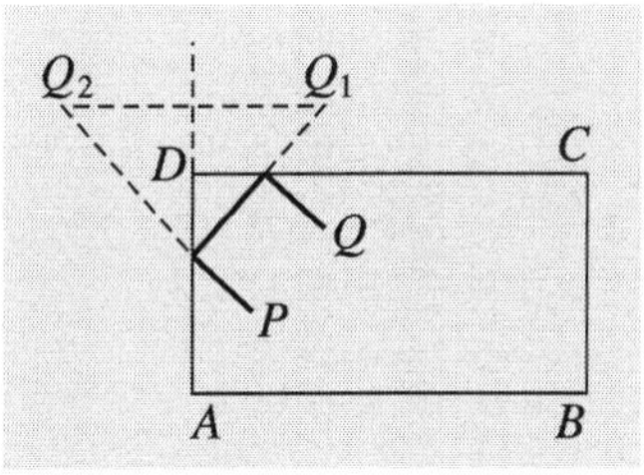

Find the *image* Q_1 of Q in DC, the *image* Q_2 of Q_1 in AD. Join as in the figure. P must be hit in the direction PQ_2.

Miscellaneous examples on the construction of triangles

EXAMPLE 7. *Construct* $\Delta\, ABC$, *given* $a + b$, c, A.

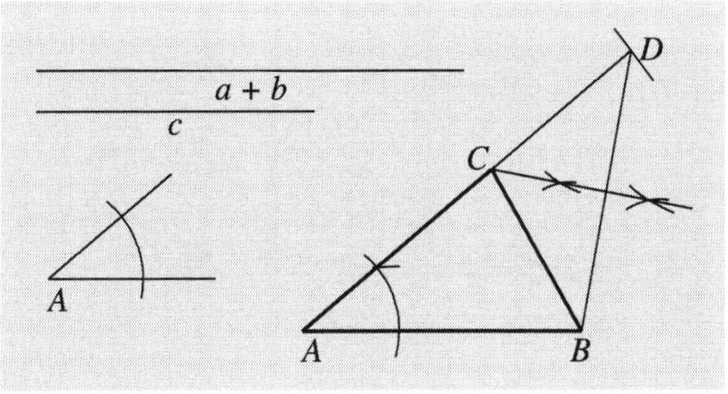

Let $AB = c$, make $\angle\, BAD = A$, and set off $AD = a + b$. Join BD. Draw a line bisecting BD at rt. $\angle$s and meeting AD in C. Join BC. Prove that BAC is the triangle required.

EXAMPLE 8. *Construct Δ ABC, given the perimeter and two angles B, C.*

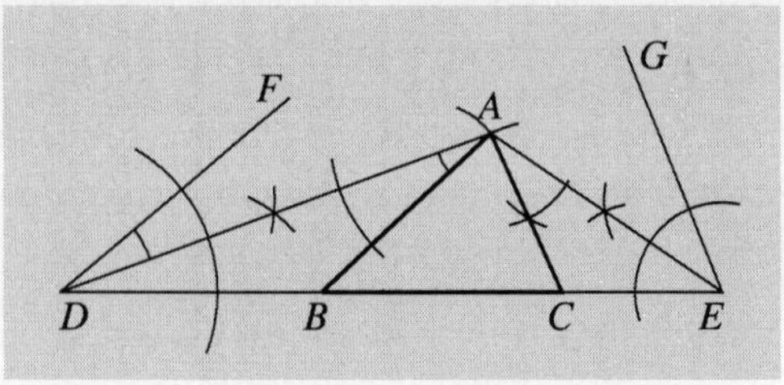

Let *DE* = given perimeter. Let ∠ *EDF* = *B*, ∠ *DEG* = *C*. Bisect ∠s *D*, *E* by lines meeting in *A*. Draw *AB* || *FD*, *AC* || *GE* to meet *DE* in *B*, *O* respectively. Prove that *ABC* is the triangle required.

EXAMPLE 9. *Construct Δ ABC, given b + c, a, A.*

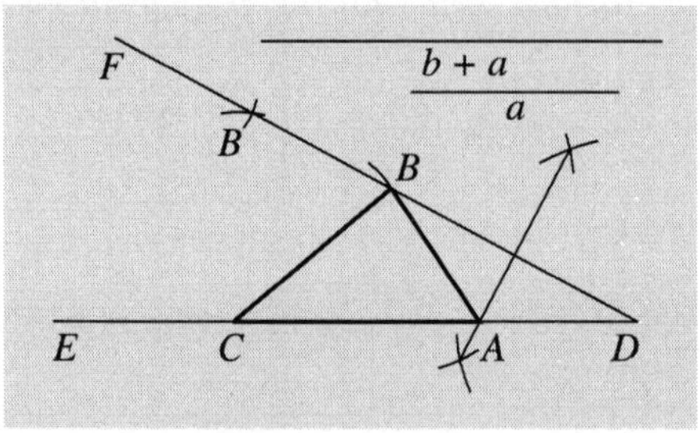

Let ∠ *EDF* = $\frac{1}{2}$*A*. Along *DE* set off *DC* = *b* + *c*.

With centre *C*, and radius *a*, draw an arc of a circle cutting. *DF* in *B*. Bisect *DB* at right angles by a line cutting *DC* in *A*. Join *AB*. Then *ABC* is the required triangle. Supply proof, and show that there are two solutions.

EXAMPLE 10. *Construct Δ ABC, given b – c, a, A.*

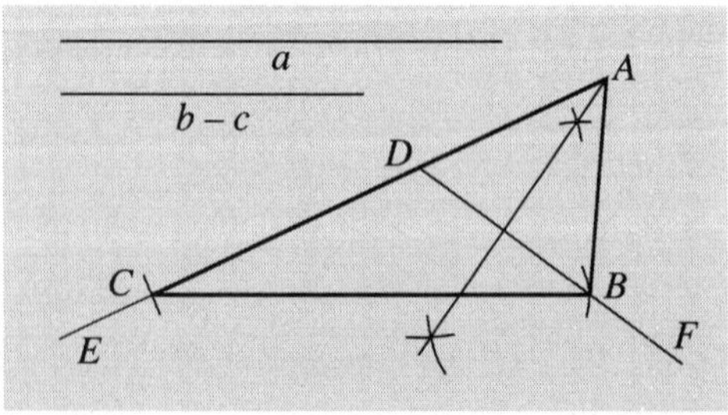

Let ∠ *EDF* = 90° + $\frac{A}{2}$. Along *DE* set off *DC* = *b* – *c*. Proceed as in the last example. Complete construction and proof.

EXERCISE XXII(a)

1. The sides AB, AC of a triangle are produced. Prove that the bisectors of the interior angle at A and the exterior angles at B and C are concurrent.
2. If the medians BE, CF of the triangle ABC are equal, show that $AB = AC$.

   [Let BE, CF meet in G. Prove $GB = GC$, and hence show that Δs BEC, CFB are congruent.]
3. Construct the triangle ABC, given the lengths of BC and the medians BE, CF. [First construct the triangle BGC]
4. Construct a triangle, having given the lengths of the three medians.

   [Referring to the figure of Theorem 36, we can construct the parallelogram $BGCH$, for the lengths of BG, GC, GH are known.]
5. Given the base BC of the triangle ABC in position and the length of the median BE, prove that the locus of A is a circle.

   [The locus of G is a circle with centre B and radius $\frac{2}{3}BE$, and, if D is the middle point of BC, $DA = 3DG$. Hence, show that the locus of A is a circle. Compare Ex. XIX, 9]
6. P is any point in the straight line, drawn through the vertex A of an isosceles triangle ABC, parallel to the base BC. Prove that $PB + PC > AB + AC$.

   [Draw CN perpendicular to AP. Produce CN to meet BA produced in X. Prove that $PB + PC = PB + PX$, etc.]
7. A, B are two billiard balls, l is a cushion; find where B must be made to strike l in order to hit A.
8. Draw the *image* of an arrow PQ, as seen by reflection at a plane mirror.
9. $ABCD$ is a billiard table, AB = 12 ft, AD = 6 ft. A ball is placed at O, the centre of the table, and struck so that, after hitting the cushion CD, it goes into the pocket at A. Draw a diagram (scale 1 in = 3 ft) and construct the path of the ball. Measure the distance travelled by the ball.
10. In Example 6, p. 85, construct the course of the ball P in order that it may hit Q after striking the cushions BA, AD, DC in the order named.
11. A ball is placed anywhere on a billiard table and struck so that, after hitting all four cushions once, it returns to the point from which it started. Construct its path and prove that it must be struck in a direction, parallel to a diagonal of the table and that the total distance it travels is twice the length of a diagonal.

EXERCISE XXII(b)

Construction of Triangles (Continued)

1. Explain how to construct the triangle ABC, having given $c - b$, a, B. [Use the figure of Theorem 19]

2. From Exercise XIV, 1, deduce a construction for the triangle ABC with the following data:

(1) $a, c - b, C - B$.

(2) $a, c + b, C - B$.

3. Given a, A, $B - C$, construct the triangle. Construct the triangle ABC with the following data; in each case measure the longest constructed side:

4.	$a + b = 3$ in,	$c = 2.6$ in,	$C = 120°$
5.	$a = 4$ in,	$c - b = 1.46$ in,	$A = 60°$.
6.	$a + b = 3.77$ in,	$c = 0.75$ in,	$A = 62°$.
7.	$b - a = 0.48$ in,	$c = 2.03$ in,	$A = 33°$.
8.	Perimeter = 5.01 in,	$A = 42°$,	$B = 36°$.
9.	$c = 2.41$ in,	$b - a = 1.57$ in,	$B - A = 65°$.
10.	$b = 1.10$ in,	$c + a = 3.08$ in,	$C - A = 60°$.
11.	$a = 2$ in,	$A = 40°$,	$B - C = 20°$,

XV. THE FORMS OF SIMPLE SOLIDS

In **geometry** the space occupied by any object (say a brick) is called a solid. The boundaries of a solid are surfaces.

Def. A **solid** has length, breadth and thickness. The volume of a solid is the amount of space enclosed by its boundaries.

Any brick-shaped solid will be called a rectangular block. The plane figures which bound such a solid are called its faces. The lines in which pairs of faces intersect are the edges of the solid. The points of intersection of edges are the corners or vertices of the block.

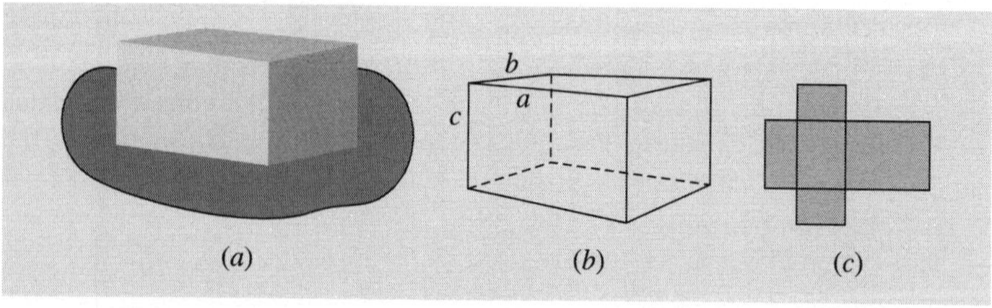

Fig. 1

A rectangular block, whose length, breadth and height are all equal, is called a **cube**.

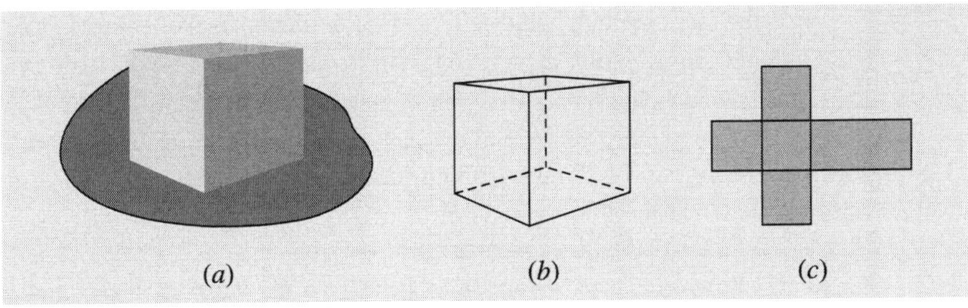

Fig. 2

NOTE: In the diagrams illustrating this section, the right-hand figure is a photograph on a reduced scale of the piece of cardboard used to make the solid in the left-hand figure; the dark lines on the right-hand figure are the V-shaped grooves cut in the cardboard to facilitate folding. The central figure shows how such a solid should be represented on paper. Note that all lines not parallel to the plane of the paper are shortened, and that, if two such lines *in the solid* are parallel to one another, the lines which represent them, in the drawing, will meet when produced. Dotted lines represent edges which would only be seen if the solid were transparent.

EXERCISE XXIII

Rectangular Blocks

1. What measurements must be taken to describe completely the shape of a brick?

2. Consider the shape of a rectangular block whose length is a in, breadth b in. and height c in.

(*i*) How many faces has the block?
(*ii*) What is the shape of each face?
(*iii*) What is the length and breadth of each face?
(*iv*) What do you observe about any pair of opposite faces?
(*v*) How many faces meet at each vertex?
(*vi*) How many edges has the block?
(*vii*) How many edges are a in. long?
(*viii*) How many edges meet at each vertex?
(*ix*) How many vertices has the block?

3. Consider the shape of a cube.

(*i*) What do you observe about the lengths of its edges?
(*ii*) How many measurements must be taken to determine a cube completely?
(*iii*) What is the shape of each face?
(*iv*) If the length of an edge is a in, what is the sum of the lengths of the edges?

(*v*) If the faces are numbered 1, 2, 3, 4, 5, 6 (as in the case of dice), how could you describe a particular edge? and how a particular vertex?

4. Draw on fairly thick paper a diagram consisting of six squares as in Fig. 2(*c*). Cut the figure out neatly. Fold the paper so as to form creases along the dark lines. This can be done by folding the paper over the edge of a thin ruler placed along the line. Now bend the paper so as to form a cube, as in Fig. 2 (*a*). Fasten the edges together with a narrow strip of gummed paper: or, when cutting out, narrow flaps may be left on the edges of the squares, where required.

5. Draw a figure of character similar to Fig. 1 (*c*), showing how a piece of paper may be cut and folded so as to form a rectangular block whose edges are 3 in, 2 in. and 1.6 in. respectively.

6. What is the smallest number of bricks, each 9 in. by $4\frac{1}{2}$ in. by $2\frac{1}{4}$ in, which can be built into a cube? Make a freehand drawing showing how the bricks are arranged?

Draw any triangle *ABC*, and take a point *D* outside the plane of your paper. (Fig. 3 (*b*))

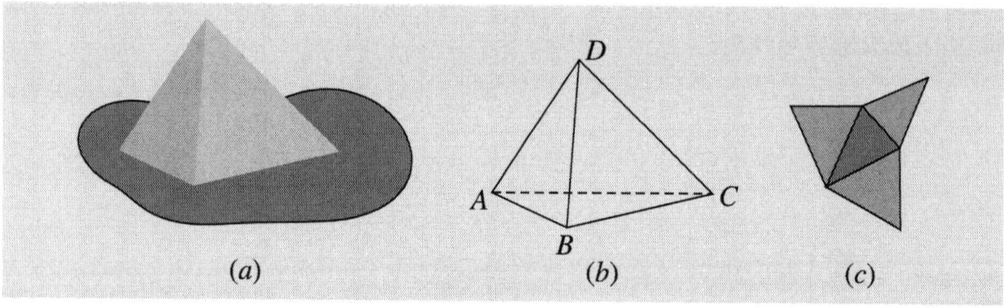

Fig. 3

Suppose the point *D* joined to *A*, *B* and *C*. The figure bounded by the four triangles *ABC*, *DBC*, *DCA*, *DAB* is called a **tetrahedron** (plural **tetrahedra**).

If all the edges of a tetrahedron are equal, the tetrahedron is said to be regular. A regular tetrahedron is therefore bounded by four equilateral triangles. (Fig. 4)

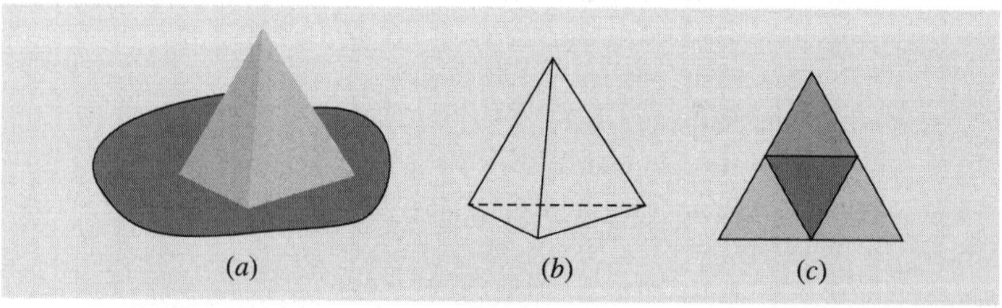

Fig. 4

Draw any polygon *ABCDEF*, and join any point *O* outside its plane to the vertices, thus forming a number of triangles.

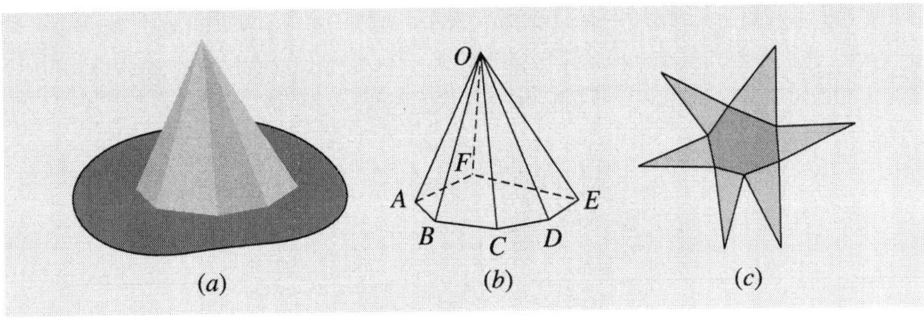

Fig. 5

The solid bounded by the polygon and the triangles is called a **pyramid**.

The polygon is the base of the pyramid, the triangles are its lateral faces, the lines *OA*, *OB*, *OC*, etc., are its lateral edges, and the point *O* is its vertex. (Fig. 5).

A pyramid is **triangular**, **quadrilateral**, etc., according as its base is a triangle, a quadrilateral, etc. Thus, a tetrahedron is a triangular pyramid, and any one of its faces may be considered to be the base.

EXERCISE XXIV

Pyramids

1. What is the smallest number of planes which can bound a solid?
2. What is the solid called which is bounded by four planes which do not all meet in a point?
3. With regard to a tetrahedron,
   (*i*) How many faces has it?
   (*ii*) How many edges?
   (*iii*) How many vertices?
   (*iv*) What is the shape of each face?
4. Draw on paper a triangle *ABC* in which *BC* = 2.5 in, *CA* = 2 in, AB = 2.3 in. Through *A*, *B*, *C* draw parallels to *BC*, *CA*, *AB* respectively, forming a triangle *A'B'C'*. What are the lengths of the sides of *A'B'C'*? Prove the correctness of your answer. Cut out the triangle *A'B'C'*. Crease the paper along the dark lines. You will now be able to construct a tetrahedron whose opposite edges are equal. What are the lengths of the pairs of opposite edges?
5. Draw a figure showing how a piece of paper may be cut and bent so as to form a tetrahedron *ABCD* in which *BC* = 3.4 in, *CA* = 2.5 in, *AB* = 3 in, *AD* = 2.84 in, *BD* = 2.9. in, *CD* = 3.15 in.
6. Draw a figure showing how a piece of paper may be cut and bent in order to form a pyramid whose base is a square of 2 in. side and each lateral face an equilateral triangle.
7. Draw a figure showing how a piece of paper must be cut and bent in order to make a pyramid whose base is a rectangle of length 3 in. and breadth 2 in, each lateral edge being 2.5 in. long.

In Fig. 6, *OA* and *OB* represent the intersection of two walls and the floor of a room, *OC* is the intersection of the walls. Observe that *OC* is perpendicular to every straight line (such as *OD*) drawn through *O* on the floor.

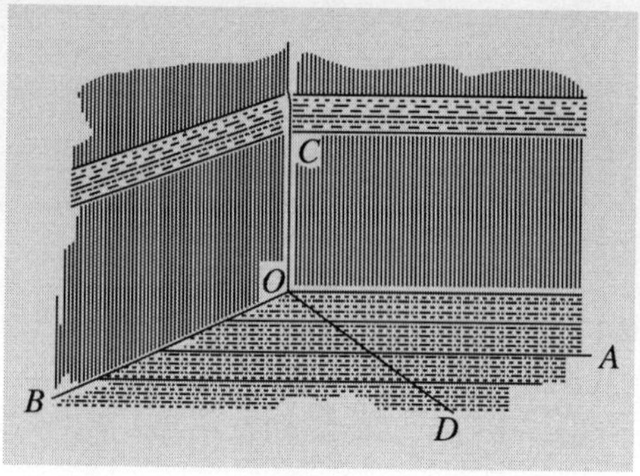

Fig. 6

Hence *OC* is said to be perpendicular to the plane of the floor. Usually, a floor is a horizontal plane, the walls are vertical planes and two vertical walls intersect in a vertical line.

Draw any two straight lines on your paper, and construct a line perpendicular to both of them as follows:

Take the stiff covers *AOCD*, *BOCE* of an exercise book from which the interior has been removed.

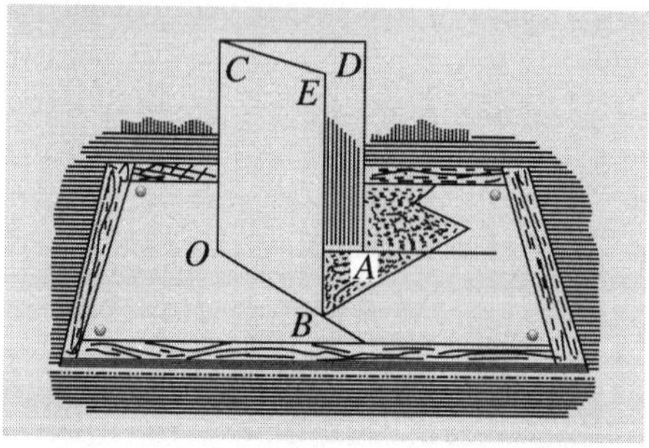

Fig. 7

Place the covers so that the edges *OA* and *OB* are along the lines drawn on your paper, then the edge *OC* is perpendicular to both *OA* and *OB*. (Fig. 7)

Now hold the cover *BOCE* firm; you will find that *AOCD* can be turned about *OC* so that the edge *OA* remains in contact with the paper.

Hence *OC* is perpendicular to every straight line through *O* in the plane of the paper, *i.e.* *OC* is the perpendicular through *O* to the plane of the paper.

Thus it will be seen that if a straight line (*OC*) is perpendicular to each of two straight lines (*OA*, *OB*) at their point of intersection, it is perpendicular to the plane which contains them.

You will now understand that any vertical line is perpendicular to every horizontal plane, and that an edge of a rectangular block is perpendicular to either of the faces which it intersects.

Planes which do not intersect, however far they are produced, are said to be parallel. Thus two opposite faces of a rectangular block are parallel planes.

EXERCISE XXV

1. Name several surfaces which you would call horizontal.
2. Name several lines which you would call vertical.
3. Give instances of pairs of parallel planes.
4. Give instances of lines and planes which are mutually perpendicular.
5. Stand your pencil, as nearly as you can judge, perpendicularly to the plane of your paper. By using a set square how can you discover if it is exactly perpendicular to the paper?
6. How would you determine by means of a spirit level whether the surface of a table is horizontal?
7. How would you use a plumb-line to determine whether a post is placed vertically?
8. How would you proceed to level the surface of a three-legged table?
9. Why is it easier to level a three-legged table than a four legged one?

Suppose a large number of triangular pieces of paper, all of the same size and shape, cut out and piled as in Fig. 8 (*a*), so that each covers exactly the one below it. The solid so formed is called a right triangular prism or wedge.

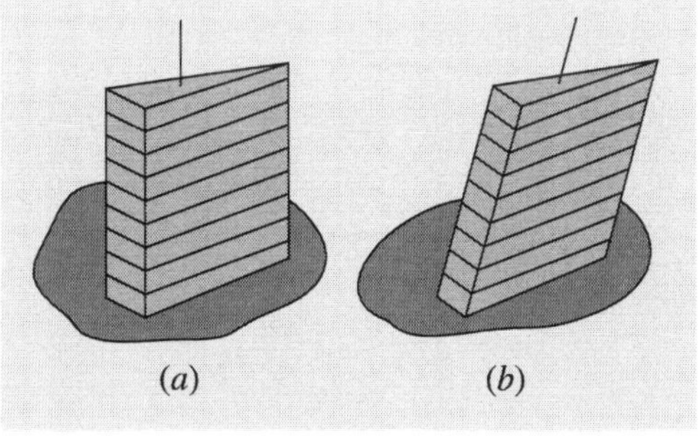

Fig. 8

Next suppose that, without moving the triangles, a hole is bored through the pile in a line perpendicular to the base and a thin metal rod, such as a knitting needle, is put through the hole.

You will see that the pile can be displaced into the shape shown in Fig. 8 (*b*), in which each lateral edge remains straight, but is no longer at right angles to the plane of the base. The solid obtained in Fig. 8 (*b*) is called an oblique triangular prism.

Observe that in the process just described, neither the volume of the solid (*i.e.* the amount of space occupied by it) nor its height (*i.e.* the perpendicular distance between the triangular faces) has been altered. Hence it is seen that the volume of an oblique triangular prism is equal to that of the right triangular prism of the same height standing on the same base.

EXERCISE XXVI

Prisms

1. What is the shape of the lateral faces of a right triangular prism?
2. What is the shape of the faces of an oblique triangular prism?
3. Why are all the lateral edges of a triangular prism equal?
4. What sort of solid can be constructed by cutting out from-paper a figure of the shape shown in Fig. 9 (*c*) consisting of three equal rectangles and two equilateral triangles and bending it about the dark lines?

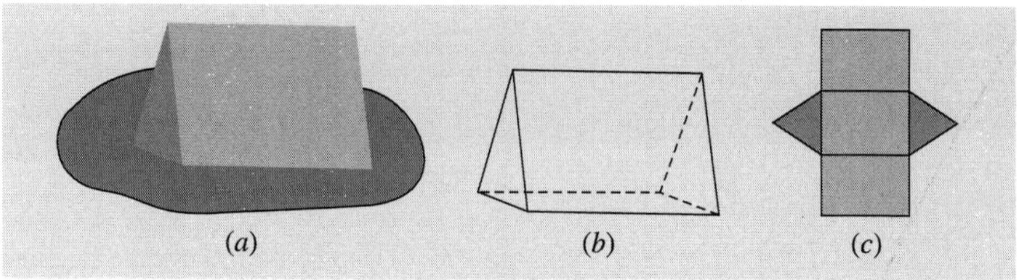

Fig. 9

5. Cut out from paper a figure of shape similar to that in Fig. 10 (*c*), but of five or six times its linear dimensions (*i.e.* each line five or six times the length of the corresponding line in Fig. 10(*c*).

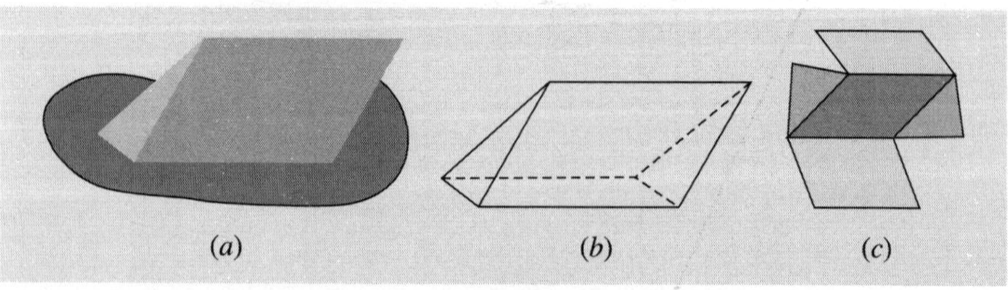

Fig. 10

Bend about the dark lines and an oblique triangular prism will be obtained.

6. Draw a figure showing how to cut and bend paper to form a right triangular prism the sides of whose base are 2 in, 1.25 in. and 1.5 in, and whose height is 2.5 in.
7. If two parallel planes are cut by a third plane, the lines of intersection are parallel. Can you give a reason for this?

Draw two congruent polygons *ABCDE*, *A'B'C'D'E'* and place them in parallel planes with their corresponding sides parallel.

Then the figures *ABB'A' BCC'B'*, etc., are parallelograms (Why?), and the solid bounded by the polygons and parallelograms is called a prism (Fig. 11).

The polygons are the bases of the prism and the perpendicular distance between the bases is the height of the prism.

A prism is triangular, quadrangular, etc., according as its bases are triangles, quadrilaterals, etc.

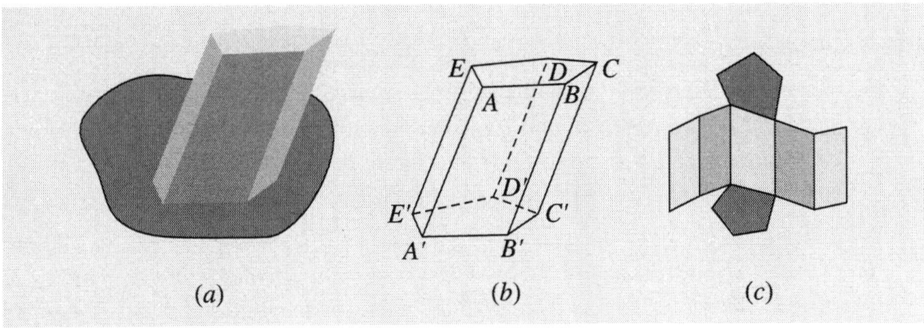

Fig. 11

If the **lateral edges** *AA'*, *BB'*, etc., are perpendicular to the plane of a base, the solid is called a **right prism**.

Figure 12 represents a right prism whose base is a regular polygon of 24 sides. If the regular polygon had a very great number of sides, the shape of the corresponding prism would approximate to that represented in Fig. 13.

A solid of the shape of a round uncut pencil is called a right circular cylinder, and the curved surface of the pencil is called a cylindrical surface.

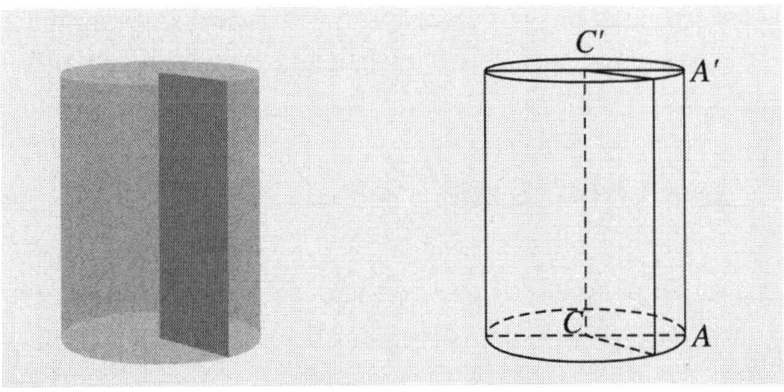

Fig. 12 Fig. 13

To be more precise, if a rectangle *ACC'A'* is made to revolve about one of its sides *CC'* as axis, it is said to generate a solid.

This solid is called a **right circular cylinder** and the curved surface of the solid is a **cylindrical surface** (Fig. 13).

The line *CC'* is called the **axis** of the cylinder and the circles generated by the revolution of *CA* and *C'A'* are its **bases**.

EXERCISE XXVII

Prisms, etc.

1. What do you observe about all the lateral edges of a prism?
2. Draw a figure representing a right prism whose base is a rectangle. What is the other name of this solid?
3. Describe exactly the shape (giving the correct name) of an uncut hexagonal pencil.
4. Explain exactly why, in Fig. 13, as *CA* revolves, it always moves in the same plane.
5. Take a cylindrical ruler and measure its circumference as follows: wrap a piece of paper tightly round the ruler until the paper overlaps. Stick a pin through the paper so as to mark two points just under one another. Unfold and measure the distance between the points. This is the circumference of the cylinder.
6. Draw a freehand sketch of the section of a right circular cylinder made by a plane which is (*i*) perpendicular to the axis, (*ii*) parallel to the axis (*i.e.* such that the axis never meets the plane if both are produced indefinitely), (*iii*) inclined to the axis.

   In both cases (*i*) and (*ii*) say exactly what figure the section is.
7. Measure the circumferences of several cylinders, such as a ruler, a pencil, a draughts-man, a cork bung. Measure also the diameters, and verify that, in all cases, the circumference is approximately $\frac{22}{7}$ times the diameter.
8. The curved surface of a tin box in the shape of a right circular cylinder, whose height is 3 in. and circumference 4 in, is cut through along a line parallel to the axis of the cylinder. Draw a figure showing the exact size and shape of the curved surface when flattened out on a table. What is the figure called?
9. A cylindrical box of height 4 in. and diameter 3 in. is to be made of tin. Draw a figure of the exact shape and size of the piece of tin required for the curved surface. What is the length and breadth of the figure?
10. A room is 30 ft. long, 20 ft. wide and 15 ft. high. By drawing a figure to scale, find, as accurately as you can, the distance of a corner of the floor from the opposite corner of the ceiling. [Scale 1 in. = 10 ft.]

Figure 14 represents a pyramid whose base is a regular polygon of 24 sides. If the polygon had a very great number of sides, the shape of the corresponding pyramid approximates to that represented in Fig. 15.

A **right circular cone** is the solid generated by the revolution of a right-angled triangle *AOC* about *OC*, one of the sides containing the right angle. The curved surface of the cone is called a conical surface. (Fig. 15)

Fig. 14

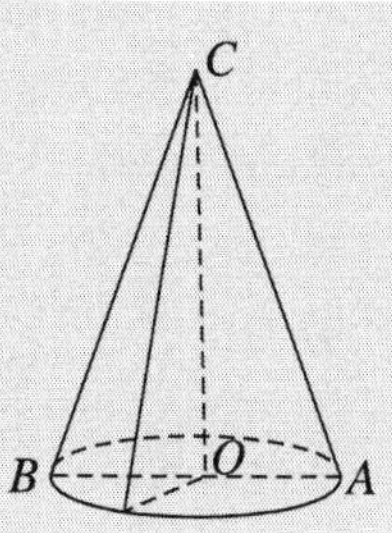

Fig. 15

During the revolution, the side *OA* generates a circle whose centre is *O*, and whose plane is perpendicular to *CO*. This circle is called the base of the cone, the line *CO* its axis and the point *O* its vertex. The length of *CO* is the **height** of the cone, and *CA* is called a slant side.

Observe that the surface of a right circular cone can be **developed** (*i.e.* rolled out flat) on a plane. If this is done, the resulting figure is a sector of a circle, for each point in the circumference of the base of the cone is at the same distance from the **vertex**.

EXERCISE XXVIII

Cones

1. Explain why the section of a cone by a plane perpendicular to the axis is a circle.
2. What is the section of a cone by a plane through the vertex?
3. A piece of wood in the shape of a right circular cone, of height 2 in. and the radius of whose base is 1.5 in, is to be covered with tin. Draw a figure showing the exact size and shape of the piece of tin required and find the angle of the sector of the circle.
4. Find, by drawing a diagram, the semi-vertical angle (*i.e.* the angle between the axis and a slant side), of the cone, the surface of which, when rolled out flat on a plane, forms a semi-circle.

A solid of the shape of a cricket ball is called a sphere, and its surface is called a **spherical** surface (Fig. 16).

A **sphere** is a solid bounded by a surface, all points of which are equidistant from a certain point within the surface, This point is called the **centre**.

A **radius** of a sphere is a straight line drawn from the centre to any point on the surface; all radii are equal.

A **diameter** of a sphere is a straight line passing through the centre and terminated both ways by the surface of the sphere.

A sphere may also be considered as the solid generated by the revolution of a semi-circle about its diameter.

The surface of a sphere cannot be flattened out on a plane without crumpling.

It is readily shown that the section of a sphere by a plane is a circle (Fig. 17).

Fig. 16

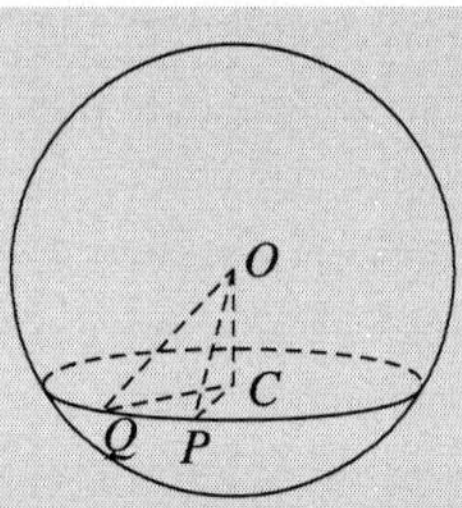

Fig. 17

Let O be the centre of the sphere, OC perpendicular to the plane of section, P and Q any two points on the curve of section;

(*i*) The $\angle$s OCP, OCQ are right angles. (Why?)

(*ii*) The Δs OCP, OCQ are congruent. (Why?)

Hence $\qquad CP = CQ.$

In the same way any two points on the curve of section may be shown to be equidistant from C.

Hence the section is a circle whose centre is C.

Note: For Revision papers on Unit I, see pp. 136-140, Nos., I–IX.

UNIT 2

Areas

XVI. AREAS OF PARALLELOGRAMS AND TRIANGLES

DEF. The area of a plane figure is the amount of surface enclosed by its boundaries.

Congruent figures have the same area, for they can be made to coincide.

For example, two rectangles, which have the same length and breadth, have also the same area. For they are obviously congruent.

It will be seen presently that figures of the same area are not necessarily of the same shape.

Measurement of Areas. In order to measure areas numerically, some unit of area must be chosen.

The area of a square whose side is the unit of length will be taken as the unit of area.

Cross-rule some paper with lines at unit distance apart, and draw a rectangle whose sides are 6 and 3 units of length respectively. The rectangle contains 3 rows of 6 squares each, and each square contains the unit of area.

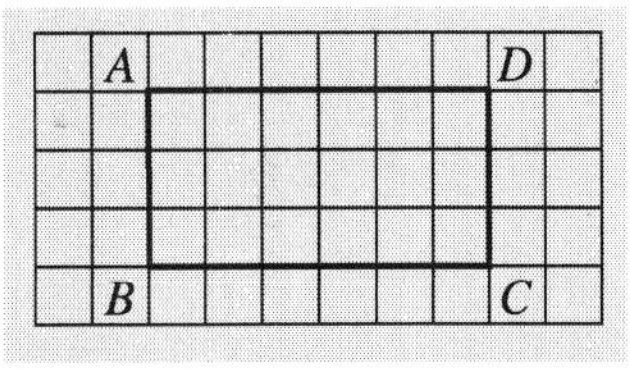

$\therefore$ Area of rectangle = 6×3 units of area.

Similarly, the area of any other rectangle whose sides contain an exact number of units is found by multiplication.

A **square inch** is the area of a square whose side is an inch.

If the unit of length is $\frac{1}{10}$ of an inch, the unit of area is the area of the small shaded square in the figure, and is $\frac{1}{100}$ of a square inch.

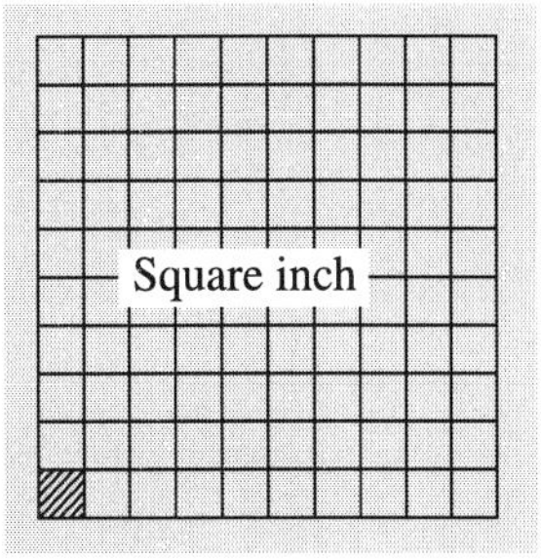

Take squared paper ruled in inches and tenths, and draw on it the rectangle *ABCD*, where $AB = 0.5$ in. and $AD = 1.3$ in. Take 0.1 in. as the unit of length

$\therefore$ $AB = 5$ units, $BC = 13$ units;

∴ Area of $ABCD$

$= 13 \times 5$ units of area

$= 13 \times 5$ hundredths of a sq. in.

$= \frac{13}{10} \times \frac{5}{10}$ sq. in.

$= 1.3 \times 0.5$ sq. in.

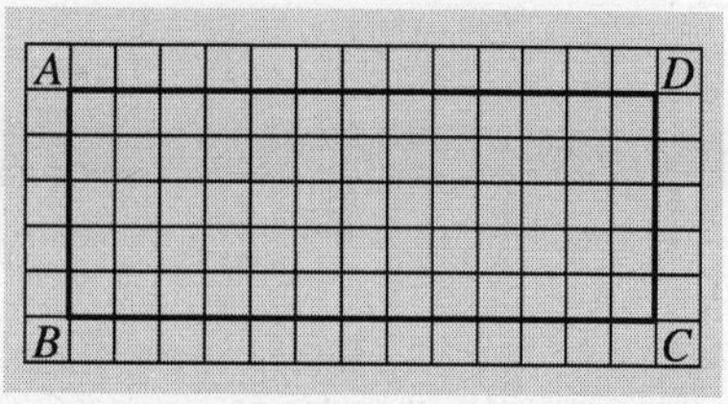

Thus, again the area is given by multiplication, although the length and breadth are both fractions of an inch.

It will thus be seen that the **area of a rectangle** is given by the formula

$$A = l \times b,$$

where A stands for the number of units of area in the rectangle and l, b for the numbers of units of length in two adjacent sides, these numbers being whole or fractional.

DEF. Any side of a parallelogram or triangle may be called the **base**.

The perpendicular distance between the base and the opposite side is called the **altitude** or **height** of the **parallelogram**.

The perpendicular distance of the vertex of a triangle from the base is called the **altitude** or **height** of the **triangle**.

In each of the figures below, if b is the base, h is the height.

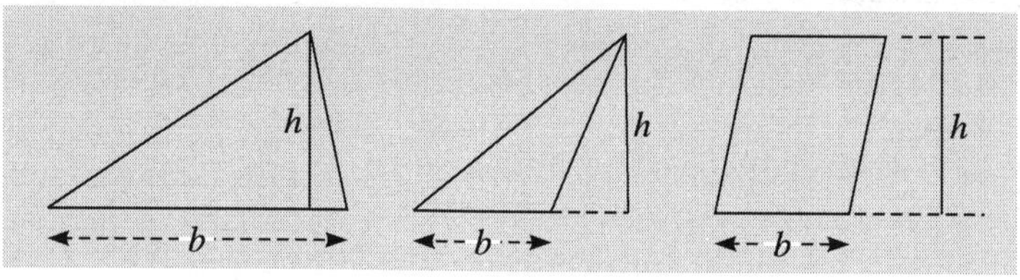

Parallelograms and triangles, situated as in the figure below, are said to be **between the same parallels**.

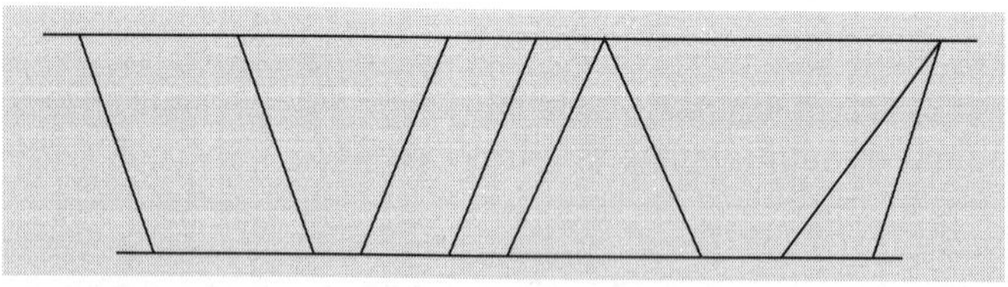

Since the perpendicular distance between two parallel straight lines is everywhere the same, these figures are of the same height.

Conversely, it is obvious that parallelograms and triangles of the same altitude can be placed between the same parallels.

For shortness, the parallelogram $ABCD$ is often called 'the parallelogram AC' or 'the parallelogram BD.'

Let $ABCD$ be a parallelogram and X any point in the diagonal BD. Through X draw HXK parallel to AB to meet AD in H and BC in K. Through X draw LXM parallel to BC to meet AB in L and CD in M.

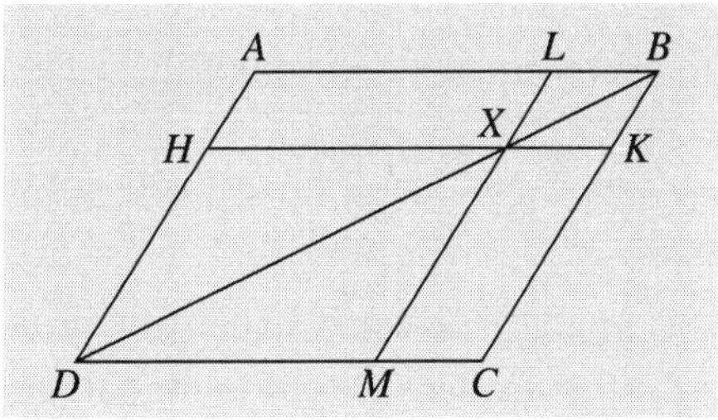

The figures *HM*, *LK* are called the **parallelograms** about the diagonal of *ABCD*, and the figures *AX*, *XC* are called the complements of the parallelograms about the diagonal of *ABCD*.

> **NOTE:** *The abbreviation* $\square^m$ *stands for 'parallelogram.'*

THEOREM 38 (Euclid I. 35)

Parallelograms on the same base and between the same parallels are equal in area.

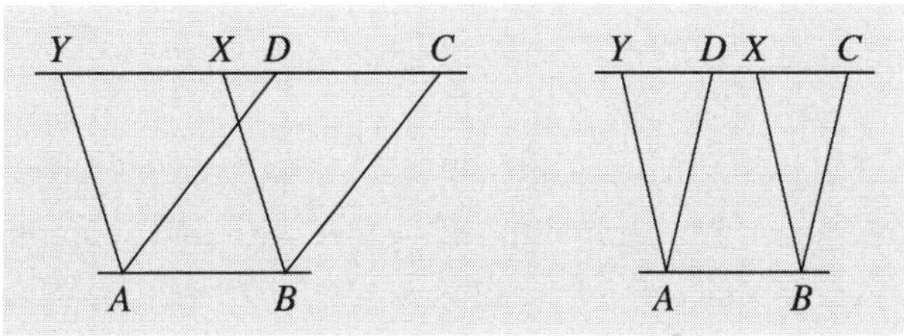

Let *ABCD*, *ABXY* be parallelograms on the same base *AB* and between the same parallels *AB*, *YC*.*

It is required to prove that *ABCD*, *ABXY* are equal in area.

Proof. Because *YC* meets the parallels *AD*, *BC*,

∴ $\angle ADY$ = The corresponding $\angle BCX$.

Again, because *YC* meets the parallels *AY*, *BX*,

∴ $\angle AYD$ = The corresponding $\angle BXC$.

Also, the opposite sides of the parallelogram *ABXY* are equal;

∴ $AY = BX$.

Hence, in the triangles *ADY*, *BCX*,

$$\angle ADY = \angle BCX$$

$$\angle AYD = \angle BXC \text{ (proved)};$$

$$AY = BX$$

∴ The triangles *ADY*, *BCX* are congruent and equal in area.

Take each of these equal triangles, *in succession*, away from the figure *ABCY*.

*The parallels are supposed to be indefinitely produced in either direction.

∴ The remaining figures *ABCD*, *ABXY* are equal in area.

Cor. 1. Parallelograms on equal bases and of the same altitude are equal in area.

For they can be placed so as to be on the same base and between the same parallels.

Cor. 2. The area of a parallelogram is equal to that of a rectangle, whose adjacent sides are equal to the base and altitude of the parallelogram, respectively.

Theorem 39 (Euclid I. 41)

If a triangle and a parallelogram are on the same base and between the same parallels, the area of the triangle is equal to half that of the parallelogram.

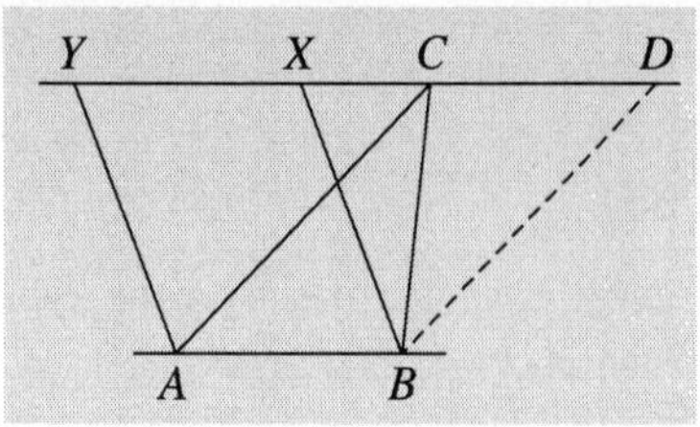

Let the triangle *ABC* and the parallelogram *ABXY* be on the same base *AB* and between the same parallels *AB*, *YC*.

It is required to prove that the area of *ABC* is half that of *ABXY*.

Construction. Draw *BD* parallel to *AC* to meet *YC*, or *YC* produced at *D*.

Proof. It is given that *AB* is parallel to *CD*, and, by construction, *BD* is parallel to *AC*;

∴ *ABDC* is a parallelogram;

and because *ABDC* is bisected by its diagonal *BC*,

$$\therefore \qquad \Delta\, ABC = \frac{1}{2} \text{▱}^{m}\, ABDC.$$

Now the parallelograms *ABDC*, *ABXY* are on the same base and between the same parallels;

$$\therefore \qquad \text{▱}^{m}\, ABDC = \text{▱}^{m}\, ABXY;$$

$$\therefore \qquad \Delta\, ABC = \frac{1}{2} \text{▱}^{m}\, ABXY.$$

Cor. 1. If a triangle and a parallelogram are on equal bases and of the same altitude, the area of the triangle is equal to half that of the parallelogram.

For they can be placed so as to be on the same base, and between the same parallels.

Cor. 2. The area of a triangle is equal to half that of a rectangle, whose adjacent sides are respectively equal to the base and altitude of the triangle.

THEOREM 40 (Euclid I. 37, 38)

Triangles on equal bases, or on the same base, and between the same parallels, are equal in area.

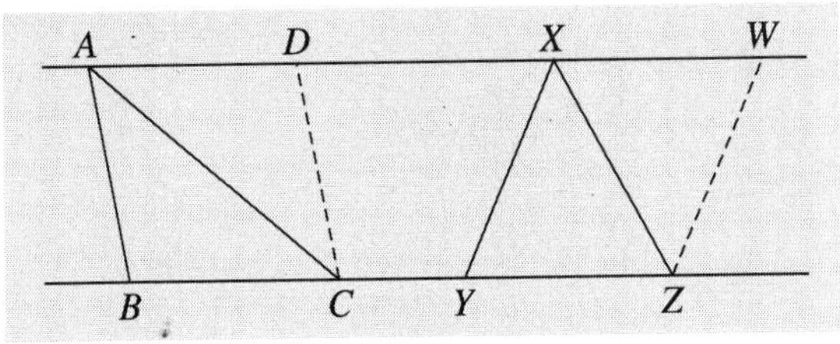

Let *ABC*, *XYZ* be triangles on equal bases *BC*, *YZ* and between the same parallels *BZ*, *AX*.

It is required to prove that *ABC*, *XYZ* are equal in area.

Construction. Through *C* draw a straight line parallel to *BA*, and through *Z* draw a straight line parallel to *YX*. Let these lines cut *AX* or *AX* produced in *D*, *W* respectively.

Proof. Each of the figures *ABCD*, *XYZW* is a parallelogram, and they are on equal bases *BC*, *YZ* and between the same parallels *BZ*, *AW*;

$$\therefore \qquad \square^{m}\ ABCD = \square^{m}\ XYZW.$$

Now a parallelogram is bisected by a diagonal;

$$\therefore \qquad \Delta\ ABC = \frac{1}{2}\ \square^{m}\ ABCD$$

$$\text{and} \qquad \Delta\ XYZ = \frac{1}{2}\ \square^{m}\ XYZW;$$

$$\therefore \qquad \Delta\ ABC = \Delta\ XYZ.$$

In the same way it can be shown that triangles on the same base and between the same parallels are equal in area.

THEOREM 41 (Euclid I. 40)

Triangles of equal area, which are on equal bases in the same straight line and on the same side of it, are between the same parallels.

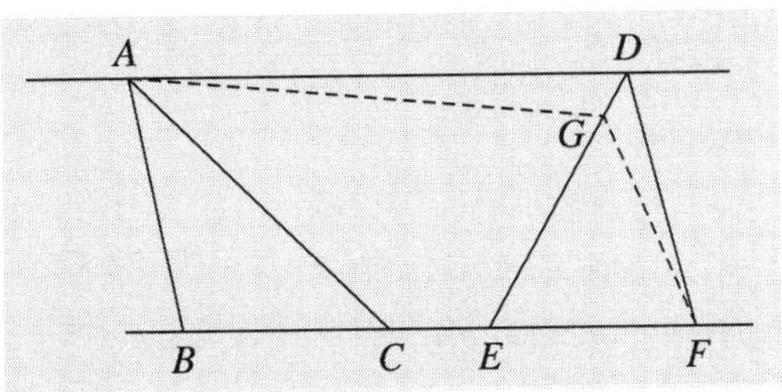

Let the triangles *ABC*, *DEF* be equal in area, and on equal bases *BC*, *EF*, the bases being in the same straight line, and the triangles on the same side of it.

It is required to prove that *AD* is parallel to *BF*.

Proof. If *AD* is not parallel to *BF*, let *AG* be drawn parallel to *BF*, to cut *ED* or *ED* produced at *G*.

It is given that the triangles *ABC*, *GEF* are on equal bases; and, by supposition, they are between the same parallels *BF*, *AG*;

$\therefore \qquad \Delta\, ABC = \Delta\, GEF.$

But it is given that

$$\Delta\, ABC = \Delta\, DEF;$$

$\therefore \qquad \Delta\, GEF = \Delta\, DEF,$

that is, a part equal to the whole, which is impossible.

Hence it is wrong to suppose that *AD* is not parallel to *BF*;

$\therefore$ *AD* is parallel to *BF*.

THEOREM 42 (Euclid I. 43)

The complements of the parallelograms which are about the diagonal of a parallelogram are equal.

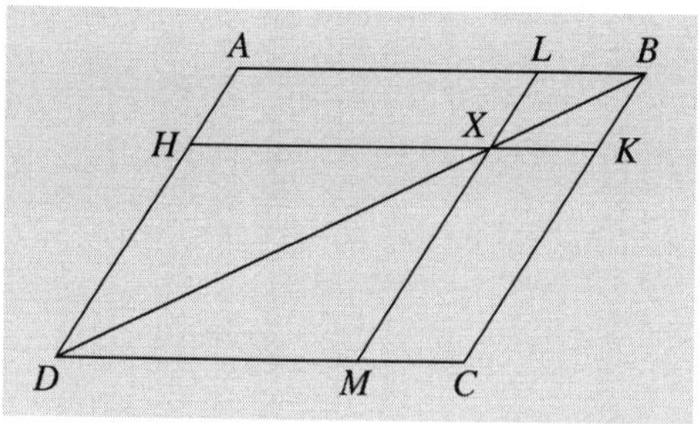

Let *ABCD* be a parallelogram, and *X* any point in the diagonal *BD*. Through *X*, let a parallel to *AB* be drawn to meet *AD* in *H*, and *BC* in *K*. Through *X*, let a parallel to *BC* be drawn to meet *AB* in *L* and *CD* in *M*.

It is required to prove that the complements *AX*, *XC* are equal.

Proof. The parallelogram *AC* is bisected by the diagonal *BD*;

$\therefore \qquad \Delta\, ABD = \Delta\, CBD;$

and since the figures *HM*, *LK* are parallelograms,

$\therefore$ *HM* is bisected by *DX*

and *LK* is bisected by *XB*;

$\therefore \qquad \Delta\, HDX = \Delta\, MDX$

and $\qquad \Delta\, LXB = \Delta\, KXB;$

$\therefore \qquad \Delta\, HDX + \Delta\, LXB = \Delta\, MDX + \Delta\, KXB.$

But $\qquad \Delta\, ABD = \Delta\, CBD;$

the remainders, the complements *AX*, *XC*, are equal.

EXERCISE XXIX (Theorems 38–42)

1. *ABCD* is a parallelogram and *E* is the middle point of *CD*; join *AE*; show that the area of the triangle *ADE* is one-fourth that of the parallelogram.
2. Of all parallelograms on equal bases, whose areas are equal, that which is a rectangle has the least perimeter.
3. Construct a rhombus equal in area to a given parallelogram, having one side common with the parallelogram.
4. If *ABCD*, *ABXY* are parallelograms on opposite sides of the common base *AB*, and *CX* is bisected by *AB* or *AB* produced, then the parallelograms are equal in area. [Draw *CL*, *XM* perpendicular to *AB*. Prove $CL = XM$.]
5. Find the locus of the vertex of a triangle of which the area is equal to that of a given triangle *ABC*, and which is on the same base *BC* and on the same side of it.
6. *ABC* is a given triangle. Construct a triangle equal to *ABC* in area, with its vertex on a given straight line.
7. Construct an isosceles triangle equal in area to a given triangle, and on the same base.
8. Given the lengths of two sides of a triangle, show that its area is greatest when the angle contained by these sides is a right angle.
9. *AB* is the base of a given parallelogram; construct a rhombus equal in area to the parallelogram, with *AB* as a diagonal.
10. *ABCD* is a given parallelogram. Find the locus of the remaining vertices of parallelograms equal in area to *ABCD*, having *AC* as diagonal.
11. *ABCD* is a quadrilateral, and *BD* is bisected in *M*. Show that the area of $AMCB$ = half that of *ABCD*.
12. *ABCD* is a quadrilateral. Through *A* and *C* draw parallels to *BD*, and through *B* and *D* draw parallels to *AC*. Show that the area of the resulting parallelogram is twice that of *ABCD*.
13. *ABCD* is a quadrilateral. Construct a triangle two sides of which are equal and parallel to *AC* and *BD* respectively, and show that its area is equal to that of *ABCD*. [Use Ex. 12]
14. *ABCD* is a trapezium, with *AB* parallel to *CD*. Construct a parallelogram equal in area to *ABCD*, with *AD* for one of its sides, and two sides along *AB*, *DC* respectively.
15. If *D* is the middle point of the base *BC* of a triangle *ABC* and *P* any point in *AD*, show that $\Delta APB = \Delta APC$.
16. *ABC* is a triangle and *P* is a point such that $\Delta PAB = \Delta PAC$. Find the locus of *P*. [Let *AP* cut *BC* in *X*. Draw *BL*, *CM* perpendicular to *AP*. Prove Δs *BXL*, *CXM* congruent.]
17. Two triangles are on the same base, and the height of one is double that of the other; show that the area of one is double that of the other.

18. *ABCD* is a parallelogram, *E* is any point in the diagonal *AC*; show that the triangles *ABE*, *ADE* are of equal area. [Join *BD*]

19. If two triangles have two sides of the one equal to two sides of the other and the included angles supplementary, they are equal in area.

20. If one diagonal *AC* of a quadrilateral bisects the other diagonal *BD*, show that *AC* also bisects the quadrilateral.

21. *D* is a point in the side *AB* of the triangle *ABC*. Find a point *E* in *BC* so that the triangles *EAD*, *CAE* are equal. [Use Ex. 20]

22. If a quadrilateral is divided into four equal triangles by its diagonals, show that it is a parallelogram.

23. *ABC* is a triangle and *D*, *E* are the middle points of *AB*, *AC*. Prove that $\Delta\ BCD = \Delta\ BCE$. Hence show that *DE* is parallel to *BC*.

24. *ABC* is a triangle and *D*, *E* arc the middle points of *AB*, *AC*. If *BE*,*CD* meet in *G*, prove that the triangles *GBC*, *GCA*, *GAB* are equal in area. Hence show that *AG* produced bisects *BC*.

25. Of all triangles on the same base whose areas are equal, that which is isosceles has the least perimeter. [See Exercise XXII *a*, 6]

26. *ABCD* is a parallelogram and *P* is any point within the angle which *AD* makes with *BA* produced. Prove that $\Delta\ PAC = \Delta\ PAB + \Delta\ PAD$. [Draw *DE*, *BF* parallel to *AP*, to cut *AC* in *E* and *F*. Join *PE*, *PF*. Prove $AE = CF$, etc.]

27. *ABCD* is a parallelogram and *P* is any point within the angle *CAD*. Prove that $\Delta\ PAC = \Delta\ PAB - \Delta\ PAD$. [Same construction as in Ex. 26, except that *E*, *F* are in *AC* produced.]

28. In the base *BC* of a triangle *ABC*, take points *D* and *E* in the order *B*, *D*, *E*, *C*, so that *BD* may equal *EC*; join *AD*, *AE*; draw*DF* parallel to *AC* to meet *AB* in *F*, and *EG* parallel to *AB* to meet *AC* in *G*; show that the triangles *ADF* and *AEG* are equal in area.

29. Draw the figure of Theorem 42. Join *HM*, *LK*, and show that each of these lines is parallel to *AC*. [Join *AM*, *CH*. Prove $\Delta\ AHM = \Delta\ CHM$.]

30. Draw the figure of Theorem 42. Join *DL*, *DK*, *LK*, *HM*, and prove that $\Delta\ DLK + \Delta\ DHM = \Delta\ DAC$.

XVII. CALCULATION OF AREAS

Area of a Parallelogram

Let *ABCD* be a parallelogram, in which the base $AB = b$ units of length, the height $BE = h$ units of length.

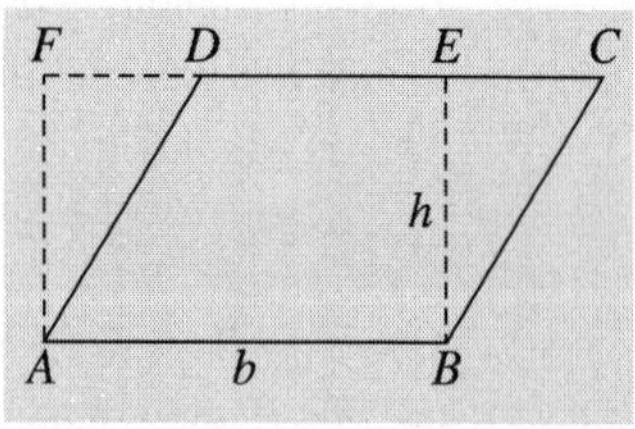

Complete the rectangle *ABEF*.

Then the parallelograms *ABCD*, *ABEF* are on the same base and between the parallels;

$\therefore$ Area of $ABCD$ = Area of $ABEF$

$= b \times h$ units of area.

Thus, **the area of a parallelogram** is the product of the base and the height.

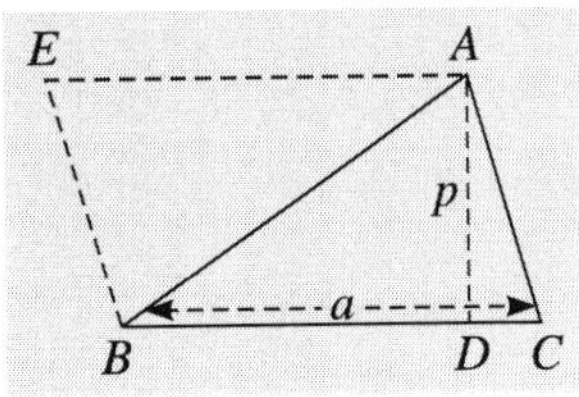

Area of a Triangle. Let ABC be a triangle, in which the base $BC = a$ units of length, the height $AD = p$ units of length.

Complete the parallelogram $BCAE$.

This is of the same height as the triangle ABC, and it is bisected by the diagonal AB;

$\therefore$ Area of ABC $= \frac{1}{2}$ area of $BCAE$

$= \frac{1}{2}\, ap$ units of area.

Thus, the area of a triangle is half the product of the base and the height.

EXAMPLE: *ABC is a triangular field in which the length of BC is 6 chains and the perpendicular distance from A to BC is 4 chains. Find the area of the field in acres.*

$$\text{Area} = \frac{1}{2} \times 6 \times 4 \text{ sq. ch.} = 12 \text{ sq. ch.} = 1.2 \text{ acres.}$$

Area of a Trapezium. Let $ABCD$ be a trapezium with AB parallel to CD.

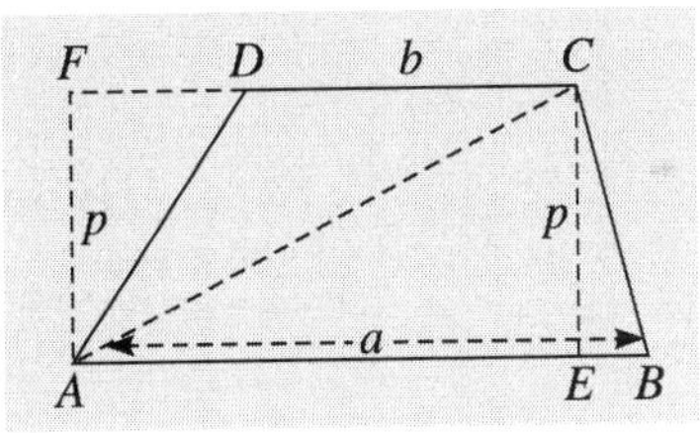

Draw CE perpendicular to AB and AF perpendicular to CD.

Join AC.

Let the lengths of AB, CD and CE, be a, b and p units respectively.

Since $AECF$ is a rectangle,

$\therefore AF = CE = p$ units;

$\therefore$ Area of $ABCD$ = Area of ABC + Area of ADC.

$= \frac{1}{2}\, ap + \frac{1}{2}\, bp$ units of area

$= \frac{1}{2}\,(a + b)p$ units of area.

Thus, the area of a trapezium is the product of half the sum of the parallel sides and the perpendicular distance between them.

To find the **Area of a Quadrilateral**, draw a diagonal, find the areas of the triangles so formed and add the results.

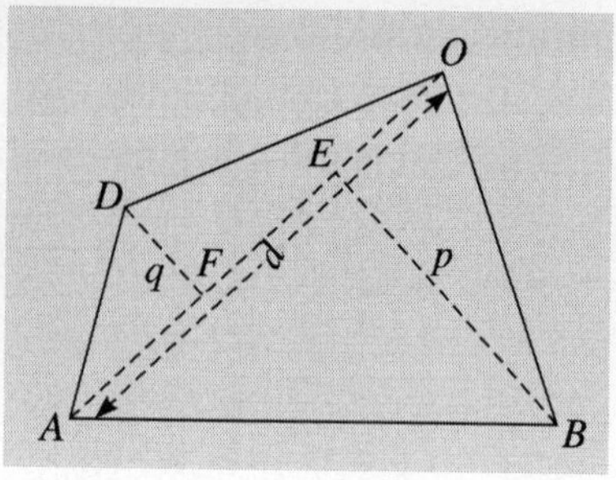

This can be done by measuring three lines.

Let *ABCD* be the quadrilateral. Join *AC*. Draw *BE*, *DF* perpendicular to *AC*.

Let the lengths of *AC*, *BE* and *DF* be *d*, *p* and *q* units respectively.

$\therefore$ Area of *ABCD* = Area of *ABC* + Area of *ADC*

$$= \frac{1}{2} dp + \frac{1}{2} dq \text{ units of area}$$

$$= \frac{1}{2} (p + q)d \text{ units of area.}$$

In connection with the area of a quadrilateral, the following example is useful:

EXAMPLE: *The area of a quadrilateral is equal to that of a triangle, two sides of which are equal and parallel to the diagonals.*

Let *ABCD* be the quadrilateral.

Through *A*, *C* draw parallels to *BD*, and through *B*, *D* draw parallels to *AC*, forming the parallelogram *PQRS*.

Join *QS*.

Prove that

quad. *ABCD* = $\frac{1}{2}$ ▱^{m} *PQRS* = Δ *PQS*.

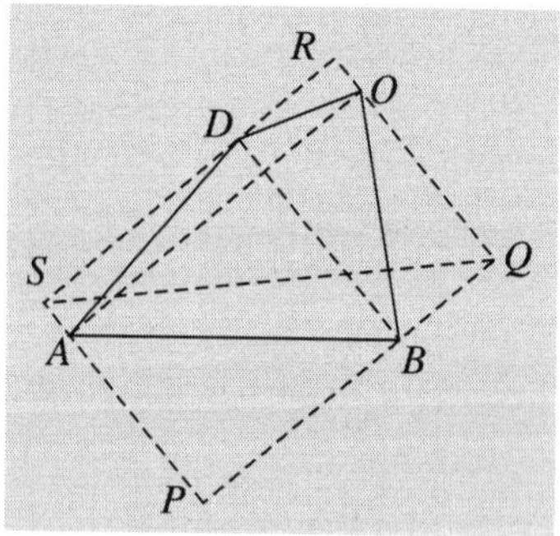

EXERCISE XXX(a)

Numerical

In Exercises 1–15, draw a sketch of the figure named, freehand and find the area by calculation.

1. A rectangle, 3 in. by 4 in.
2. A right-angled Δ, sides containing the rt. ∠, 2 in. and 3 in.
3. A triangle *BC* = 1.2 in., *CA* = 2.5 in., ∠ *C* = 90°.
4. A triangle ∠ *A* = ∠ *C* = 45°, *AB* = 1 in.

5. A parallelogram, base 3 in., height 1.2 in.
6. A triangle, base 2 in., altitude 1.3 in.
7. A parallelogram, adjacent sides 3 in., 2.02 in.; contained $\angle$ 30°.
8. A parallelogram, adjacent sides 3 in., 2.02 in.; contained $\angle$ 150°.
9. A rhombus, diagonals 2.4 in., 0.8 in.
10. A trapezium, parallel sides 3 in. and 1 in., perpendicular distance between them 1.5 in.
11. A quadrilateral *ABCD*, in which the perpendiculars from *B* and *D* on *AC* are 1 in. and 1.2 in. respectively and *AC* = 2 in.
12. A quadrilateral whose diagonals are 1.5 in. and 1.2 in. and are at right angles.
13. A quadrilateral *ABCD*, in which *AC* = 3.01 in., *BD* = 2.4 in., and *AC*, *BD* are inclined at 45°.
14. A quadrilateral *ABCD*, in which *AC* = 4 in., *BD* = 3 in., *AC*, *BD* inclined at $\angle$30°.
15. Find the area (in acres) of a field *ABCD*, given *AB* = 5.35 chains, *CD* = 8.65 chains, *AB* is parallel to *CD*, and the perpendicular distance between *AB*, *CD* = 4.2 chains.

EXERCISE XXX(b)

Numerical

In Exs. 1.13, find the areas of the figures by drawing to a convenient scale and measurement.

1. A rectangle, one side 0.8 in., and diagonal 1.7 in.
2. A rectangle *ABCD*, where *AB* = 2 in., $\angle$ *CAB* = 30°.
3. A right-angled Δ, hypotenuse 0.73 in., one side 0.55 in.
4. A triangle, $\angle A = \angle B = \angle C$, *AB* = 2 in.
5. A triangle, $\angle A$ = 30°, *AB* = 3 in., *BC* = 2 in.
6. A triangle, $\angle A$ = 33°, $\angle B$ = 113°, *AB* = 3 in.
7. A parallelogram, base 1.2 in., one angle 45°, opposite diagonal 1 in.
8. A parallelogram, perpendiculars between pairs of parallel sides 1 in. and 2 in.; one $\angle$ 60°.
9. A rhombus, side 1.4 in., one $\angle$ 40°.
10. A trapezium *ABCD*, in which *AB* = 2 in., *AD* = *DC* = 1 in.; $\angle A$ = 42°.
11. A quadrilateral *ABCD*, in which *AB* = 1.68 in., *BC* = 0.26 in., *CD* = 0.72 in., *AD* =1.54 in. and $\angle B$ = 90°.
12. A quadrilateral whose diagonals are 2 in. and 3 in. and are inclined at 38°.
13. A quadrilateral *ABCD*, where *AB* = 2.4 in., *BC* = 2 in., *CD* = 1.5 in., *DA* = 0.7 in., *BD* = 2.5 in.

Area of any Rectilineal Figure

The following are convenient practical methods of determining the area of a given rectilineal figure *ABCDE.*

First Method: 'Triangulation'

Divide the figure *ABCDE* into triangles by joining any vertex *A* to the remaining vertices.

Find the area of each triangle by taking suitable measurements.

If the figure is convex, as in the diagram, the required area is the sum of the areas of the triangles.

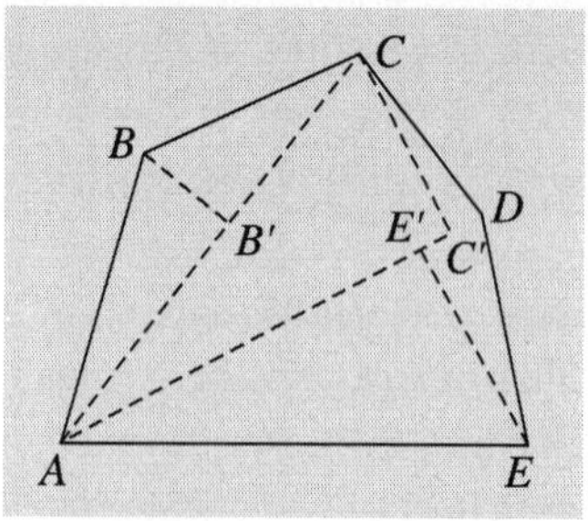

If the figure is not convex, and has an angle *B* greater than two right angles, it will be found that the area of *ABC* must be *subtracted* from the sum of the other triangles. (The student should draw a figure to illustrate this case.)

Let *BB'*, *CC'*, *EE'* be perpendiculars from *B*, *C*, *E* to *AC*, *AD*, *AD* respectively.

Then area of *ABCDE*

$$= \frac{1}{2} AC \cdot BB' + \frac{1}{2} AD \cdot CC' + \frac{1}{2} AD \cdot EE'.*$$

Second Method: 'Field-Book'

Join *AD*, and take this as *base line.*

Draw *BB'*, *CC'*, *EE'* perpendicular to *AD*. Measure *AB'*, *AE'*, *AC'*, *AD* and the *offsets B'B, E'E, C'C.*

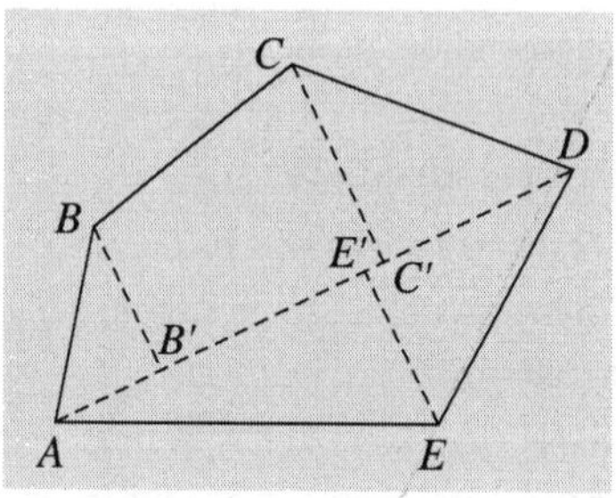

* The notation $AC \cdot BB'$ for an area is to be understood thus : If the lengths of *AC*, *BB'* are *x*, *y* units respectively, then the area denoted by $AC \cdot BB'$ is $x \times y$ units of area.

Then the area of $ABCDE$ is the sum of the areas of the triangles ABB', AED, $CC'D$ and the trapezium $BB'C'C$, and these can be found by the formulae on pp. 107, 108.

Let $ABCDE$ be a plan of a field, in which $AD = 500$ links, $AC' = 400$ links, $AE' = 380$ links, $AB' = 100$ links, $E'E = 200$ links, $B'B = 190$ links, $C'C = 210$ links.

This information would be entered in a Surveyor's field-book in the following form:

Left offsets	Links	Right offsets
	To D	
	500	
To. C 210	400	
	380	200 to E
To B 190	100	
	From A	

the offsets being placed to the right or left of the central column, according as the corresponding points are to the right or left as you walk from A to D.

$$\text{Then} \qquad \Delta\, ABB' = \frac{1}{2} AB' \cdot BB' = \frac{1}{2} \times 100 \times 190 = 9{,}500$$

$$\Delta\, AED = \frac{1}{2} AD \cdot EE' = \frac{1}{2} \times 500 \times 200 = 50{,}000$$

$$\Delta\, CC'D = \frac{1}{2} C'D \cdot CC' = \frac{1}{2} \times 100 \times 210 = 10{,}500$$

$$\text{fig. } BB'C'C = \frac{1}{2} B'C'\,(BB' + CC') = \frac{1}{2} \times 300 \times 400 = 60{,}000$$

$$1{,}30{,}000$$

$$\therefore \qquad \text{Area of } ABCDE = 1{,}30{,}000 \text{ sq. links}$$
$$= 13 \text{ sq. chains}$$
$$= 1.3 \text{ acres.}$$

EXERCISE XXXI

1. Calculate the area of a regular hexagon whose side is 1 in., by dividing it into triangles by joining each vertex to the centre of the figure.

Also calculate its area by joining any vertex to all the other vertices and finding the areas of the triangles into which the hexagon is divided.

2. Draw a plan of the field $ABCDE$ from the field-book entry on the previous page, to scale 1 in. = 1 chain, and find the area by the first method, making any necessary measurements.

3. From the following extracts from a surveyor's field-book, representing measurements of certain fields, draw plans to scale 2 in. = 1 chain, and find the areas of the fields, in acres, by both methods:

(*i*)

	Links	
	To *B*	
	230	
	200	160 to *C*
	170	75 to *D*
	From *A*	

(*ii*)

	Links	
	To *D*	
	260	
	115	28 to *C*
	90	72 to *B*
To *E* 112	40	
	From *A*	

XVIII. AREA CONSTRUCTIONS

Construction 14

Draw a triangle equal in area to a given quadrilateral.

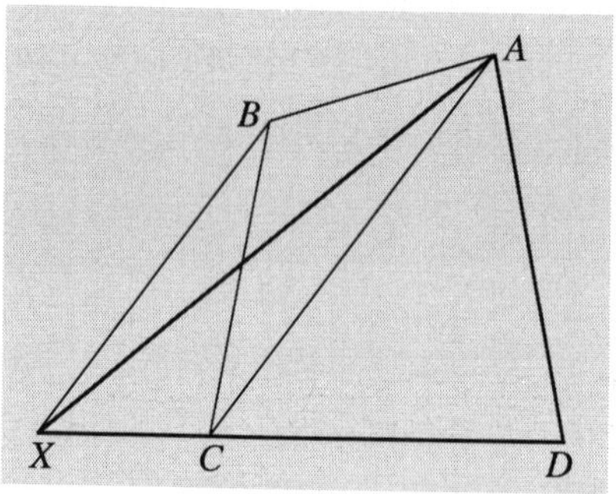

Let *ABCD* be the given quadrilateral.

It is required to draw a triangle equal in area to *ABCD*.

Construction. Join *AC*.

Draw *BX* parallel to *AC* to cut *DC* produced at *X*. Join *AX*. Then *AXD* is the required triangle.

Proof. The triangles *ACX*, *ACB* are on the same base *AC* and between the same parallels *AC*, *BX*;

$\therefore$ $$\Delta\ ACX = \Delta\ ACB.$$

To each add the triangle *ACD*;

$\therefore$ $$\Delta\ AXD = \text{The figure } ABCD.$$

Construction 15

Make a rectilineal figure equal to a given convex rectilineal figure and having fewer sides by one than the given figure.

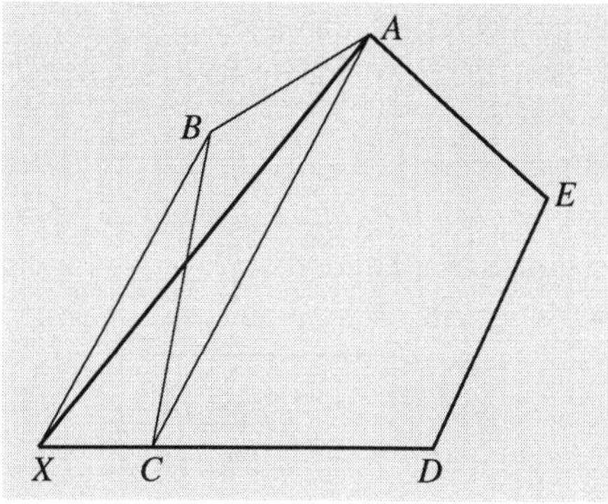

Let *ABCDE* be a rectilineal figure with (say) five sides.

It is required to make a rectilineal figure equal to *ABCDE*, with four sides.

Construction. Join *AC*.

Draw *BX* parallel to *AC* to cut *DC* produced at *X*,

Join *AX*.

Then *AXDE* is the required figure.

Proof. The triangles *ACX*, *ACB* are on the same base *AC* and between the same parallels;

$\therefore \qquad \Delta\, ACX = \Delta\, ACB.$

To each add the figure *ACDE*;

∴ The figure *AXDE* = The figure *ABCDE*.

NOTE: By a succession of steps, similar to the above, we can construct a triangle equal in area to any given rectilineal figure.

EXERCISE XXXII (Constructions 14, 15)

1. Bisect a given triangle *ABC* by a straight line through *A*.
2. Divide a triangle into 5 equal parts by straight lines through *A*.
3. Reduce a trapezium to a triangle of equal area, by Construction 14, and show that the rule already given for the area of a trapezium follows.
4. Draw a pentagon *ABCDE* and reduce it to a triangle of equal area, one vertex of the triangle being at *A* and the others in *CD*, produced both ways.
5. Draw a straight line *AB* one inch long. On *AB* describe a regular hexagon *ABCDEF* (see Ex. XIII. 17). Draw a triangle equal in area to the hexagon, with one vertex at *A* and the others in *CD* produced. Find by measurement the area of the triangle.
6. *P* is a point in the side *BC* of a triangle *ABC*. Find a point *X*, in *CA* produced through *A*, such that the triangle *PCX* may be equal in area to the triangle *ABC*. [Suppose *X* to be in the required position, Join *PA*. Prove that Δ *PAX* must be equal to Δ *PAB*. Hence *BX* must be parallel to *PA*.]

7. *ABCDE* is the plan of a field. Show how to replace the fence *ABC* by a straight fence *AX*, *X* being some point in *CD*, or in *CD* produced, without altering the area of the field.

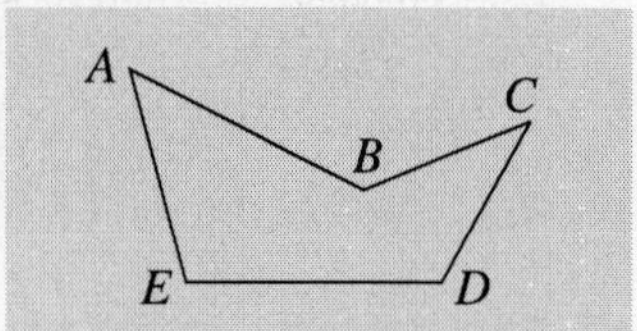

8. *ABCDEF* is the plan of a field. Show how to replace the fence *ABCD* by a straight fence *AY*, *Y* being some point in *ED*, or in *ED* produced, without altering the area of the field. [First find *X* in *CD*, such that the area *AXDEF* may be equal to the area *ABCDEF*.]

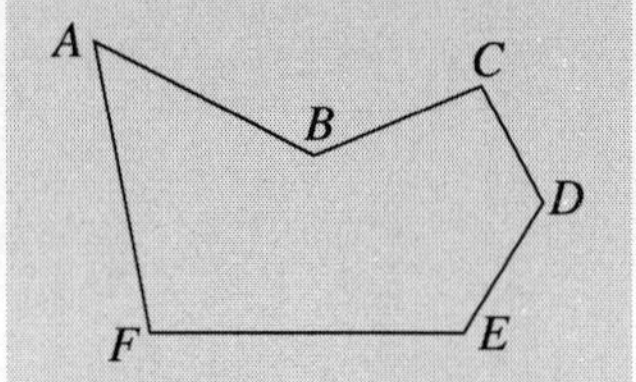

9. *Bisect a triangle ABC by a straight line through a given point P in the side BC.*

[If *BP* < *PC*, by Ex. 6, find *X*, in *CA* produced, such that Δ *PCX* = Δ *ABC*. Bisect *CX* at *Y*.

Join *PY*. Prove that Δ *PCY* = $\frac{1}{2}$ Δ *ABC*, and that *Y* lies between *A* and *C*.]

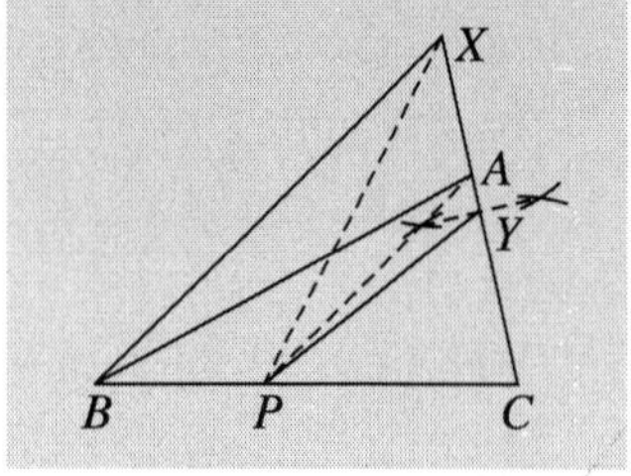

10. Prove also that the following construction is correct: If *BP* < *PC*, bisect *AC* at *X*. Join *PX*. Draw *BY*, parallel to *PX*, to cut *AC* in *Y*. Join *PY*. Then Δ *CPY* = $\frac{1}{2}$ Δ *ABC*.

11. *Bisect a quadrilateral ABCD by a straight line through A.*

[If Δ *ADC* < Δ *ABC*, find *X* in *BC* produced, such that Δ *ABX* = the quad. *ABCD*. Bisect *BX* at *Y*. Join *AY*.]

Prove that Δ *ABY* = $\frac{1}{2}$ quad. *ABCD*, and that *Y* lies between *B* and *C*.

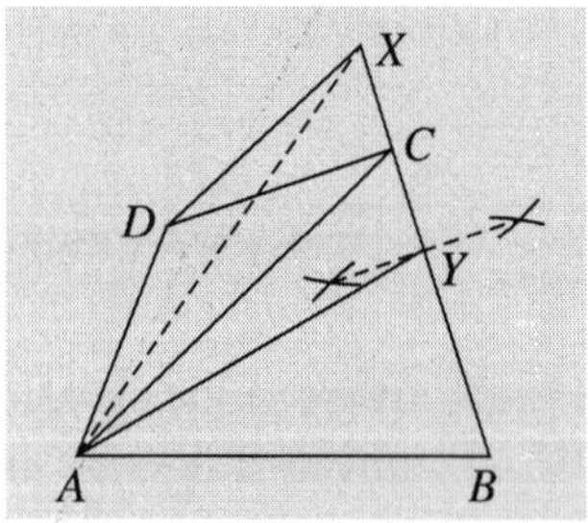

12. Prove also that the following construction is correct : Bisect the diagonal *BD* at *E*. If Δ *ADC* < Δ *ABC*, through *E* draw *EY* parallel to *AC* to cut *BC* in *Y*. Then Δ *ABY* = $\frac{1}{2}$ quad. *ABCD*.

13. Divide a square into three equal parts by straight lines drawn through a vertex.

14. From a given triangle *ABC*, cut off any fractional part, say two-fifths, by a straight line through a given point *P* in *BC*. [If *BP* < $\frac{2}{5}$ *PC*, by Ex. 6, find *X*, in *CA* produced, such that Δ *PCX* = Δ *ABC*. Along *CX* set off *CY* = $\frac{2}{5}$ *CX*. Join *PY*. Prove Δ *PCY* = $\frac{2}{5}$ Δ *ABC*.]

15. Give an alternative construction for Ex. 14, similar to the construction of Ex. 10.

16. Give constructions for Exs. 9, 10 when $BP > PC$; for Exs. 11, and 12, when $\Delta\, ADC > \Delta\, ABC$; and for Ex. 14, when $BP > \frac{2}{5} PC$.

17. *Divide the rectilineal figure ABCDE into five equal parts, by straight lines through A.*

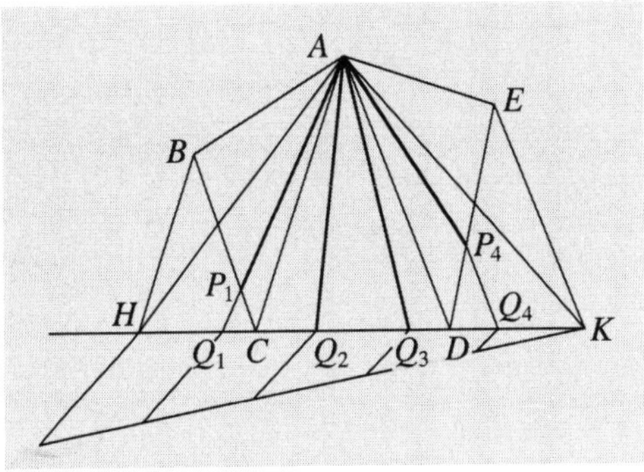

Make $\Delta\, AHK$ = fig. $ABCDE$. Divide HK into 5 equal parts at Q_1, Q_2, Q_3, Q_4. Join A to Q_2, Q_3, the points in CD. Draw $Q_1P_1 \parallel AC$, and $Q_4P_4 \parallel AD$. The lines AP_1, AQ_2, AQ_3, AP_4 are the required lines of division. Supply proof.

Construction 16

Draw a parallelogram equal in area to a given triangle and having one of its angles equal to a given angle.

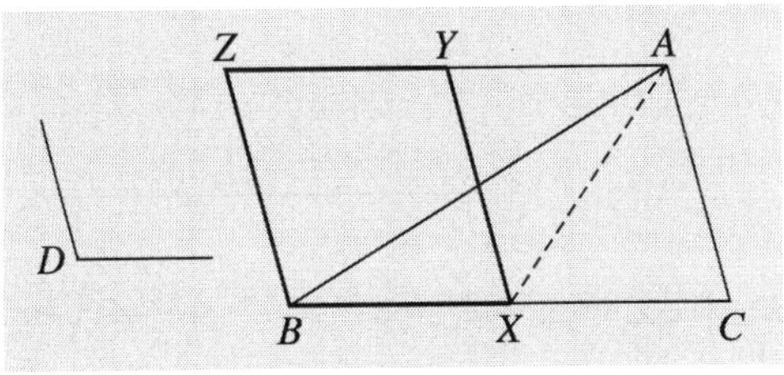

Let ABC be the given triangle and D the given angle. It is required to make a parallelogram equal in area to the triangle ABC with one of its angles equal to the angle D.

Construction. Bisect BC at X,

Make the angle XBZ equal to the angle D.

Draw XY parallel to BZ.

Draw AYZ parallel to BC to cut XY, BZ at Y, Z.

Then $BXYZ$ is the required parallelogram.

Proof. Join AX.

Because the parallelogram BY and the triangle ABX are on the same base BX and between the same parallels;

$$\therefore \quad \square^m\ BY = 2\Delta\ ABX$$

Again, the triangles ABX, ACX are on equal bases and between the same parallels;

$$\therefore \quad \Delta\ ABX = \Delta\ ACX;$$

$$\therefore \quad \Delta\ ABC = 2\Delta\ ABX.$$

But $\quad \square^m\ BY = 2\Delta\ ABX$ (*proved*);

$$\therefore \quad \square^m\ BY = \Delta\ ABC.$$

Also $\angle XBZ$ was made equal to $\angle D$;

$\therefore$ BY is the required parallelogram.

> **NOTE:** *A particular case of the above is to construct a* rectangle *equal in area to a given triangle.* By Constructions 14, 15, 16 we can therefore draw a rectangle equal to any given rectilineal figure.

CONSTRUCTION 17

Draw a rectangle with one of its sides equal to a given straight line and its area equal to that of a given rectangle.

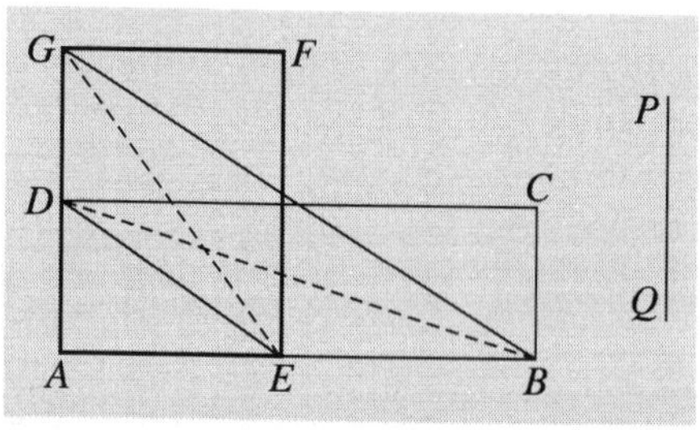

Let $ABCD$ be the given rectangle and PQ the given straight line.

It is required to draw a rectangle with one side equal to PQ and its area equal to that of $ABCD$.

Construction. Along AB set off AE equal to PQ.

Join DE.

Draw BG parallel to ED to cut AD or AD produced in G.

Complete the rectangle $AEFG$.

Then $AEFG$ is the required rectangle.

Proof. Join BD, EG.

The triangles GED, BED are on the same base ED and between the same parallels ED, BG;

$$\therefore \quad \Delta\ GED = \Delta\ BED,$$

$\therefore$ $\Delta\ GAE = \Delta\ BAD$.*

But, since a rectangle is bisected by a diagonal,

$\therefore$ rect. $AEFG = 2\Delta\ GAE$

and rect. $ABCD = 2\Delta\ BAD$;

$\therefore$ rect. $AEFG$ = rect. $ABCD$.

Also AE was made equal to PQ;

$\therefore$ $AEFG$ is the required rectangle.

EXERCISE XXXIII(a) (Constructions 16, 17)

1. *Draw a parallelogram ABCD with its sides AB, AD equal to given straight lines and its area equal to that of a given rectangle AEFG.*

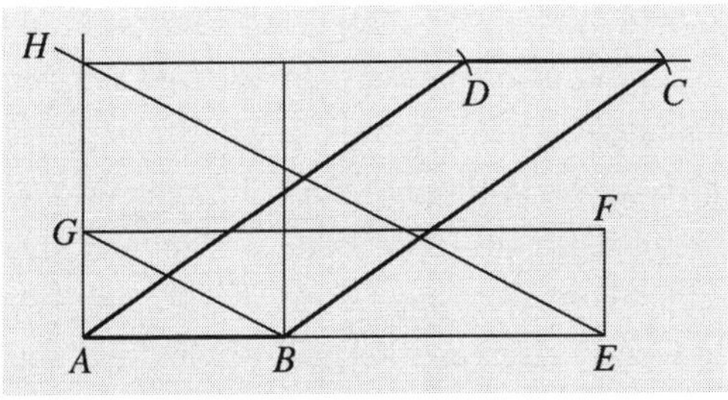

Along AE or AE produced set off AB of the proper length. Join BG. Draw EH parallel to BG to cut AG or AG produced in H. Draw HDC parallel to AE. Complete the construction and supply proof.

2. Draw a rhombus of given area and altitude.

3. *Construct the triangle ABC, given a, B, and that the area is equal to that of a given triangle A'BC'.*

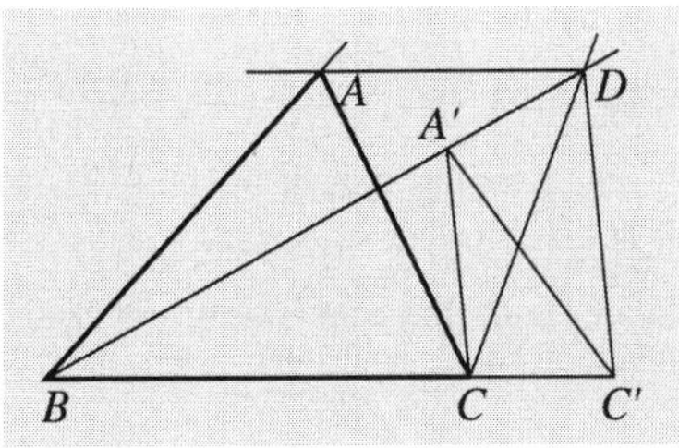

Along BC', set off BC, equal to a. Join CA'. Draw $C'D \parallel CA'$, to meet BA', produced if necessary, in D. Draw $DA \parallel BC'$, and make $\angle\ CBA$ equal to B. ABC is the triangle required. Supply proof.

* If, as in the figure, G is in AD produced, this follows by adding $\Delta\ ADE$ to each of the Δs GED, BED. If G is within AD, the Δs GED, BED must be taken away, in succession, from $\Delta\ ADE$.

4. Construct the triangle ABC, given a, b, and area equal to that of a given triangle. When is this impossible?
5. Construct a rhombus $ABCD$ of area equal to that of a given triangle PQR, and with its diagonal AC of given length.

   [Find a point S in QR, such that $QS = \frac{1}{4} QR$, and use Ex. 3 to construct a triangle AQB, having a right angle at Q, QB equal to $\frac{1}{2} AC$, and its area equal to that of $\Delta\ PQS$. AQB is a quarter of the rhombus. Complete the construction and supply proof.]

EXERCISE XXXIII(b)

Numerical

In Exercises 5–10, *measure, in each case, the length of the constructed side of the parallelogram or rectangle.*

5. Construct a rectangle, given one side = 1.53 in., area = 1 sq. in.
6. Construct a rectangle, given one side = 2 in. and area equal to that of a parallelogram, whose sides are 1.5 in., 2.4 in. and included angle 30°.
7. Construct a rectangle, given one side = 2 in., and area equal to that of a rhombus whose side is 1.5 in. and one of whose angles is 75°.
8. Construct a rectangle, one side 1.5 in., and area equal to that of a triangle whose sides are 3 in., 4 in., 5 in.
9. Construct a rectangle, one side 1.2 in., and area equal to that of a trapezium whose parallel sides are 3 in., 1 in. and altitude 1.5 in.
10. Construct a parallelogram, one side 2 in., one angle = 45°, area = 2.4 sq. in.

In Exercises 11–13, construct the triangle ABC with the given data. Measure the parts named.

11. a = 2 in., B = 45°, area equal to that of a triangle whose sides are 3 in., 2.6 in., 2.8 in. Measure AB.
12. a = 2.5 in., b = 1.5 in., area that of a triangle whose sides are 2 in., 3.4 in., 1.8 in. Measure AB.
13. On a line of length = 2 in., as hypotenuse, construct a right-angled triangle, having its area equal to that of a triangle whose sides are 1.3 in., 1.4 in., 1.5 in. Measure the perpendicular from the right angle to the hypotenuse. [Use the method of Ex. 3 to construct, on a base of 2 in., any triangle having an area equal to that of the given triangle, and then use the Theorem on p. 47.]

XIX. SQUARES ON THE SIDES OF RIGHT-ANGLED TRIANGLES

THEOREM 43 (Theorem of Pythagoras. Euclid I. 47)

In any right-angled triangle, the square on the hypotenuse is equal to the sum of the squares on the sides containing the right angle.

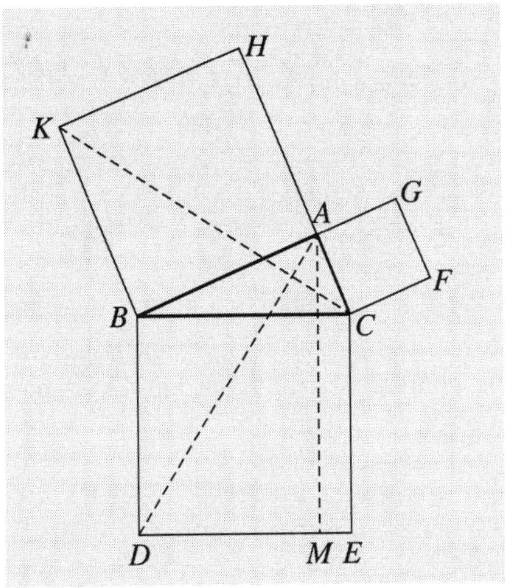

Let *ABC* be a triangle with the angle *A* a right angle. It is required to prove that sq. on *BC* = sq. on *CA* + sq. on *AB*.

Construction. Draw the squares *BCED*, *CAGF*, *ABKH*. Draw *AM* parallel to *BD* to meet *DE* in *M*.

Join *KC*, *AD*.

Proof. The adjacent angles *BAC*, *BAH* are right angles;

∴ *CA*, *AH* are in the same straight line.

Similarly, *BA*, *AG* are in the same straight line.

Again, $\angle KBA = \angle CBD$.

To each add $\angle ABC$;

∴ $\angle KBC = \angle ABD$.

Hence, in the triangles *KBC*, *ABD*,

$$\begin{cases} KB = AB \text{ (sides of a square)}, \\ BC = BD \text{ (sides of a square)}, \\ \angle KBC = \angle ABD \text{ (proved)}; \end{cases}$$

∴ The triangles are congruent and equal in area.

Now the square *AK* and the triangle *KBC* are on the same base *BK*, and between the same parallels *BK*, *CH*;

∴ sq. $AK = 2\Delta\ KBC$.

Also the rectangle *BM* and the triangle *ABD* are on the same base *BD*, and between the same parallels *BD*, *AM*.

∴ rect. $BM = 2\Delta\ ABD$.

But $\Delta\ KBC = \Delta\ ABD$ (*proved*);

∴ sq. AK = Rect. BM.

Similarly it can be shown that

sq. AF = Rect. CM;

∴ Sq. AK + Sq. CG = Rect. BM + Rect. CM = Sq. BE.

EXERCISE XXXIV

1. By means of the figures below, show that the square on the side opposite the angle *A* of the triangle *ABC* is greater than, or less than, the sum of the squares on the sides containing the angle *A*, according as the angle *A* is obtuse or acute.

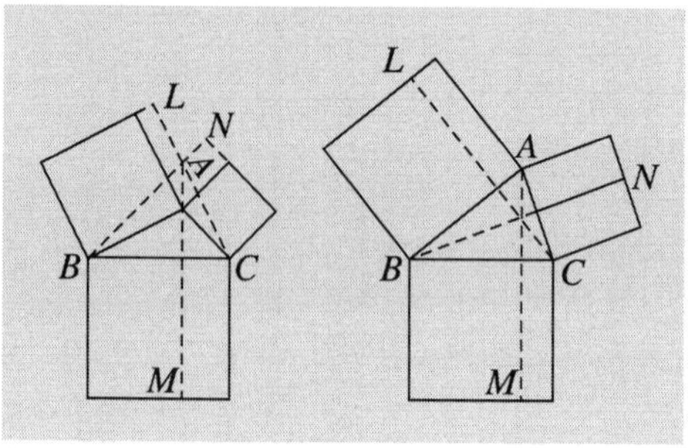

[In each case, prove, by the method of Theorem 43, that
Rect. CM = Rect. CN,
Rect. BM = Rect. BL,
Rect. AL = Rect. AN.]

2. State the converse of Theorem 43, and prove it by the method of 'reductio ad absurdum.' [Use Ex. 1.]

Theorem 44 (Euclid I. 48)

If the square on one side of a triangle is equal to the sum of the squares on the other sides, then the angle contained by these sides is a right angle.

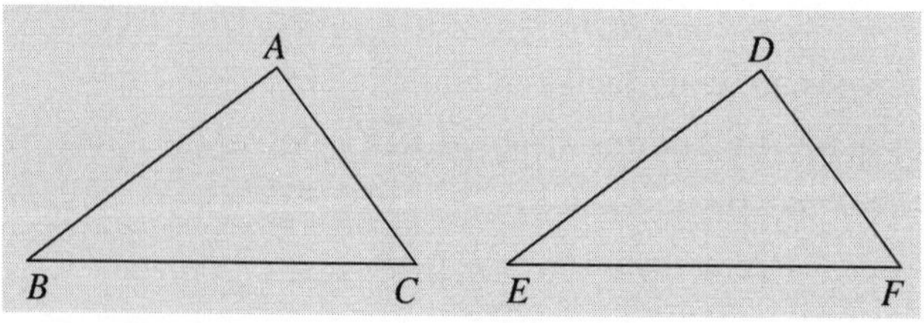

In the triangle ABC, let the square on BC be equal to the sum of the squares on AB, AC.

It is required to prove that the angle A is a right angle.

Construction. Draw a triangle DEF, in which the angle D is a right angle, $DE = AB$ and $DF = AC$.

Proof. In the triangle DEF, the angle D is a right angle;

$\therefore$ Sq. on EF = Sq. on DE + Sq. on DF.

Now, by construction,

$$DE = AB \text{ and } DF = AC;$$

$\therefore$ Sq. on EF = Sq. on AB + Sq. on AC.

But it is given that

Sq. on BC = Sq. on AB + Sq. on AC;

$\therefore$ Sq. on BC = Sq. on EF,

$\therefore$ $BC = EF$.

Hence, in the triangles ABC, DEF,

$$\begin{cases} AB = DE \ (\textit{construction}), \\ AC = DF \ (\textit{construction}), \\ BC = EF \ (\textit{proved}). \end{cases}$$

$\therefore$ The triangles are congruent,

$\therefore$ $\angle A = \angle D$.

But $\angle D$ is a right angle (*construction*);

$\therefore$ $\angle A$ is a right angle.

EXERCISE XXXV (Theorems 43, 44)

1. If N is any point in a straight line AB, or in AB produced, and P any point on the perpendicular at N to AB, then the difference of the squares on AP, BP is equal to the difference of the squares on AN, BN.
2. In the figure of Theorem 43, prove that
   (*i*) F, A, K are in a straight line.
   (*ii*) The triangles HAG, KBD, ECF, ABC are all equal in area,
   (*iii*) AD is at right angles to CK.
   (*iv*) BH and CG are parallel.
3. If a quadrilateral has its diagonals at right angles to each other, show that the sum of the squares on two opposite sides is equal to the sum of the squares on the other two opposite sides.
4. ABC is an equilateral triangle'and AD is perpendicular to BC. Prove that the square on AD is equal to three times the square on BD.

5. Construct a square equal in area to three times a given square. [Use Ex. 4]

6. In the triangle *ABC*, $\angle A = 30°$, $\angle B = 45°$. If *CD* is drawn perpendicular to *AB*, prove that the square on *AD* is three times the square on *BD*.

7. Divide a straight line into two parts so that the square on one part may be equal to three times the square on the other part. [Use Ex. 6]

8. If *AB*, *CD* are chords of a circle, prove the following by means of Theorem 43.

(*i*) If *AB* = *CD*, then *AB* and *CD* are equidistant from the centre.

(*ii*) If *AB* and *CD* are equidistant from the centre, then *AB* = *CD*.

(*iii*) If *AB* > *CD*, then *AB* is nearer the centre than *CD*.

(*iv*) If *AB* is nearer the centre than *CD*, then *AB* > *CD*.

[Let *O* be the centre; draw *OM*, *ON* perp. to *AB*, *CD*. Then *AB*, *CD* are bisected at *M*, *N*. Next show that sq. on *AM* + sq. on *OM* = sq. on *CN* + Sq. on *ON*.]

9. In a given straight line *AB*, or in *AB* produced, find a point *X* such that the difference of the squares on *AX* and *XB* may be equal to the difference between two given squares. [Use Ex. 1]

10. From a point *O*, within a triangle *ABC*, *OD*, *OE*, *OF* are drawn perpendicular to *BC*, *CA*, *AB* respectively; show that the sum of the squares on *BD*, *CE*, *AF* is equal to the sum of the squares on *CD*, *AE*, *BF*. [Use Ex. 1]

11. *ABC* is a triangle right angled at *C*: draw *CD* at right angles to *AB*; in *CB* take *CE* equal to *BD*, and in *CA* take *CF* equal to *AD*; join *AE* and *BF*; show that *AE* and *BF* are equal.

EXAMPLE 1. *Given a piece of paper AEFGCD, in the shape of two squares ABCD, BEFG, explain how to divide the figure into three pieces, by two straight cuts, which will fit together into one square.*

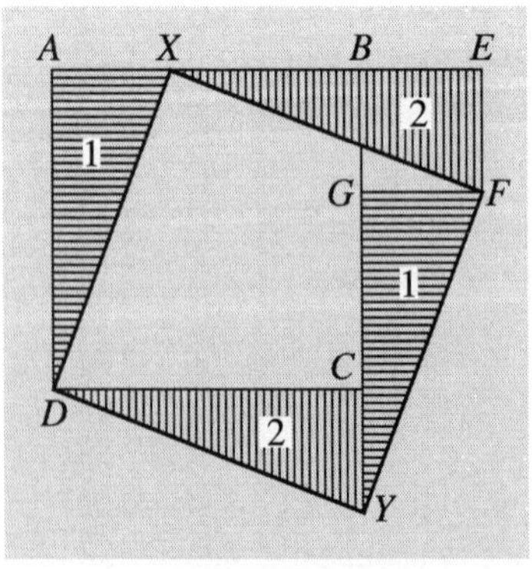

Along *AE* set off *AX* equal to *BE*.

Join *XD*, *XF*.

Cut the figure along XD, XF.

Then the pieces can be fitted to form a square.

Outline of proof. Produce BC to Y, making $CY = BE$. Join YD, YF.

By construction, $AX = BE$, $CY = BE$;

$\therefore$ $AX = CY = GF = EF$.

Now prove that $AD = DC = GY = XE$.

Hence show that Δs AXD, CYD, GFY, EFX are congruent, and therefore all the sides of the figure $XDFY$ are equal.

Finally, show that all the angles of $XDFY$ are rt. $\angle$s.

Hence $XDFY$ is a square.

Also, since $\Delta\ AXD$ can be made to coincide with $\Delta\ GFY$ and $\Delta\ EFX$ with $\Delta\ CYD$, the square $XDFY$ can be made out of the pieces of the given figure.

EXAMPLE 2. *Find, by calculation, the length of a chord of a circle of radius 5 feet, if the distance of the chord from the centre is 3 feet.*

[As this has to be done by calculation, it is unnecessary to draw a figure to scale. A freehand sketch is sufficient.]

Let PQ be the chord, O the centre of the circle, ON perpendicular to PQ.

Let $PQ = 2x$ ft.;

$\therefore$ $PN = x$ ft. (See Ex. VII. 8)

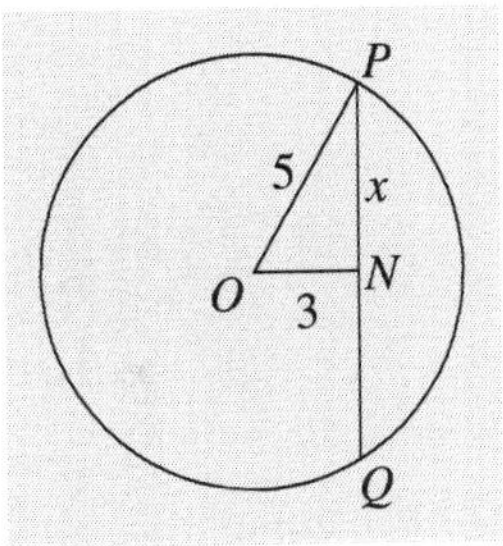

Also, because ONP is a right angle,

$\therefore$ Sq. on PN + Sq. on ON = Sq. on OP;

$\therefore$ $x^2 + 3^2 = 5^2$;

$\therefore$ $x^2 = 16$,

$\therefore$ $x = 4$,

$\therefore$ $2x = 8$;

$\therefore$ $PQ = 8$ feet.

EXAMPLE 3. *In the triangle ABC, if BC=13 in., CA= 12 in., AB = 5 in., prove that $\angle$ A is a right angle.*

The area of the square on $BC = 13^2$ sq. in.,

.... $CA = 12^2$ sq. in.,

.... $AB = 5^2$ sq. in.

Now $12^2 + 5^2 = 144 + 25 = 169 = 13^2$;

$\therefore$ Sq. on BC = Sq. on CA + Sq. on AB,

$\therefore$ $\angle A$ is a right angle.

EXAMPLE 4. *If ABC is an equilateral triangle, and AB = 2 in., find the length of AD, the perpendicular from A to BC.*

Let $AD = x$ in. Because $\angle\ ADB$ is a rt. $\angle$.

$\therefore$ Sq. on AB = Sq. on AD + Sq. on BD.

But AD bisects BC;

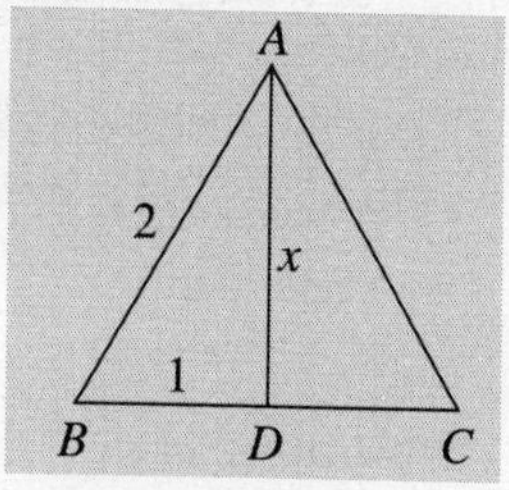

$\therefore \quad BD = 1$ in.

$\therefore \quad 2^2 = x^2 + 1^2,$

$\therefore \quad x^2 = 3,$

$\therefore \quad x = \sqrt{3};$

$\therefore \quad AD = \sqrt{3}$ inches.

EXAMPLE 5. *The length of a rectangular box is 2 in., its breadth 1 in., its height 2 in. Find the length of a diagonal of the box, drawn from a bottom corner to the opposite top corner.*

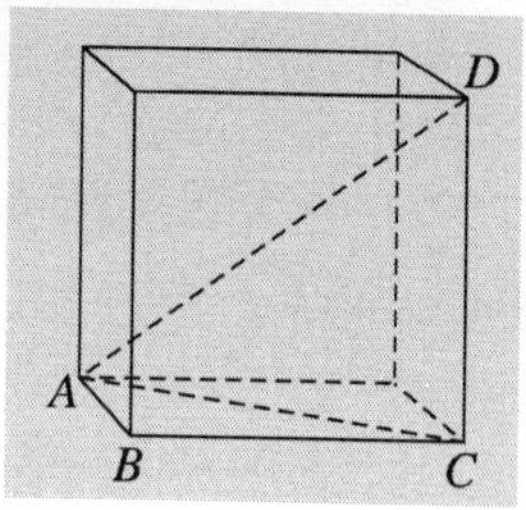

The edge CD of the box is perpendicular to every straight line drawn through C in the plane of the bottom ABC of the box;

$\therefore \angle ACD$ is a rt. $\angle$,

$\therefore$ Sq. on AD = Sq. on AC + Sq. on CD.

Again, because ABC is a rt. $\angle$.

$\therefore$ Sq. on AC = Sq. on AB + Sq. on BC,

$\therefore$ Sq. on AD = Sq. on AB + Sq. on BC + Sq. on CD.

Hence, if $AD = x$ in.,

$$x^2 = 1^2 + 2^2 + 2^2 = 9;$$

$\therefore \quad x = 3;$

$\therefore \quad AD = 3$ in.

EXERCISE XXXVI(a) (Theorems 43, 44)

Numerical

1. Without actually drawing the figure, show that a triangle, whose sides are 8, 15 and 17 inches long, is a right-angled triangle.
2. If m, n are any two numbers, show that a triangle whose sides, measured in inches, are

   $$m^2 + n^2, m^2 - n^2, 2mn,$$

   is a right-angled triangle.

   What values must m and n have to produce the triangle in Ex. 1?
3. Use the equality $13 = 2^2 + 3^2$ to draw a straight line of length $\sqrt{13}$ inches.
4. Use the equality $7 = 4^2 - 3^2$ to draw a straight line $\sqrt{7}$ inches long.
5. Use the equality $6 = 1^2 + 1^2 + 2^2$ to draw a straight line $\sqrt{6}$ inches long.

Exercises 6–16 are to be done by calculation.

6. Find the length of a chord of a circle of radius 1.7 in. which is distant 0.8 in. from the centre.

7. The diameter of a circular lake is 14.6 miles. A and B are places on the shore of the lake distant 11 miles apart. I sail straight from A to B. Find my shortest distance from the centre.

8. Two points A, B are 54 ft. apart. Find the radius of a circle, which passes through A, B, whose centre is 120 ft. from the straight line AB.

9. A gun, whose range is 2.65 miles, is situated at a point 1.4 miles from a straight road. What length of the road is commanded by the gun?

10. A ladder 20 ft. long rests with its top against a vertical wall and its base, on level ground, distant 12 feet from the wall. Find the height of the top above the ground.

11. Find the length of a ladder which, with its base on level ground, 8 ft. from a vertical wall, will just reach a point on the wall 31.5 feet from the ground.

12. The diagonals of a rhombus are 112 ft. and 66 ft. Find the length of a side.

13. A side of a rhombus is 34 in., one diagonal is 60 in. Find the length of the other diagonal.

14. The radii of two intersecting circles are respectively 15 in. and 13 in., and the common chord of the circles is 24 in. What length of the line joining the centres lies within both circles?

15. In a field, in the form of a quadrilateral $ABCD$, B is North of A and D is East of A. Also $AB = 7.5$ chains, $BC = 8.4$ chains, $CD = 1.3$ chains, DA = 4 chains. Show that $\angle BCD$ is a right angle, and find the area of the field in acres.

16. Two men start from the same point A, to walk by different routes to B, where the routes cross. One walks due North from A a distance of 2300 yards, and turns off to the right through an angle of 45 degrees. The other walks due East from A a distance of 3700 yards, and then due North to B. Find the distance travelled by each.

EXERCISE XXXVI(b)

Solid Figures

1. A room is 30 ft. long, 20 ft. wide and 12 ft. high. Find the distance from a corner of the floor to the opposite corner of the ceiling.

2. Find, correct to two places of decimals, the length of a diagonal of a cube whose edge is 1 inch.

3. A point P is taken, in the line of intersection of two walls of a room 6 feet above the floor. A point Q is taken in the floor 2 feet from one of the walls and 3 feet from the other. Find the distance from P to Q.

4. The height of a cone is 6 in. and the diameter of the base is 5 in. Find the length of a slant side.

5. A bar of rectangular section is to be cut from a cylindrical log of wood. The diameter of the log is 15.7 inches and the width of the bar is to be 13.2 inches. What is the greatest possible thickness of the bar?

6. A slice is cut off from a spherical ball of radius 2.5 in. The extreme thickness of the slice is 1.8 in. Find the radius of the circular face ofthe slice. [In Fig. 17, p. 98, we should have in Δ OCQ, $\angle OCQ = 90°$, $OQ = 2.5$ in., $OC = (2.5 - 1.8)$ in.]

7. The radius of the top of a spherical bowl is 2 ft., the extreme depth of the bowl is 1 ft.: find the radius of the sphere of which the bowl is a part. [If the radius of the sphere is r ft., we should have in Fig. 17, p. 98, in Δ OCQ, $\angle OCQ = 90°$, $OQ = r$, $OC = r - 1$, $QC = 2$.]

8. Find how many square feet of material are required to make a cycle-camper's tent, which is to have a square base and a single pole. A side of the base is to be 6.6 feet and the height of the pole 5.6 feet.

9. A rectangular block 2.6 in. by 2.2 in. by 1.9. in. can just be cut out of a spherical ball of iron. What is the diameter of the ball?

XX. TRIGONOMETRICAL RATIOS

Let AB and CD be two straight lines.

In the expression $\frac{AB}{CD}$, AB stands for a *number*, namely the number of units of length in AB, and CD has a similar meaning.

The fraction $\frac{AB}{CD}$ is often called the ratio of AB to CD, and is written in the form $AB : CD$.

Thus $AB : CD$ means ths same as $\frac{AB}{CD}$.

If four straight lines AB, CD, PQ, RS are such that

$$\frac{AB}{CD} = \frac{PQ}{RS},$$

the four straight lines are said to be proportional.

The above equation is often written in the form

$$AB : CD = PQ : RS,$$

which is read thus:

AB is to CD as PQ is to RS.

If AB and CD are straight lines, and a straight line XY can be found such that

$$AB = m \cdot XY \text{ and } CD = n \cdot XY$$

where m and n are whole numbers, then XY is called a common measure of AB and CD.

In practice, all straight lines may be regarded as having a common measure; for example, if

$$AB = 1.234 \text{ in. and } CD = 2.345 \text{ in.,}$$

the length 0.001 in. is a common measure of AB and CD. For it is contained 1234 times in AB and 2345 times in CD.

Sides of equiangular triangles, which are opposite equal angles in each, are called corresponding sides.

We shall prove that

If two triangles are equiangular, their corresponding sides are proportional. Let ABC, DEF be two equiangular triangles with

$$\angle A = \angle D,\ \angle B = \angle E \text{ and } \angle C = \angle F.$$

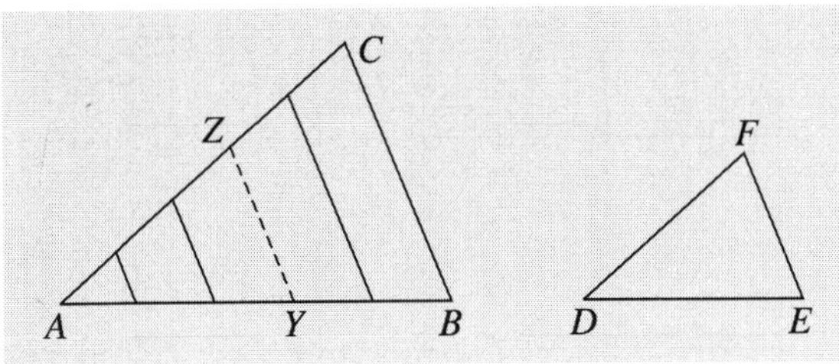

We have to prove that $\dfrac{AB}{DE} = \dfrac{AC}{DF} = \dfrac{BC}{EF}$.

Place $\Delta\, DEF$ so that D falls at A, E falls at some point Y in AB and F falls at some point Z.

Because $\angle D = \angle A$.

$\therefore$ The point Z is in AC.

Again, AB meets YZ and BC so as to make

$$\angle AYZ = \text{The corresponding } \angle ABC;$$

$\therefore$ YZ is parallel to BC.

Now suppose that AB and DE have a common measure.

Suppose, for example, that the common measure is contained 5 times in AB and 3 times in DE.

Mark off lengths equal to this common measure along AB, and through the points of division, draw parallels to BC.

These parallels divide AC into 5 equal parts, and YZ is one of the parallels.

$$\therefore \qquad \frac{AC}{DF} = \frac{AC}{AZ} = \frac{5}{3};\ \text{and}\ \frac{AB}{DE} = \frac{5}{3},$$

$$\therefore \qquad \frac{AB}{DE} = \frac{AC}{DF}.$$

Similarly, by placing $\Delta\, DEF$ so that the angles E and B coincide, we can show that

$$\frac{AB}{DE} = \frac{BC}{EF}.$$

Hence $$\frac{AB}{DE} = \frac{AC}{DF} = \frac{BC}{EF}.$$

Trigonometrical Ratios. Let P be any point in either arm of an acute angle A. Draw PM perpendicular to the other arm of the angle.

We shall prove that no matter where P is taken in either arm, each one of the fractions

$$\frac{MP}{AP}, \frac{AM}{AP}, \frac{MP}{AM}$$

always has the same value.

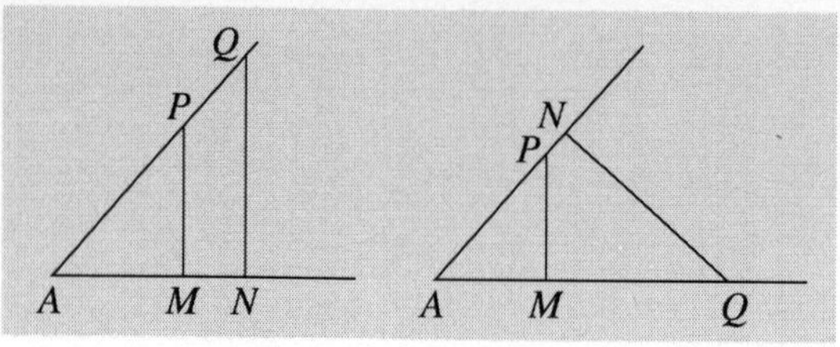

Take any point Q in either arm of $\angle A$, and draw QN perpendicular to the other arm.

Then, in the triangles APM, AQN,

$$\begin{cases} \angle A \text{ is common,} \\ \angle AMP = \angle ANQ \text{ (being rt. } \angle \text{ s).} \end{cases}$$

∴ The triangles are equiangular.

Hence, by the last article,

$$\frac{AM}{AN} = \frac{MP}{NQ} = \frac{AP}{AQ};$$

$$\therefore \qquad \frac{MP}{AP} = \frac{NQ}{AQ}, \frac{AM}{AP} = \frac{AN}{AQ}, \frac{MP}{AM} = \frac{NQ}{AN}.$$

which proves the theorem.

The ratios, or fractions, $\frac{MP}{AP}, \frac{AM}{AP}, \frac{MP}{AM}$ are called the **sine**, **cosine** and **tangent** of the angle A, respectively.

The names of the ratios are abbreviated into sin A, cos A, tan A

Thus, $\sin A = \frac{MP}{AP}$, $\cos A = \frac{AM}{AP}$, $\tan A = \frac{MP}{AM}$.

The reciprocals* of the sine, cosine and tangent of an angle A are called the **cosecant**, **secant** and **cotangent** of the angle A respectively. The names of these ratios are abbreviated into cosec A, sec A, cot A.

---

* *The reciprocal of x is* $\frac{1}{x}$.

Thus, $$\operatorname{cosec} A = \frac{1}{\sin A} = \frac{AP}{MP},$$

$$\sec A = \frac{1}{\cos A} = \frac{AP}{AM},$$

$$\cot A = \frac{1}{\tan A} = \frac{AM}{MP}.$$

Example 1. *Draw an acute angle whose cosine is* $\frac{2}{3}$.

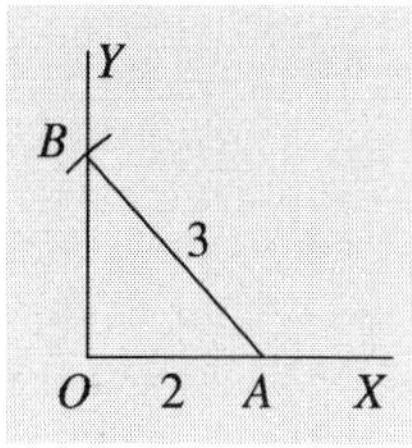

Draw two straight lines OX, OY at rt. ∠ s. Along OX set off $OA = 2$ units. With centre A and radius equal to 3 units, draw an arc cutting OY at B. Then OAB is the angle required.

For $$\cos OAB = \frac{AO}{AB} = \frac{2}{3}.$$

Example 2. *Draw an acute angle whose sine is* $\frac{2}{3}$.

Make the same construction as in Ex. 1. Then OBA is the required angle.

For $$\sin OBA = \frac{AO}{AB} = \frac{2}{3}.$$

Example 3. *Draw an acute angle whose tangent is* $\frac{2}{3}$.

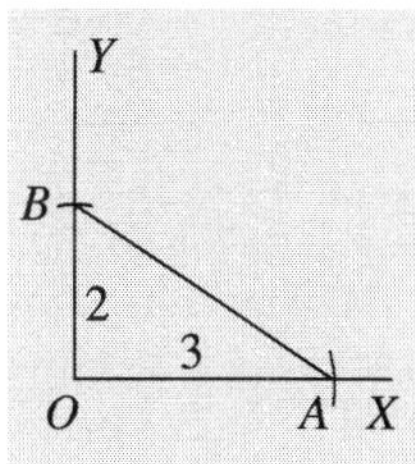

The construction will be obvious from the figure. The required angle is OAB.

For $$\tan OAB = \frac{OB}{AO} = \frac{2}{3}.$$

Example 4. *Use the tables on p. 132 to find, as accurately as possible, the angle whose sine is* $\frac{1}{3}$.

We have $$\frac{1}{3} = 0.3333 \textit{ nearly.}$$

Looking at the numbers in the column of the tables headed *sine*, we see that the nearest of these to 0.3333 is 0.3256. The corresponding angle (in the first column) is 19°. Hence

the required angle = 19° *to the nearest degree.*

EXERCISE XXXVII

Trigonometrical Ratios

1. Draw an acute angle whose sine is $\frac{3}{4}$. Measure the angle with your protractor, and check the result by the tables.

2. Draw an acute angle whose cosine is $\frac{3}{4}$. Measure the angle, and check the result by the tables.

3. Draw an acute angle whose tangent is $\frac{3}{4}$. Measure the angle, and check the result by the tables.

4. Draw an acute angle whose sine is $\frac{3}{5}$. Prove from your figure that the cosine of the angle is $\frac{4}{5}$ and its tangent $\frac{3}{4}$.

5. Prove that the cosine of an angle is the sine of the complement of the angle.

6. If the angle A is supposed to increase from 0 to 90°, prove that

(*i*) sin A increases from 0 to 1.

(*ii*) cos A decreases from 1 to 0.

(*iii*) tan A increases from zero, and, by taking A sufficiently near to 90°, tan A can be made greater than any positive number that we may choose, however great.

[To do (*i*) and (*ii*), draw OX, OY at right angles. Draw an arc of a circle with unit radius to cut OX at A and OY at B. Take any point P on the arc AB. Draw PM perpendicular to OA. Prove that sin $AOP = MP$, cos $AOP = OM$.]

EXAMPLE 5. *Find the inclination of a road to the horizon which rises 1 in 3.*

Let x be the number of degrees in the angle. Referring to the sketch, it is seen that

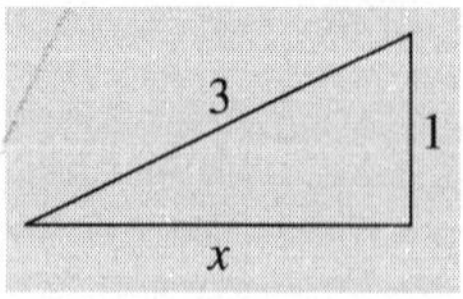

$$\sin x = \frac{1}{3}$$
$$= 0.3333 \textit{ nearly};$$
$\therefore$ $x = 19°$ *to the nearest degree.* (See Ex. 4)

EXAMPLE 6. *Find the remaining parts of* ΔABC, *given* $A = 90°$, $B = 40°$, $c = 10$.

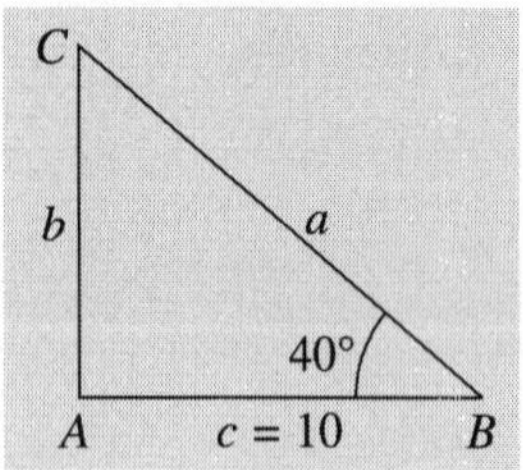

Since $A + B + C = 180°$,

$\therefore$ $90° + 40° + C = 180°$;

$\therefore$ $C = 50°$.

Referring to the sketch, we have

$$\frac{b}{10} = \tan 40° = 0.8391;$$

$\therefore$ $b = 0.8391 \times 10 = 8.391$ *nearly.*

Also $\frac{10}{a} = \cos 40° = 0.7660;$

$\therefore$ $10 = 0.7660 \times a;$

$\therefore$ $a = \frac{10}{0.7660} = 13.054$ *nearly.*

> **NOTE:** If we have tables of *secants*, we can proceed thus :
>
> $\frac{a}{10} = \sec 40° = 1.3054;$
>
> $\therefore$ $a = 1.3054 \times 10 = 13.054$ *nearly*

We are thus saved the trouble of division.

EXAMPLE 7. *From the top of a hill, I observe a horizontal road running straight away from the foot of the hill. The angles of depression of two consecutive milestones along this road are 20° and 10°. Calculate the height of the hill in feet.*

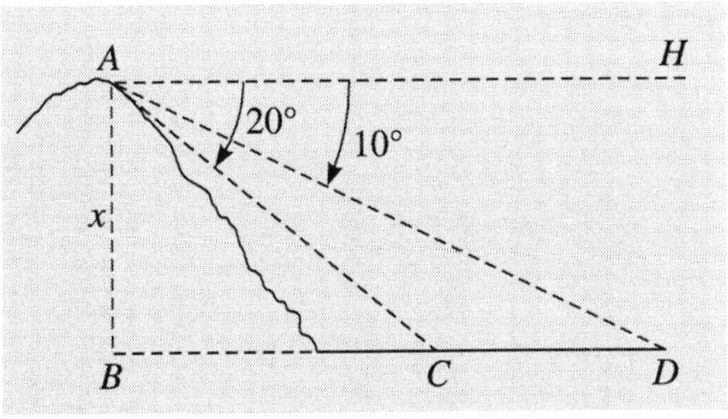

The figure is a rough sketch in which A is the top of the hill, C and D the nearer and further milestones, AB perpendicular to DC produced and AH parallel to CD. Thus AH is horizontal, and it is given that

$$\angle HAC = 20°, \angle HAD = 10°;$$

$\therefore$ $\angle BAC = 70°, \angle BAD = 80°.$

Also $CD = 1$ mile $= 5280$ ft.

Let $AB = x$ ft.

Then $\frac{BD}{x} = \tan BAD = \tan 80° = 5.6713,$

$\frac{BC}{x} = \tan BAC = \tan 70° = 2.7475;$

Hence, by subtraction, $\frac{CD}{x} = 2.9238;$

$\therefore$ $CD = 2.9238 \times x;$

$\therefore$ $x = \frac{5280}{2.9238} = 1805$ *nearly;*

$\therefore$ Height of hill $= 1805$ ft. *nearly.*

Table of Trigonometrical Ratios

Angle	*Sine*	*Cosine*	*Tangent*	*Angle*	*Sine*	*Cosine*	*Tangent*
0°	0	1	0	45°	.7071	.7071	1.0000
1	.0175	.9998	.0175	46	.7198	.6947	1.0355
2	.0349	.9994	.0349	47	.7314	.6820	1.0724
3	.0523	.9986	.0524	48	.7431	.6691	1.1106
4	.0698	.9976	.0699	49	.7547	.6561	1.1504
5	.0872	.9962	.0875	50	.7660	.6428	1.1918
6	.1045	.9945	.1051	51	.7771	.6298	1.2849
7	.1219	.9925	.1228	52	.7880	.6157	1.2799
8	.1392	.9903	.1405	53	.7986	.6018	1.3270
9	.1504	.9877	.1584	54	.8090	.5878	1.3764
10	.1736	.9348	.1763	55	.8192	.5736	1.4281
11	.1908	.9816	.1944	56	.8290	.5592	1.4826
12	.2079	.9781	.2126	57	.8387	.5446	1.5899
13	.2250	.9744	.2309	58	.8480	.5299	1.6003
14	.2419	.9703	.2493	59	.8572	.5150	1.6643
15	.2588	.9659	.2679	60	.8660	.5000	1.7821
16	.2756	.9613	.2867	61	.8746	.4848	1.8040
17	.2924	.9563	.3057	62	.8829	.4695	1.8807
18	.3000	.9511	.3249	63	.8910	.4540	1.9626
19	.3256	.9455	.3443	64	.8988	.4384	2.0508
20	.3420	.9397	.3840	65	.9063	.4226	2.1445
21	.3584	.9336	.3839	66	.9135	.4067	2.2460
22	.3746	.9272	.4040	67	.9205	.3907	2.3559
23	.3907	.9205	.4245	68	.9272	.3746	2.4751
24	.4067	.9135	.4452	69	.9336	.3584	2.6051
25	.4226	.9003	.4663	70	.9397	.3420	2.7475
26	.4384	.8938	.4877	71	.9455	.3256	2.9042
27	.4540	.8910	.5095	72	.9511	.3090	3.0777
28	.4695	.8829	.5317	73	.9563	.2924	3.2709
29	.4848	.8746	.5543	74	.9618	.2756	3.4874
30	.5000	.8660	.5774	75	.9659	.2588	3.7321
31	.5150	.8572	.6009	76	.9703	.2419	4.0108
32	.5299	.8430	.6249	77	.9744	.2250	4.8315
33	.5440	.8337	.6494	78	.9781	.2079	4.7046
34	.5592	.8290	.6745	79	.9816	.1908	5.1446
35	.5736	.8192	.7002	80	.9848	.1736	5.6713
36	.5378	.8090	.7265	81	.9877	.1564	6.8138
37	.6018	.7986	.7536	82	.9903	.1392	7.1154
38	.6157	.7880	.7813	83	.9925	.1219	8.1443
39	.6293	.7771	.8098	84	.9945	.1045	9.5144
40	.6428	.7660	.8391	85	.9962	.0872	11.4301
41	.6561	.7547	.8693	86	.9976	.0698	14.3007
42	.6691	.7431	.9004	87	.9986	.0528	19.0811
43	.6820	.7314	.9325	88	.9994	.0349	28.6363
44	.6947	.7193	.9657	89	.9993	.0175	57.2900
45	.7071	.7071	1.0000	90	1	0	∞

EXERCISE XXXVIII

The following examples are to be done by calculation and the use of the tables on p. *132*.

1. A road is inclined at 10° to the horizon; how much does it rise in 100 yards, measured along the slope?
2. What is the inclination of a road to the horizon that rises 1 ft in 6 ft.?
3. Calculate the length of the diagonal of a square whose side is 2 in.
4. Calculate the lengths of the diagonals of a rhombus of 1 in. side, one angle of which is 50°.
5. Calculate the lengths of the sides of a right-angled triangle, whose hypotenuse is 1.5 in., one angle being 50°.
6. Calculate the remaining parts of Δ *ABC* with the following data:
   (1) $A = 90°$, $b = 4.8$ ft., $c = 5.5$ ft.
   (2) $A = 90°$, $b = 1.6$ yd., $a = 6.5$ yd.
   (3) $A = 90°$, $c = 4$, ch. 50 lk., $a = 5$ ch. 30 lk.
   (4) $A = 90°$, $B = 15°$, $c = 8$ in.
   (5) $A = 90°$, $C = 50°$, $a = 8$ in.
7. A chimney casts a shadow 200 ft. long when the sun's altitude is 20°. Calculate the height of the chimney.
8. A chimney 98 ft. high casts a shadow 138 ft long. Calculate the sun's altitude.
9. From the top of a cliff I observe a house at sea-level, which I know to be two miles distant from the foot of the cliff. The angle of depression of the house is 15°. Calculate the height of the cliff in feet.
10. A lighthouse facing *E.* throws out a fan-shaped beam of light covering an angle of 20°. A vessel sailing due *S* at 6 miles an hour sees the light for 10 minutes. Calculate the ship's shortest distance from the lighthouse.
11. Calculate the length of the common chord of two equal circles, each of which passes through the centre of the other, the distance between their centres being 2 in.
12. A tower stands on a level plane, and, from a point *A* in the plane the elevation of the top of the tower is 12°; from a point *B* in the plane, 100 yd. nearer the foot of the tower, the elevation of the top is 22°. Find the height of the tower.
13. In the triangle *ABC*, $B = 40°$, $C = 70°$, and the length of the perpendicular *AN* from *A* to *BC* is 2 in. Calculate the lengths of *BN*, *CN*, and hence that of *a*.
14. In the triangle *ABC*, $B = 35°$, $C = 50°$, $a = 3$ in. Calculate the length of the perpendicular from *A* to *BC*.
15. In the triangle *ABC*, $A = 30°$, $B = 20°$, $a = 2$ in. Find the length of the perpendicular from *A* to *BC*.

16. A railway cutting, 40 ft deep, is 118 ft wide at the top and 18 ft wide at the bottom. Calculate the angle of slope at the sides, these being equally inclined to the horizon.

17. I start from the most southerly point of a circular lake, whose diameter is 20 miles, and sail straight across in a direction 15° East of North. Find, by calculation, how far I sail, and my shortest distance from the centre of the lake.

18. I observe the elevation of the top of a chimney from a certain point and find it to be 20°; I walk straight towards the chimney a distance of 50 yds., and then find the elevation of the top to be 30°. Calculate the height of the chimney in feet.

19. Shooting through a loop-hole which faces North, I can only cover 10° horizontally with my rifle. A man, running West, crosses my line of fire at the rate of 10 yds. a second. Find by calculation how long he is under fire, if the range when he is nearest is 500 yds.

20. The bore of a pistol is 0.45 in. in diameter and the length of the barrel is 6 in.; calculate the amount of angular space which you can see through it, if you place your eye at one extremity.

21. From the top of a mountain, I observe the angles of depression of two houses which are one mile apart and due North of the top of the mountain. I find the angles of depression to be 24° and 13°. Find the height of the mountain in feet above the plain in which the houses lie.

Trigonometrical Expressions for the Areas of Parallelograms, Triangles, etc.

Area of a Parallelogram: Let *ABCD* be a parallelogram, of which $\angle A$ is an acute angle.

Draw *DE* perpendicular to *AB*.

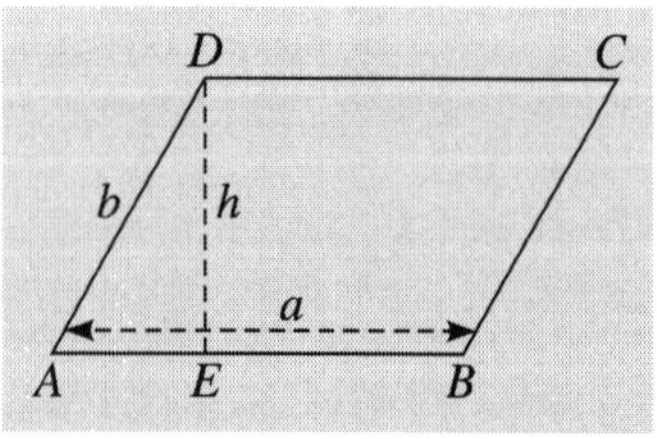

Let $AB = a$ units of length,

$AD = b$,, ,,

$DE = h$,, ,,

Then area of ▱^{m} $= ah$ units of area

But $\sin A = \dfrac{h}{b}$,

$\therefore$ $h = b \sin A$.

$\therefore$ Area of ▱^{m} $= ab \sin A$ units of area.

Thus, the area of a parallelogram is the product of two adjacent sides and the sine of the included angle.

Area of a Triangle: Let ABC be a triangle, in which $\angle A$ is an acute angle.

Draw CD perpendicular to AB.

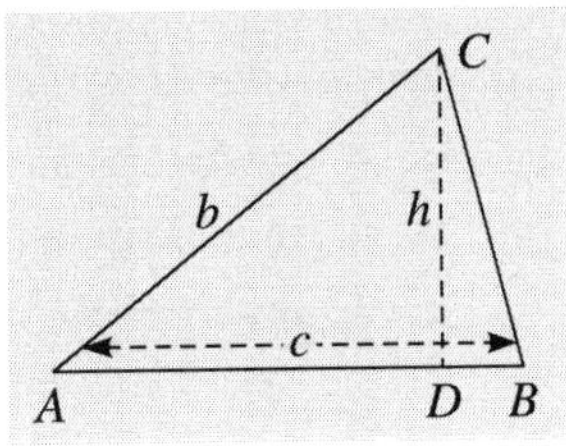

Let $AB = c$ units of length,

$AC = b$,, ,,

$CD = h$,, ,,

Then area of $\Delta\, ABC = \frac{1}{2} ch.$

But $\sin A = \frac{h}{b},$

$\therefore$ $h = b \sin A;$

$\therefore$ Area of $\Delta\, ABC = \frac{1}{2} bc \sin A.$

Thus, the area of a triangle is given by half the product of two sides and the sine of the included angle.*

> **NOTE:** *The area of the triangle ABC is generally denoted by the symbol Δ(delta).*

Area of a Quadrilateral: Let $ABCD$ be a quadrilateral, of which the diagonals are inclined at α.

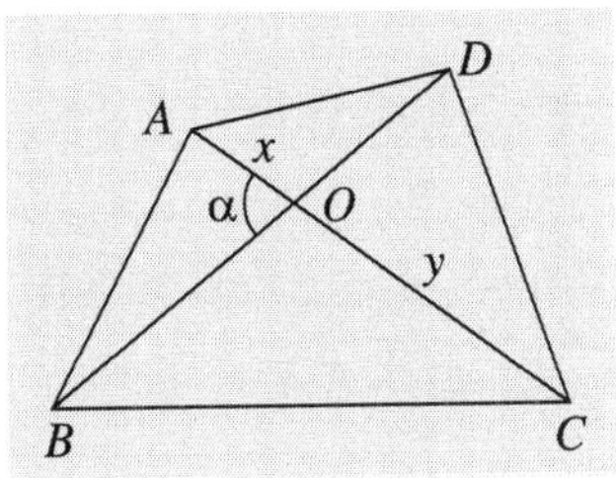

Let $AO = x$ units of length,

$OC = y$,, ,,

Regarding BD as the base of Δs ABD, CBD, the altitudes of these Δs are $x \sin \alpha$ and $y \sin \alpha$.

$$\therefore \quad \text{Area of } ABCD = \text{Area of } \Delta\, ABD + \text{Area of } \Delta\, CBD$$

$$= \frac{1}{2} BD \cdot x \sin \alpha + \frac{1}{2} BD \cdot y \sin \alpha$$

$$= \frac{1}{2} BD \cdot (x + y) \sin \alpha$$

$$= \frac{1}{2} BD \cdot AC \sin \alpha.$$

Thus, the area of a quadrilateral is given by half the product of its diagonals and the sine of the included angle.

---

* If the given angle A is obtuse, produce BA to D : their area of $\Delta\, ABC = \frac{1}{2} bc \sin CAD$. The student should draw a figure for this case and supply a proof.

EXERCISE XXXIX

Areas of parallelograms and triangles

Calculate, *by the use of the tables on* *p. 132*, the areas of the following figures:

1. A parallelogram, adjacent sides 3 in., 2.02 in; contained ∠, 30º.
2. A parallelogram, adjacent sides 3 in., 2.02 in.; contained ∠, 150º.
3. A parallelogram, perpendiculars between pairs of parallel sides 1 in. and 2 in.; one angle, 60°.
4. A rhombus, side 1.4 in.; one angle, 40°.

Find *by calculation* the area of the triangle *ABC* in the following cases:

5. $a = b = c = 2$ in.
6. $a = 3$ in., $b = c = 2$ in.
7. $a = 3$ in., $b = 2$ in., $C = 50°$.
8. $b = c = 3$ in., $B = 40°$.
9. $b = c = 2$ in., $A = 70°$.
10. $a = 5$ in., $B = 55°$, $C = 35°$ (use a method similar to that of Example 7, p. 131).
11. $B = 40°$, $C = 60°$, and the perpendicular from *A* to *BC* = 4 in.

MISCELLANEOUS EXERCISES

Arranged in Sets for Homework or Revision

PAPER I (to Section VII)

1. *ABCDEF* is a regular hexagon; prove that *A*, *C*, *E* are the vertices of an equilateral triangle.
2. If the bisector *AX* of the angle *A* of a triangle *ABC* cuts the base *BC* in *X*, show that *X* is equidistant from *AB* and *AC*.
3. *ABC* is a triangle; *P* is a point on the bisector of the angle *B*; *PQ*, is drawn parallel to *CB* to meet *AB*, produced if necessary, in *Q*. Prove that *BQ* = *QP*.
4. *CA*, *CB* are radii of a circle at right angles to each other, and *AM*, *BN* are drawn perpendicular to any diameter, meeting it in *M* and *N*; prove that *AM* is equal to *CN* and *BN* is equal to *CM*.
5. Draw figures to show into how many regions three straight lines divide a plane. Consider the different cases that may arise.
6. *ABCDE* is a pentagon, in which *AB* = *BC* = *CD* and *AE* = *ED*; also, each of the angles at *A*, *B*, *C*, *D* is double the angle at *E*.

   Find the angle at *E*: and, if *AB* is 1 inch, construct the pentagon and calculate the length of *AE*.

PAPER II (to Section VIII)

1. *ABCDE* is a regular pentagon. Join *AC*, *AD*, and produce *BC*, *ED* to meet at *F*. Find the number of degrees in the angles *ACD*, *FCD*, and prove that the triangles *ACD*, *FCD* are congruent.

2. Draw a circle with centre O. Draw a straight line to cut the circle at A and B. Produce AB to C, making BC equal to OA. Join CO and produce it to E. Prove that $\angle AOE = 3\angle ACE$.
3. Draw an isosceles triangle ABC, in which $AB = AC$. Produce BA to D, making AD equal to AB. Prove that the angle BCD is a right angle.
4. Prove that the diagonals of a rhombus meet at a point which is equidistant from the sides.
5. Using ruler and compasses only, draw two straight lines OX, OY inclined at an angle of 45°; mark off $OX = 1$ in. and $OY = 2$ in. Find a point P within the angle XOY, distant 1 in. from Y and 2 in. from X. Measure OP.
6. A chimney is 98 ft. high. What is the angular elevation of its top at a point at the same level as its base and 138 ft distant?

Paper III (to Section IX)

1. ABC is a triangle; BO, CO bisect the angles B, C and intersect in O; through O a straight line is drawn parallel to BC, cutting AB in D and AC in E. Prove that $DE = BD + CE$.
2. $ABCD$ is a quadrilateral such that the sum of the angles at A and B is equal to the sum of the angles at C and D. Prove that two sides of the quadrilateral are parallel to one another.
3. AOB, COD are intersecting straight lines, and each of the figures $AOCE$, $BODF$ is a rhombus. Show that the straight line EF passes through O, and that AC, BD are parallel.
4. ABC is a triangle in which AB is greater than AC. From B and C draw two equal straight lines BD, CE, perpendicular to AB and AC respectively, so that C, D are on opposite sides of AB, and B are on opposite sides of AC. Join BE and CD. Prove that $BE > CD$.
5. Draw four straight lines at random and produce them indefinitely. In the most general case, state (*a*) in how many points the lines intersect, (*b*) how many different triangles are formed, (*c*) into how many regions the lines divide the plane.
6. The inclination of a road to the horizon is 10 degrees. Find by diagram (scale 1 in. = 25 yards) how many feet the road rises in 100 yards.

Paper IV (to Section X)

1. Squares $ABHK$, $ACLM$ are described on the sides AB, AC of any triangle ABC, externally* to the triangle. Show that $BM = KC$.
2. ABC is an acute-angled triangle, in which the angle C is less than the angle B ; take a point D in AB; with centre D and radius DB, describe a circle cutting BC in E; produce ED and CA to meet at F. Prove that the angle AFD is equal to $B - C$.

---

* That is, $ABHK$ and ABC are on opposite sides of AB, and so on.

3. *ABC* is a triangle, and through *D* the middle point of *AB*, *DE* is drawn parallel to *BC* ; *BE* is drawn to bisect the angle *ABC* and to meet *DE* in *E*. Prove that *AEB* is a right angle. [Use Ex. 3 in Paper I]

4. *ADEF* is a rhombus, having the angle at *A* a little greater than a right angle, and *B* and *C* are points in *DE* and *EF* respectively, such that *ABC* is an equilateral triangle. If a side of the rhombus is equal to a side of the equilateral triangle, prove that the angle *DEF* is ten-ninths of a right-angle.

5. With ruler and compasses only, make an isosceles triangle *ABC*, of which the base *AB* is 2 inches and the vertical angle *C* is 120°. Measure *CA*.

6. From the top of a hill, I observe a level road running straight away from the foot of the hill. The angular depressions of two consecutive milestones along this road are 20° and 10°. Find, by diagram, the height of the hill in feet.

PAPER V (to Section XI)

1. *ABC* is an equilateral triangle; *Q* is a point in *BC* produced in the direction *B* to *C*. Prove that *Q* is nearer to *A* than to *B*.

2. The vertical angle *A* of an isosceles triangle *ABC* is half a right angle, and the perpendiculars *AD*, *BE*, from *A*, *B* to the opposite sides, meet in *F*. Prove that *FE* = *EC*.

3. Two angles which are such that the arms of one are respectively perpendicular to the arms of the other are either equal or supplementary.

4. Any point *O* is taken within a square *ABCD*, and on *OA*, and on the same side of it as *D*, a square *AOEF* is described. Prove that *BO* is equal to *DF*.

5. *B* is a point 2 miles *E*. of another point *A*. *C* is a tower N.E. of *B* and E.N.E. of *A*. Find the distances *AC*, *BC* in yards. [Scale 1 in. = 1000 yards]

6. Draw two straight lines *OX*, *OY* making an angle of 45° with one another. Find a point *P* distant 1 in. from *OX* and 2 in. from *OY*.

PAPER VI (to Section XII)

1. *AB* and *CD* are any two diameters of a circle ; prove that *CB*, *DB* bisect the angles made with *AB* by a straight line through *B* parallel to *CD*.

2. From a point *P* in the base *BC* of an isosceles triangle *ABC*, perpendiculars *PL*, *PM* are drawn to the sides *AB*, *AC*. Prove that *PL* + *PM* is constant for all positions of *P*.

   [Complete the parallelogram *BACD*, and show that *PL* + *PM* = distance between opposite sides.]

3. *AD* bisects the angle *A* of a triangle *ABC*. The perpendicular from *C* to *AD* meets *AD* in *N*. *O* is the middle point of *BC*. Prove that *ON* is half the difference between *AB* and *AC*.

   [Produce *CN* to meet *AB*]

4. *AB* is a straight line, *AP* any other straight line. Along *AP* mark off three equal lengths *AC*, *CD*, *DE*. Join *EB* and produce it to *F*, so that *BF* = *EB*. Prove that *FC* bisects *AB*. [Join *DB*]
5. Construct an isosceles triangle *ABC*, of which the perimeter is 6 inches and the vertical angle *C* is 40°. Measure *AB*.
6. A bridge is thrown across a ravine through which a stream 17 ft wide runs. The sides of the ravine slope upwards right from the banks of the stream at angles of 80° and 74° with the horizontal. The bridge is 85 feet long. What is its height above the stream? [Scale 1 in. = 25 ft.]

Paper VII (to Section XIII)

1. *M* and *N* are the middle points of the sides *AB* and *AC* of a triangle *ABC*. *L* is the foot of the perpendicular from *A* to *BC*. Prove that the triangles *AMN*, *LMN* are congruent.
2. If *P* is any point within a parallelogram *ABCD*, prove that the sum of its distances from the angular points is greater than the sum of the diagonals, but less than the perimeter of the parallelogram.
3. Two triangles are drawn, one with two sides equal and parallel to the sides *AB*, *DC* of a quadrilateral *ABCD*, the other with two sides equal and parallel to *AD*, *BC*. Prove that their bases are equal.
4. *ABCD* is a square, and equal lengths *AE*, *BF*, *CG*, *DH* are marked off in order round the sides of the square. Prove that *EFGH* is a square.
5. Draw two straight lines *AB*, *AC* inclined at an angle of 60°. Find a point *P*, within the arms of the angle, half-an-inch from *AC* and an inch from *AB*. Through *P* draw a straight line *QPR*, to cut *AB*, *AC* in *Q* and *R*, such that *QP* = *PR*. Measure *QR*.
6. A lighthouse facing due *E*. throws out a fan-shaped beam of light covering an angle of 20°. A vessel sailing due *S*. at 6 miles an hour sees the light for 10 minutes. What was the ship's shortest distance from the lighthouse? [Scale 1 in. = 1 mile]

Paper VIII (to Section XIV)

1. Draw any triangle *ABC*. Bisect *AC* at *E* and *AB* at *F*. Join *BE* and produce it to *G*, so that *EG* = *BE*. Join *CF* and produce it to *H* so that *FH* = *CF*. Make a careful construction, with ruler and compasses only, showing all the work, and *prove* that the points *H*, *A*, *G* *should* lie on a straight line parallel to *BC*.
2. A triangle *ABC* is rotated in its own plane about the point *A* into a position *A'B'C'*. If *AC* bisects *BB'*, prove that *AB'* (produced if necessary) bisects *CC'*.
3. Draw a parallelogram *ABCD*. With centre *B* and radius *BC*, draw a circle to cut *DC* or *DC* produced in *E*. Join *CA*, *AE*, *BD*. Prove that the angle *CAE* is equal to the difference between the angles *DCA* and *DBA*. [Prove Δs *BED*, *ADE* congruent]

4. *AB* is a straight line, *CBD* is any other straight line such that *CB* = *BD* = *AB*. Join *CA* and produce it to *E*, so that *AE* = *AB*. Let *AD*, *BE* meet in *F*. Along *EB* mark off *EH* = *FB*. Prove that *HAB* is a right angle.
5. Given the middle points of the sides *AB*, *AC* of a triangle *ABC* and the foot of the perpendicular from *A* to *BC*, construct the triangle.
6. *ABC* is a triangle, in which *AB* is $2\frac{1}{2}$ in., *BC* is 3 in., *CA* is $3\frac{1}{2}$ in. Find a point *X* in *BC* which is half-an-inch further from *AB* than from *AC*. Verify by measuring the perpendiculars from *X* to *AB*, *AC* to the nearest hundredth of an inch.

Paper IX (to Section XV)

1. *ABC* is a triangle *L*, *M*, *N* are the middle points of the sides *BC*, *CA*, *AB*. *BM* cuts *LN* in *P*, *CN* cuts *LM* in *Q*. Prove that $PQ = \frac{1}{4} BC$, and that *PQ* is parallel to *BC*.
2. The sides *BC*, *CA*, *AB* of a triangle are divided in *P*, *Q*, *R*, so that *BP* = 2*PC*, *CQ* = 2*QA*, *AR* = 2*RB*. *BQ* cuts *AP* in *X* and *CR* cuts *AP* in *Y*. Prove that *AX* = *XY* = 3*YP*.

   [If *P*′ is the middle point of *BP*, draw parallels to *BQ* through *P*′, *P*: on a second figure draw parallels to *CR* through *P*′, *P*.]
3. *A*, *B*, *C* are three points in a straight line. On *AB*, *AC*, squares *ABDE*, *ACFG* are described, so as to lie on the same side of the straight line. Show that the straight line through *A* at right angles to *BG* will bisect *EC*.

   [Produce *ED* to meet *CF* in *K*, and prove *AK* is at right angles to *BG*.]
4. *AOB*, *COD* are two straight lines intersecting at *O*. *P* is a point such that the sum of the perpendiculars from *P* to these lines is equal to a given length. Show that the locus of *P* consists of the four sides of a certain rectangle.
5. *If ABCD is a quadrilateral, in which AB + CD = BC + AD, prove that the bisectors of the angles of the quadrilateral meet in a point which is equidistant from the sides of the quadrilateral.*

   Proceed thus:* Let *AD* be the shortest side, so that *BC* is the longest. (Why?) Along *AB* cut off *AE* equal to *AD*. Along *CB* cut : off *CF* equal to *CD*. Join *DE*, *EF*, *FD*. Prove that *BE* = *BF*, and that the bisectors of ∠s *A*, *B*, *C* are the perpendicular bisectors of the sides of Δ *DEF*. The bisectors of ∠s *A*, *B*, *C* therefore meet at a point *O* which is equidistant from the sides of *ABCD*. (Explain this) Finally prove that *OD* bisects ∠*D*.
6. *OA*, *OB*, *OC* are three edges of a rectangular block. If *OA* = 4 in., *OB* = 3 in., *OC* = 2 in., find by drawing and measurement, as accurately as you can, the angles of the triangle *ABC*.

---

* This proof was given by A. H. Teagle, a pupil at Derby School.

Paper X (to Section XVII)

1. Prove that, if *ABCD* is a square, and the bisector of the angle *BAC* meets *BC* in *E*, then *BE* is equal to the difference between *AC* and *AB*, and *CE* is equal to the difference between 2*AB* and *AC*.
2. *ABC*, *DBC* are two triangles on opposite sides of the common base *BC*, such that the area of the triangle *ABC* is twice the area of the triangle *DBC*. *BC*, produced if necessary, cuts *AD* in *K*. Prove $AK = 2KD$.
3. *ABCD* is a parallelogram. *PQR* is a straight line, parallel to *AB*, which cuts *AD* in *P*, *AC* in *Q*, and *BC* in *R*. Prove that the triangles *APR*, *AQD* are equal in area.
4. *P* is a point on the diagonal *AC* of a parallelogram *ABCD*, such that $AC = n \cdot AP$. Through *P* draw parallels to the sides of the parallelogram. Prove that the areas of the complements to *BP*, *DP* are each $(n - 1)/n^2$ of the area of the parallelogram.
5. Construct a parallelogram *ABCD* of area equal to 16 sq. cm, and having *AB* = 5 cm, *AD* = 4 cm. Construct complements such that the area of each is 3 sq. cm.
6. *ABCD* is a field in the form of a quadrilateral. *B* is 100 yd. E of *A* and 250 yd. N. of *A*; *C* is 300 yd. E. of *A* and 350 yd, N. of *A*; *D* is 400 due E. of *A*. Draw a plan of the field, scale 1 in. = 100 yd. and calculate its area.

Paper XI (to Section XX)

1. *ABC* is a triangle ; *P* is a point in *BC* such that $m \cdot BP = n \cdot PC$. What fraction is the triangle *ABP* of the triangle *ABC* ?
2. In the triangle *ABC*, *AD* and *AE* are drawn from the vertex *A* to the base *BC*, making the angle *BAD* equal to the angle *ACB*, and the angle *CAE* equal to the angle *ABC* ; prove that the perpendicular from *A* to *BC* bisects the angle *DAE*.
3. *ABC* is a triangle, in which *B* is not a right angle. Draw a triangle *PQR*, such that *PQ* = *AB*, *QR* = *BC*, and the angle *Q* is a right angle. Prove that
   (*i*) *AC* is greater or less than *PR*, according as the angle *B* is greater or less than a right angle.
   (*ii*) Sq. on *AC* is greater or less than the sum of the squares on *AB* and *BC*, according as the angle *B* is obtuse or acute.
4. A corner is sawn off a rectangular block so as to leave a triangular section. Prove that the triangle must be acute angled.
5. Starting with a straight line *AB*, 1 in. long, give a construction for obtaining in succession lines whose lengths are $\sqrt{2}, \sqrt{3}, \sqrt{4}, \sqrt{5}$,... inches. Hence find $\sqrt{5}$ and $\sqrt{7}$ to two decimal places.
6. A railway, running North and South, passes through a cutting 40 feet deep, whose sides make an angle of 30° with the horizon. Starting from a point at

the bottom of a face of the cutting, I walk in a straight line to a point at the top, 50 feet North of my starting point. Find, by diagram and measurement, how far I have walked.

Paper XII (to Section XX)

1. *ABCDEFGH* is a regular octagon, and *AF*, *BE*, *CH*, *DG* are drawn. Prove that their intersections are the angular points of a square.
2. *ABC* is a triangle, right angled at *C* ; *AB*, *BC*, CA are 25, 20, 15 in. in length respectively; *P*, *Q* are points on *BC*, *CA* distant 10 in. from *O*. Join *AP* and *BQ*, intersecting in *O*.
   (*i*) Prove that $AO = OP$ and $BO = 3OQ$.
   (*ii*) Calculate the areas of the triangles *ABC*, *AOB*, *BOP*, *AOQ*, and that of the quadrilateral *OPCQ*. [Use Example 1 in Paper XI]
3. From *O*, a point within a triangle *ABC*, perpendiculars *OD*, *OE* and *OF* are drawn to *BC*, *CA* and *AB* respectively. Prove that the sum of the squares on *BD*, *CE*, *AF* is equal to the sum of the squares on *CD*, *AE*, *BF*.
4. A figure consists of five equal squares in the form of a cross. Show how to divide it, by two straight cuts, into four equal figures which will fit together to form a square.
5. *ABCD* is a quadrilateral and X is a given point in *AD*. Find a point *Y* in *AB* such that the area of the triangle *AXY* is equal to that of *ABCD*. Hence show how to divide the quadrilateral *ABCD* into three equal parts by straight lines drawn through *X*.
6. A balloon starts 1000 yd. away from me and rises vertically. I observe it when its elevation is 42°, and 5 minutes later its elevation is 63°. If its motion is uniform, find, by means of trigonometrical tables, the rate of ascent in feet per second.

UNIT 3

The Circle

XXI. CHORDS OF A CIRCLE

DEF. A **circle** is a plane figure bounded by one line called the **circumference**, and is such that all straight lines drawn from a certain point within the figure to the circumference are equal.

These straight lines are called **radii**, and the point is called the **centre**.

DEF. A **diameter of a circle** is a straight line drawn through the centre and terminated both ways by the circumference.

DEF. A **semi-circle** is the figure bounded by a diameter of a circle and part of the circumference cut off by the diameter.

DEF. A **chord of a circle** is a straight line joining any two points on the circumference.

DEF. An **arc of a circle** is any part of the circumference.

DEF. A **sector** of a circle is the figure bounded by two radii and the arc intercepted between them.

DEF. A **segment** of a circle is the figure bounded by any straight line and one of the arcs into which it divides the circumference.

DEF. Circles which have the same centre are said to be **concentric**.

The following properties of circles are obvious:

(*i*) Circles with equal radii are congruent figures.

(*ii*) A point is without, upon, or within the circumference of a circle according as its distance from the centre is greater than, equal to, or less than the radius.

(*iii*) Concentric circles, whose radii are unequal, do not cut one another.

DEF. If a straight line divides a plane figure in such a way that, when the figure is folded about the line, one part coincides with the other, the figure is said to be **symmetrical with regard to the line**.

The line is called an axis of symmetry.

DEF. A figure is said to be **symmetrical with regard** to a point O, if, corresponding to every point P of the figure, there is another point P' such that the straight line PP' is bisected at O.

EXERCISE XL

Symmetry

1. A rhombus is symmetrical with regard to a diagonal.
2. If a parallelogram is symmetrical about a diagonal, it is a rhombus.
3. A rectangle is symmetrical about the straight line which bisects a pair of opposite sides.
4. If a parallelogram is symmetrical about a straight line which bisects a pair of opposite sides, it is a rectangle.
5. How many axes of symmetry has a square? What are they?
6. A circle is symmetrical with regard to its centre.
7. A parallelogram is symmetrical with regard to the intersection of its diagonals.
8. If a quadrilateral is symmetrical with regard to any point, it is a parallelogram.

THEOREM 45 (Euclid III. 3)

(*i*) The diameter of a circle which bisects a chord (not a diameter) is perpendicular to the chord;

(*ii*) Conversely, the diameter which is perpendicular to a chord bisects it.

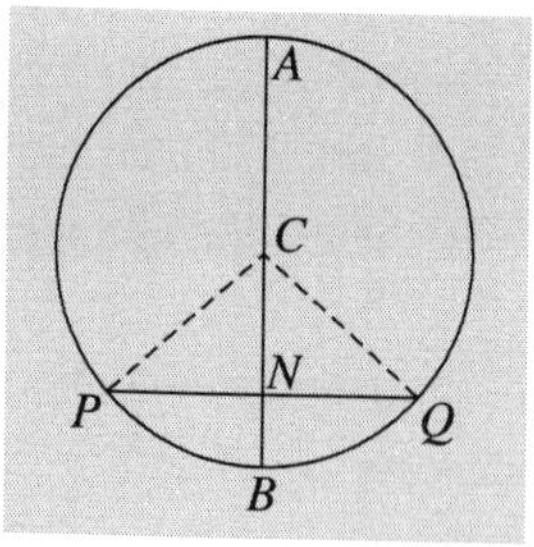

Let AB be a diameter and PQ a chord of a circle of which C is the centre.

(*i*) Let AB bisect PQ at N.

It is required to prove that AB is perpendicular to PQ.

Construction. Join CP, CQ.

Proof. In the triangles CNP, CNQ,

$$\begin{cases} NP = NQ \ (given), \\ CN \text{ is common}, \\ CP = CQ. \end{cases}$$

∴ The triangles are congruent;

∴ $\angle CNP = \angle CNQ$;

∴ AB is perpendicular to PQ.

(*ii*) Let AB be perpendicular to PQ.

It is required to prove that $NP = NQ$.

Proof. In the right-angled triangles *CNP*, *CNQ*,

the hypotenuse *CP* = The hypotenuse *CQ*,
and *CN* is common.

∴ The triangles are congruent;

∴ $NP = NQ$.

THEOREM 46

The perpendicular bisector of a chord of a circle passes through the centre.

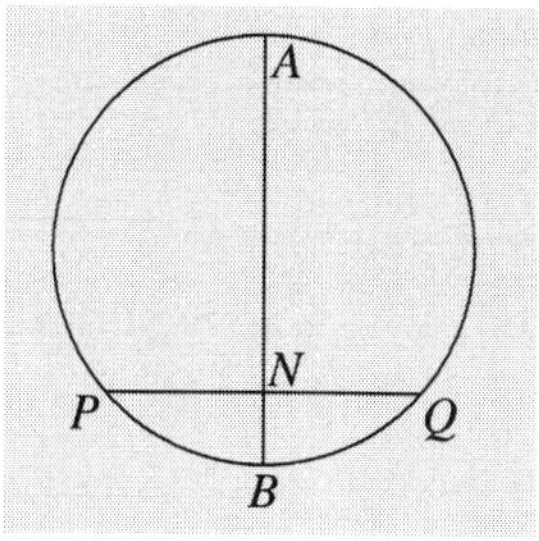

Let *PQ* be a chord of a circle of which *C* is the centre, and let *AB* be the perpendicular bisector of *PQ*.

It is required to prove that *AB* passes through *C*.

Proof. Since *AB* is the perpendicular bisector of *PQ*, it is the locus of points which are equidistant from *P* and *Q*.

But $CP = CQ$;

∴ *AB* passes through *C*.

THEOREM 47

A circle is symmetrical with regard to any diameter.

Let *AB* be a diameter of a circle (*see figure of Theorem* 46).

It is required to prove that the circle is symmetrical with regard to *AB*.

Construction. Take any point *P* on the circumference. Draw *PN* perpendicular to *AB* and produce it to meet the circle again at *Q*.

Proof. Because the diameter *AB* is perpendicular to the chord *PQ*.

∴ $PN = QN$.

Hence, if the figure is folded about *AB*, any point *P* on one semi-circumference will fall on some point *Q* on the other semi-circumference.

∴ The circle is symmetrical with regard to *AB*.

THEOREM 48

One circle, and one only, can be drawn through three given points not in the same straight line.

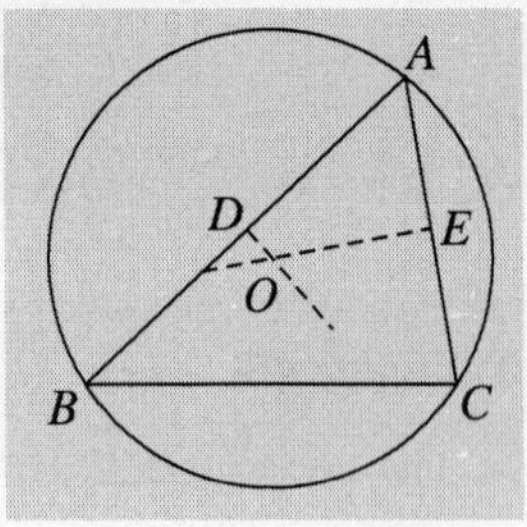

Let A, B, C be the given points. It is required to prove that one circle, and one only, can be drawn through A, B, C.

Construction. Bisect AB at D and AC at E. Through D and E, draw perpendiculars to AB and AC respectively.

Proof. Because AB and AC are not in the same straight line, these perpendiculars will meet at some point O.

And because DO is the perpendicular bisector of AB,

∴ DO is the locus of points which are equidistant from A and B.

Similarly, EO is the locus of points which are equidistant from A and C.

Hence O is the point, and the only point, which is equidistant from A, B and C.

∴ the circle with centre O and radius OA passes through A, B and C, and there is no other circle which passes through these three points.

COR. 1. Circles which have three points common coincide.

COR. 2. Two circles cannot cut in more than two points.

For if they have a third common point, they coincide.

COR. 3. If O is a point within a circle from which three equal straight lines OA, OB, OC can be drawn to the circumference, O is the centre of the circle.

THEOREM 49 (Euclid III. 14)

Equal chords of a circle are equidistant from the centre, and conversely, chords of a circle which are equidistant from the centre are equal.

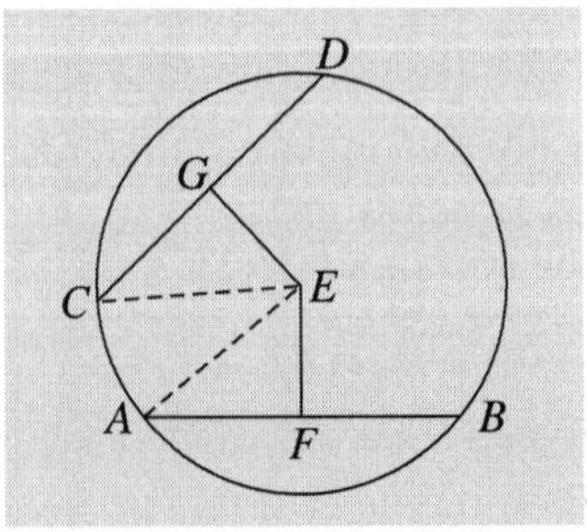

Let AB, CD be chords of a circle of which E is the centre, and let EF, EG be the perpendiculars from E to AB, CD respectively.

---

* Exercise on Theorems 45–48 are given on p. 148.

(i) If $AB = CD$, it is required to prove that $EF = EG$.

Construction. Join EA, EC.

Proof. Because EF is perpendicular to AB,

∴ $AF = FB$;

∴ $AB = 2AF$.

Similarly, $CD = 2CG$.

But it is given that $AB = CD$;

∴ $AF = CG$.

Hence, in the right-angled triangles AEF, CEG,

the hypotenuse AE = The hypotenuse CE,
and $AF = CG$ *(proved)*;

∴ The triangles are congruent;

∴ $EF = EG$.

(ii) If $EF = EG$, it is required to prove that $AB = CD$.

In the right-angled triangles AEF, CEG,

the hypotenuse AE = The hypotenuse CE,
and $EF = EG$ *(given)*;

∴ The triangles are congruent;

∴ $AF = CG$.

But, as before, $AB = 2AF$ and $CD = 2CG$;

∴ $AB = CD$.

Theorem 50 (Euclid III. 15)

Of any two chords of a circle, the one which is nearer the centre is greater than the one more remote.

Conversely, the greater chord is nearer the centre than the less.

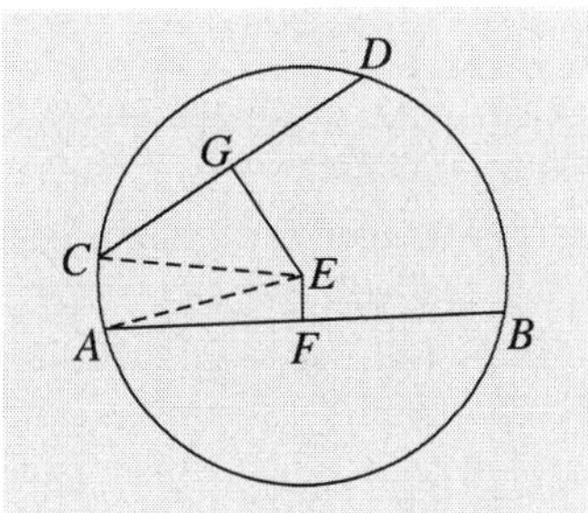

Let AB, CD be chords of a circle of which E is the centre, and let EF, EG be the perpendiculars from E to AB, CD respectively.

(i) If $EF < EG$, it is required to prove that $AB > CD$.

Construction. Join EA, EC.

Proof. Because EF is perpendicular to AB,

$\therefore$ $AB = 2AF$;

and because EG is perpendicular to CD,

$\therefore$ $CD = 2CG$.

Because AFE, CGE are right angles,

$\therefore$ Sq. on AE = Sq. on AF + Sq. on EF,

and Sq. on CE = Sq. on CG + Sq. on EG.

But $AE = CE$;

$\therefore$ Sq. on AE = Sq. on CE;

$\therefore$ Sq. on AF + Sq. on EF = Sq. on CG + Sq. on EG.

But, Sq. on EF < Sq. on EG, for $EF < EG$ *(given)*;

$\therefore$ Sq. on AF > Sq. on CG;

$\therefore$ $AF > CG$;

$\therefore$ $AB > CD$.

(*ii*) If $AB > CD$, it is required to prove that $EF < EG$.

Proof. As before, $AB = 2AF$ and $CD = 2CG$.

But $AB > CD$ *(given)*;

$\therefore$ $AF > CG$.

As before, Sq. on AF + Sq. on EF = Sq. on CG + Sq. on EG.

But Sq. on AF > Sq. on CG, for $AF > CG$.

$\therefore$ Sq. on EF < Sq. on EG;

$\therefore$ $EF < EG$.

EXERCISE XLI(a) (Theorems 45–48)

1. From a point O, two equal straight lines OA, OB are drawn to the circumference of a circle whose centre is C. Prove that $\angle AOC = \angle BOC$.
2. If any two points are taken on the circumference of a circle, the chord which joins them lies within the circle.
3. Prove that a straight line cannot cut a circle in more than two points.
4. If a straight line cuts a circle in one point, in general, it will cut the circle in a second point, if produced far enough. Prove this, and state the exceptional case.
5. The radii of two circles are r and r', and d is the distance between their centres. If the circles cut, prove that $r - r' < d < r + r'$.
6. The straight line joining the centres of two intersecting circles bisects their common chord at right angles.
7. Explain how to draw the chord of a circle which is bisected at a given point within the circle.

8. A chord PQ of a circle cuts a concentric circle in P', Q'. Prove that $PP' = QQ'$.
9. The straight line joining the middle points of two parallel chords of a circle passes through the centre.
10. Find the locus of the middle point of a chord of a circle drawn parallel to a given straight line.
11. Two chords of a circle cannot bisect each other unless they are both diameters.
12. If two circles cut at one point, prove that they must also cut at a second point.
13. Every circle which can be drawn through a given point A, with its centre on a given straight line PQ, passes through another fixed point.
14. Two circles intersect in A and B. Through these points two parallel straight lines CAD, EBF are drawn, meeting the circles again in C, D, E, F. Prove that $CD = EF$. [Draw perpendiculars to CD, EF from the centres.]
15. Two circles whose centres are C and D intersect at A and B. M is the middle point of CD, and a straight line PAQ is drawn through A, perpendicular to AM, to cut the circles again in P, Q. Prove that $AP = AQ$.
16. Through a point of intersection of two circles, draw a straight line such that the chords intercepted on it by the circles shall be equal. [Use the last example.]
17. Through a point O two pairs of equal straight lines OP, OP' and OQ, OQ' are drawn to the circumference of a circle whose centre is C. Prove that PQ, $P'Q'$ meet on OC (or OC produced). [Let PQ, $P'Q'$ cut OC in T, T', First take Δs OPQ, $OP'Q'$, then Δs OPT, $OP'T'$. Hence show that T, T' coincide.]
18. A circle cuts two concentric circles in P, Q and P', Q'. By considering the symmetry of the figure, prove that (*i*) the chords PP', QQ' are equal; (*ii*) the arcs PP', QQ', are equal.

EXERCISE XLI(b) (Theorems 49, 50)

1. A diameter of a circle is greater than any chord which is not a diameter. [Let C be the centre and AB a chord. Consider Δ CAB.]
2. Find the locus of the middle points of equal chords of a circle.
3. Explain how to draw the least chord of a circle which passes through a given point within the circle.
4. Chords of a circle which make equal angles with the straight line joining their point of intersection to the centre are equal.
5. AB is a chord of a circle. In AB take a point O. Draw a chord through O equal to AB.
6. POQ, XOY are chords of a circle of which C is the centre. If $\angle COP > \angle COX$ and both of these angles are acute, prove that $PQ < XY$.

7. Construct a chord of a circle equal to a given chord and making an angle of 30 degrees with it.
8. Construct a chord of a circle of given length, parallel to a given straight line.
9. Calculate the length of the least chord of a circle of radius 5 in. which can be drawn through a point distant 3 in. from the centre.
10. *AB* and *CD* are parallel chords of a circle of radius 2.5 in., on the same side of the centre. If $AB = 3$ in. and $CD = 1.4$ in., calculate the distance between *AB'* and *CD*. Verify by drawing and measurement.
11. A circular field is crossed by two parallel paths. The distance between these paths is 93 yards and their lengths are 112 and 50 yards respectively. Calculate the diameter of the field. [Let the distance from the centre to the longest path be *x* yards.]

XXII. ANGLES IN SEGMENTS

If any point *P* is taken on the arc of a segment of which *AB* is the chord, it will be proved that the angle *APB* is always of the same size, no matter where the point *P* is taken on the arc. The angle *APB* is called the angle in the segment.

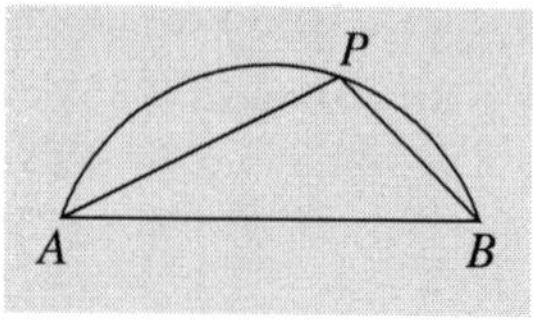

Def. The **angle in a segment** is the angle subtended by the chord of the segment at any point on the arc of the segment. The segment is said to contain the angle.

Def. Similar **segments of circles** are those which contain equal angles.

Def. If all the vertices of a rectilineal figure lie on the circumference of a circle, the figure is said to be inscribed in the circle, and the circle is said to be **described about the figure** or **circumscribed** to it.

The circle which passes through the vertices of a triangle is called the circum-circle of the triangle and its radius the circum-radius.

It was proved in Theorem 48 that one and only one circle can be drawn through three given points, *A*, *B*, *C*.

Hence *in general* it is not possible to draw a circle to pass through four given points *A*, *B*, *C*, *D*.

Def. A **rectilineal figure** which is such that a circle can be described about it is said to be cyclic.

Def. Four or more points such that a circle can be drawn through them are said to be **concyclic**.

THEOREM 51 (Euclid III. 20)

The angle which an arc of a circle subtends at the centre is double that which it subtends at any point on the remaining part of the circumference.

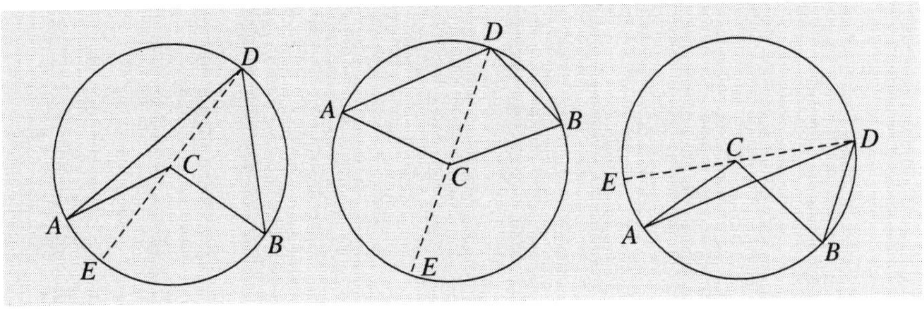

Fig. 1 Fig. 2 Fig. 3

Let AB be an arc of a circle of which C is the centre, and let D be any point on the remaining part of the circumference.

It is required to prove that $\angle ACB = 2\angle ADB$.

Join DC, and produce it to meet the circle at E.

Proof. First suppose that the point C is within the angle ACB, as in Figs. 1 and 2.

Because $CA = CD$,

$\therefore$ $\angle CAD = \angle CDA$.

Now the side DC of the triangle CAD is produced to E;

$\therefore$ $\angle ECA = \angle CAD + \angle CDA$;

$\therefore$ $\angle ECA = 2\angle CDA$.

Similarly, $\angle ECB = 2\angle CDB$;

$\therefore$ The sum of $\angle$s ECA, ECB = Twice the sum of $\angle$s CDA, CDB;

that is, $\angle ACB = 2\angle ADB$.

Next, suppose that the point C is not within the angle ADB, as in Fig. 3.

Then, as before, $\angle ECA = 2\angle CDA$,

and $\angle ECB = 2\angle CDB$.

$\therefore$ Difference of $\angle$s ECA, ECB = Twice the diffcrence of $\angle$s CDA, CDB;

that is,

$$\angle ACB = 2\angle ADB.$$

NOTE: In Fig, 2, the angle subtended at C by the arc AEB is greater than two right-angles.

THEOREM 52 (Euclid III. 21)

Angles in the same segment of a circle are equal

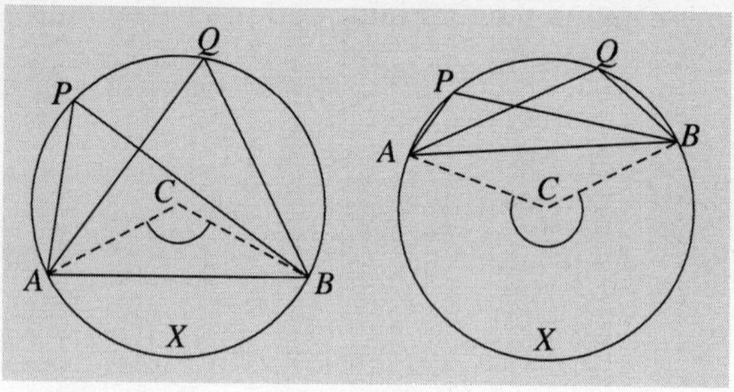

Let APB be a segment of a circle $APBX$, and let APB, AQB be any angles in the segment.

It is required to prove that

$$\angle APB = \angle AQB.$$

Construction. Let C be the centre. Join CA, CB.

Proof. The angle ACB subtended by the arc AXB at the centre is double that which it subtends at any point on the remaining part of the circumference;

$\therefore$ $$\angle ACB = 2\angle APB,$$

and $$\angle ACB = 2\angle AQB;$$

$\therefore$ $$\angle APB = \angle AQB.$$

THEOREM 53

This is the converse of Theorem 52

If a straight line subtends equal angles at two points on the same side of it, a circle can be drawn through these points and the extremities of the line.

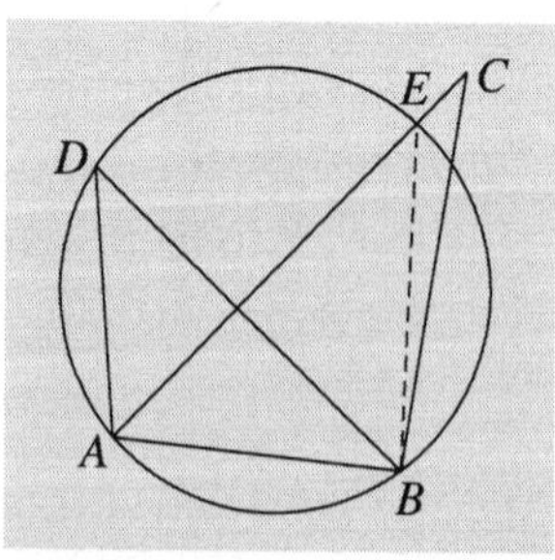

Let AB be a straight line, and C, D points on the same side of it such that the angles ACB, ADB are equal.

It is required to prove that a circle can be drawn through C, D, A, B.

Proof. A circle can be drawn through A, D, B.

If this circle does not pass through C, it will cut AC, or AC produced, at some point E.

Join BE.

Because angles in the same segment are equal,

$\therefore$ $\angle ADB = \angle AEB$.

But it is given that $\angle ADB = \angle ACB$;

$\therefore$ $\angle AEB = \angle ACB$;

that is, the exterior angle of the triangle BEC is equal to the interior opposite angle, and this is impossible.

Hence the circle through A, D, B passes through C.

THEOREM 54 (Euclid III. 31)

The angle in a semi-circle* is a right angle; the angle in a segment greater than a semi-circle is less than a right angle; the angle in a segment less than a semi-circle is greater than a right angle.

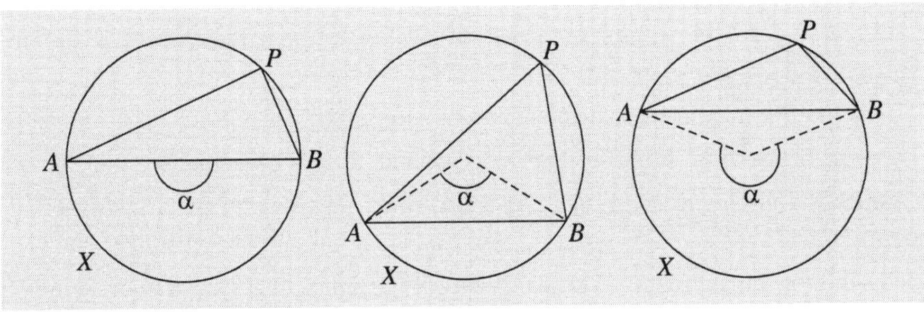

Let APB be a segment of a circle $APBX$.

It is required to prove that

(*i*) if the segment APB is a semicircle, $\angle APB$ is a right angle.

(*ii*) if the segment APB is greater than a semi-circle, $\angle APB$ is less than a right angle.

(*iii*) if the segment APB is less than a semi-circle, $\angle APB$ is greater than a right angle.

Proof. The angle which an arc subtends at the centre is double that which it subtends at the circumference.

Hence, if α is the angle which the arc AXB subtends at the centre,

$$\angle \alpha = 2\angle APB.$$

(*i*) If APB is a semi-circle, AB is a diameter and $\angle \alpha$ = 2rt. $\angle$s;

$\therefore$ $\angle APB$ is a right angle.

(*ii*) If APB is greater than a semi-circle, the arc AXB is less than the semi-circumference; and $\angle\alpha$ is less than 2 rt. $\angle$s;

$\therefore$ $\angle APB$ is less than a right angle.

* For an alternative proof of this part, see p. 47.

(*iii*) If *APB* is less than a semi-circle, the arc *AXB* is greater than the semi-circumference; and $\angle\alpha$ is greater than 2rt. $\angle$s.

$\therefore$ $\angle APB$ is greater than a right angle.

THEOREM 55 (Euclid III. 23)

On the same chord, and on the same side of it, there cannot be two similar segments of circles not coinciding with one another.

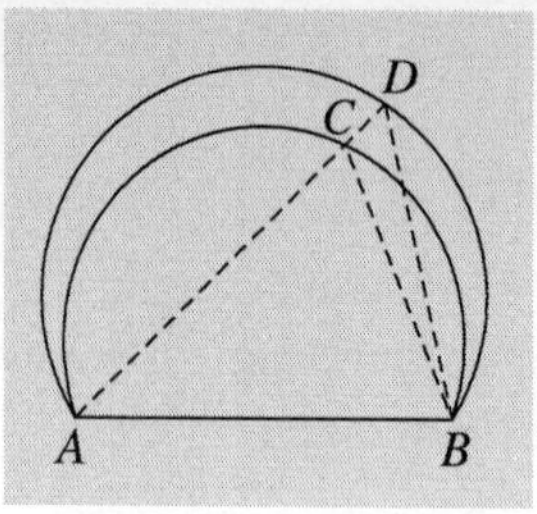

Let *ACB*, *ADB* be similar segments of circles on the same chord *AB* and on the same side of it.

It is required to prove that the segments coincide.

Proof. If possible, suppose that the segments do not coincide.

Since the circles *ACB*, *ADB* intersect at *A*, *B*, they cannot meet at any other point.

$\therefore$ One of the segments is entirely within the other.

Let *ACB* be within *ADB*.

Join *AC*, and produce it to cut the arc *ADB* at *D*.

Join *BC*, *BD*.

Because the segments *ACB*, *ADB* are similar,

$\therefore$ $$\angle ACB = \angle ADB;$$

that is, the exterior angle *ACB* of the triangle is equal to the interior opposite angle *ADB*, and this is impossible.

$\therefore$ the segments must coincide.

COR. Two similar segments of circles on equal chords are equal to one another.

For if the segments are placed on the same chord and on the same side of it, by the above theorem, they will coincide.

EXERCISE XLII (Theorems 51–55)

1. If *O* is the centre of the circle described about the triangle *ABC*, show that the angle *OBC* is the complement of the angle *A*. [Draw *OL* perpendicular to *BC*.]

2. If ABC, $A'B'C'$ are two triangles in which $BC = B'C'$ and $\angle A = \angle A'$, show that the circles circumscribing the triangles are equal. [Let O, O' be the centres. Use Ex. 1 to prove Δs OBC, $O'B'C'$ congruent.]
3. Similar segments on equal chords are parts of equal circles. [Use Ex. 2.]
4. AC, BD are chords of a circle which cut at a point E within the circle. If O is the centre, show that
$$\angle AOB + \angle COD = 2\angle AEB.$$
5. $ABCD$ is a quadrilateral inscribed in a circle whose centre is O. If the diagonals AC, BD are at right angles, show that the opposite sides AB, CD subtend supplementary angles at O. [Use Ex. 4]
6. A point P moves at a uniform rate round the circumference of a circle of which AB is a diameter and C is the centre. Prove that the angle BCP increases twice as fast as the angle BAP. [Suppose that, in 1 second, P moves to Q, etc.]
7. If A, B are fixed points, find the locus of a point P which moves so that the angle APB is a right angle.
8. A, B are given points and XY is a given straight line. Find a point in XY at which AB subtends a right angle.
9. Inscribe a rectangle in a given circle, with one of its sides equal to a given straight line.
10. Any parallelogram inscribed in a circle is a rectangle. [Prove that the centre is at the intersection of the diagonals.]
11. Circles are drawn on AB, AC, two sides of the triangle ABC, as diameters; show that the circles meet on BC.
12. Two circles intersect in A, B, and through B the diameters BP, BQ are drawn; show that P, A, Q are in the same straight line.
13. Find the locus of the intersection of the diagonals of a rhombus, one of whose sides is given in magnitude and position.
14. AP is a chord of a circle of which C is the centre. Prove that the circle on AC as diameter bisects AP.
15. Find the locus of middle points of chords of a circle which pass through a fixed point.
16. Find a point X on the circle circumscribing the triangle ABC, obtuse-angled at A, such that AX is bisected by BC. [Consider the locus of the middle points of chords through A. See also Ex. XXI(a) 11]
17. Find the locus of the vertex of a triangle which stands on the same base AB as a given triangle ABC, and which has a vertical angle equal to the angle C.
18. ABC is a given triangle and XY a given straight line. Find a point in XY at which AB subtends an angle equal to the angle C. Draw figures showing that there may be two, four, or no solutions.
19. CAB, DAB are triangles on the same base AB. Construct a triangle on AB as base, equal in area to CAB and having a vertical angle equal to the angle D.

20. Given the base and the vertical angle of a triangle, show that its area is greatest when it is isosceles.

21. X and Y are any points on the arcs of two segments on the same chord AB and on the same side of it. The bisectors of the angles XAY, XBY meet in Z. Prove that the angle AZB is constant.

[Show that, if X is on the inner arc,

$$\angle Z = \angle X + \frac{1}{2}\angle A - \frac{1}{2}\angle B, \text{ and } \angle Z = \angle Y + \frac{1}{2}\angle B - \frac{1}{2}\angle A]$$

22. AD, BE are the perpendiculars from A, B to the opposite sides of the triangle ABC. Prove that the quadrilateral $ABDE$ is cyclic.

23. Two circles intersect at A and B; PAQ is any straight line through A, cutting the circumferences at P and Q. Prove that the angle PBQ is constant.

24. AD, BE are the perpendiculars from A and B to the opposite sides of the triangle ABC. If AD meets BE in P and the circum-circle of the triangle in Q, show that $DP = DQ$. [Prove $\angle DBP = \angle DBQ$]

25. *ABC is a triangle and X, Y, Z are the middle points of BC, CA, AB. Draw AD perpendicular to BC. Prove that* $\angle YDC = \angle YZX$.

Hence show that the circle through X, Y, Z also passes through D.

26. Two circles intersect in A, B; PAQ, is any straight line through A, cutting the circumferences in P, Q. Show that PQ is greatest when it is parallel to the straight line joining the centres of the circles. [Let C, D be the centres. Draw CM, DN perpendicular to PQ and CX perpendicular to DN. Prove that $PQ = 2CX$.]

27. Two circles intersect in A, B; show how to draw a straight line of given length through A with its extremities on the given circles. [See hint to the last example, and note that X is on the circle on CD as diameter.]

THEOREM 56 (Euclid III. 22)

The opposite angles of a quadrilateral inscribed in a circle are supplementary.

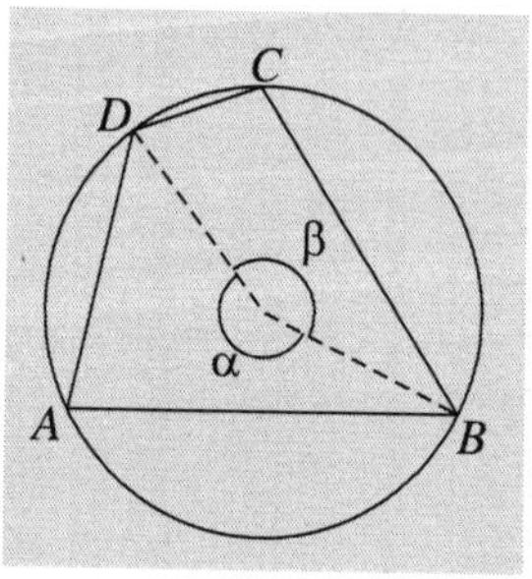

Let $ABCD$ be a quadrilateral inscribed in a circle.

It is required to prove that the angles A and C are supplementary and the angles B and D are supplementary.

Proof. Join B and D to the centre.

Let α, β be the angles which the arcs BAD, BCD subtend at the centre.

Because the angle which an arc subtends at the centre is double that which it subtends at the circumference,

$\therefore \quad \angle\alpha = 2\angle C$ and $\angle\beta = 2\angle A$;

$\therefore \quad \angle\alpha + \angle\beta = 2\angle C + 2\angle A.$

Now $\quad \angle\alpha + \angle\beta = 4$ right angles,

$\therefore \quad \angle C + \angle A = 2$ right angles.

Similarly, $\quad \angle B + \angle D = 2$ right angles.

Cor. If a side of a quadrilateral inscribed in a circle is produced, the exterior angle is equal to the interior opposite angle.

Theorem 57

***This is the converse of Theorem* 56**

If two opposite angles of a quadrilateral are supplementary, the quadrilateral is cyclic.

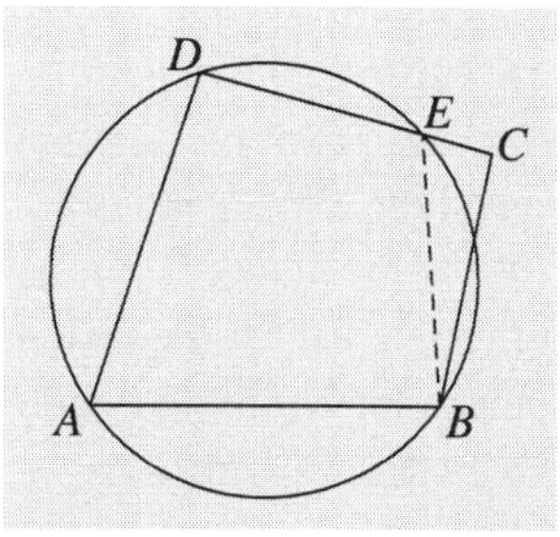

Let $ABCD$ be a quadrilateral in which the angles A and C are supplementary.

It is required to prove that $ABCD$ is a cyclic quadrilateral.

Proof. A circle can be described through the points A, B, D.

If this circle does not pass through C, it will cut DC, or DC produced, at some point E.

Join EB.

Because $ABED$ is a cyclic quadrilateral,

$\therefore \quad \angle BED =$ The supplement of $\angle A$.

But $\quad \angle C =$ The supplement of $\angle A$ *(given);*

$\therefore \quad \angle BED = \angle C$;

that is, an exterior angle of the triangle BEC is equal to an interior opposite angle, which is impossible.

$\therefore$ The circle through A, B, D passes through C; that is, $ABCD$ is a cyclic quadrilateral.

Cor. If one side of a quadrilateral is produced and the exterior angle so formed is equal to the interior opposite angle, then the quadrilateral is cyclic.

EXERCISE XLIII (Theorems 56, 57)

1. *ABCD* is a quadrilateral inscribed in a circle, and *AB*, *CD* are produced to meet at *P*. Prove that the triangles *PAD*, *PCB* are equiangular.
2. The bisectors of the angles *B*, *C* of the triangle *ABC* meet at I, and the bisectors of the exterior angles at *B*, *C* meet at *J*. Prove that *BICJ* is a cyclic quadrilateral.
3. If the bases of two triangles are equal and their vertical angles are supplementary, prove that their circum-circles are equal. [Place the triangles on opposite sides of a common base.]
4. If *P* is any point in the base *AB* of an isosceles triangle *ABC*, prove that the circum-circles of the triangles *APC*, *BPC* are equal.
5. *AD*, *BE* are the perpendiculars from *A*, *B* to the opposite sides of the triangle *ABC*. If *AD*, *BE* meet at *P*, show that the quadrilateral *DPEC* is cyclic. Hence show that the angle *APB* is the supplement of *C*, and that the circles *APB*, *ACB* are equal.
6. Let *O* be the centre, *AB* a diameter, and *AC* a chord of a circle; draw *OD* at right angles to *AB*, meeting *AC*, or *AC* produced, in *D*; show that the circle about *AOD* is equal to the circle through *O*, *B*, *C*, *D*.
7. *ABCD* is a quadrilateral inscribed in a circle; the angle *ABC* is bisected by a straight line *BE* meeting the circumference in *E*; if the side *AD* is produced to *F*, show that *DE* bisects the angle *CDF*.
8. Two circles intersect in the points *B*, *D*; a straight line *ABC* cuts the circles in *A*, *C*; *AD*, *CD* cut the circles again in *P*, *Q*; *AQ*, *CP* meet in *R*; prove that *DPQR* is a cyclic quadrilateral. [Join BD]
9. The opposite sides of a cyclic quadrilateral are produced to meet in *P* and *Q*, and, about the triangles so formed, outside the quadrilateral, circles are described intersecting in *R*. Show that *P*, *R*, *Q*, are in the same straight line.
10. *ABCD* is a quadrilateral, and the bisectors of the angles *A*, *B*; *B*, *C*; *C*, *D*; *D*, *A* meet in *E*, *F*, *G*, *H* respectively. Show that the figure *EFGH* is cyclic.
11. If the exterior angles of a quadrilateral are bisected by four straight lines, prove that the quadrilateral formed by these lines is cyclic.
12. *If any points X, Y, Z are taken on the sides BC, CA, AB of a triangle ABC, prove that the circum-circles of the triangles AYZ, BZX, CXY meet at a point. (Let the circles AYZ, BZX cut at O. Prove that CXOY is a cyclic quadrilateral.]*
13. *A*, *B*, *C* are three points in a straight line, and *P* any other point. *AFE*, *BFD*, *CED* are perpendicular to *PA*, *PB*, *PC* respectively; prove that *P*, *D*, *E*, *F* lie on a circle. [Prove $\angle PDC = \angle PBC = \angle PFE$]
14. *ABCD* is a cyclic quadrilateral; *AB*, *DC* are produced to meet at *E* and *AD*, *BC* are produced to meet at *F*. Show that the bisectors of the angles *BEC*,

CFD are at right angles. [Let the bisectors of *BEC*, *CFD* meet at *X*. Prove that

$$\angle ECF = \angle C = \angle EXF + \frac{1}{2}\angle BEC + \frac{1}{2}\angle CFD;$$

also $$\angle A = \angle EXF - \frac{1}{2}\angle BEC - \frac{1}{2}\angle CFD.]$$

15. Let *ABC* be a triangle. Draw *AD*, *BE* perpendicular to *BC*, *CA*, meeting at *P*. Join *CP* and produce it to cut *AB* at *F*. Prove that

$$\angle DPC = \angle DEC = \angle FBD.$$

Hence show that *BDPF* is a cyclic quadrilateral, and that *CF* is perpendicular to *AB*.

16. If *P* is the orthocentre of the triangle *ABC*, prove that the circum-circles of the triangles *BPC*, *CPA*, *APB* are equal. [See Ex. 5]

17. *If P is the orthocentre of the triangle ABC, and X, Y, Z are the middle points of BC, CA, AB, respectively, prove that the circle through X, Y, Z passes through the middle point of AP.*

[Bisect *AP* at *L*. Prove that $\angle YLZ = \angle BPC =$ suppl. of $\angle YXZ$.]

18. *Take any point X on the circum-circle of the triangle ABC. Draw XM, XN perpendicular to CA, AB. Let MN (or MN produced cut BC at L. Join XL, XA, XB. Prove that*

$\angle XNL = \angle XAC = \angle XBL$ (or its supplement, if *X* is on arc *AB*).

Hence show that XNBL is a cyclic quadrilateral and that XL is *perpendicular to BC.*

XXIII. TANGENTS

Let a circle and a straight line be drawn at random, and let the straight line be produced to any length.

In general, the line will not meet the circle at all, or else it will cut the circle in two points.

The exceptional case is that in which the straight line meets the circle in one point only.

Def. Any straight line which cuts a circle is called a **secant**.

Def. A straight line which meets a circle and which, when produced, does not cut it, is called a **tangent**.

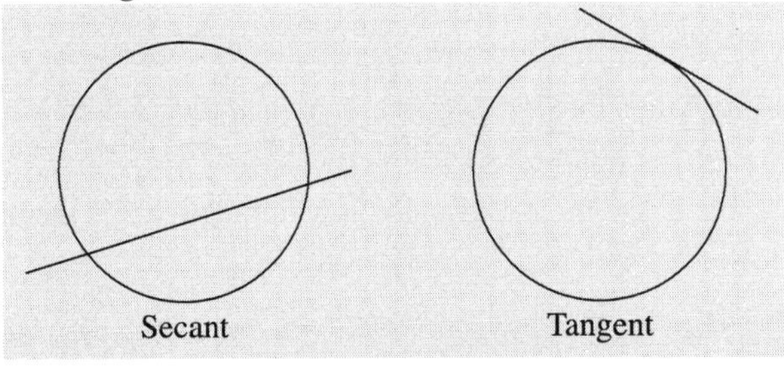

The point at which a tangent meets the circle is called the point of contact, and the tangent is said to touch the circle at this point.

Def. Two circles are said to **touch one another** when they meet at a point and have a common tangent at this point.

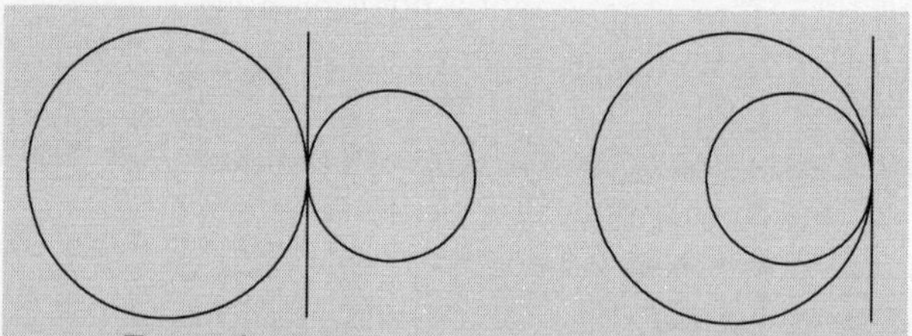

External contact   Internal contact

Circles may touch either **externally** (when they are on opposite sides of the common tangent) or **internally** (when they are on the same side of the common tangent).

Def. If all the sides of a rectilineal figure touch a circle, the figure is said to be **described** about the circle or **circumscribed** to it, and the circle is said to be inscribed in the figure.

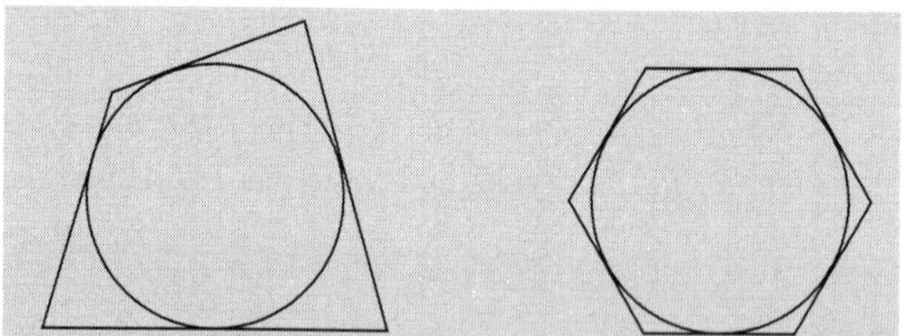

Circumscribe quadrilateral   Circumscribed hexagon

A circle which touches the three sides of a triangle is called the **inscribed** circle, and a circle which touches one side and the other two sides produced is called an **escribed** circle.

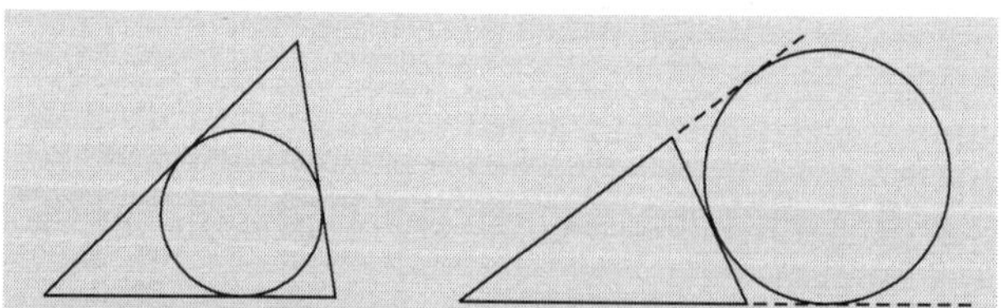

Inscribed circle   Escribed circle

If a straight line cuts a circle at a point A, the angle at which it cuts the circle is the angle which the line makes with the tangent at A.

If two circles cut at a point A, the angle at which they cut is the angle between the tangents at A.

> **Note:** The definition of a *tangent to a circle*, given on p. 159, is Euclid's. Another definition, which (in the opinion of the authors) is unsuitable for the beginner, is given on p. 198.

Theorem 58 (Euclid III. 16)

(*i*) The straight line drawn through a point on a circle perpendicular to the radius touches the circle.

(*ii*) Any other straight line drawn through the point cuts the circle.

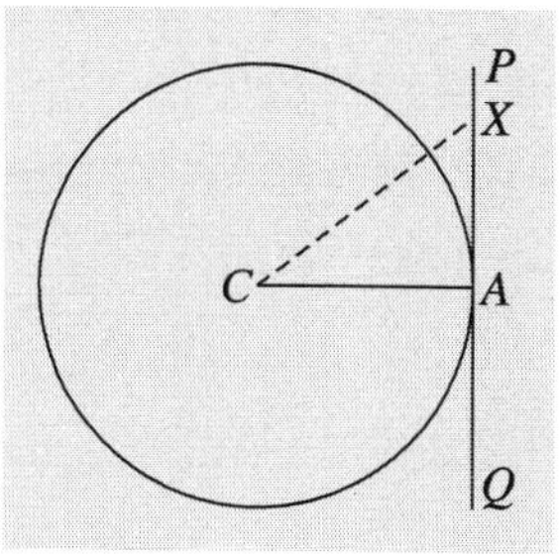

Let A be a point on the circumference of a circle, of which C is the centre. Let PAQ be a straight line perpendicular to the radius CA.

(*i*) It is required to prove that PAQ is a tangent.

Construction. In PQ take any point X, distinct from A.

Join CX.

Proof. In the triangle CAX,

$\angle CAX$ is a right angle;

$\therefore \angle CXA$ is less than a right angle;

$$\therefore \quad \angle CAX > \angle CXA,$$

and the greater angle is subtended by the greater side;

$$\therefore \quad CX > CA;$$

that is, CX is greater than a radius;

$\therefore X$ is outside the circle.

In the same way it can be shown that any other point in PQ, except A, is outside the circle;

$\therefore PQ$ meets the circle and, if produced, does not cut it;

$\therefore PQ$ is a tangent.

(*ii*) Let HAK be any other straight line through A.

It is required to prove that HAK cuts the circle.

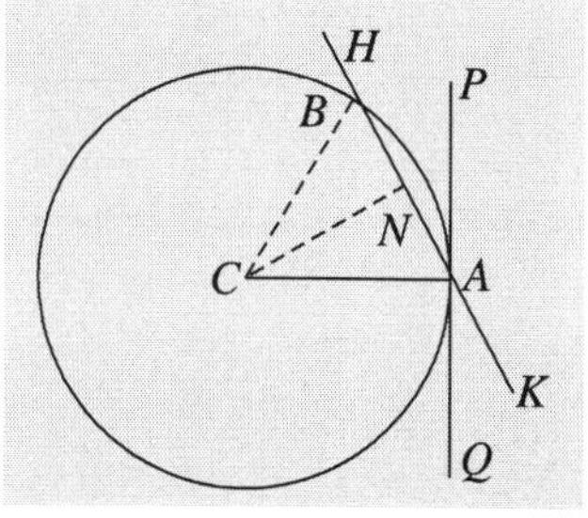

Construction. Draw CN perpendicular to HK, meeting it in N. Along NH cut off NB equal to AN.

Join CB.

Proof. In the triangles CNA, CNB,

$$\left\{\begin{aligned} NA &= NB \text{ (construction)},\\ &\text{CN is common},\\ \angle CNA &= \angle CNB \text{ (construction)}; \end{aligned}\right.$$

$$\therefore \qquad CA = CB;$$

$\therefore$ B is on the circumference:

$\therefore$ HK cuts the circle at B.

> **NOTE:** This theorem may be stated as follows:
> One tangent, and one only, can be drawn to a circle at any point on the circumference, and this tangent is perpendicular to the radius through the point of contact.

COR. 1. The straight line joining the centre of a circle to the point of contact of a tangent is perpendicular to the tangent.

COR. 2. The perpendicular to a tangent to a circle, at the point of contact, passes through the centre.

These corollaries follow from the fact that one perpendicular only can be drawn to a straight line from a given point.

THEOREM 59

If two tangents are drawn to a circle from an external point,

(*i*) the tangents are equal;

(*ii*) they subtend equal angles at the centre of the circle;

(*iii*) they make equal angles with the straight line joining the given point to the centre.

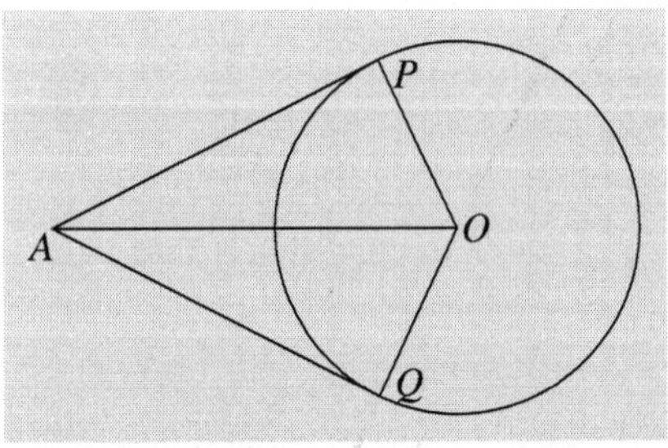

Let AP, AQ be tangents drawn to a circle whose centre is O from an external point A.

It is required to prove that

(*i*) $AP = AQ$,

(*ii*) $\angle AOP = \angle AOQ$,

(iii) $\angle PAO = \angle QAO$.

Proof. Because AP, AQ are tangents at P, Q.

∴ The angles APO, AQO are right angles.

In the right-angled triangles APO, AQO,

$$\begin{cases} \text{the hypotenuse } AO \text{ is common,} \\ OP = OQ \text{ (radii).} \end{cases}$$

∴ The triangles are congruent;

∴ $$AP = AQ,$$
$$\angle AOP = \angle AOQ,$$
$$\angle PAO = \angle QAO.$$

THEOREM 60

If two circles touch one another, the point of contact lies in the straight line joining the centres, or in that line produced.

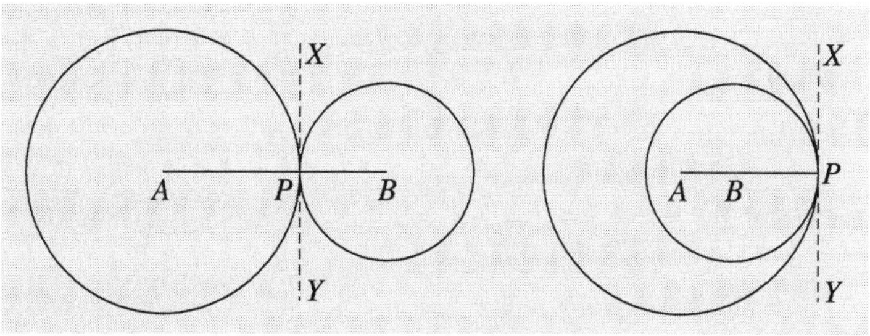

Let two circles with centres A and B touch one another, and let P be the point of contact.

It is required to prove that the points A, P, B are in the same straight line.

Proof. Because the circles touch at P, by definition, they have a common tangent at P.

Let XPY be the common tangent.

Because XPY touches both circles at P,

∴ XPY is perpendicular to each of the radii PA, PB.

∴ Both A and B lie on the straight line through P, perpendicular to XY.

COR. If two circles touch one another, the distance between their centres is equal to the sum or the difference of their radii, according as they touch externally or internally.

EXERCISE XLIV(a) (Theorems 58–60)

1. If a straight line cuts a circle at A and B, it cuts the circle at the same angle at each of these points.

2. If two circles cut one another at *A* and *B*, they cut at the same angle at each of these points.

3. If two circles cut at right angles, the square on the straight line joining their centres is equal to the sum of the squares on their radii.

4. Draw the tangents to a circle which are parallel to a given straight line.

5. What is the locus of the centre of a circle which touches a given straight line at a given point?

6. Describe a circle to touch a given straight line *AB* at a given point *C* and to pass through another given point *D*.

7. What is the locus of the centre of a circle which touches two given intersecting straight lines?

8. Describe a circle of given radius to touch two given straight lines, *OA* and *OB*.

9. Describe a circle to touch two given straight lines *AOB*, *COD*, with its centre on a given straight line *XY*. Show that, in general, two such circles can be drawn. What is the exceptional case?

10. What is the locus of the centre of a circle which touches two given parallel straight lines?

11. Describe a circle to touch two given parallel straight lines and to pass through a given point.

12. If two circles are concentric, any chord of the outer which touches the inner is bisected at the point of contact.

13. What is the locus of the centres of circles which touch a given circle at a given point?

14. Draw a circle to pass through a given point *A* and to touch a given circle at a given point *B*.

15. What is the locus of the centres of circles of given radius which touch a given circle?

16. Draw a circle of given radius to touch a given circle and have its centre on a given straight line. Draw figures showing that there may be four, three, two, one, or no solutions.

17. Draw a circle of given radius to touch a given circle and a given straight line.

18. Given two non-intersecting circles, each outside the other. Explain how to draw a circle of given radius to touch both circles (*i*) externally; (*ii*) internally.

19. Given two non-intersecting circles, each outside the other. Explain how to draw a circle of given radius to touch one of the circles externally and the other internally.

20. If three circles are such that each touches the other two, the common tangents at the points of contact meet in a point. [Consider the triangle formed by joining the centres, and its inscribed circle.]

21. Four circular coins of different sizes lie on a table, and each touches two and only two of the others; show that the four points of contact lie on a circle.

22. *If a quadrilateral circumscribes a circle, the sum of one pair of opposite sides is equal to the sum of the other pair.*

For the converse of this theorem see p. 140, Paper IX. 5.

23. If a parallelogram is circumscribed to a circle, it is a rhombus.

24. If a rhombus circumscribes a circle, the lines joining the opposite points of contact pass through the centre.

25. If a rectangle is circumscribed about a circle, it is a square.

26. If a hexagon *ABCDEF* circumscribes a circle, prove that the sum of the sides *AB*, *CD*, *EF* is equal to the sum of the sides *BC*, *DE*, *FA*.

27. If a rectilineal figure of an even number of sides is described about a circle, show that the sum of one set of alternate sides is equal to the sum of the other set.

28. Two circles whose centres are *C* and *D* touch one another at *A*. Through *A* any straight line is drawn, cutting the circles at *P* and *Q*. Prove that the radii *CP*, *DQ* are parallel.

29. Two circles intersect in *A*. *A* straight line *PAQ* is drawn through *A*, terminated by the circumferences of the circles, and tangents are drawn at *P*, *Q*, intersecting in *R*; show that the angle *PRQ* is equal to the angle between the tangents at *A*.

30. If a straight line cuts two concentric circles in *A*, *A'* and *B*, *B'*, the four intersections of a tangent at an *A* and a *B* lie on another concentric circle. [Let the tangent at *B* meet the tangents at *A*, *A'* in *Q*, *Q'*. Let *C* be the common centre. Prove that the quads. *ABQC*, *A'CBQ'* are cyclic. Hence show *CQ* = *CQ'*.]

EXERCISE XLIV(b)

Graphical and Numerical

1. Draw a triangle *ABC* in which *AB* = *AC* = 1.6 in., *BC* = 1.2 in. Draw a circle to touch *BC* at *B* and to pass through *A*. Measure the radius and verify by calculation.

2. Draw two straight lines *OA*, *OB*, making $\angle AOB = 50^\circ$. Draw a circle of radius 1 in. to touch *OA*, *OB*. Measure the distance of its centre from *O*.

3. Draw a triangle *ABC*, in which *BC* = 4 cm., *CA* = 6 cm., *AB* = 5 cm. Draw two circles with their centres *X*, *Y* on *BC*, or *BC* produced, to touch *AB*, *AC*, or these lines produced. Measure *XY*.

4. Draw two parallel straight lines *AB*, *CD* 2 in. apart. Take a point *E* between the lines and distant 0.4 in. from *AB*. Draw two circles through *E* to touch *AB* and *CD*. Measure the distance between their centres. Verify by calculation.

5. Draw a circle with centre O and radius 1 in. Draw radii OA, OB at an angle of 30°. Produce OB to C, making BC = 1 in. Draw a circle to pass through C and to touch the first circle at A. Measure its radius.
6. Given a circle of radius 1 in. and a straight line at a perpendicular distance of 1.4 in. from the centre. Draw two circles, each of radius 0.5 in., to touch the given circle and line. Measure the distance between the centres of the required circles. Verify by calculation.
7. Given a circle of radius 0.5 in. and a straight line distant 3 in. from the centre. Draw a circle of radius 2 in. to touch the given straight line and the given circle, having internal contact with the given circle. Explain the construction.
8. Given two circles of radii 0.8 in. and 0.5 in., the distance between their centres being 1.7 in. Draw a circle of radius 0.6 in. to touch both the given circles externally. Explain the construction.
9. Given two circles as in Ex. 8. Draw a circle of radius 1.4 in. to touch the smaller of the given circles internally and the larger externally. Explain the construction.
10. Given two circles as in Ex. 8. Draw a circle of radius 1.4 in. to touch the larger of the given circles internally and the smaller externally. Explain the construction.
11. Given two circles of radii 0.8 in. and 0.6 in, the distance between their centres being 1 in. Draw a circle of radius 2.4 in. to touch both the given circles internally. Explain the construction.

XXIV. CONSTRUCTION OF TANGENTS

CONSTRUCTION 18

Draw a tangent to a circle from an external point.

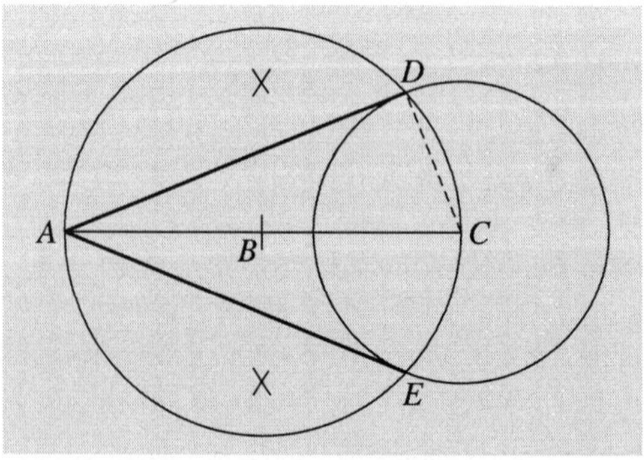

Let A be the given point and C the centre of the given circle.

Construction. Join AC, and bisect it at B. With centre B and radius BA, draw a circle, cutting the given circle at D and E.

Join AD.

Then AD is a tangent to the given circle.

Proof. Join CD.

By construction, ADC is a semi-circle on AC as diameter;

$\therefore$ $\angle ADC$ is a right angle;

$\therefore$ AD is perpendicular to the radius CD;

$\therefore$ AD touches the given circle.

> **NOTE:** Similarly it can be shown that the straight line AE is a tangent. Thus two tangents can be drawn to a circle from an external point.
>
> **NOTE:** *The circle described with centre c and radius r is often called the circle (C, r).*

CONSTRUCTION 19

Draw a direct common tangent to two given circles.

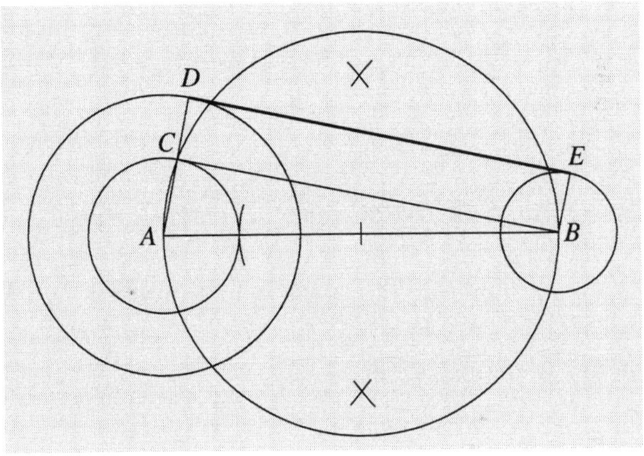

Let A, B be the centres and r, r' the radii of the circles, and suppose that $r > r'$.

It is required to draw a direct common tangent.

Construction. Join AB.

With centre A and radius $r - r'$, draw a circle.

Draw a tangent BC to this circle. Join AC, and produce it to cut the circle (A, r) at D.

Draw BE parallel to AD, to cut the circle (B, r') at E.

Join DE.

Then DE is the required common tangent.

Proof. Because $AD = r$ and $AC = r - r'$.

$\therefore$ $$CD = r - (r - r') = r';$$

$\therefore$ $$CD = BE;$$

$\therefore$ CD, BE are equal and parallel;

$\therefore$ $CBED$ is a parallelogram.

Now, by construction, BC is a tangent at C;

$\therefore \angle BCD$ is a right angle;

$\therefore$ $CBED$ is a rectangle.

$\therefore$ The angles ADE, BED are right angles;

$\therefore$ DE touches each of the given circles.

CONSTRUCTION 20

Draw a transverse common tangent to two given non-intersecting circles.

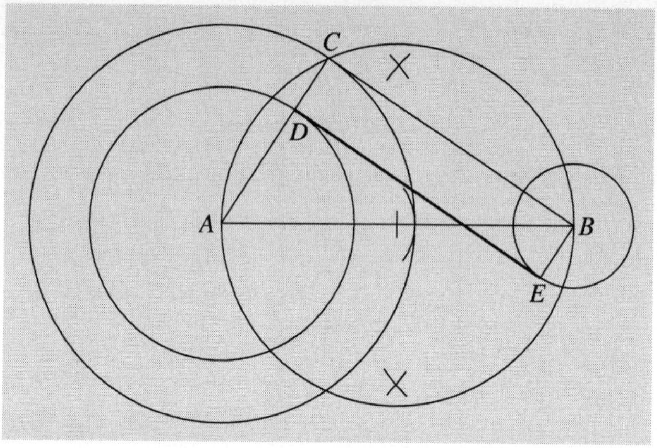

The construction is similar to the preceding, except that the circle described with centre A is of radius $r + r'$.

Proof. Because $AD = r$ and $AC = r + r'$,

$\therefore$ $$DC = (r + r') - r = r';$$

$\therefore$ $$CD = BE.$$

The rest of the proof is the same as in the preceding construction.

EXAMPLE 1. *Through a given point A, draw a straight line to meet a given circle, whose centre is O, in H, K, so that HK may be equal to a given straight line BC.*

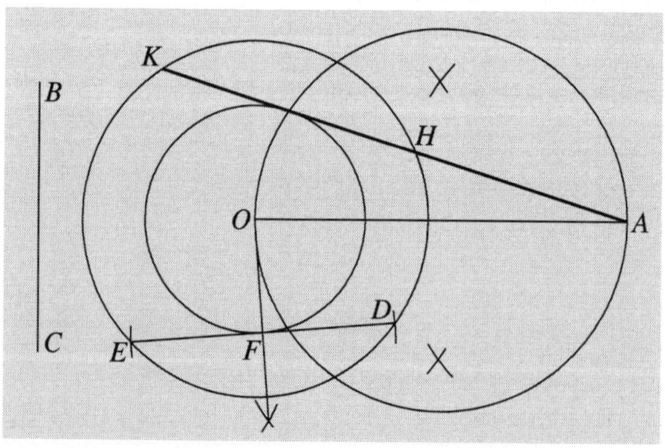

Place a chord *DE*, equal to *BC*, in the circle, and proceed as indicated in the figure. Complete the explanation and give a proof.

EXAMPLE 2. *Draw a circle to touch a given circle and a given straight line at a given point P.*

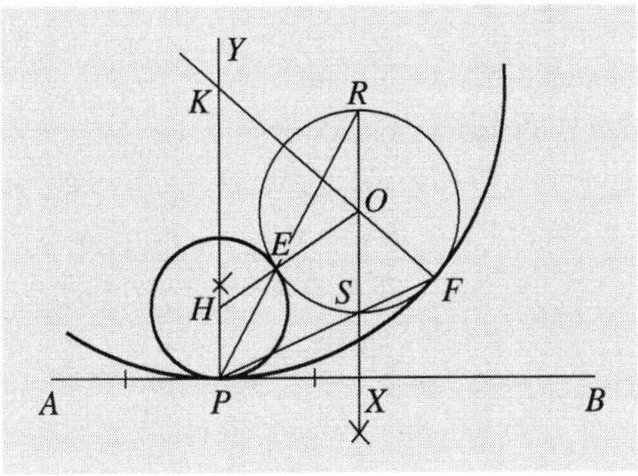

Let *O* be the centre of the given circle. Draw *OX*, *PY* each perpendicular to *AB*. Let *OX* cut the given circle in *R* and *S*. Join *PR*, *PS*, cutting the given circles in *E*, *F* respectively. Join *OE*, *OF* and produce to cut *PY* in *H*, *K*. Then *H*, *K* are centres of circles which fulfil the conditions.

Supply proof by showing $\angle HPE = \angle HEP$ and $\angle KPF = \angle KFP$.

EXAMPLE 3. *Draw a circle to touch a given line AB, and a given circle at a given point P.*

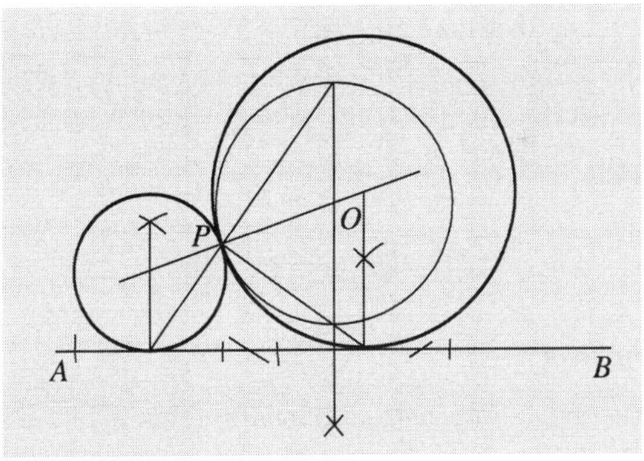

The construction is indicated in the above figure. Supply explanation and proof.

EXERCISE XLV(a) (Constructions 18–20)

1. Draw a straight line through a given point, so that the perpendicular on it from another given point may be equal to a given straight line.
2. Construct the triangle *ABC*, given the angle *A*, the side *AB* and the length of the perpendicular from *A* to *BC*.

3. A and B are given points; draw a circle, centre A, so that the tangent to it from B is of given length (less than AB).
4. Two circles whose centres are A and B touch externally at C. The common tangent at C cuts another common tangent DE at F. Prove the following:
   (*i*) DE is bisected at F.
   (*ii*) The angle DCE is a right angle.
   (*iii*) The angle AFB is a right angle.
   (*iv*) The perpendicular to DE through F bisects AB.
   (*v*) The circle on AB as diameter touches DE at F.
5. If r, r' are the radii of two circles and d is the distance between the centres,
   (*i*) Prove that the length of a direct common tangent is
   $$\sqrt{\{d^2 - (r - r')^2\}},$$
   and deduce the condition that one circle is entirely within the other.
   (*ii*) Prove that the length of a transverse common tangent is
   $$\sqrt{\{d^2 - (r + r')^2\}},$$
   and deduce the condition that each circle is entirely outside the other.
6. Two circles whose centres are A and B touch externally at C. DE is a direct common tangent and DAG, EBH are the diameters through D, E. Prove that DH and GE pass through C.
7. In a given circle, place a chord of given length. How many such chords can be drawn? Prove that they all touch a fixed concentric circle.
8. Given two non-intersecting circles, explain how to draw a straight line $PQRS$ such that the chords PQ, RS intercepted on it by the circles may be of given lengths. [Use Ex. 7]

EXERCISE XLV(b)

Numerical and Graphical

1. Calculate the length of a tangent drawn to a circle of radius 3.3 in. from a point distant 6.5 in. from the centre.
2. Draw the tangents to a circle of radius 1 in. from a point distant 2 in. from the centre. Measure the angle between the tangents.
3. Draw a circle of radius 1 in. Draw two tangents inclined at 50° to one another. Measure their lengths and verify by calculation.
4. Draw a circle of radius 1 in. Construct the locus of a point P which moves so that the length of the tangent from P to the circle is always 2 in.
5. Draw a circle of radius 1 in. Construct the locus of a point P such that the circle subtends an angle of 45° at P, *i.e.* such that the tangents from P contain an angle of 45°.

6. Draw circles of radii 1 in. and 1.5 in., the distance between their centres being 2 in. Find a point *P* at which the first circle subtends an angle of 45° and the second circle subtends an angle of 60°. Measure the distance of *P* from the line of centres. [Use Ex. 5]
7. Draw a circle of radius 5 cm and take a point *P* distant 8.5 cm from the centre. Draw a straight line *PQR* through *P*, to cut the circle at *Q*, *R* such that *QR* = 6 cm. Measure *PQ*. Verify by calculation.
8. Find by calculation the lengths of the common tangents to two circles of radii 1.1 in. and 0.4 in. with their centres at a distance of 2.5 in. Verify by measurement.
9. Take two points, *A*, *B*, 2 in. apart. Draw a straight line *PQ* such that the perpendiculars from *A*, *B* on it are 0.9 in., 0.4 in. respectively. Show that there are four positions of *PQ*, and measure the angles they make with *AB*.

XXV. CONSTRUCTION OF INSCRIBED AND ESCRIBED CIRCLES

CONSTRUCTION 21

Inscribe a circle in a given triangle.

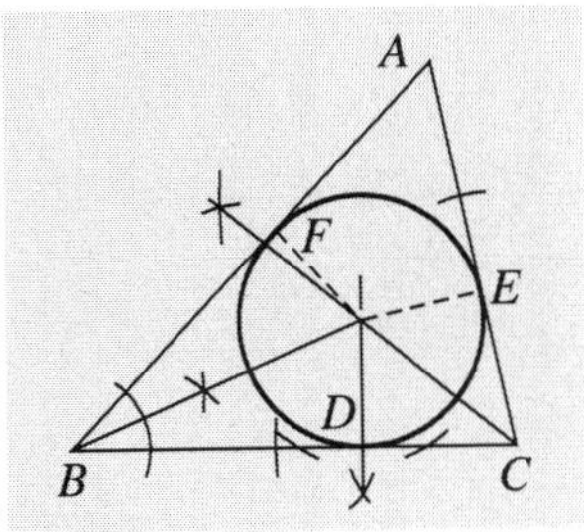

Let *ABC* be the given triangle.

It is required to inscribe a circle in *ABC*.

Construction. Bisect the angles *B* and *C* by straight lines meeting at I.

Draw *ID* perpendicular to *BC*.

With centre *I* and radius *ID*, draw a circle.

This is the required circle.

Proof. Draw *IE*, *IF* perpendicular to *CA*, *AB* respectively.

In the triangles *IBD*, *IBF*,

$$\begin{cases} \angle IBD = \angle IBF \text{ (construction)}, \\ \angle IDB = \angle IFB \text{ (right angles)}, \end{cases}$$

IB is common.

∴ The triangles are congruent.

∴ $ID = IF.$

Similarly $ID = IE.$

∴ The circle drawn with centre I and radius ID passes through E and F.

Also, because the angles at D, E, F are right angles, the circle touches BC, CA, AB.

CONSTRUCTION 22

Draw an escribed circle of a triangle.

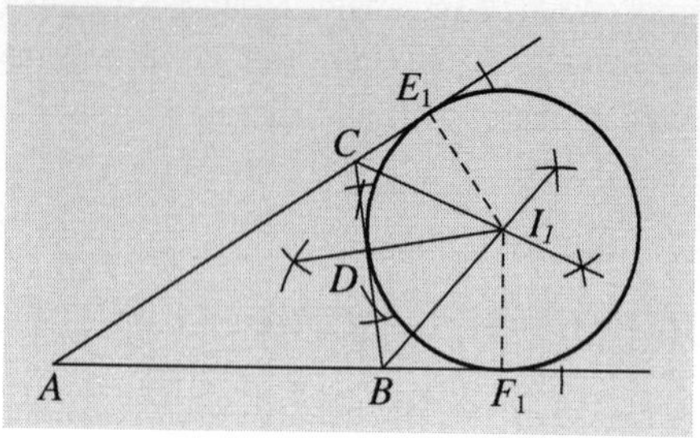

Let ABC be the given triangle.

It is required to draw a circle to touch the side BC and the sides AC, AB, produced.

Construction. Produce AB to F_1 and AC to E_1.

Bisect the angles CBF_1, BCE_1 by straight lines meeting at I_1.

Draw I_1D_1 perpendicular to BC.

With centre I_1 and radius I_1D_1, draw a circle.

This is the required circle.

Proof. Draw I_1E_1, I_1F_1 perpendicular to AC, AB respectively. In the triangles I_1BD_1, I_1BF_1,

$$\begin{cases} \angle I_1BD_1 = \angle I_1BF_1 \text{ (construction)}, \\ \angle I_1D_1B = \angle I_1F_1B \text{ (right angles)}, \\ I_1B \text{ is common.} \end{cases}$$

∴ The triangles are congruent;

∴ $I_1D_1 = I_1F_1;$

Similarly $I_1D_1 = I_1E_1;$

∴ The circle drawn with centre I_1 and radius I_1D_1 passes through E_1 and F_1.

Also, because the angles at D_1, E_1, F_1 are right angles, the circle touches BC, AE_1, AF_1.

Lengths of Certain Lines connected with the In- and E-scribed Circles of a Triangle.

1. *Find the radius of the circle inscribed in the triangle ABC.*

Let I be the centre of this circle; D, E, F its points of contact with the sides and r its radius.

Then $\angle$s at D, E, F are rt. $\angle$s.

Hence, if Δ = Area of $\Delta\, ABC$, and $2s = a + b + c$,

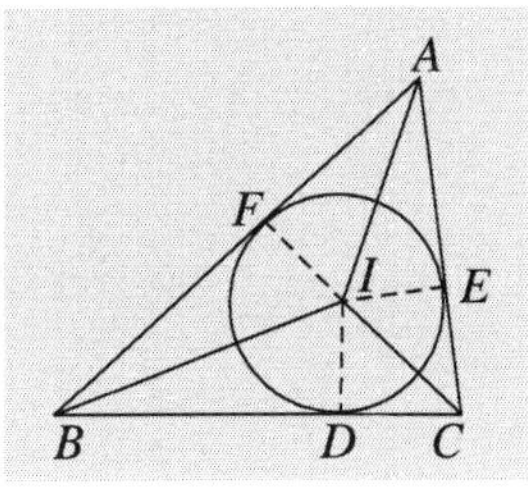

$$\Delta = \Delta\, IBC + \Delta\, ICA + \Delta\, IAB$$

$$= \frac{1}{2}ar + \frac{1}{2}br + \frac{1}{2}cr$$

$$= \frac{1}{2}r(a + b + c) = rs;$$

$\therefore$ $$r = \frac{\Delta}{s}.$$

2. *Find the radii of the escribed circles of the triangle ABC.*

Let I_1 be the centre of the circle escribed to $\Delta\, ABC$, opposite $\angle\, A$, D_1, E_1, F_1 its points of contact with the sides and r_1 its radius.

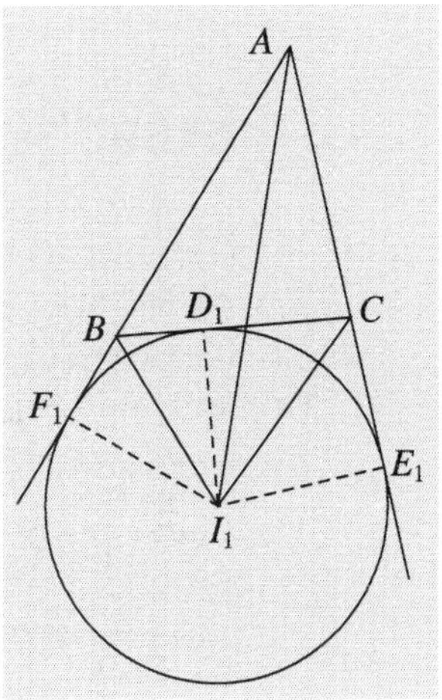

Then $\angle$s at D_1, E_1, F_1 are rt. $\angle$s.

Hence, if Δ = area of $\Delta\, ABC$, and $2s = a + b + c$,

$$\Delta = \Delta\, I_1CA + \Delta\, I_1AB - \Delta\, I_1BC$$

$$= \frac{1}{2}br_1 + \frac{1}{2}cr_1 - \frac{1}{2}ar_1$$

$$= \frac{1}{2}r_1(b + c - a) = r_1(s - a);$$

$\therefore$ $$r_1 = \frac{\Delta}{s - a}.$$

Similarly, if r_2, r_3 are the radii of circles escribed to $\Delta\, ABC$, opposite $\angle$s B, C,

$$r_2 = \frac{\Delta}{s - b},\ r_3 = \frac{\Delta}{s - c}.$$

3. *If the inscribed circle of the triangle ABC touches BC, CA, AB at D, E, F respectively, then*

$$AE = AF = s - a,$$

where s is the semi-perimeter of the triangle.

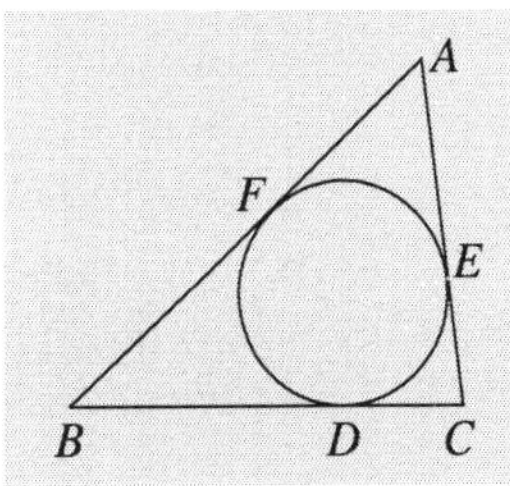

Because AE, AF are the tangents from A,

$\therefore$ $$AE = AF.$$

Similarly $$BD = BF$$

and $CD = CE$;

$\therefore$ $AE + BD + CD =$ Semi-perimeter;

$\therefore$ $AE + BC = s$;

$\therefore$ $AE = s - a$.

4. *If the circle escribed to the triangle ABC touches BC at D, and AC, AB produced at E, F respectively, then*

$$AE_1 = AF_1 = s \text{ and } BD_1 = s - c.$$

We have $AE_1 + AF_1$

$$= AC + CE_1 + AB + BF_1$$
$$= AC + CD_1 + AB + BD_1$$
$$= AC + AB + BC$$
$$= 2s.$$

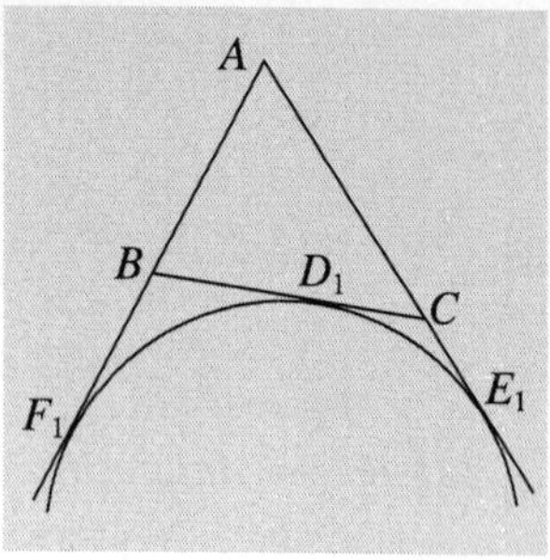

Now $AE_1 = AF_1$;

$\therefore$ $AE_1 = AF_1 = s$.

Also $BD_1 = BF_1 = AF_1 - AB = s - c$.

EXERCISE XLVI(a) (Constructions 21, 22)

1. If in the figure of Construction 21 the straight line AI is drawn, prove that it bisects the angle A.
2. If in the figure of Construction 22 the straight line AI_1 is drawn, prove that it bisects the angle A.
3. In the figures of Constructions 21, 22, prove that
$$\angle BIC = 90 + \frac{A}{2} \text{ and } \angle BI_1C = 90 - \frac{A}{2}.$$
4. Construct the triangle ABC, given the side AB, the angle A and the radius of the inscribed circle.
5. If I is the centre of the circle inscribed in the triangle ABC and I_1 is the centre of the circle escribed to the triangle opposite A, prove that A, I, I_1 are in the same straight line.
6. Draw two straight lines AB, CD which would meet, if produced at some point X outside the sheet of paper. Construct the bisector of the angle X between the lines. [Draw any straight line cutting AB in E, and CD in F Bisect $\angle$s AEF, CFE by lines meeting in G. Bisect $\angle$s BEF, DFE by lines meeting in H. Join GH. This is the line required. Supply proof.]
7. A circle is inscribed in a right-angled triangle; show that the sum of the hypotenuse and the diameter of the circle is equal to the sum of the sides containing the right angle.

8. ABC is a triangle right-angled at A; prove that the hypotenuse BC is equal to the difference between the radius of the circle inscribed in the triangle and the radius of the circle which touches BC and the other sides produced.

9. The inscribed circle of the triangle ABC touches BC at D, and the circle escribed to the triangle, opposite A, touches BC at D_1. If $AB > AC$ and X is the middle point of BC, prove that

$$XD = XD_1 = \frac{1}{2}(AB - AC).$$

10. AX, AY are fixed tangents to a given circle. A variable tangent to the circle meets AX at B and AY at C. Prove that either $AB + AC + BC$ or $AB + AC - BC$ is constant.

11. If I is the centre of the circle inscribed in the triangle ABC and I_1, I_2, I_3 are the centres of circles escribed to the triangle opposite A, B, C respectively, prove that

(*i*) the straight line AII_1 is perpendicular to I_2I_3.

(*ii*) I is the orthocentre of the triangle $I_1I_2I_3$.

12. AP, BQ, CR are the perpendiculars from A, B, C to the opposite sides of the triangle ABC, meeting in the orthocentre O. Prove that O, A, B, C are the centres of the inscribed and escribed circles of the triangle PQR.

In general, a circle can be drawn to touch three given straight lines, but in order that it may be possible to draw a circle to touch the, four sides of a quadrilateral, a certain condition must be satisfied.

13. *If the sum of one pair of opposite sides of a quadrilateral is equal to the sum of the other pair; a circle can be inscribed in it.*

[Let $ABCD$ be the quadrilateral. Draw a circle to touch three sides AB, BC, AD. If DC does not touch this circle, draw DC' to touch it, meeting BC or BC produced at C'. Use Ex. XLIV(*a*), 22 to prove that C' coincides with C.]

For a direct proof, see Paper IX, Ex. 5, p. 140.

14. If a circle can be inscribed in a quadrilateral, prove that the bisectors of the angles of the quadrilateral meet in a point

15. In the accompanying figure, the sides of the quadrilateral $ABDC$, when produced, touch a circle; prove that

$$AB - CD = BD - AC.$$

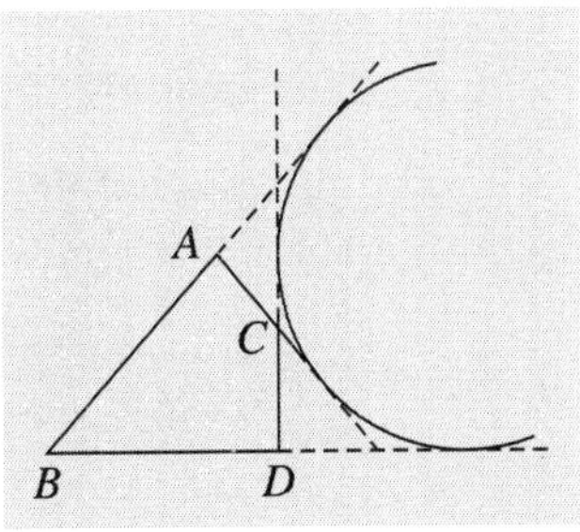

16. If A, B, C, D are any four points such that $AB + BC = AD + DC$, prove that a circle can be drawn to touch AB, BC, AD, DC, produced if necessary.

17. Four circular coins of different sizes lie on a table, and each touches two and only two of the others; show that a circle can be inscribed in the quadrilateral formed by joining the centres.

Exercise XLVI(b)

Numerical and Graphical

1. Draw triangles whose sides are (*i*) 3 in., 2.6 in., 2.8 in.; (*ii*) 3.4 in., 2 in., 1.8 in., Inscribe circles in them and measure their radii.
2. Draw a triangle *ABC* in which a=1.5 in., b = 1.4 in., c = 1.3 in. Draw the inscribed circle and the escribed circle which touches *AB* and *AC* produced. Measure their radii.
3. Draw an isosceles right-angled ΔABC, whose hypotenuse *AB* is 4 in. Along *AC*, *BC* set off *AP*, *BQ* equal to 2 in., 1 in. respectively. Join *PQ*, and construct, as in Ex. 6 of the preceding Exercise, the bisector of the angle between *PQ* and *AB*. Verify by producing the lines.
4. Draw a quadrilateral *ABCD*, in which *AB* = 3 in., *BC* = 3 in. *CD* = 1.5 in., *DA* = 2 in., *AC* = 3 in. Draw, as in Ex. 6 of the preceding Exercise, the bisectors of the angles between *AD*, *BC* and *CD*, *BA*. Verify by producing the sides.
5. The sides of a triangle are 3 in., 4 in., 5 in. Find by calculation the radii of the inscribed circle and of the escribed circle which touch the shorter sides produced.
6. In the triangle *ABC*, *BC* = 13 in., *CA* = 11 in., *AB* = 10 in. If the inscribed circle touches *BC* at *D*, calculate the length of *BD*.
7. Draw any quadrilateral *ABCD* in which *AB* = 3 in., *BC* = 2 in., *CD* = 1 in., *DA* = 2 in. Explain why a circle can be inscribed in it, and draw this circle. *(See Ex.* 13 of the preceding Exercise.)
8. Draw two straight lines *AD*, *BC* crossing one another at *X*, such that *AD* = 3 in., *BC* = 2 in., *AB* = 2 in., *CD* = 1 in. Produce *AB*, *CD* to meet in *Y*. Explain why a circle can be inscribed in the quadrilateral *BXDY*, and draw this circle. *(See Ex.* 16 of the preceding Exercise.)

XXVI. EQUAL ANGLES, CHORDS AND ARCS

Theorem 61 (Euclid III. 26)

In equal circles (or in the same circle), if two arcs subtend equal angles at the centres, or at the circumferences, they are equal.

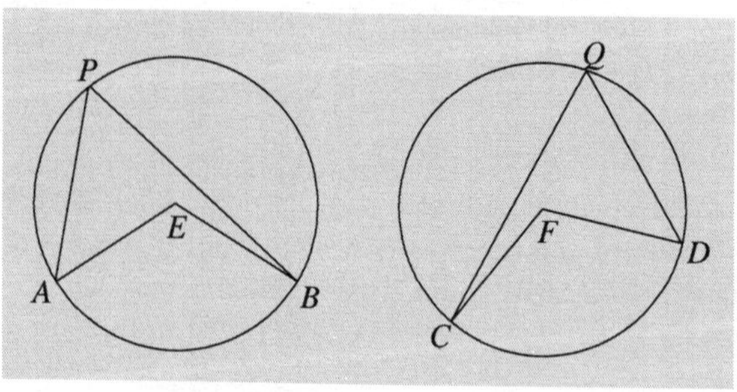

Let AB, CD be arcs of equal circles, of which the centres are E and F.

(*i*) If $\angle AEB = \angle CFD$, it is required to prove that the arc AB = the arc CD.

Proof. Apply the circle AB to the circle CD so that E falls on F and EA falls along FC.

Because $\quad \angle AEB = \angle CFD$ (*given*),

$\therefore$ EB falls along FD.

Because the circles are equal, their radii are equal;

$\therefore$ A falls on C and B on D, and the circumferences coincide.

$\therefore$ The arcs AB, CD coincide;

$\therefore \quad$ The arc AB = The arc CD.

(*ii*) If the angles APB, CQD which the arcs AB, CD subtend at points P, Q on the circumferences are equal, it is required to prove that the arc AB = the arc CD.

Proof. The angle which an arc subtends at the centre is twice that which it subtends at the circumference;

$\therefore \quad \angle AEB = 2\angle APB$

and $\quad \angle CFD = 2\angle CQD$

But $\quad \angle APB = \angle CQD$ (*given*);

$\therefore \quad \angle AEB = \angle CFD$;

$\therefore \quad$ arc AB = arc CD.

Theorem 62 (Euclid III. 27)

In equal circles (or in the same circle), if two arcs are equal, they subtend equal angles at the centres and at the circumferences.

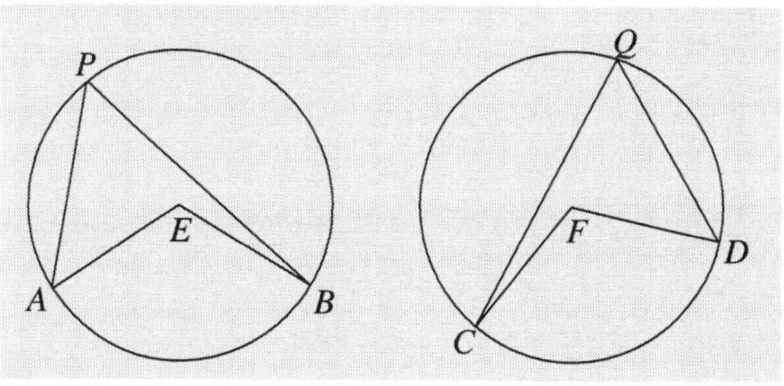

Let AB, CD be equal arcs of equal circles, subtending the angles AEB, CFD at the centres and angles APB, CQD at the circumferences.

It is required to prove that

$$\angle AEB = \angle CFD$$

and $\quad \angle APB = CQD$.

Proof. Apply the circle AB to the circle CD so that E falls on F and EA falls along FC.

Because the circles are equal, their radii are equal;

∴ A falls on C,

and the circumferences coincide.

Also, because The arc AB = The arc CD,

∴ B falls on D;

∴ EB falls along FD;

∴ $$\angle AEB = \angle CFD.$$

Again, the angle which an arc subtends at the centre is twice that which it subtends at the circumference;

∴ $$\angle AEB = 2\angle APB$$

and $$\angle CFD = 2\angle CQD;$$

∴ $$\angle APB = \angle CQD.$$

Cor. If the arcs AB, CD of two equal circles are equal, the sectors AEB, CFD which stand on these arcs are equal.

Theorem 63 (Euclid III. 28)

In equal circles (or in the same circle), if two chords are equal, the arcs which they cut off are equal, the greater to the greater and the less to the less.

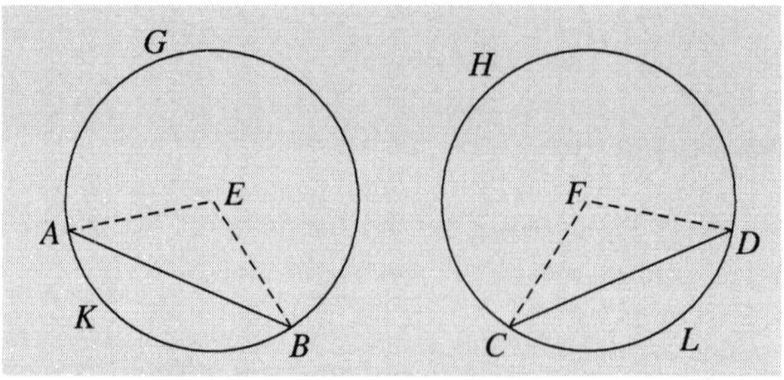

Let AB, CD be equal chords of equal circles, of which the centres are E and F.

It is required to prove that the arcs which they cut off are equal, namely that

$$\text{arc } AGB = \text{arc } CHD$$

and $$\text{arc } AKB = \text{arc } CLD.$$

Construction. Join EA, EB, FC, FD.

Proof. Because the circles are equal, their radii are equal; Therefore, in the triangles AEB, CFD,

$$\begin{cases} EA = FC, \\ EB = FD, \\ AB = CD \ (\textit{given.}); \end{cases}$$

∴ The triangles are congruent:

∴ $$\angle AEB = \angle CFD.$$

Hence the arcs AKB, CLD subtend equal angles at the centres;

$\therefore$ arc AKB = arc CLD.

But the circumferences of the circles are equal.

$\therefore$ the remaining arc AGB = the remaining arc CHD.

THEOREM 64 (Euclid III. 29)

In equal circles (or in the same circle), if two arcs are equal, the chords of the arcs are equal.

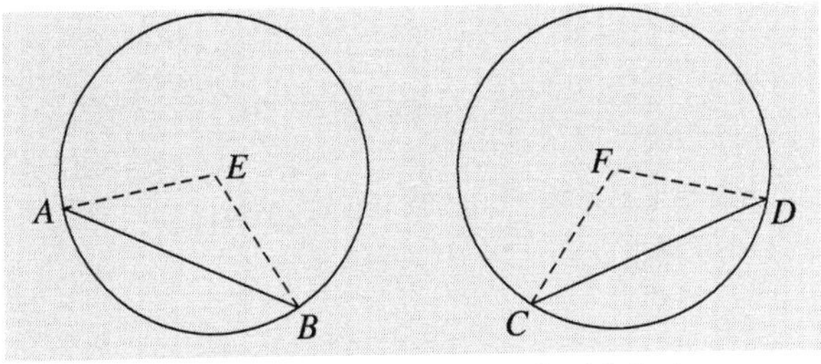

Let AB, CD be equal arcs of equal circles, of which the centres are E and F.

It is required to prove that the chord AB = The chord CD.

Construction. Join EA, EB, FC, FD.

Proof. Because AB, CD are equal arcs of equal circles,

$\therefore$ they subtend equal angles at the centres,

that is, $\angle AEB = \angle CFD$.

Now, because the circles are equal, their radii are equal.

Therefore, in the triangles AEB, CFD,

$$\begin{cases} EA = FC, \\ EB = FD, \\ \angle AEB = \angle CFD. \end{cases}$$

$\therefore$ The chord AB = The chord CD.

CONSTRUCTION 23

Bisect an arc of a circle.

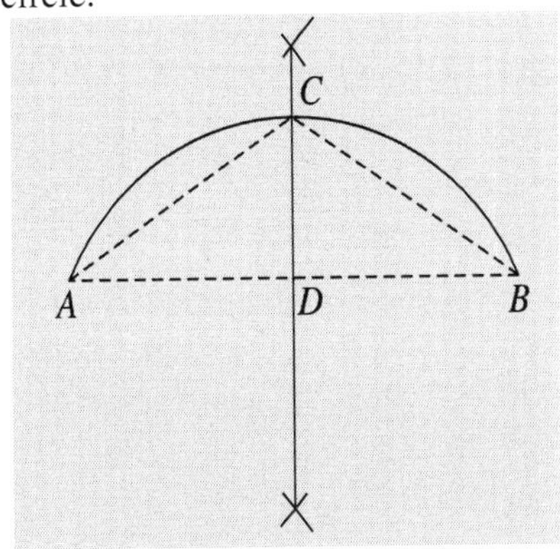

Let AB be the given arc which it is required to bisect.

Construction. Draw the perpendicular bisector of the straight line joining A, B, meeting the arc AB at C.

Then the arc AB is bisected at C.

Proof. Join AB, AC, CB. Let AB cut the perpendicular bisector at D.

In the triangles ADC, BDC,

$$\begin{cases} AD = BD \text{ (construction)}, \\ CD \text{ is common}, \\ \angle ADC = \angle BDC \text{ (construction)}. \end{cases}$$

∴ The triangles are congruent;

∴ $$AC = BC.$$

Now the equal chords AC, BC cut off equal arcs.

∴ The arc AC = The arc BC.

EXERCISE XLVII (Theorems 61–64)

1. The arcs intercepted by two parallel chords of a circle are equal.
2. State and prove the converse of the preceding.
3. The perpendiculars from A, B to the opposite sides of the triangle ABC meet the circum-circle of the triangle at X and Y. Prove that the arcs CX, CY are equal.
4. The bisector of the angle A of the triangle ABC bisects the arc BC of the circum-circle of the triangle.
5. C is any point on the arc AB of a circle and BC is produced to D. Prove that the bisector of the angle ACD bisects the arc AB.
6. Given the base AB and the magnitude of the angle C of a triangle ABC, show that the bisectors of the interior and exterior angles at C pass each through a fixed point. [Use Exs. 4, 5]
7. PAQ, PBQ, PCQ are three equal angles on the same side of PQ, and the bisectors of the angles PAQ, PBQ meet in R. Prove that CR bisects the angle PCQ.
8. Two circles intersect in A, B and any straight line through A meets the circumferences in P, Q. Show that the angles of the triangle PBQ are constant.
9. $ABCD$ is a quadrilateral inscribed in a circle whose centre is O; if the angles BAD and BOD are together equal to two right angles, show that the arc BAD is double the arc BCD.

10. *ABCD* is a trapezium inscribed in a circle; show that the non-parallel sides are equal.

11. *OA*, *OB* are perpendicular radii of a circle and *AX*, *BY* are parallel chords; show that *BX*, *AY* are at right angles. [Prove $\Delta\ BOX = \Delta\ AOY$ and $\therefore\ \angle\ OBX = \angle\ OAY$.]

12. *AP*, *AQ* are any two chords of a circle; the arcs *AP*, *AQ* are bisected at *M*, *N*; show that *MN* makes equal angles with *AP*, *AQ*.

13. *AB*, *CD* are arcs of the same circle. If the arc *AB* is twice the arc *CD*, show that the chord *AB* is less than twice the chord *CD*.

14. *AB* is an arc of a circle bisected in *C*. Construct a chord *AD* equal to half the chord *AB*, and show that the arc *AD* is less than the arc *AC*.

15. Two equal circles intersect in *A*, *B*; any straight line through *A* meets the circumferences in *P*, *Q*; show that $BP = BQ$.

16. Two equal circles intersect in *A*, *B* and any straight line through *A* meets the circles again in *P*, *Q*. Find the locus of the middle point of *PQ*. [Use the preceding example.]

17. If, in two circles, equal chords subtend equal or supplementary angles at points on the circumferences, the circles are equal.

18. *ABC* is a triangle, and *D* any point in *AC*. Show that, if the circles round *BAD* and *BCD* are equal, the triangle is isosceles.

19. From the ends of a diameter *BC* of a circle, parallel chords *BE*, *CF* are drawn, meeting the circle again in *E* and *F*; prove that *EF* passes through the centre of the circle.

20. *ACB*, *ADB* are two arcs of circles. If *C* is any point in the arc *ACB* and *AC*, *BC* cut the arc *ADB* in *D* and *E*, prove that the chord *DE* is of constant length.

21. A chord *PQ* of a given circle subtends a constant angle at a point *A* on the circumference. Show that *PQ* touches a fixed circle.

22. A straight line of given length subtends given equal angles at two fixed points; prove that the straight line always touches a fixed circle.

23. *A*, *B* are two fixed points on a circle and *C*, *D* the extremities of a chord of constant length; prove that the intersections of *AD*, *BC* and of *AC*, *BD* lie on fixed circles.

24. Three equal circles pass through the point *P* and meet one another, two and two, at *A*, *B*, *C*. Prove that *P* is the orthocentre of the triangle *ABC*. [Let *AP*, *BP*, *CP* cut *BC*, *CA*, *AB* at *L*, *M*, *N*. Prove $\angle\ PAB = \angle\ PCB$, $\angle\ PAC = \angle\ PBC$, $\angle\ PCA = \angle\ PBA$. Hence show that Δs *ALB*, *BMC*, *CNA* are right-angled triangles.]

XXVII. ANGLES IN SEGMENTS *(Continued)*

THEOREM 65 (Euclid III. 32)

If a straight line touches a circle and, from the point of contact, a chord is drawn, the angles which the chord makes with the tangent are equal to the angles in the alternate segments of the circle.

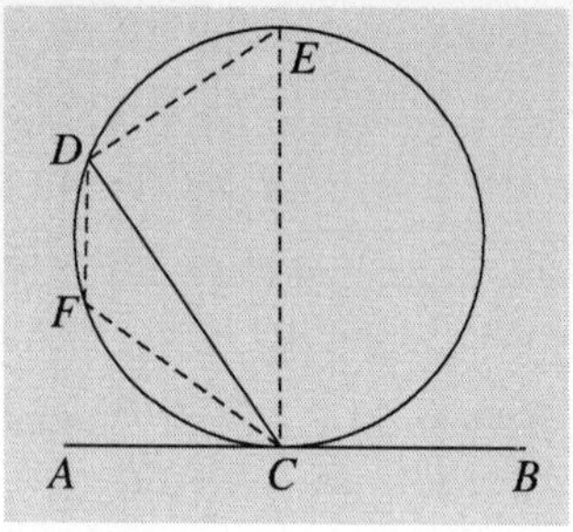

Let the straight line *AB* touch a circle at *C*, and let *CD* be a chord making an acute angle with *CA*. Let *CD* divide the circle into the segments *CED*, *CFD*, the points *E* and *A* being on opposite sides of *CD*.

(*i*) It is required to prove that

$$\angle ACD = \text{The angle in the alternate segment } CED.$$

Construction. Draw the diameter *CE*. Join *DE*.

Proof. Because *ACB* is a tangent and *CE* a diameter,

$\therefore$ $\angle ACE$ is a right angle.

Again, the angles of the triangle *CDE* are together equal to two right angles.

But the angle *CDE* is the angle in a semi-circle;

$\therefore$ $\angle CDE$ is a right angle;

$\therefore$ $\quad \angle DCE + \angle CED = \text{A right angle};$

$\therefore$ $\quad \angle ACE = \angle DCE + \angle CED.$

From each of these equals take the angle *DCE*;

$\therefore$ $\quad \angle ACD = \angle CED.$

(*ii*) It is required to prove that

$$\angle BCD = \text{The angle in the alternate segment } CFD.$$

Construction. Join *CF*, *FD*.

Proof. Because *CEDF* is a quadrilateral inscribed in a circle,

$\therefore$ $\quad \angle CFD = \text{The supplement of } \angle CED.$

Also; because *ACB* is a straight line,

$\therefore$ $\quad \angle BCD = \text{The supplement of } \angle ACD.$

But $\quad \angle ACD = \angle CED$ *(proved);*

$\therefore$ $\quad \angle BCD = \angle CFD.$

NOTE: The converse of Theorem 65 is as follows:
If through an extremity of a chord of a circle a straight line is drawn, making with the chord an angle equal to the angle in the alternate segment, the straight line touches the circle.

Let CD be the chord and CA a straight line through C, making $\angle ACD$ equal to $\angle CXD$ in the alternate segment.

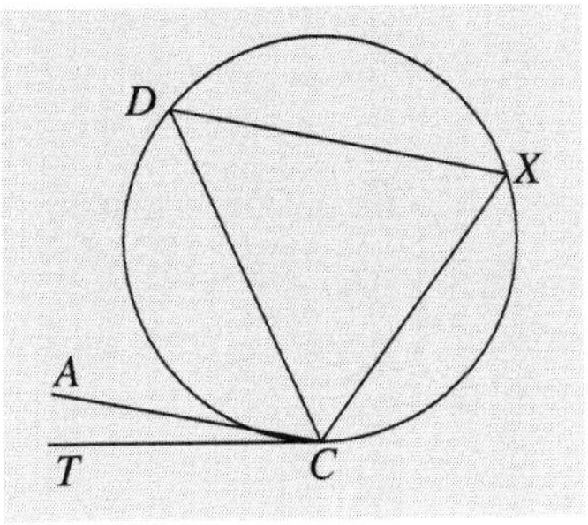

It is required to prove that CA touches the circle.

Proof. Suppose CT to be the tangent at C, T and A being on the same side of CD.

Then, because CT is a tangent and CD a chord,

$\therefore$ $\angle TCD = \angle CXD$, in the alternate segment.

But $\angle ACD = \angle CXD$ *(given)*;

$\therefore$ $\angle ACD = \angle TCD$;

$\therefore$ CA coincides with CT, which is the tangent at C;

$\therefore$ CA touches the circle.

EXERCISE XLVIII (Theorem 65)

1. A tangent to a circle is drawn parallel to a chord; show that the point of contact bisects the arc cut off by the chord.
2. Draw a tangent at a given point on a circular arc of large radius without finding the centre, showing your construction.
3. AB is an arc of a circle and AT is the tangent at A. Prove that the bisector of the angle BAT bisects the arc AB.
4. If AB, AC are tangents at B, C to a circle, and if D is the middle point of the arc BC, prove that D is the centre of the circle inscribed in the triangle ABC.
5. In the side AB of a triangle ABC a point D is taken so that the angle CDB is equal to the angle C. Show that CB touches the circle ACD.
6. P is any point on an arc AB of a circle, and the tangent at P meets the chord AB produced, at C. Prove that the angle PCA is equal to the difference between the angles PAB and PBA.

7. In a right-angled triangle *ABC*, of which *C* is the right angle, *CD* is drawn perpendicular to *AB*, meeting it in *D*; prove that *CB* touches the circle circumscribing the triangle *CDA*.

8. Three points *A*, *B*, *C* are taken on a circle, and a straight line parallel to the tangent at *A* intersects *AB*, *AC* in *D*, *E*; prove that a circle can be described through the four points *B*, *C*, *D*, *E*.

9. Two circles touch at *A* and *PAQ*, *XAY* are straight lines through *A*, cutting the circles at *P*, *Q*, *X*, *Y*. Show that *PX*, *QY* are parallel. [Draw the common tangent at *A*.]

10. If two circles touch, any straight line through the point of contact cuts off similar segments.

11. *ABCD* is a parallelogram whose diagonals meet in *O*; prove that the circles described about the triangles *AOB*, *COD* touch each other.

12. *AB* is a chord, and *CAD* a tangent to a circle; take *P* and *Q*, two points on the circumference, such that *PA* and *QA* bisect the angles *CAB*, *DAB*. Show that *PQ* is the diameter perpendicular to *AB*.

13. Two circles cut at *A*, *B* and a straight line *PAQ* cuts the circles at *P*, *Q*. If the tangents at *P*, *Q* meet in *T*, prove that the points *P*, *B*, *Q*, *T* are concyclic. [Join *AB*]

14. Two circles intersect at *A* and *PAQ*, *XAY* are straight lines through *A*, cutting the circles at *P*, *Q*, *X*, *Y*. Show that *PX*, *QY* are inclined to each other at the same angle as the tangents at *A*.

15. From a fixed point *A* straight lines *ABC*, *AEF* are drawn to meet two fixed lines in *B*, *C* and *E*, *F*. Prove that the circles *ABE*, *ACF* cut at a constant angle.

16. A straight line touches a circle at the point *P* and *QR* is a chord of a second circle, parallel to this tangent. *PQ*, *PR* cut the first circle in *S*, *T* and the second circle in *U*, *V*; prove that *ST* and *UV* are parallel to each other.

17. The bisector of the angle *C* of the triangle *ABC* meets *AB* at *D* and the perpendicular bisector of *CD* meets *AB* produced at *E*. Prove that *CE* touches the circum-circle of the triangle.

18. In the triangle *ABC*, the angles *B*, *C* are each double the angle *A*; prove that a circle, drawn through *A* to touch *BC* at *B*, intersects *AC* in a point *D* such that *AD*, *BD*, *BC* are all equal.

19. *ABCDE* is a regular pentagon, *O* the centre of the circum-scribing circle, *G* the point of intersection of *AC* and *BE*; prove that the circle *AOG* touches *AB*.

20. Two circles touch internally at *X* and a straight line cuts them at *A*, *B*, *C*, *D* (in order from left to right). Prove that *AB*, *CD* subtend equal angles at *X*. [Draw the common tangent at *X*]

21. Two circles touch externally at X and a straight line cuts them in A, B, C, D (in order from left to right). Prove that AD, BC subtend supplementary angles at X. [Draw the common tangent at X]

22. AB is the hypotenuse of a right-angled triangle ABC; the lines bisecting the internal and external angles at C cut AB and AB produced in D and E; if M is the middle point of AB, show that MC touches the circle that passes through C, D, E.

23. The tangent at A to the circumscribing circle of the triangle ABC meets BC produced in D, and the internal and external bisectors of the angle BAC meet BC in E and F respectively; show that EF is bisected in D. [Prove $\angle AED = \angle EAD$]

24. A triangle circumscribes a circle, and from each point of contact a straight line is drawn perpendicular to the line joining the other two; prove that the straight lines joining the feet of these perpendiculars are parallel to the sides of the original triangle.

CONSTRUCTION 24

Cut off from a given circle a segment containing an angle equal to a given angle.

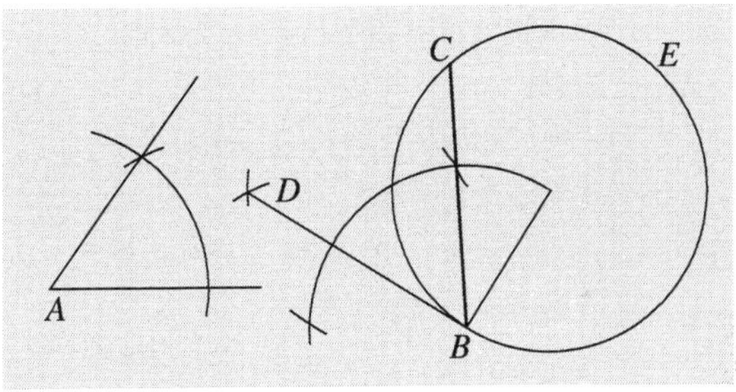

Let BCE be the given circle, and A the given angle.

It is required to cut off from the circle BCE a segment containing an angle equal to A.

Construction. Take any point B on the circumference.

Draw BD, the tangent at B.

Draw the chord BC, making the angle DBC equal to A.

Then the alternate segment BEC is the one required.

Proof. Because BD is a tangent and BC is a chord of the circle,

$\therefore \angle DBC$ = The angle in the alternate segment BEC.

But $\angle DBC = \angle A$;

$\therefore$ The segment BEC contains an angle equal to the angle A.

CONSTRUCTION 25

On a given straight line draw a segment of a circle containing an angle equal to a given angle.

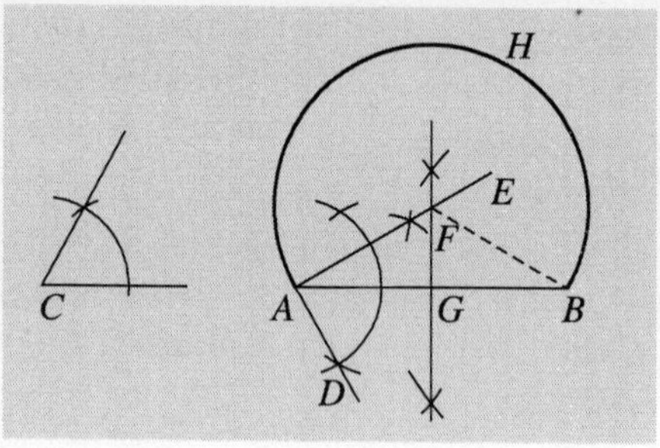

Let AB be the given straight line and C the given angle.

It is required to draw on AB a segment of a circle, containing an angle equal to the angle C.

Construction. At A, make an angle BAD equal to the angle C.

Draw AE perpendicular to AD.

Draw the perpendicular bisector of AB, meeting AE at F.

With centre F and radius FA, draw a circle.

Then the segment of this circle alternate to the angle BAD is the one required.

Proof. Let AB cut its perpendicular bisector at G. Join BF.

In the triangles AGF, BGF,

$$\begin{cases} AG = BG \text{ (construction)}, \\ GF \text{ is common}, \\ \angle AGF = \angle BGF \text{ (construction)}; \end{cases}$$

$\therefore$ $AF = BF$.

$\therefore$ The circle passes through B.

But, by construction, AD is perpendicular to the radius AF;

$\therefore$ AD is the tangent at A;

$\therefore$ $\angle BAD =$ The angle in the alternate segment AHB;

$\therefore$ AHB is the required segment.

EXERCISE XLIX(a) (Constructions 24, 25)

1. Draw a chord of a circle dividing it into two segments, one of which contains an angle equal to twice that contained by the other.
2. Construct a triangle, having given the base, vertical angle and altitude.
3. Construct a triangle on the same base BC and of the same area as a given triangle ABC, and having a vertical angle equal to a given angle D.
4. A, B are given points and CD is a given straight line. Find a point X in CD such that the angle AXB may be equal to a given angle E.

5. Give a construction to draw the tangent to the circum-circle of the triangle *ABC* at *A*, without actually drawing the circle.

6. In a given circle inscribe a triangle *ABC* so that *A* is a given angle, and the sides *AB*, *AC* pass each through a given point.

7. Show how to construct the triangle *ABC*, given the base *BC*, the vertical angle *A* and the point *D* where the bisector of the angle *A* meets *BC*.

8. Construct a square with two adjacent sides passing through two given points, and the intersection of the diagonals at a third given point. Show that there are generally two solutions.

9. *A* and *B* are the extremities of a given arc of a circle; show how to find the position of a point *C* on the arc such that the chords *AC* and *BC* may be together equal to a given straight line. [If *C* is the required point and *AC* is produced to *X* so that $CX = CB$, prove that $\angle AXB = \frac{1}{2}\angle ACB$; find the centre of the arc *AXB*.]

10. Explain how to construct a triangle, given the base, the sum of the sides and the radius of the circum-circle.

11. Explain how to find a point *C* on a given arc *AB* of a circle such that *AC* may exceed *BC* by a given length. [If *C* is the required point and along *CA* a length *CX* is cut off equal to *CB*, prove that $\angle AXB = 90° + \frac{1}{2}C$; find the centre of the arc *AXB*.]

12. Explain how to construct a triangle, given the base, the difference of the sides and the radius of the circum-circle.

13. Draw a triangle *ABC* and, along *BC* produced both ways, cut off *BX* equal to *BA* and *CY* equal to *CA*. Join *AX*, *AY*, and prove that $\angle XAY = 90° + \frac{1}{2}A$.

    Hence deduce a construction for a triangle, having given its perimeter, altitude and vertical angle.

14. Draw a triangle *ABC*. On *BC* and *CA* draw segments of circles containing angles equal to the supplements of the angles *C* and *A* respectively. Let the arcs meet at a point *O* inside the triangle.

    Prove that $\angle OBC = \angle OCA = \angle OAB$.

15. *If O is a point within the triangle ABC such that*

    $$\angle OBC = \angle OCA = \angle OAB,$$

    prove that the angles BOC, COA, AOB are the supplements of the angles C, A, B respectively.

16. *Given three straight lines OA, OB, OC, drawn from a point O. Draw a straight line to cut OA, OB, OC at X, Y, Z respectively such that XY and YZ may be each of given length.*

    [Draw a straight line *X'Y'Z'* such that *X'Y'*, *Y'Z'* are of the given lengths. Find a point *O'* such that $\angle X'O'Y' = \angle AOB$, $\angle Y'O'Z' = \angle BOC$ etc.]

EXERCISE XLIX(b)

Numerical

1. Draw a circle of radius 1.5 in. From it cut off a segment containing an angle of 50°. Measure the chord of the segment.
2. Draw a circle of radius 1 in. From it cut off a segment containing an angle of 150°. Measure the chord of the segment.
3. Draw a square *ABCD* in which *AB* = 2 in. Find points *X*, *Y* in *CD* produced such that ∠ *AXB* = ∠ *AYB* = 30°. Measure *XY*.
4. Draw an equilateral triangle *ABC* in which *AB* = 3 in. Bisect *AB* at *D*. Find points in *BC* or *BC* produced at which *AD* subtends an angle of 25°. Measure the distance between the points.
5. Draw a triangle *ABC* in which the base *BC* = 3 in., altitude = 3.2 in., ∠ *A* = 50°. Measure the greater of the constructed sides.
6. Draw a triangle *ABC* of area equal to 3 square inches, in which *BC* = 2 in. and *A* = 30°.

   In the following examples, R denotes the radius of the circum-circle of the triangle.
7. Construct the triangle *ABC*, with the following data, and in each case measure the constructed sides:
   (*i*) $b + c = 4.2$ in., $a = 1.4$ in., $R = 1.5$ in.
   (*ii*) $c - b = 2$ in., $a = 2.5$ in., $R = 1.6$ in.
   (*iii*) $a = 2$ in., $R = 1.33$ in., $C - B = 29°$.
8. Draw a quadrilateral *ABCD* in which *AB* = 1.3 in., *BC* = 1 in., *AD* = 2 in., ∠ *A* = 60°, ∠ *C* = 120°. Explain why a circle can be described about *ABCD*; draw this circle and measure the radius.

XXVIII. INSCRIBED AND CIRCUMSCRIBED FIGURES

CONSTRUCTION 26 (Euclid IV. 2)

In a given circle, inscribe a triangle equiangular to a given triangle.

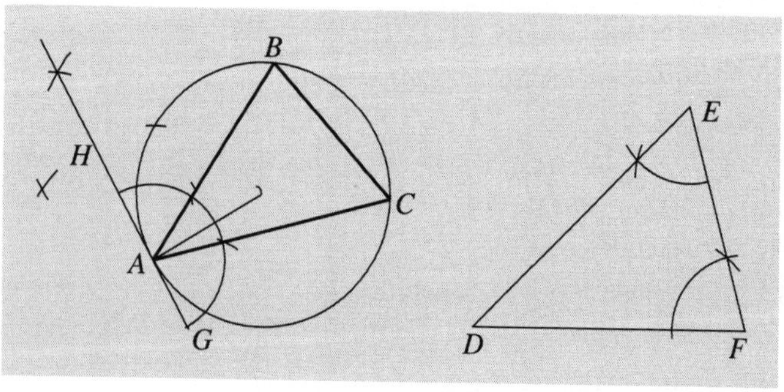

Let ABC be the given circle and DEF the given triangle.

It is required to inscribe in the circle ABC a triangle equiangular to the triangle DEF.

Construction. Take any point A on the circle.

At A draw the tangent GAH.

Draw chords AB, AC, making the angles HAB, GAC equal to the angles F, E respectively. Join B, C.

Then ABC is the required triangle.

Proof. Because HAG is a tangent and AC a chord of the circle

∴ $\angle GAC = \angle B$ in the alternate segment.

But $\angle GAC = \angle E$ *(construction);*

∴ $\angle B = \angle E$.

Similarly, $\angle C = \angle F$;

∴ The remaining $\angle A$ = The remaining $\angle D$.

Construction 27 (Euclid IV. 3)

About a given circle to describe a triangle equiangular to a given triangle.

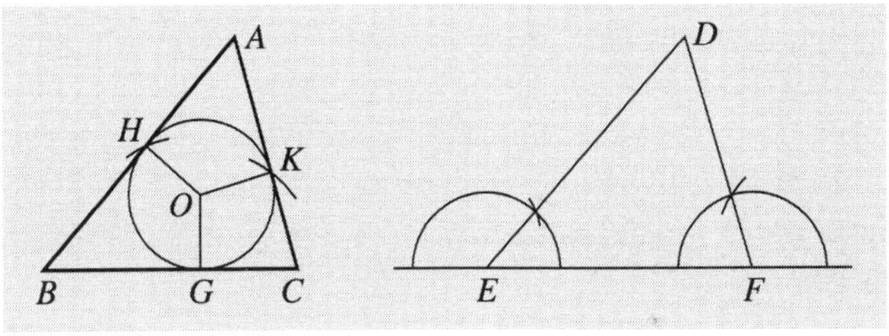

Let GHK be the given circle and DEF the given triangle.

It is required to describe a triangle about the circle GHK, equiangular to the triangle DEF.

Construction. Let O be the centre of the circle.

Take any point G on the circumference.

Join OG, and draw radii OH, OK, making the angles GOH, GOK equal to the supplements of the angles DEF, DFE respectively.

Draw the tangents BGC, CKA, AHB at G, K, H.

Then ABC is the required triangle.

Proof. Because GB is a tangent and OG a radius,

∴ $\angle OGB$ is a right angle.

Similarly, $\angle OHB$ is a right angle.

Hence, the quadrilateral $OGBH$ is cyclic;

∴ $\angle GOH$ = The supplement of $\angle B$.

But $\angle GOH$ = The supplement of $\angle DEF$ (*construction*);

∴ $\angle B = \angle DEF$.

Similarly, $\angle C = \angle DFE$;

∴ the remaining $\angle A$ = the remaining $\angle D$.

EXERCISE L

Inscribed and Circumscribed Figures

1. Draw a circle of radius 1 in., and inscribe an equilateral triangle in it. Measure a side of the triangle.

2. Draw a circle of radius 1 in. and, about it, describe an equilateral triangle. Measure a side of the triangle.

3. An equilateral triangle is inscribed in a circle and another equilateral triangle is described about it. Prove that a side of the first triangle is equal to half a side of the second.

4. Draw a circle of 1.5 in. radius. Inscribe in it and describe about it triangles whose angles are 50°, 60°, 70°. Measure the greatest side of each triangle.

5. Draw a circle of radius 1 in. Inscribe in it and describe about it triangles equiangular to a triangle whose sides are 1.3 in., 1.4 in., 1.5 in. Measure the greatest side of each.

6. Explain how to construct a triangle, given two angles (B, C) and the radius (R) of the circum-circle.

Construct the triangle when R = 1.5 in., B = 60°, C = 40°. Measure the longest side.

7. Explain how to construct a triangle, given two angles (B, C) and the radius (r) of the inscribed circle.

Construct the triangle when r = 0.7 in., B = 60°, C = 40°. Measure the longest side.

Explain how to effect the constructions in Exs. 8–12.

8. Inscribe a circle in a given square. (See Fig. *a*)

9. Circumscribe a circle about a given square. (Fig. *a*)

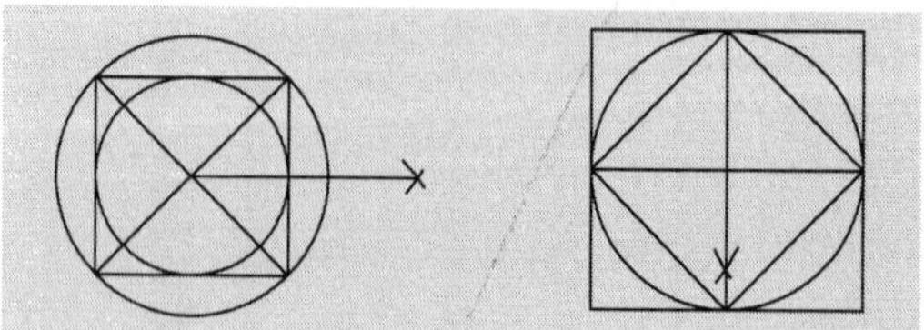

(*a*) (*b*)

10. Inscribe a square in a given circle. (Fig. *b*)

11. Circumscribe a square about a given circle. (Fig. *b*)

12. Inscribe a circle in a given rhombus *ABCD*.

13. Draw a quadrilateral *ABCD* in which *AB* = 2 in., *BC* = 1.7 in., *CD* = 1 in., *DA* = 1.3 in., *AC* = 2 in. Explain why a circle can be inscribed in *ABCD*; draw the circle and measure its radius.

NOTE: *The beginner is advised to learn the facts stated in the enunciations of Theorems 66–68, and to omit the proofs.*

THEOREM 66

If the circumference of a circle is divided into any number of equal parts, the points of division are the vertices of a regular polygon.

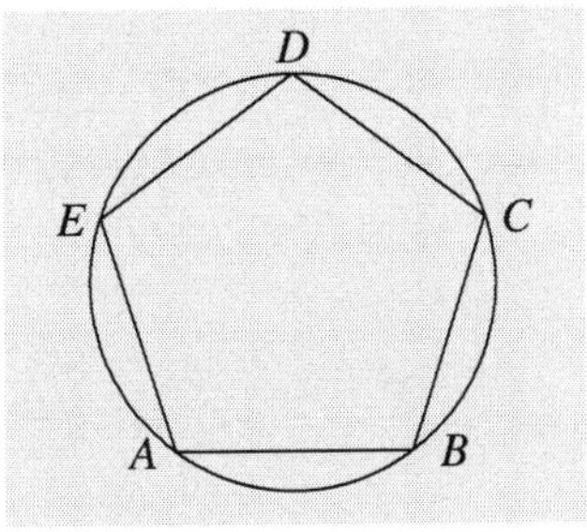

Let the circumference of a circle be divided into any number of equal parts, *five* for example, at *A*, *B*, *C*, *D*, *E*.

Join *AB*, *BC*, *CD*, *DE*, *EA*.

It is required to prove that *ABCDE* is a regular polygon.

Proof. Because the arcs *AB*, *BC*, *CD*, *DE*, *EA* are equal *(given)*.

∴ The chords *AB*, *BC*, *CD*, *DE*, *EA* are equal.

∴ The figure *ABCDE* is equilateral.

Again, the arc *BC* = The arc *EA* *(given)*.

To each add the arc *CDE*;

∴ The arc *BCDE* = The arc *CDEA*;

∴ Angles at the circumference, standing on these arcs, are equal;

∴ $\angle BAE = \angle CBA$.

Similarly, any two consecutive angles of *ABCDE* are equal.

∴ The figure *ABCDE* is equiangular;

and it has been proved to be equilateral;

∴ *ABCDE* is a regular polygon.

THEOREM 67

If the circumference of a circle is divided into any number of equal parts, the tangents at the points of division are the sides of a regular polygon.

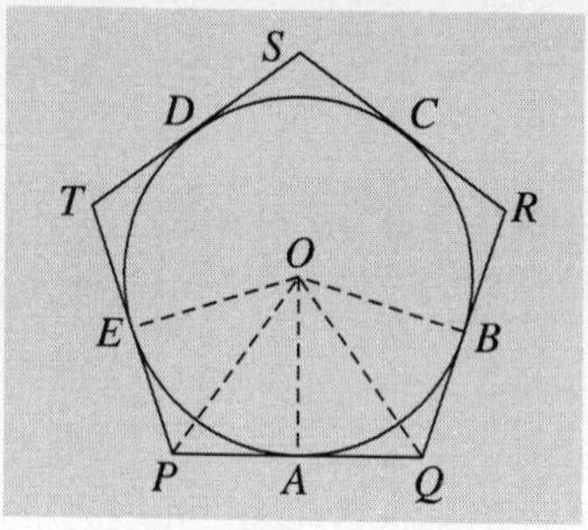

Let the circumference of a circle be divided into any number of equal parts. *five* for example, at A, B, C, D, E.

Let the tangents PAQ, QBR, RCS, SDT, TEP be drawn.

It is required to prove that $PQRST$ is a regular polygon.

Construction. Join the centre O to E, P, A, Q, B.

Proof. Because PA, PE are tangents,

$\therefore$ $PE = PA$,

$\angle POE = \angle POA$,

$\angle EPO = \angle APO$;

$\therefore$ $\angle EOA = 2\angle POA$

and $\angle EPA = 2\angle APO$.

Similarly, $\angle AOB = 2\angle QOA$

and $\angle AQB = 2\angle AQO$.

Now the arc EA = The arc AB (*given*);

$\therefore$ $\angle EOA = \angle AOB$;

$\therefore$ $\angle POA = \angle QOA$;

and, because PAQ is the tangent at A,

$\therefore$ $\angle$s PAO, QAO are right $\angle$s.

Hence, in the triangles POA, QOA,

$\angle POA = \angle QOA$,

$\angle PAO = \angle QAO$, OA is common.

$\therefore$ The triangles are congruent;

$\therefore$ $PA = QA$

and $\angle APO = \angle AQO$.

Hence, from the above, $\angle EPA = \angle AQB$.

Similarly, it may be shown that any two consecutive angles of *PQRST* are equal; that is, *PQRST* is equiangular.

Again, because $PA = QA$,

$\therefore$ $PQ = 2PA$.

Similarly, $TP = 2PE$.

But $PE = PA$;

$\therefore$ $TP = PQ$.

Similarly, it may be shown that any two consecutive sides of *PQRST* are equal; that is, *PQRST* is equilateral;

and it has been proved to be equiangular;

$\therefore$ *ABCDE* is a regular polygon.

THEOREM 68

(*i*) The bisectors of the angles of a regular polygon meet in a point. (*ii*) This point is the centre of the circle circumscribing the polygon. (*iii*) It is also the centre of the circle inscribed in the polygon.

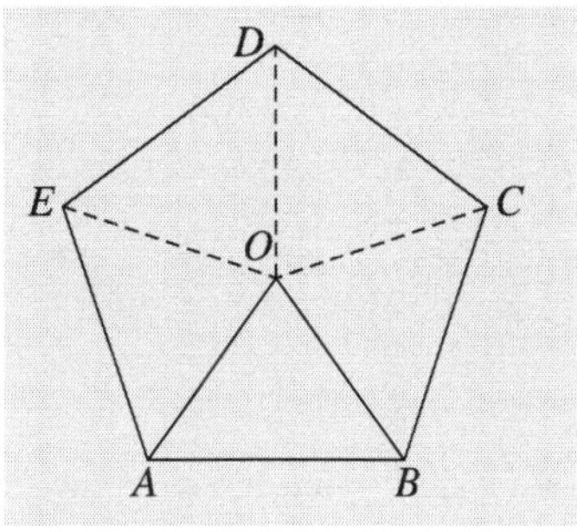

Take a regular polygon with any number of sides, *five* for example. Call it *ABCDE*.

(*i*) It is required to prove that the bisectors of the angles *A*, *B*, *C*, *D*, *E* meet in a point.

Construction. Bisect the angles *EAB*, *ABC* by straight lines meeting at *O*. Join *OC*, *OD*, *OE*.

Proof. In the triangles *EAO*, *BAO*,

$$\begin{cases} EA = BA, \text{ being sides of a regular polygon,} \\ AO \text{ is common,} \\ \angle EAO = \angle BAO \text{ (construction).} \end{cases}$$

$\therefore$ The triangles are congruent;

$\therefore$ $\angle AEO = \angle ABO$.

Now *BO* bisects $\angle ABC$ (*construction*);

$\therefore \qquad \angle ABO = \frac{1}{2} \angle ABC;$

$\therefore \qquad \angle AEO = \frac{1}{2} \angle ABC.$

But $\angle ABC = \angle AED$, being angles of a regular polygon;

$\therefore \qquad \angle AEO = \frac{1}{2} \angle AED;$

$\therefore$ *EO* bisects $\angle AED$.

Similarly, *DO*, *CO* bisect $\angle$s *CDE*, *BCD*.

$\therefore$ The bisectors of $\angle$s *A*, *B*, *C*, *D*, *E* meet at *O*.

(*ii*) It is required to prove that *O* is the centre of the circle circumscribing the polygon.

Because *OA*, *OB* bisect the equal angles *EAB*, *ABC*,

$\therefore \qquad \angle OAB = \angle OBA;$

$\therefore \qquad OA = OB.$

Similarly, $\qquad OB = OC = OD = OE.$

$\therefore$ The circle with centre *O* and radius *OA* passes through *B*, *C*, *D*, *E*, and therefore circumscribes the polygon.

(*iii*) It is required to prove that *O* is the centre of the circle inscribed in the polygon.

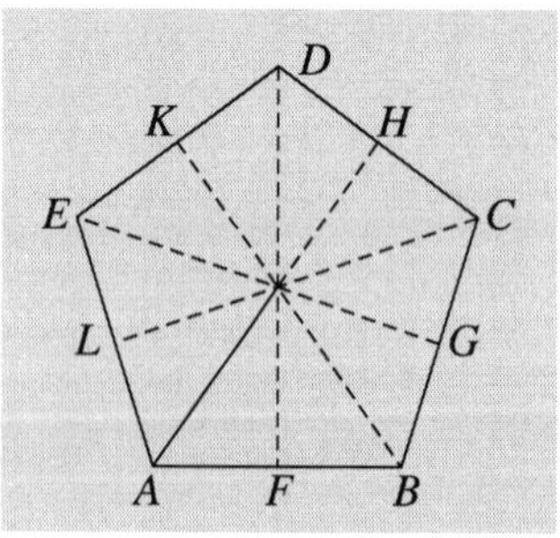

Construction. Draw *OF*, *OG*, *OH*, *OK*, *OL* perpendicular to *AB*, *BC*, *CD*, *DE*, *EA*.

Proof. In the triangles *ALO*, *AFO*,

$$\begin{cases} \angle LAO = \angle FAO, \\ \angle ALO = \angle AFO, \\ OA \text{ is common;} \end{cases}$$

$\therefore$ The triangles are congruent;

$\therefore \qquad OL = OF.$

Similarly, $OF = OG = OH = OK$;

∴ The circle with centre O and radius OF passes through F, G, H, K, L.

Also, because the angles at F, G, H, K, L are right angles.

∴ The circle touches the sides of the polygon and is the circle inscribed in it.

Construction 28

Inscribe a regular hexagon in a given circle.

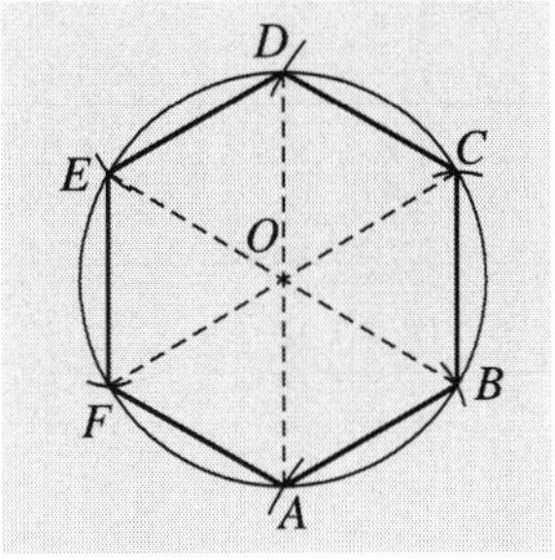

Let O be the centre of the circle in which it is required to inscribe a regular hexagon.

Construction. Take any point A on the circumference.

In the circle place five chords AB, BC, CD, DE, EF, each equal to the radius.

Join AF.

Then $ABCDEF$ is a regular hexagon.

Proof. Join O to A, B, C, D, E, F.

Because $OA = OB = AB$ *(construction)*,

∴ OAB is an equilateral triangle;

∴ $$\angle AOB = 60°.$$

Similarly, ∠s BOC, COD, DOE, EOF each = 60°.

But the sum of the angles at O is 360°;

∴ $$\angle AOF = 60°.$$

Now, the angles of the triangle OAF are together equal to 180°;

∴ $$\angle OAF + \angle OFA = 120°.$$

But $$OA = OF;$$

∴ $$\angle OAF = \angle OFA = 60°;$$

∴ The triangle OAF is equilateral;

∴ $$AF = \text{The radius } OA.$$

∴ The figure $ABCDEF$ is equilateral,

and, since each of its angles is 120°, it is also equiangular;

∴ $ABCDEF$ is a regular hexagon.

Constructions with the Protractor

EXAMPLE 1. *Using the protractor, explain how to inscribe, in a given circle, a regular polygon with a given number of sides, say with five sides.*

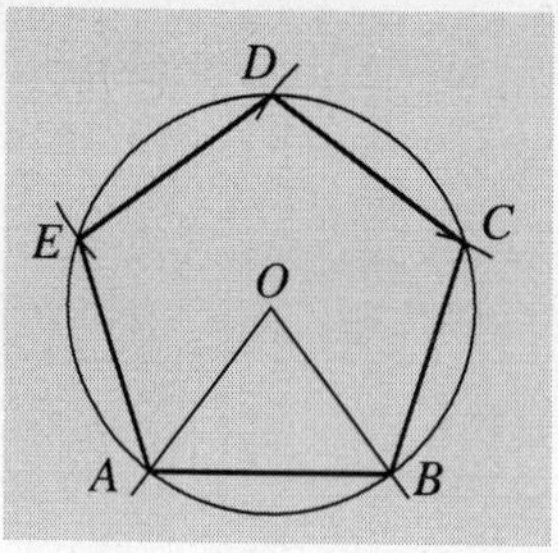

Let O be the centre of the circle. Draw two radii OA, OB, making $\angle AOB = \frac{1}{5}$ of $360°$.

Join AB. In the circle, place chords BC, CD, DE, each equal to AB. Join EA. Then $ABCDE$ is the required polygon.

Supply proof by Theorem 66.

EXAMPLE 2. *Using the protractor, explain how to describe about a given circle a regular polygon with a given number of sides.*

By Ex. 1, find the vertices A, B, C, etc., of an inscribed polygon, with the given number of sides. Draw tangents at A, B, C, etc. These are the sides of the required polygon.

EXAMPLE 3. *Using the protractor, explain how to describe, on a given straight line AB, a regular polygon with any number of sides, say one with* seven *sides.*

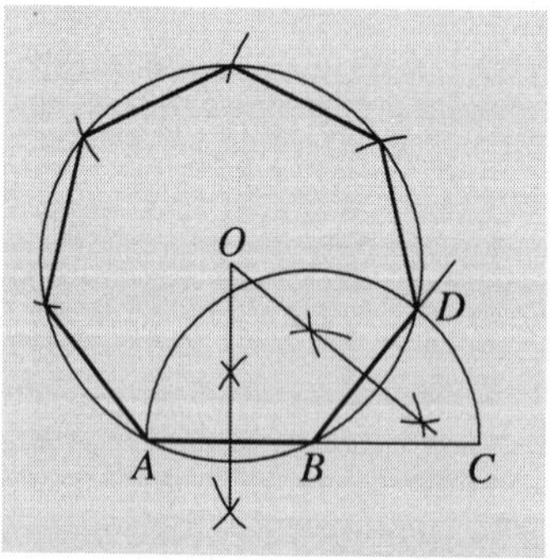

With centre B, and radius BA, draw a semi-circle ADC. Using the protractor, make an angle CBD* equal to $\frac{2}{7}$ of 2 right angles.

---

* It is advisable to check, and if necessary, to correct, the angle 'by completing the circle on AC as diameter,' and 'stepping out' the arc CD round the circumference.

Let BD cut the arc of the semi-circle in D.

Then BD is a side of the required heptagon, and if a circle is drawn through A, B, D, the heptagon can be completed by placing in it chords equal to AB.

Proof. The angle $CBD = \frac{2}{7}$ of 2 right angles

$= \frac{1}{7}$ of 4 right angles

= Exterior angle of a regular heptagon;

$\therefore$ BD is a side of the heptagon, and ABD the circumscribing circle.

EXERCISE LI

Regular Figures

1. Draw a circle of radius 1 in. Describe about it a regular hexagon. Measure a side.
2. Describe a regular hexagon on a given straight line AB.
3. An equilateral triangle and a regular hexagon are described about a circle. Prove that a side of the triangle is equal to three times a side of the hexagon.
4. Inscribe a regular octagon in a given circle. [Draw two diameters AOB, COD, at right angles; bisect the arcs AC, CB, etc.]
5. Complete and prove the following construction for drawing a regular octagon on a given straight line AB:

   Draw CX, bisecting AB at right angles. Along CX mark off $CD = CA$, $DO = DA$. Then O is the centre of the circle circumscribing the octagon, etc.
6. Draw a circle of radius 1 in. Use the protractor to inscribe in the circle a regular pentagon. Measure a side.
7. Draw a straight line AB of length 1 in. Use the protractor to describe on AB a regular pentagon. Measure the radius of the circle circumscribing the pentagon.
8. Draw a straight line AB of length 1 in. On AB describe a regular hexagon $ABCDEF$. Make a triangle DXY equal in area to the hexagon, the points X, Y being in AB produced both ways. Measure XY and verify your result by geometrical reasoning.
9. Draw a straight line AB of length 1 in. On AB describe a regular heptagon $ABCDEFG$. Make a triangle EXY equal in area to the heptagon, X and Y being in AB produced. By suitable measurements, find the area of the triangle.

10. If a is a side of a regular polygon, R the radius of the circle circumscribing the polygon, and r the radius of the circle inscribed in it, prove that

$$R = \frac{1}{2} a \operatorname{cosec} \frac{180°}{n} \text{ and } r = \frac{1}{2} a \cot \frac{180°}{n}.$$

[For, in the figure of Theorem 67,

$$AQ = \frac{1}{2} a,\ OQ = R,\ OA = r,\ \angle AOQ = \frac{180°}{n}.]$$

11. Referring to Ex. 10, if $a = 2$, calculate the values of R and r, for (*i*) a regular pentagon; (*ii*) a regular heptagon.

12. If a is a side of a regular n-sided polygon, Δ the area of the polygon, and r the radius of the inscribed circle, prove that

$$\Delta = \frac{1}{2} nar.$$ [*See figure of Theorem 67, p. 192*]

13. If Δ is the area of a regular n-sided polygon inscribed in a circle of radius R, prove that

$$\Delta = \frac{1}{2} nR^2 \sin \frac{360°}{n}.$$

14. If Δ is the area of a regular n-sided polygon circumscribing a circle of radius r, prove that

$$\Delta = nr^2 \tan \frac{180°}{n}.$$

15. Use the above formulae to calculate the area of

(*i*) a regular pentagon inscribed in a circle of radius 2 in.

(*ii*) a regular hexagon described about a circle of radius 2 in.

(*iii*) a regular heptagon of which a side is 2 in.

16. Prove that the sum of the perpendiculars drawn to the sides of a regular polygon from any point within it is constant.

XXIX. TANGENT TO A CURVE*

Continuity. Lines which are not straight are called **curves**.

In figure as given below, the curve APB may be traced by a point which moves from A to B *without leaving the paper.* We therefore say that the curve is continuous from A to B.

Let Q, R be two points on the curve APB, as close together as we can possibly take them. The idea of continuity leads us to assert that however near Q is to R, there are points on the curve between Q and R.

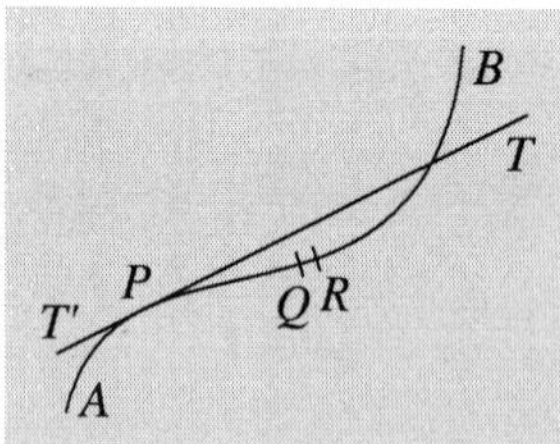

* *Omit on first reading.*

In other words, *there is no such thing as two consecutive points on a continuous curve.*

Tangent to a Curve. The reader will probably agree that the straight line $T'PT$ in figure is a fair representation of the tangent at P.

Now, $T'PT$ cuts the curve in a point distinct from P. It is therefore clear that there are curves for which the tangent cannot be defined in the same way as in the case of the circle.

DEF. Let P be any point on a curve, and let Q be a point which is supposed to move along the curve so as to approach P from either side.

It is supposed that the curve is continuous near P, so that Q may be as near to P as we like.

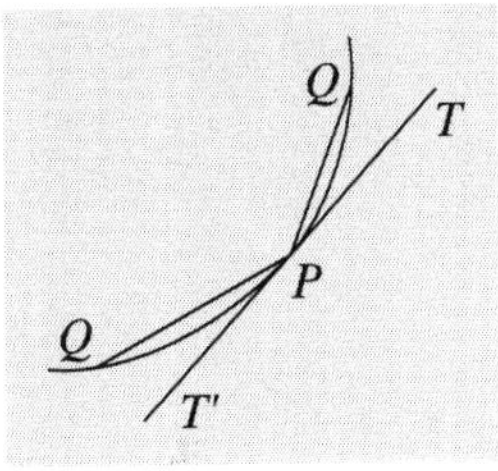

If a straight line $T'PT$ exists such that, as Q, approaches P from either side, the angle which PQ, makes with $T'PT$ becomes and (as Q continues to approach P) remains less than any angle we may choose, however small, then $T'PT$ is called the **tangent** to the curve at P.

This, what is meant by saying that the tangent to a curve at P, is the limiting position of the chord PQ, as Q tends to coincidence with P.

The reader may consider this definition unnecessarily long, but let him be assured that there is no other way of putting the matter without talking nonsense.

He is particularly warned against the definitions given in many of the text-books, of which the following is typical: "A straight line which meets a curve in two points which are indefinitely close, and which ultimately coincide, is called a tangent."

This is sheer nonsense, for so long as the points are distinct, the straight line joining them does not touch the curve; and, when the two points coincide, they become one and cease to define a straight line.

EXAMPLE 1. *Taking the 'limit' definition of a tangent, prove that the tangent to a circle is perpendicular to the diameter through the point of contact.*

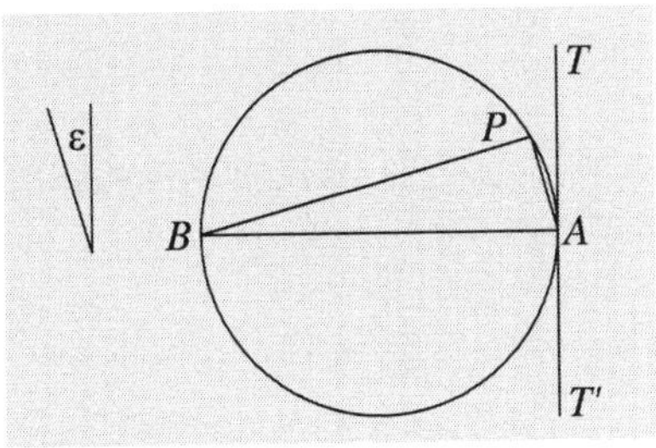

Let A be any point on the circle and AB the diameter through A.

Let $T'AT$ be perpendicular to AB. It is required to prove that $T'AT$ is the tangent at A.

Proof. Choose any angle ε (epsilon), no matter how small.

On either side of *AB*, make an angle *ABP* less than ε, and let *BP* meet the circle again at *P*. Join *AF*. Because *APB* is a semi-circle,

∴ $\angle APB$ is a right angle;

∴ $\angle BAP$ is the complement of $\angle ABP$;

∴ $$\angle PAT = \angle ABP < \varepsilon.$$

If the point *P* is supposed to move along the circumference, so as to approach *A*, the angle *ABP* decreases. Hence the angle which *AP* makes with *T'AT* becomes, and remains, less than the angle ε.

Now, this is true from whichever side *P* approaches *A*;

∴ *T'AT* is a tangent to a circle.

MISCELLANEOUS EXERCISES

Arranged in Sets for Homework or Revision

Paper XIII (to Section XXII)

1. *O* is the orthocentre of the triangle *ABC*, and the parallelogram *BOCD* is completed; show that *AD* is a diameter of the circle circumscribing *ABC*.
2. *XY* is a chord of a circle perpendicular to a diameter *AB*; *YMZ* is any other chord meeting *AB* in *M*. Prove that either *XA* or *XB* bisects the angle *ZXM*. [Join *AY* or *BY*]
3. *A, B, C, D* are four given points; *P, Q* are points on the circles *ADB, ADC* respectively, such that *P, A, Q* are in one straight line. Find the locus of the intersection of *BP, CQ*. Consider also the case in which *B, C, D* are in one straight line. [Consider the angles *PQ*]
4. *P, Q*, are the points of intersection of two circles; a straight line cuts one of the circles at *A* and *C*, and the other at *B* and *D*: prove that the angle *APB*, *CQD* are equal or supplementary.
5. *ABC* is an equilateral triangle, with *AB* = 3 in. Construct an equilateral triangle *PQR*, with *PQ* = 2.5 in., so that *P, Q, R* lie on *BC, CA, AB* respectively. Measure *AP*.
6. A tunnel is constructed in the form of a circular tube, and a level roadway is made by filling up a small segment of the circular section throughout its length. The breadth of the roadway is 24 feet and the height above the roadway to the top of the arch is 18 feet. Calculate the diameter of the tube.

PAPER XIV (to Section XXII)

1. In the circle *ABCD*, draw two chords *AC*, *BD* intersecting at right angles in *E*; join *CD*; draw *AF* perpendicular to *CD*; let *AF* intersect *BD*, or *BD* produced, in *G*. Show that $BE = EG$. [Join *AB*]

2. *ABCD* is a quadrilateral in which the angles *B* and *D* are right angles. Show that the difference between the angles *BAC*, *DAC* is equal to the difference between the angles *BCA*, *DCA*. What is this difference if the diagonals cut at 60°?

3. *AB* is a given chord of a circle whose centre is *O*; *P* is any given point within the circle. Draw through *P* a straight line cutting the circle in *C* and *D*, such that *CD* is bisected by *AB*. Show that the problem admits of two solutions or no solution, according as the sum of the distances of *P* and *O* from *AB* is less or greater than *OP*.

4. Through *A*, one of the points of intersection of two given circles, a straight line *PAQ* is drawn to meet one of the circles again in *P* and the other, in *Q*. If *X* is the middle point of *PQ*, and *C* is the middle point of the line joining the centres of the circle, prove that $CX = CA$. What is the locus of the point *X*?

5. Construct a rhombus, of which two sides lie along two given parallel straight lines, whilst the other two pass each through a given fixed point. Of how many solutions does the problem admit?

   [Note that perpendicular distances between the pairs of parallel sides of a rhombus are equal.]

6. Draw a circle, 3 in. in diameter, and take a point *O*, 2.5 in. from the centre of the circle. Draw a straight line through *O* to meet the circle in *P* and *Q* so that *OQ* is bisected in *P*. Measure *PQ*.

   [If *R* is any point on the circle construct the locus of the middle point of *OR*.]

PAPER XV (to Section XXIII)

1. Two equal circles cut at right angles. Show that the area common to the two circles together with the square on the radius is equal to half the area of either circle.

2. *A* and *B* are the points of intersection of two given circles. Draw through *B* a line *CBD* terminated by the circumferences, such that the area of the triangle *ACD* may be the greatest possible.

   [See Ex. XLII, 26]

3. A chord *AB* of a circle, centre *O*, is bisected at *D* by a chord *EF*, and the tangents at *E* and *F* meet *AB* produced in *G* and *H*. Prove that $AG = BH$. [Show that the triangles *OEG*, *OFH* are congruent.]

4. *ABP*, *ABCD* are two circles, *O* being the centre of the former; *P* is any point on *ABP*; *PA*, *PB* cut the circle *ABCD* again in *C* and *D*. Prove that *OP* is perpendicular to *CD*.
5. Given an arc of a circle of very large radius, whose centre is not available; *P* is a point near the middle point of the arc but not on the arc. Draw a circle, with centre *P*, to touch the given arc.
   [Use Ex. VII, 7 and Theorem 60]
6. Draw two circles of radii 1 in. and 1.5 in., the distance between their centres being 2 in. Find a point *P* such that the lengths of the tangents from *P* to the first and second circles are 2.5 in. and 2 in. respectively. Measure the distance of *P* from the line of centres.

PAPER XVI (to Section XXIV)

1. *BD*, *CE* are the transverse common tangents to two given circles and *AF* is a direct common tangent. Show that the part of *AF*, intercepted between *BD* and *CE*, is equal to *BD*.
2. A quadrilateral inscribed in a circle has its diagonals at right angles, and from their point of intersection a perpendicular is drawn to one of the sides. Show that, when produced, it bisects the opposite side of the quadrilateral.
3. Two equal circles touch one another at *A*; a circle, of double the radius, is drawn having internal contact with one of them at *B* and cutting the other at *P* and *Q*. Prove that the straight line joining *A*, *B* passes through *P* or *Q*.
4. Find the locus of the points of contact of tangents from a given point to a system of concentric circles.
5. Given a straight line *MN* and two points *A*, *B* on the same side of it, find a point *P* on *MN* such that *PA* bisects one of the angles which *PB* makes with *MN*.
6. Draw a straight line $AB = 7$ cm. Draw a circle with centre *A* and radius 3 cm. Draw a circle with centre *B* and radius 2 cm. Draw a straight line cutting *AB* at *X*, the first circle at *P*, *Q*, and the second circle at *R*, *S*, such that $PQ = 3$ cm. and $RS = 2$ cm. Measure *AX*.

PAPER XVII (to Section XXV)

1. A straight line *AB* is trisected in *C*, *D*, and an equilateral triangle *PCD* is described on *CD*. Show that the circle *BCP* touches *AP*.
   [Show that $\angle APD$ is a right angle]
2. Two tangents *AP*, *AQ* are drawn to a circle from a fixed point *A*, and a variable tangent meets *AP*, *AQ* in *B*, *C*. Show that the perimeter of the triangle *ABC* is constant, and *BC* subtends a constant angle at the centre of the circle.

3. H, K are the centres of two of the circles which touch three given straight lines. Prove that the circle on HK as diameter passes through two of the vertices of the triangle formed by the lines.
4. The circle inscribed in the triangle ABC touches BC in D. Show that the circles inscribed in the triangles BAD, CAD touch one another.
5. Given the base and the vertical angle of a triangle, find the loci of (*i*) the in-centre, (*ii*) the e-centres.
6. The base of a triangle is 3 in., the altitude is 2.24 in., the radius of the inscribed circle is 0.8 in. Construct the triangle and measure the other two sides. [Construct the semi-vertical angle.]

Paper XVIII (to Section XXVI)

1. The centre of the inscribed circle of a triangle is within each of the circles on the sides of the triangle as diameters.
2. Prove that, if the chord of half the arc of a circle is greater than the chord of the whole arc, the arc must be greater than two-thirds of the whole circumference.
3. Two circles touch internally at A, and a straight line APQ is drawn cutting the inner circle at P and the outer at Q. If the tangent at P meets the outer circle at H and K, prove that Q is the middle point of the arc HK.
4. Two circles intersect in A and B and a variable point P on one circle is joined to A and B. PA, PB, produced if necessary, meet the second circle in Q and R. Prove that
   (*i*) QR is of constant length;
   (*ii*) the radius of the circle PQR is constant;
   (*iii*) the locus of the centre of this circle is a circle.
5. Through a given point A, draw two straight lines AP, AQ to make equal angles with a given straight line AB, such that the part PQ intercepted on another given straight line may be bisected at a given point S on the line.
   [Suppose the construction effected and describe a circle through A, P, Q.]
6. Draw a triangle SII_1, in which $II_1 = 3.4$ in., $SI_1 = 3.2$ in., $SI = 0.6$ in. Draw a triangle ABC of which S is the circumcentre, I the in-centre and I_1 an e-centre. Measure the longest side of ABC.
   [Prove that the middle point of II_1 lies on the circumcircle, and use Ex. 3 of Paper XVII.]

Paper XIX (to Section XXVII)

1. The circle described, with centre A and radius AB, cuts the circle circumscribing the rectangle $ABCD$ in E. Show that $CE = AD$ and that DE is parallel to AC.

2. If the angle ACB of the triangle ACB is bisected by CE, cutting AB in E, and another point D is taken in AB produced, such that the $\angle ECD = \angle CED$, prove that CD touches the circle ACB.
3. A circle touches one side BC of a triangle, and the other two sides AB, AC produced, the points of contact being D, F, E. If I is the centre of the inscribed circle, prove that
$$\Delta IAE = \Delta IAF = \frac{1}{2}\Delta ABC.$$
4. B is a given point on the bisector of the angle between two given straight lines AX, AY. A variable circle through A and B cuts AX, AY in P, Q respectively. Prove that $AP + AQ$ is constant.
[If BR, BS are perpendicular to AX, AY, prove $PR = QS$.]
5. AB, AC are straight lines, inclined at 75°, drawn from the centre, of a given circle, whose diameter is 3 in. Draw a tangent to the circle, so that the part PQ intercepted between AB and AC is 2.5 in. Measure AP, AQ.
6. OA, OB, OC are three straight lines meeting in a point, so that $AOB = 105°$, $BOC = 120°$, $COA = 135°$. Construct an equilateral triangle PQR, with each side of length 2 in., and with its vertices P, Q, R on OA, OB, OC respectively. Measure OP.

PAPER XX (to Section XXIX)

1. Through the vertex A of a square $ABCD$, a straight line is drawn meeting CB, CD, DB, each being produced at E, F, K respectively. Prove that CK touches the circle through C, E, F.
2. A rhombus is circumscribed to a given rectangle $ABCD$. If P is the intersection of the sides passing through A and B respectively, prove that P lies on a fixed straight line or on a fixed circle passing through A, B, and the intersection of the diagonals of the rectangle.
3. Points P, Q, R are taken on three circles, BOC, COA, AOB respectively, so that PBR, PCQ are straight lines. Show that QAR is a straight line, and that the tangents to the circles at P, Q, R form a triangle whose angles are constant for all positions of P.
4. In a given circle a quadrilateral $ABCD$ is inscribed such that its angles A and D are each half a right angle. The perpendicular from B to AD is produced to meet the circumference in E. Show that CE is the diameter of the circle parallel to the tangent at A.
5. A hexagon is described about a given circle and an equilateral triangle is inscribed in the circle. Prove that the area of the hexagon is eight-thirds of the area of the triangle.
6. Construct an equilateral triangle, given its centre and one point on each of its sides.

UNIT 4

Areas of Rectangles

XXX. RECTANGLES CONTAINED BY SEGMENTS OF LINES

DEF. A **rectangle is a parallelogram**, of which one angle is a right angle.

DEF. A **square is a rectangle**, of which two adjacent sides are equal.

DEF. If *AB*, *CD* are two straight lines, any rectangle, which has one side equal to *AB* and an adjacent side equal to *CD*, is called the **rectangle contained** by *AB* and *CD*.

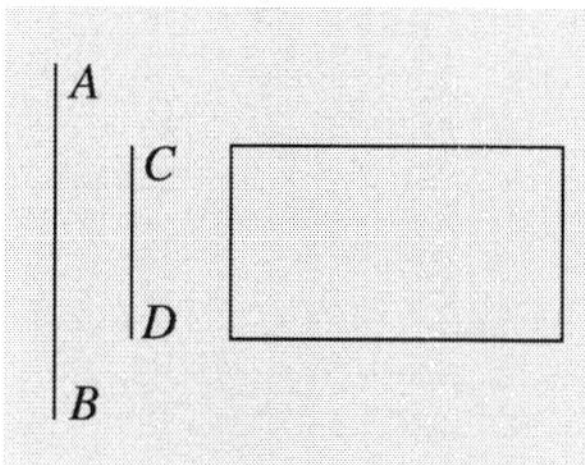

Observe that all such rectangles are congruent and therefore equal in area. "The rectangle contained by *AB* and *CD*" is written shortly thus,—The rect. *AB*, *CD*.

In working examples the following notation is commonly used,— "The rectangle contained by *AB* and *CD*" is denoted by $AB \cdot CD$.

Here the dot is the multiplication sign of algebra, *AB* stands for the number of units of length in *AB* and *CD* has a similar meaning.

In the bookwork, "the square on *AB*" is written thus,—the sq. on *AB* : in examples, it is denoted by AB^2.

DEF. If *M*, *N* are the feet of the perpendiculars from any two points *A*, *B* on a straight line *CD*, then *MN* is called the projection of the straight line *AB* on *CD*.

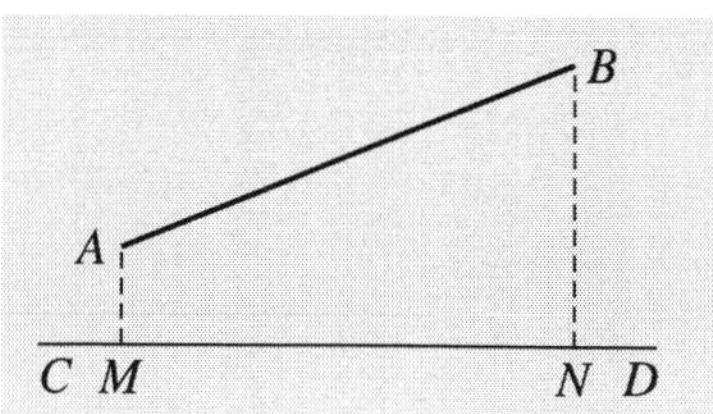

THEOREM 69 (Euclid II. 1)

If there are two straight lines, one of which is divided into any number of parts, the rectangle contained by the two straight lines is equal to the sum of the rectangles contained by the undivided line and the several parts of the divided line.

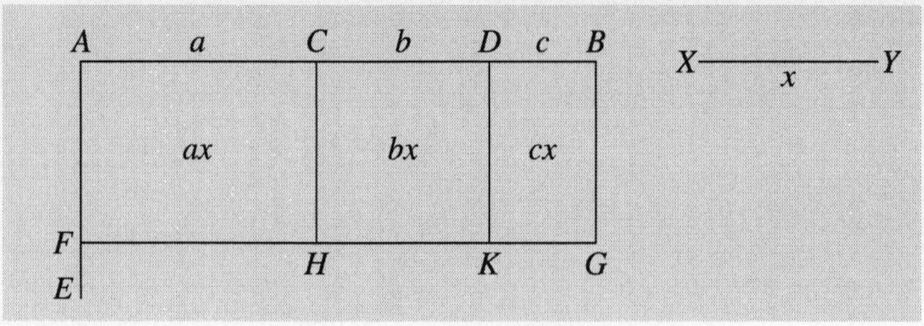

Let *AB*, *XY* be straight lines, and let *AB* be divided into any number of parts, three for example, at *C*, *D*. It is required to prove that

rect. *AB*, *XY* = rect. *AC*, *XY* + rect. *CD*, *XY* + rect. *DB*, *XY*.

Construction. Draw *AE* perpendicular to *AB*. Along *AE* cut off *AF* equal to *XY*. Complete the rectangle *ABGF*. Draw *CH*, *DK* parallel to *AF*, meeting *FG* in *H*, *K*.

Proof. By construction, the figures *AG*, *AH*, *CK*, *DG* are rectangles,

and $AF = CH = DK = XY$;

∴ fig. *AG* = rect. *AB*, *XY*, fig. *AH* = rect *AC*, *XY*,

fig. *CK* = rect. *CD*, *XY*, fig. *DG* = rect. *DB*, *XY*.

But fig. *AG* = fig. *AH* + fig. *CK* + fig. *DG*;

∴ rect. *AB*, *XY* = rect. *AC*, *XY* + rect. *CD*, *XY* + rect *DB*, *XY*.

Note: Let a, b, c, x stand for the number of units of length in *AC*, *CD*, *DB*, *XY* respectively. Then $(a + b + c)x$ is the number of units of area in *AG*. Also ax, bx, cx are the numbers of units of area in *AH*, *CK*, *DG* respectively. Hence the above geometrical theorem corresponds to the algebraical identity

$$(a+b+c)x = ax+bx+cx.$$

Theorem 70 (Euclid II. 2)

If a straight line is divided into any two parts, the square on the whole line is equal to the sum of the rectangles contained by the whole line and each of the parts.

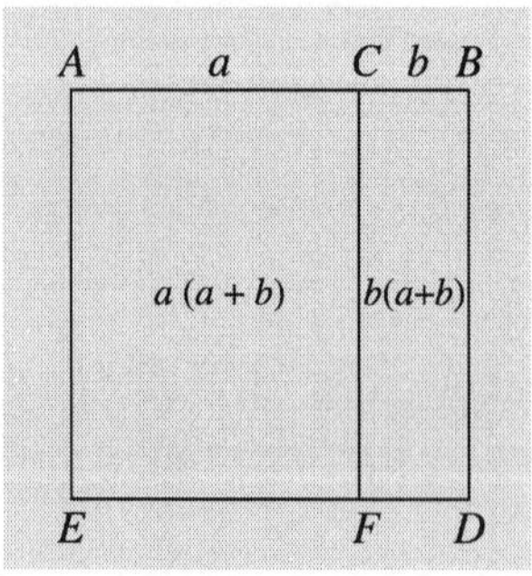

Let the straight line AB be divided into any two parts at C. It is required to prove that

$$\text{sq. on } AB = \text{rect. } AB, AC + \text{rect. } AB, CB.$$

Construction. Describe the square $ABDE$. Draw CF parallel to AE, meeting DE at F.

Proof. By construction, the figures AF, CD are rectangles,

and $AE = BD = AB$;

∴ The fig. AF = rect. AB, AC,

fig. CD = rect. AB, CB.

But fig. AD = fig. AF + fig. CD;

∴ sq. on AB = rect. AB, AC + rect. AB, CB.

NOTE: Let a, b stand for the number of units of length in AC, CB respectively.
Then fig. AD contains $(a + b)^2$ units of area.
fig. AF .... $a(a + b)$ ....
fig. CD .... $b(a + b)$ ....
Hence the above geometrical theorem corresponds to the algebraical identity

$$(a + b)^2 = a(a + b) + b(a + b).$$

THEOREM 71 (Euclid II. 3)

If a straight line is divided into any two parts, the rectangle contained by the whole line and one part is equal to the square on that part together with the rectangle contained by the two parts.

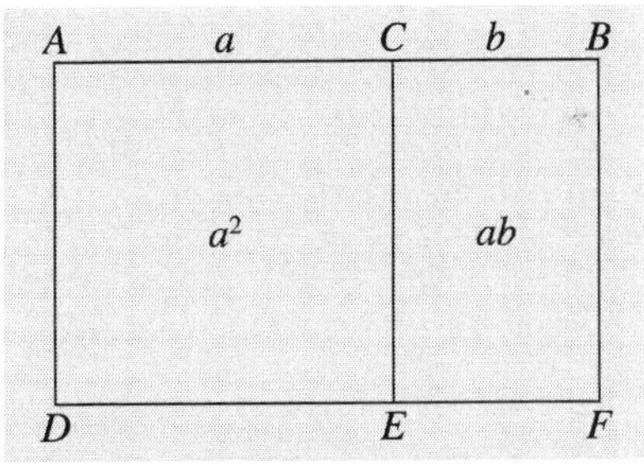

Let the straight line AB be divided into any two parts at C. It is required to prove that

$$\text{rect. } AB, AC = \text{sq. on } AC + \text{rect. } AC, CB.$$

Construction. Describe the square $CADE$.

Draw BF parallel to CE to meet DE produced at F.

Proof. By construction, the figures AF, CF are rectangles,

and $CE = AD = AC$;

∴ The fig. AF = rect. AB, AC,

The fig. CF = rect. AC, CB.

But fig. AF = fig. AE + fig. CF;

∴ rect. AB, AC = sq. on AC + rect. AC, CB.

> **NOTE:** Let a, b stand for the number of units of length in AC, CB respectively.
> Then fig. AF contains $a(a + b)$ units of area,
> fig. AE .... a^2 ....
> fig. CF .... ab ....
> Hence, the above geometrical theorem corresponds to the algebraical identity
> $$a(a + b) = a^2 + ab.$$

THEOREM 72 (Euclid II. 4)

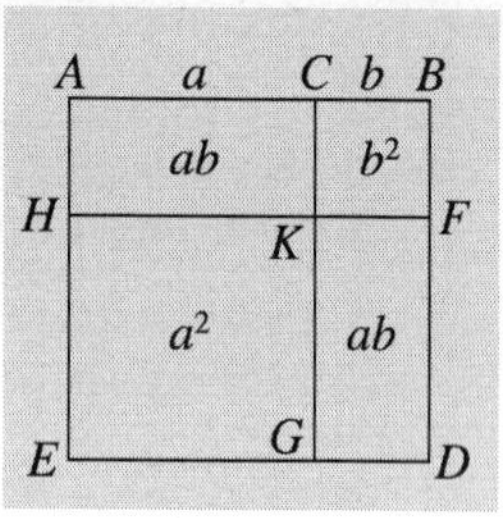

If a straight line is divided into any two parts, the square on the whole line is equal to the sum of the squares on the two parts together with twice the rectangle contained by the parts.

Let the straight line AB be divided into any two parts at C. It is required to prove that

sq. on AB = sq. on AC + sq. on CB + 2 rect. AC, CB.

Construction. Describe the square $ABDE$. Draw CG parallel to AE, meeting DE at G. Along BD, cut off BF equal to CB, Draw FK, parallel to AB, meeting AE at H and CG at K.

Proof. By construction, all the quadrilaterals in the figure are rectangles, and $BF = BC$;

∴ The fig. CF is the sq. on CB.

Also, because $KG = FD = AC = HK$;

∴ The fig. HG is the sq. on HK= the sq. on AC;

and, because $FD = AC$ and $KF = CB = CK$,

∴ fig. AK = fig. KD = rect. AC, CB.

Now, fig. AD= fig. HG + fig. CF + fig. AK + fig. KD;

∴ sq. on AB = sq. on AC + sq. on CB + 2 rect. AC, CB.

> **NOTE:** Let a, b stand for the number of units of length in AC, BC respectively.
> Then fig. AD contains $(a + b)^2$ units of area,
> fig. HG .... a^2 ....
> fig. CF .... b^2 ....
> figs. AK, KD each contain ab ....
> Hence, the above geometrical theorem corresponds to the algebraical identity
> $$(a + b)^2 = a^2 + 2ab + b^2.$$

THEOREM 73 (Euclid II. 5)

If a straight line is divided equally, and also unequally, the rectangle contained by the unequal parts, together with the square on the line between the points of section, is equal to the square on half the line.

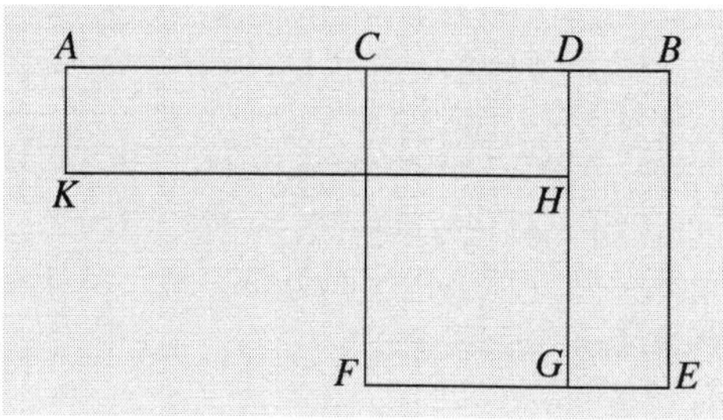

Let the straight line AB be divided equally at C and unequally at D.

It is required to prove that

$$\text{rect. } AD, DB + \text{sq. on } CD = \text{sq. on } CB.$$

Construction. Describe the square $CBEF$. Draw DG parallel to CF, meeting EF at G. Along DG, cut off DH equal to DB. Complete the rectangle $ADHK$.

Proof. By construction, all the quadrilaterals in the figure are rectangles, and $DH = DB$;

$\therefore$ fig. AH = rect. AD, DB.

Also, because $DG = BE = CB$;

$\therefore$ $DG - DH = CB - DB$, that is, $HG = CD$;

$\therefore$ $HG = CD = FG$;

$\therefore$ The fig. HF is the sq. on FG = the sq. on CD.

Again, $AK = DH = DB$

and $AC = CB = BE$;

$\therefore$ The rect. KC = the rect. DE.

Now, fig. AH + fig. HF = fig. KC + fig. CH + fig. HF

= fig. DE + fig. CH + fig. HF

= fig. CE;

$\therefore$ rect AD, DB + sq. on CD = sq. on CB.

NOTE ON THEOREM 73:

Let $AC = CB = a$ units of length,

$CD = b$ ....

$\therefore$ $AD = (a + b)$ ....

$DB = (a - b)$ ....

$\therefore$ the rect. $AH = (a + b)(a - b)$ units of area,

the sq. $HF = b^2$ ....

the sq. $CE = a^2$ ....

(Contd.)

Now, fig. AH + fig. HF = fig. CE.

Hence, Theorem 73 corresponds to the algebraical identity

$$(a + b)(a - b) + b^2 = a^2,$$

which is the same as

$$(a + b)(a - b) = a^2 - b^2.$$

Thus, Theorem 73 may also be enunciated as follows:

The rectangle contained by the sum and difference of two straight lines is **equal** to the difference between the squares on the lines.

Observe that Theorems 72–75 may be deduced from Theorems 69, 70 and 71, without any construction whatever.

EXAMPLE. *Prove Theorem 73 by means of Theorems 69, 70 and 71.*

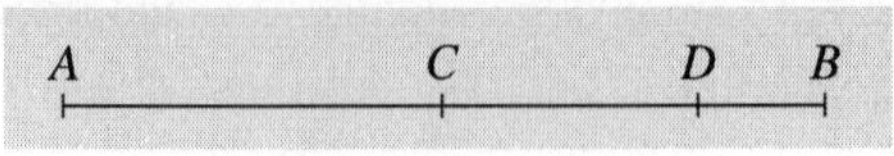

If AB is divided equally at C and unequally at D, it is required to prove that

$$AD \cdot DB + CD^2 = CB^2.$$

Because AD is divided into two parts at C and DB is another straight line,

$$\therefore \quad AD \cdot DB = AC \cdot DB + CD \cdot DB \ (Th.\ 69);$$

$$\therefore \quad AD \cdot DB + CD^2 = AC \cdot DB + CD \cdot DB + CD^2$$
$$= AC \cdot DB + CD \cdot CB \ (Th.\ 71)$$
$$= CB \cdot DB + CB \cdot CD$$
$$= CB^2 \ (Th.\ 70).$$

THEOREM 74 (Euclid II. 6)

If a straight line is bisected and produced to any point, the rectangle contained by the whole line thus produced and the part produced, together with the square on half the given line, is equal to the square on the line made up of the half and the part produced.

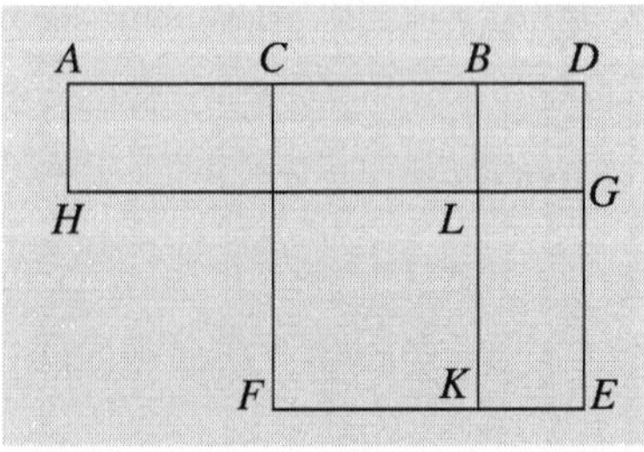

Let the straight line AB be bisected at C and produced to D. It is required to prove that

rect. AD, BD + sq. on CB = sq. on CD.

Construction. Draw the square $CDEF$. Along DE, cut off DG equal to BD. Complete the rectangle $DGHA$. Draw BLK parallel to DE, cutting EF at K and GH at L.

Proof. By construction, all the quadrilaterals in the figure are rectangles, and $DG = BD$;

∴ fig. AG = rect. AD, BD.

Also $DE - DG = CD - BD$, that is, $GE = CB$;

∴ $LK = GE = CB = FK$;

∴ the fig. FL is the sq. on FK= the sq. on CB.

Again, $AH = DG = BD = LG$ and $AC = CB = GE$;

∴ rect. HC = rect. KG.

Now, fig. AG + fig. FL = fig. HC + fig. CG + fig. FL

= fig. KG + fig. CG + fig. FL

= fig. CE;

∴ rect. AD, BD + sq. on CB = sq. on CD.

NOTE: Theorem 74 (as well as 73) corresponds to the algebraical identity $(a + b)(a - b) = a^2 - b^2$. To show this, let $CD = a$ units of length, $AC = b$ units of length.

THEOREM 75 (Euclid II. 7)

If a straight line is divided into any two parts, the sum of the squares on the whole line and on one of the parts is equal to twice the rectangle contained by the whole line and that part, together with the square on the other part.

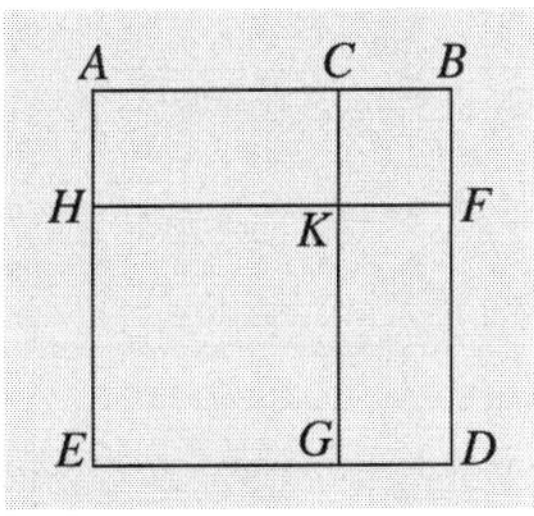

Let the straight line AB be divided into any two parts at C. It is required to prove that

sq. on AB + sq. on CB = 2 rect. AB, CB + sq. on AC.

Construction. Describe the square $ABDE$. Draw CG parallel to AE, meeting DE at G. Along BD, cut off BF equal to CB. Draw FKH parallel to AB, meeting AE at H and CG at K.

Proof. By construction, all the quadrilaterals in the figure are rectangles, and $BF = CB$;

∴ the fig. CF is the sq. on CB.

Also, because $AB = BD$ and $BF = CB$,

$\therefore$ fig. AF = fig. CD = rect. AB, CB.

Again, $KG = FD = AC = HK$;

$\therefore$ the fig. HG is the sq. on HK = the sq. on AC.

Now, fig. AD = fig. KD + fig. AF + fig. HG;

to each add the fig. CF;

$\therefore$ fig. AD + fig. CF = fig. CF + fig., KD + fig. AF + fig. HG

= fig. CD + fig. AF + fig. HG

= 2 fig. AF + fig. HG;

$\therefore$ sq. on AB + sq. on CB = 2 rect. AB, CB + sq. on AC.

NOTE ON THEOREM 75.

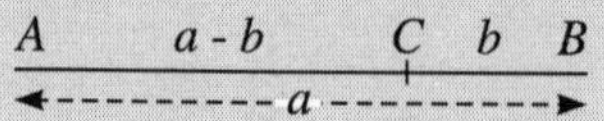

Let $AB = a$ units of length,

$CB = b$ " "

$\therefore$ $AC = (a - b)$ " "

$\therefore$ the sq. on $AB = a^2$ units of area,

the sq. on $CB = b^2$ " "

the rect. AB, CB = ab " "

the sq. on $AC = (a - b)^2$ " "

Now, it has been shown that

sq. on AB + sq. on CB = 2 rect. AB, CB + sq. on AC.

Hence, Theorem 75 corresponds to the algebraical identity

$$a^2 + b^2 = 2ab + (a - b)^2$$

which is the same as

$$(a - b)^2 = a^2 - 2ab + b^2.$$

EXAMPLE. If a straight line AB is bisected at C, and P is any point in AB, or in AB produced, then

$$PA^2 + PB^2 = 2PC^2 + 2AC^2. \text{ (Euclid II. 9, 10)}$$

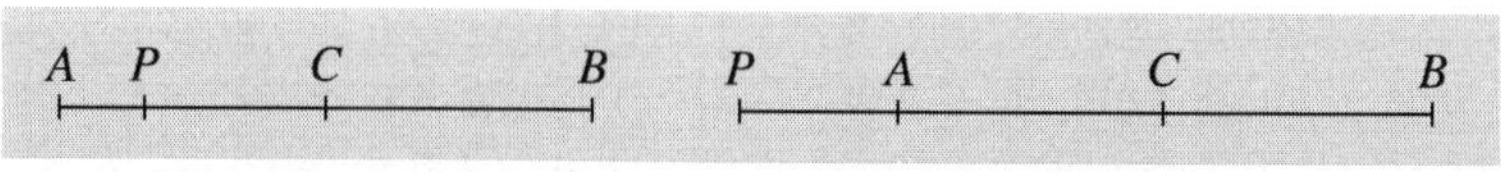

Fig. (a) **Fig. (b)**

Because in Fig. (a), AC is divided at P, and in Fig. (b), PC is divided at A,

$\therefore$ in both cases

$$PA^2 = PC^2 + AC^2 - 2PC \cdot AC \text{ (Th. 75).}$$

Also, because in both figures PB is divided at C,

$$\therefore \quad PB^2 = PC^2 + CB^2 + 2PC \cdot CB \text{ (Th. 72);}$$

$$\therefore \quad PB^2 = PC^2 + AC^2 + 2PC \cdot AC;$$

whence, by addition,

$$PA^2 + PB^2 = 2PC^2 + 2AC^2.$$

EXERCISE LII (Theorems 69–75)

1. Prove Theorems 72, 74, 75 by means of Theorems 69, 70 and 71, employing methods similar to that of the Example on p. 209.
2. Show that Theorem 74 corresponds to the identity

$$(a + b)(a - b) = a^2 - b^2.$$

Draw figures to illustrate the identities in Exs. 3–7. In each case give an explanation.

3. $(2a)^2 = 4a^2$.
4. $a(b - c) = ab - ac$.
5. $(a + b + c)^2 = a^2 + b^2 + c^2 + 2ab + 2ac + 2bc$.
6. $(a + b)(x + y) = ax + ay + bx + by$.
7. $(a - b)(x - y) = ax - ay - bx + by$.
8. *Explain how the accompanying figure illustrates the identity*

$$(a + b)^2 - (a - b)^2 = 4ab.$$

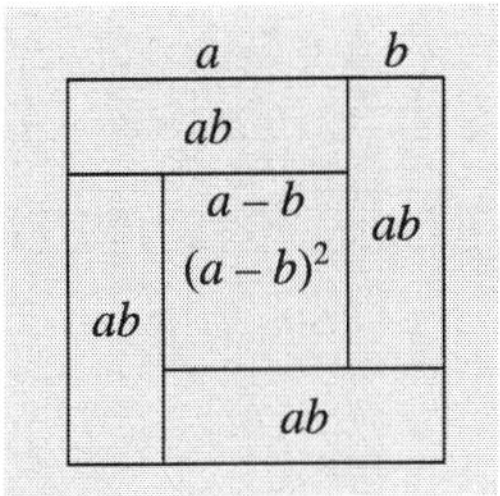

State the corresponding geometrical theorem. (Euclid II. 8).

9. If A, B, C, D are four points, in order, in a straight line, prove that

$$AC \cdot BD = BC \cdot AD + AB \cdot CD.$$

[Let $AB = a$, $BC = b$, $CD = c$; $\therefore AD = a + b + c$, etc.]

10. If A, B, C, D are four points taken in order along a straight line, show that

$$AC^2 + BD^2 = AB^2 + CD^2 + 2AD \cdot BC.$$

11. Four points A, B, C, D are taken in this order on a line; show that

$$AD^2 + BC^2 = AC^2 + BD^2 + 2AB \cdot CD.$$

12. A straight line is divided equally and also unequally, and the rectangle contained by the unequal parts is eight times the square on the line between the points of section. Compare the lengths of the unequal parts.
13. Divide a straight line into two parts such that the square on one part may be four times the square on the other part.

14. Divide a straight line into two parts such that the rectangle contained by the whole line and one part may be equal to twelve times the square on the other part. [Let the length of the line be a units, and let the length of one part be $(a - x)$ units;

$\therefore\ a(a - x) = 12x^2$, etc.]

15. Produce a given straight line AB to a point C such that

$$AC \cdot BC = 12AB^2.$$

16. *C is the middle point of a straight line AB and D is any point in AB or in AB produced. If AD > BD, prove that*

$$AD^2 - BD^2 = 2CD \cdot AB.$$

17. *D is the middle point of the side AB of the triangle ABC, and CN is the perpendicular from C to AB. If AC > BC, prove that*

$$AC^2 - BC^2 = 2AB \cdot DN.$$

18. In the triangle ABC, if $AB = AC$, and D is any point in BC produced, prove that

$$AD^2 = AB^2 + BD \cdot CD.$$

19. AB is a straight line divided into two parts at the point C. $ABDE$ is the square on AB, and the straight line drawn through C parallel to BD meets the diagonal BE in F; prove that

$$EB \cdot BF = 2AB \cdot BC. \text{ [Use Theorem 74.]}$$

XXXI. CONSTRUCTION OF A SQUARE EQUAL TO A POLYGON

Construction 29 (Euclid II. 14)

Describe a square equal to a given rectangle.

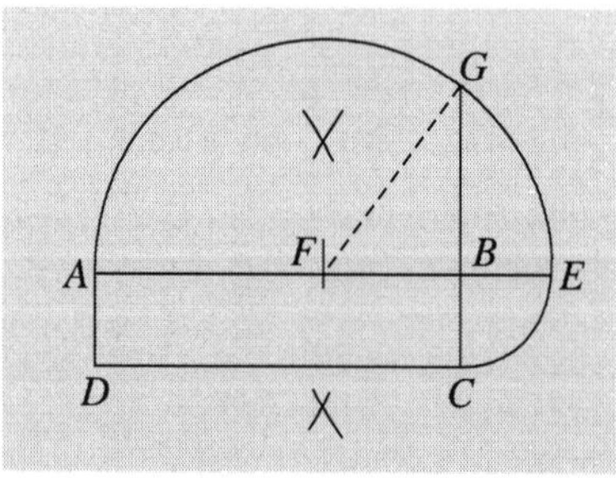

Let $ABCD$ be the given rectangle. It is required to describe a square equal to $ABCD$.

Construction. Produce AB to E, making BE equal to BC.

Bisect AE at F.

With centre F and radius FA, describe a circle.

Produce CB to cut the circle at G.

Then BG is a side of the required square.

Proof. Join FG.

Because AE is divided equally at F and unequally at B,

$\therefore$ rect. AB, BE + sq. on FB = sq, on FE.

Also, because $FE = FG$ and $\angle$ FBG is a right angle,

$\therefore$ sq. on FE = sq. on FG = sq. on BG + sq. on FB;

$\therefore$ rect. AB, BE + sq. on FB = sq. on BG + sq. on FB;

$\therefore$ rect. AB, BE = sq. on BG.

Now, $BE = BC$;

$\therefore$ rect. AB, BE = rect. AC;

$\therefore$ rect. AC = sq. on BG.

NOTE: To draw a square equal to a given rectilineal figure, the steps are as follows:

(*i*) Make a triangle equal to the figure (*Constructions* 13, 14).

(*ii*) Make a rectangle equal to the triangle (*Construction* 16).

(*iii*) Make a square equal to the rectangle.

EXERCISE LIII(a)

1. A chord AB of a circle is bisected at N by the diameter CD, Prove that $CN \cdot ND = AN^2$.

2. *If AD is the perpendicular from A to the hypotenuse BC of a right-angled triangle ABC, prove that*

   (*i*) $AD^2 = BD \cdot DC$; (*ii*) $BA^2 = BD \cdot BC$. [Draw the circle on BC as diameter.]

3. *Divide a given straight line AB at X, so that the rect. AX, XB may be equal to a given square.*

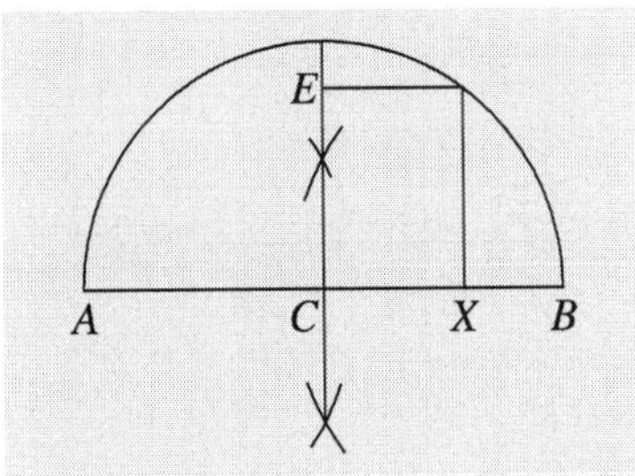

   When is the construction impossible?

   For what position of X has AX · XB its greatest value?

   [In the figure, CE is equal to a side of the given square.]

4. Prove that of all rectangles which have a given perimeter, the one of greatest area is a square.

5. Construct two straight lines, given their sum and that the rectangle contained by them is equal to a given square.

6. Prove that, if the sum of two straight lines is given, the rectangle contained by them is greatest when they are equal.

7. *AB is a given straight line. Explain how to find a point Y in AB produced, such that the rect. AY, BY may be equal to a given square whose side is c.*

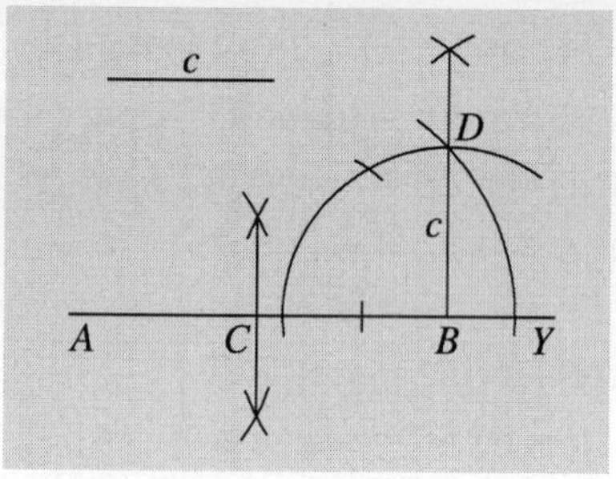

Analysis. Let Y be a point in AB produced, such that $AY \cdot BY = c^2$.

Bisect AB at C.

Then, because AB is bisected at C and produced to Y,

∴ rect. AY, $BY + CB^2 = CY^2$ (*Th.* 74).

∴ $c^2 + CB^2 = CY^2$.

Hence the following construction,— Bisect AB at C. Draw BD perp. to AB, equal to c. With centre C and radius CD, draw a circle cutting AB produced at Y. Then Y is the required point. Supply proof.

8. Explain how to draw two straight lines, given their difference and that the rectangle contained by them is equal to a given square.

9. *Find a point X in a given straight line AB, or in AB produced, such that the sum of the squares on AX, XB may be equal to a given square, whose side is c.*

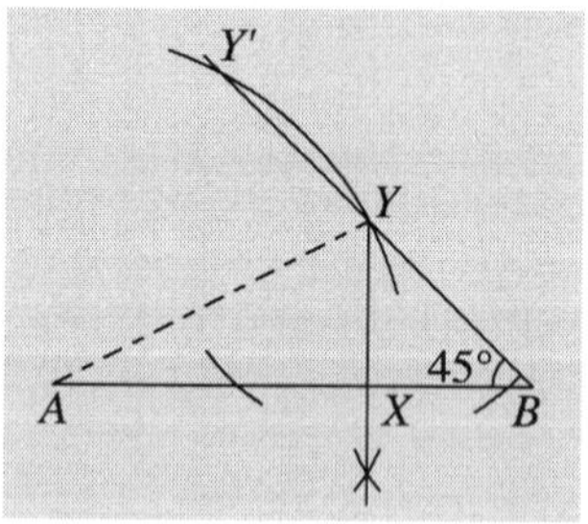

Analysis. Let X be a point in AB (or in AB produced), such that $AX^2 + XB^2 = c^2$.

Draw XY perp. to AB and equal to XB.

Join AY, BY;

∴ $\angle XBY = 45°$,

and $AY^2 = AX^2 + XY^2 = AX^2 + XB^2 = c^2$;

∴ $AY = c$.

The construction should now be evident. Supply construction and proof, showing that there are two solutions.

10. If X is any point within a given straight line AB, prove that $AX^2 + BX^2$ is least when X bisects AB.

11. Given the sum *or* the difference of two straight lines, and the sum of their squares, construct them.

12. *Find a point X in a given straight line AB, or in AB produced, such that* $AX^2 - BX^2$ *may be equal to a given square* (c^2).

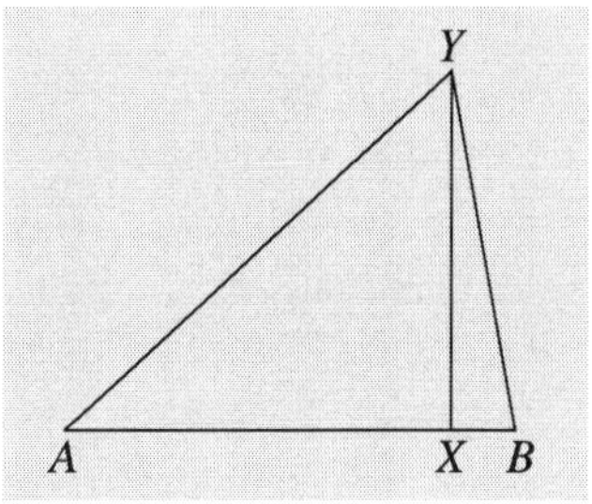

Analysis. Let X be the required point, such that $AX^2 - BX^2 = c^2$. Draw XY perpendicular to AX, and let Y be any point in this line.

Prove that $AY^2 - BY^2 = c^2$, and deduce the following construction:

Draw any ΔPQR having $\angle Q = 90°$ and $PQ = c$. Then circles with centres A, B, and radii PR, QR, respectively, meet in a point Y, such that YX is perpendicular to AX.

13. Given the sum *or* the difference of two straight lines, and the difference of their squares, construct them.

14. Find a point X in a given straight line AB, or in AB produced, such that $AB \cdot AX$ is equal to a given square. Hence deduce alternative constructions for Exs. 12, 13 above.

15. Find the radius of the circle inscribed in a rhombus whose diagonals are $2a$ and $2b$.

16. If b and c are the sides of a right-angled triangle containing the right angle and h the perpendicular from the vertex to the hypotenuse, prove the relation.

$$\frac{1}{h^2} = \frac{1}{b^2} + \frac{1}{c^2}$$

EXERCISE LIII(b)

Numerical Examples

Construct squares equal in area to the figures named in Exs. 1–8. In each case measure a side of the square.

1. A rectangle, length 2 in., breadth 1 in.

2. An equilateral triangle, side 2 in.

3. The triangle ABC in which $\angle A = 90°$, $AB = AC = 2$ in.

4. A triangle whose sides are 2.1 in., 1.7 in., 1 in.
5. A triangle whose sides are 1.3 in., 1.4 in., 1.5 in.
6. A quadrilateral *ABCD* in which *AC* is perpendicular to *BD*, *AC* = 2 in., *BD* = 3 in.
7. A quadrilateral *ABCD* in which *AB* = 2 in., *BC* = 1.3 in., *CD* = 1 in., *DA* = 1.7 in., *AC* = 2.1 in.
8. A regular hexagon, side 1 in.
9. Draw a straight line 3 in. long. Divide it so that the rectangle contained by its segments may be equal to a square whose side is 1 in. Measure the greater segment. Verify by calculation.
10. Draw a straight line *AB*, 1 in. long. Find a point *Y*, in *AB* produced through *B*, so that the rect. *AY*, *YB* may be equal to a square whose side is 2 in. Measure *BY*. Verify by solving a quadratic equation.
11. Construct two straight lines with the following data, and in each case measure the greater line:
    (*a*) Sum of lines = 1.8 in., sum of squares on lines = square whose side is 1.29 in.
    (*b*) Difference of lines = 0.62 in., sum of squares on lines = square whose side is 2.60 in.
    (*c*) Difference of lines = 1.14 in., difference of squares on lines = square whose side is 2.07 in.
    (*d*) Sum of lines = 3.71 in., difference of squares on lines = square whose side is 2.40 in.

XXXII. SQUARES ON THE SIDES OF A TRIANGLE

THEOREM 76 (Euclid II. 12)

In an obtuse-angled triangle, the square on the side opposite the obtuse angle is greater than the sum of the squares on the sides containing the obtuse angle, by twice the rectangle contained by either of these sides and the projection on this side produced of the other side adjacent to the obtuse angle.

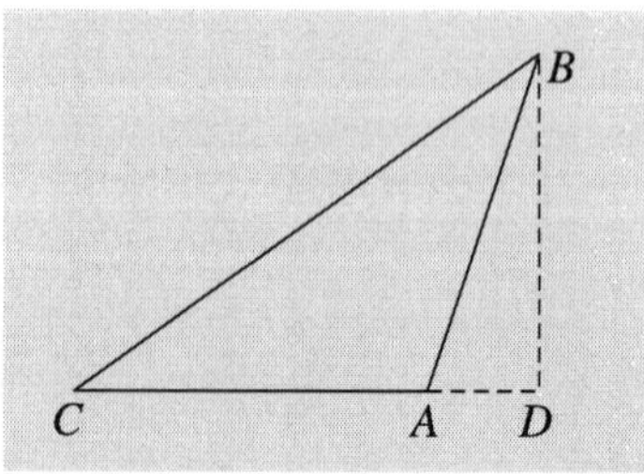

Let *ABC* be a triangle in which the angle *A* is an obtuse angle, and let *BD* be the perpendicular from *B* to *CA* produced. It is required to prove that

sq. on *BC* = sq. on *CA* + sq. on *AB* + 2 rect. *CA*, *AD*.

Proof. Because *CD* is divided at *A*,

∴ sq. on *CD*= sq. on *CA* + sq. on *AD* + 2 rect. *CA*, *AD*.

To each add the sq. on *DB*;

∴ sq. on *CD* + sq. on *DB* = sq. on *CA* + sq. on *AD* + sq. on *DB* + 2 rect. *CA*, *AD*.

But, since ∠ *CDB* is a right angle,

∴ sq. on *CD* + sq. on *DB* = sq. on *BC*,

and sq. on *AD* + sq. on *DB* = sq. on *AB*;

∴ sq. on *BC* = sq. on *CA* + sq. on *AB* + 2 rect. *CA*, *AD*.

THEOREM 77 (Euclid II. 13)

In any triangle, the square on the side opposite an acute angle is less than the sum of the squares on the sides containing the acute angle, by twice the rectangle contained by either of these sides and the projection on it of the other side adjacent to the acute angle.

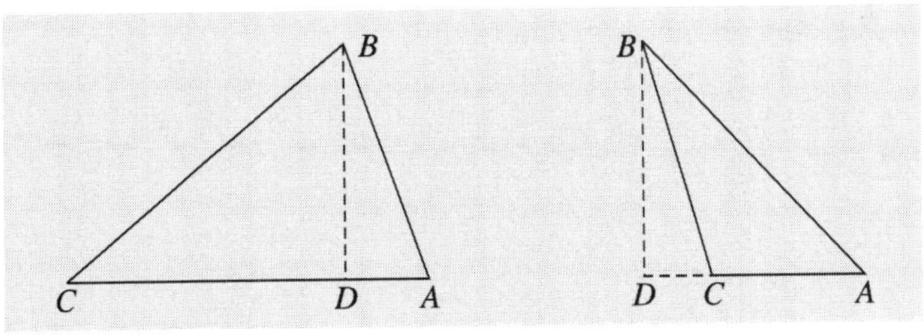

Fig. (a) Fig. (b)

Let *ABC* be a triangle in which ∠ *A* is an acute angle, and let *BD* be the perpendicular from *B* to *AC*, or to *AC* produced. It is required to prove that

sq. on *BC* = sq. on *CA* + sq. on *AB* – 2 rect. *CA*, *DA*.

Proof. Because in Fig. (*a*), *CA* is divided at *D*, and in Fig. (*b*), *DA* is divided at *C*, in both cases

sq. on *CD* = sq. on *CA* + sq. on *DA* – 2 rect. *CA*, *DA*.

To each add the sq. on *DB*;

∴ sq. on *CD* + sq. on *DB* = sq. on *CA* + sq. on *DA* + sq. on *DB* – 2 rect. *CA*, *DA*,

But, since ∠s *CDB*, *ADB* are right angles,

∴ sq. on *CD* + sq. on *DB* = sq. on *BC*,

and sq. on *DA* + sq. on *DB* = sq. on *AB*;

∴ sq. on *BC* = sq. on *CA* + sq. on *AB* – 2 rect *CA*, *DA*.

EXAMPLE 1: *In the triangle ABC, AB = 20 in., BC= 11 in., CA = 13 in. Prove that ∠ C is obtuse. Also, if AD is the perpendicular from A to BC produced, calculate the lengths of CD, AD, and find the area of the triangle ABC.*

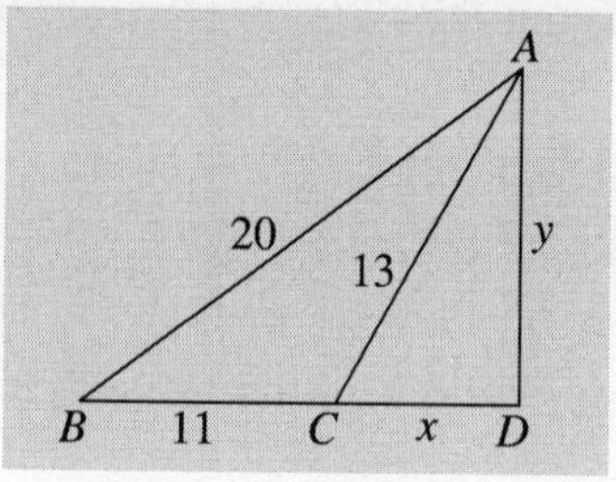

[As the lengths have to be *calculated,* the figure need not be drawn to scale.]

(*i*) We have $AB^2 = 20^2 = 400,$

$$BC^2 + CA^2 = 121 + 169 = 290;$$

$\therefore$ $AB^2 > BC^2 + CA^2,$

$\therefore$ $\angle C$ is obtuse.

(*ii*) By Theorem 76, since $\angle C$ is obtuse,

$$AB^2 = BC^2 + CA^2 + 2BC \cdot CD.$$

Hence, if $CD = x$ in., we have

$$20^2 = 11^2 + 13^2 + 2 \cdot 11 \cdot x;$$

$\therefore$ $110 = 22x,$

$\therefore$ $x = 5.$

(*iii*) Let $AD = y$ in.

Then, because $\angle CDA$ is a right angle,

$\therefore$ $CA^2 = CD^2 + AD^2;$

$\therefore$ $13^2 = 5^2 + y^2;$

$\therefore$ $144 = y^2,$

$\therefore$ $y = 12,$

(*iv*) Area of ΔABC $= \frac{1}{2} BC \cdot AD = \frac{1}{2} \cdot 11 \cdot 12.$

$\therefore$ Area = 66 sq. in.

EXAMPLE 2: *Find the area of a triangle whose sides are 13 in., 14 in., 15 in.*

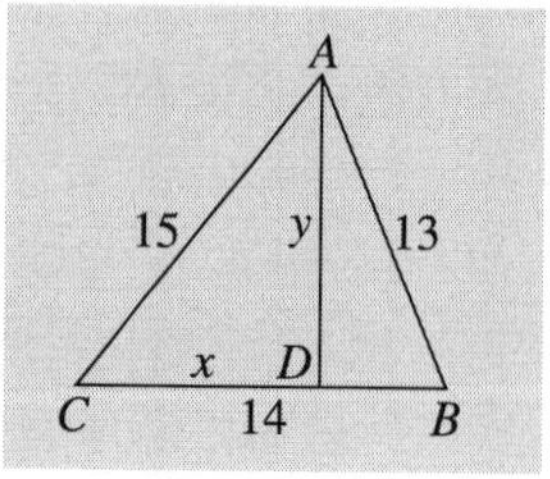

Let ABC be the triangle, in which

$$AB = 13 \text{ in.}, BC = 14 \text{ in.}, CA = 15 \text{ in.}$$

Since AB is the least side, $\angle C$ is acute.

Let AD be perpendicular to BC and let $DC = x$ in., $AD = y$ in.

Then, because $\angle C$ is acute,

$$\therefore \quad AB^2 = BC^2 + CA^2 - 2BC \cdot DC \ (Th.\ 77);$$

$$\therefore \quad 13^2 = 14^2 + 15^2 - 2 \cdot 14 \cdot x;$$

$$\therefore \quad 252 = 28x,$$

$$\therefore \quad x = 9.$$

Again, because $\angle ADC$ is a right angle,

$$\therefore \quad AC^2 = AD^2 + DC^2;$$

$$\therefore \quad 15^2 = y^2 + 9^2;$$

$$\therefore \quad 144 = y^2,$$

$$\therefore \quad y = 12.$$

$$\therefore \text{ Area of } \Delta ABC = \frac{1}{2} \cdot 14 \cdot 12 \text{ sq. in.};$$

$$\therefore \quad \text{area} = 84 \text{ sq. in.}$$

EXAMPLE 3: *In the triangle of Ex. 2, find the length of the perpendicular from B to AC.*

Let the perpendicular from B to $AC = z$ in.

$$\therefore \quad \text{Area of } \Delta ABC = \frac{1}{2} \cdot 15 \cdot z \text{ sq. in.}$$

But the area = 84 sq. in.

$$\therefore \quad \frac{1}{2} \cdot 15 \cdot z = 84,$$

$$\therefore \quad z = \frac{168}{15} = 11 \cdot 2;$$

$$\therefore \text{ Required perpendicular} = 11.2 \text{ in.}$$

EXAMPLE 4: *A piece of ground is in the form of a trapezium, the lengths of the parallel sides are 12 and 26 yards, the lengths of the other sides are 15 and 13 yards; find the area.*

Let $ABCD$ be the trapezium, with sides as shown in the figure.

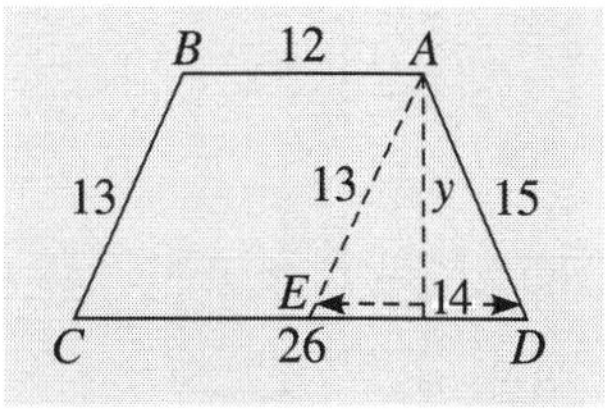

Complete the ▱^{m} $ABCE$, and let the perpendicular from A to $CD = y$ yards.

The sides of ΔAED are 13, 14, 15, and, as in Ex. 2, $y = 12$.

$\therefore$ Area of $\Delta AED = 84$ sq. yd.,

and area of ▱^{m} $ABCE = 12 \cdot 12 = 144$ sq. yd.;

$\therefore$ area of trapezium $= 144 + 84 = 228$ sq. yd.

EXAMPLE 5: *Find the radius of the circle inscribed in a triangle whose sides are 13 in., 14 in., 15 in.*

If r is the radius, we have

$$r = \frac{\Delta}{s},$$

where Δ is the area and $2s$ the perimeter of the triangle.

Now, $2s = 13 + 14 + 15 = 42$,

$\therefore$ $s = 21$;

$\Delta = 84$ (*by Ex.* 2),

$\therefore$ $r = 84/21 = 4$.

TRIGONOMETRICAL FORMULA

In any triangle ABC, if A and B are acute angles,

$$\cos A = \frac{b^2 + c^2 - a^2}{2bc}, \quad \cos B = \frac{c^2 + a^2 - b^2}{2ca}.$$

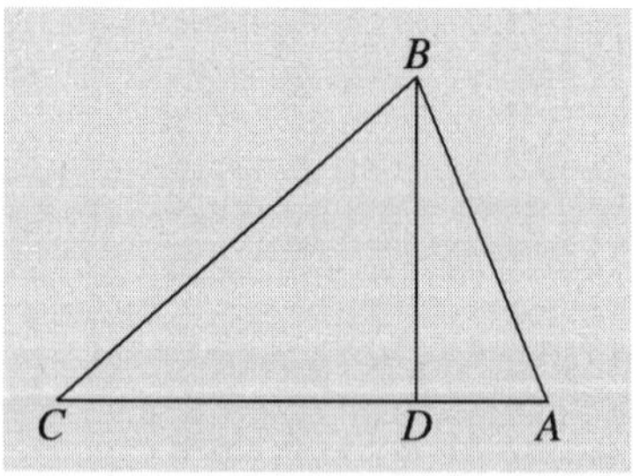

In ΔABC, let $\angle A$ be acute. Draw BD perpendicular to CA or CA produced. Then, by Theorem 77,

$$BC^2 = CA^2 + BA^2 - 2CA \cdot DA.$$

Now, $\dfrac{DA}{BA} = \cos A$,

$\therefore$ $DA = c \cos A$;

$\therefore$ $a^2 = b^2 + c^2 - 2bc \cos A$;

$\therefore$ $$\cos A = \frac{b^2 + c^2 - a^2}{2bc}.$$

We are not concerned here with the case in which A is obtuse.

The angles opposite the two smaller sides of a triangle must be acute. We can use the above formulae, with a table of cosines, to find these acute angles. The third angle can be found from the formula,

$$A + B + C = 180°.$$

EXAMPLE. *Find the least angle of the triangle ABC (to the nearest degree), given AB = 4, BC=5, CA=6.*

The least angle is C,

and $$\cos C = \frac{a^2+b^2-c^2}{2bc} = \frac{5^2+6^2-4^2}{2\cdot5\cdot6} = 0.75.$$

Hence, from the tables, $C = 41°$ *nearly.*

EXERCISE LIV (Theorems 76, 77)

These examples are intended to be done by calculation.

1. Discover by calculation whether the following are right angled, acute-angled or obtuse-angled triangles;
   (*i*) A triangle whose sides are 2, 3, 4;
   (*ii*) A triangle whose sides are 5, 6, 7.
2. In the triangle ABC, AD is the perpendicular from A to BC. Find the lengths of CD and AD, and also the area of the triangle in the following cases:
   (*i*) $BC = 4$, $CA = 13$, $AB = 15$;
   (*ii*) $BC = 21$, $CA = 13$, $AB = 20$.
3. Find the areas of the triangles whose sides are
   (*i*) 10, 17, 9;
   (*ii*) 52, 41, 15.
4. Find the radii of the circles inscribed in triangles whose sides are
   (*i*) 10, 17, 21;
   (*ii*) 25, 17, 12.
5. Find the radius of that escribed circle, which touches the two smaller sides produced, of triangles whose sides are
   (*i*) 25, 29, 36;
   (*ii*) 30, 25, 11.
6. If A is an acute angle of the triangle ABC, and R is the radius of the circum-circle, prove that

$$R = \frac{a}{2\sin A} = \frac{abc}{4\Delta}.$$

   Hence find the radii of the circum-circles of triangles whose sides are
   (i) 50, 41, 21;
   (ii) 25, 29, 6.

7. Find the lengths of the three perpendiculars from the vertices to the opposite sides of the triangle whose sides are 35, 44, 75.

8. By using a trigonometrical formula, find, to the nearest degree, the angles of the triangles whose sides are

(*i*) 2, 3, 4;

(*ii*) 8, 9, 10.

9. *Find an expression for the area of a triangle, in terms of the lengths of the sides.*

Let ABC be the triangle, and B one of its acute angles. Draw AN perpendicular to BC. Let $AN = p$, $BN = x$.

Then $$b^2 = a^2 + c^2 - 2ax, \text{ and } p^2 = c^2 - x^2;$$

$$\therefore \quad p^2 = c^2 - \frac{(a^2+c^2-b^2)^2}{4a^2} = \frac{4a^2c^2-(a^2+c^2-b^2)^2}{4a^2}$$

$$= \frac{(2ac+a^2+c^2-b^2)(2ac-a^2-c^2+b^2)}{4a^2}$$

$$= \frac{(a+c+b)(a+c-b)(b+a-c)(b-a+c)}{4a^2}$$

$$= \frac{16s(s-a)(s-b)(s-c)}{4a^2}, \text{ where } 2s = a+b+c.$$

Therefore, if Δ is the area of the triangle ABC, $\Delta = \frac{1}{2}ap$;

$$\therefore \quad \Delta = \sqrt{s(s-a)(s-b)(s-c)}.$$

10. By using the formula of Ex. 9, find the areas of triangles whose sides are

(*i*) 35, 29, 8;

(*ii*) 55, 51, 26;

11. Calculate the area of a trapezium whose parallel sides are 30 and 13, the other sides being 26 and 25.

EXERCISE LV (Theorems 76, 77)

Theoretical

1. ACB is a straight line, and on AC is drawn an equilateral triangle ACD; show that $DB^2 = AC^2 + CB^2 + AC \cdot CB$.

2. If one of the angles of a triangle is half a right angle, show that the squares on the sides containing that angle are together greater than the square on the side opposite to that angle by four times the area of the triangle.

3. If ABC be a triangle with C an obtuse angle, D and E the feet of the perpendiculars from A and B to the opposite sides, prove that $AB^2 = BC \cdot BD + AC \cdot AE$. [Write down two equations for AB^2 by Theorem 76, and add.]

4. In the quadrilateral *ABCD*, $AC = CD$, $AD = BC$, and the angle *ACB* is supplementary to the angle *ADC*; show that the square on AB = the sum of the squares on *BC*, *CD*, *DA*. [Apply Theorems 76, 77 to the triangles *ABC*, *ACD* to find AB^2, AC^2 respectively.]

5. If squares *ABDE*, *ACFG* are described externally on the sides *AB*, *AC* of a triangle *ABC*, show that the sum of the squares on *EG* and *BC* is twice the sum of the squares on *AB* and *AC*. [Apply Theorems 76, 77 to the triangles *ABC*, *AEG* to find EG^2, BC^2.]

6. The diagonal *AC* of a square *ABCD* is produced to *E* so that $CE = BC$. Prove that $BE^2 = AC \cdot AE$.

7. Squares are described externally on the sides of a triangle and their adjacent corners joined. Prove that the sum of the squares on the joining lines is three times the sum of the squares on the sides of the triangle. [Use Ex. 5.]

8. *ABCD is a quadrilateral and BM, DN are the perpendiculars from B, D to AC. Prove that*

$$(AB^2 + CD^2) \sim (AD^2 + BC^2) = 2AC \cdot MN.$$

[Let *AC*, *BD* meet at *O*, and let $\angle AOB$ be acute. From Δs *AOB*, *BOC*, *COD*, *DOA*, we have

$$AB^2 = OA^2 + OB^2 - 2OA \cdot OM, \text{ etc.]}$$

9. If the squares on the first and third sides of a quadrilateral are together equal to those on the second and fourth, the diagonals intersect at right angles. [Use Ex. 8.]

10. On each of the sides of an acute-angled triangle a square is described external to the triangle; from the angular points of the triangle perpendiculars are drawn to the opposite sides, and produced so as to divide each of the squares into two parts. Show that each square is equal to the sum of the adjacent parts of the other two squares. [Proceed as in Theorem 43.]

THEOREM 78

In any triangle, the sum of the squares on two sides is equal to twice the square on half the base, together with twice the square on the median which bisects the base.

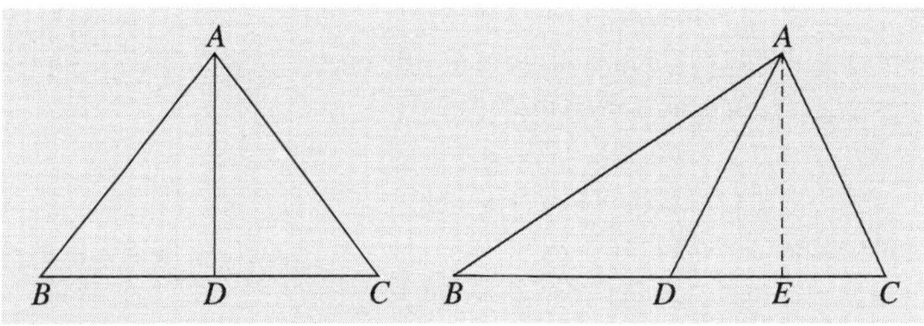

Let *ABC* be a triangle, and let *AD* be the median which bisects *BC*.

It is required to prove that

sq. on AB + sq. on AC = 2 sq. on BD + 2 sq. on AD.

First Case. If $\angle ADB$ is a right angle,

then $\angle ADC$ is also a right angle;

∴ sq. on AB = sq. on BD + sq. on AD,

and sq. on AC = sq. on DC + sq. on AD.

But $BD = DC$;

∴ sq. on AB + sq. on AC = 2 sq. on BD + 2 sq. on AD.

Second Case. Let $\angle ADB$ be obtuse.

Draw AE perpendicular to BC.

Then, because $\angle ADB$ of the triangle ADB is obtuse,

∴ sq. on AB = sq. on BD + sq. on AD + 2 rect. BD, DE:

and, because $\angle ADC$ of the triangle ADC is acute,

∴ sq. on AC = sq. on DC + sq. on AD – 2 rect. DC, DE.

But $BD = DC$.

Hence, by addition,

sq. on AB + sq. on AC = 2 sq. on BD + 2 sq. on AD.

EXAMPLE 1: *In the triangle ABC, AB = 18 in., BC = 8 in., CA = 14 in., and X is the middle point of AB. Find the length of CX.*

We have $CA^2 + CB^2 = 2AX^2 + 2CX^2$.

Hence, if $CX = x$ in.,

$$14^2 + 8^2 = 2.9^2 + 2x^2;$$

∴ $x^2 = 49$;

∴ $x = 7$;

∴ $CX = 7$ in.

EXERCISE LVI (Theorem 78)

Numerical and Theoretical

1. If D is the middle point of the side BC of the triangle ABC, find the length of AD for triangles in which
   (*i*) $AB = 7$, $BC = 8$, $CA = 9$.
   (*ii*) $AB = 13$, $BC = 16$, $CA = 11$.
2. Two adjacent sides of a parallelogram are 13 and 19, and one diagonal is 24. Find the length of the other diagonal.
3. G is the centroid of a triangle ABC, in which $AB = 6$, $BC = 11$, $CA = 7$. Find the length of GA.

4. Find the sum of the squares on the medians of a triangle whose sides are 5, 6, 7.

5. Prove that the sum of the squares on the diagonals of a parallelogram is equal to the sum of the squares on the sides.

6. A, B are given points and C is the middle point of AB. Show that the locus of a point P, which moves so that $PA^2 + PB^2$ is constant, is a circle with C as centre.

7. Let A and B be two fixed points, and CD a given straight line. Find the point P on CD, such that the sum of the squares on PA, PB has its least possible value.

[Bisect AB at X. Draw XP perpendicular to CD; then P is the required point. Supply proof.]

8. The sum of the squares on the four sides of a quadrilateral is equal to the sum of the squares on the two diagonals with four times the square on the line joining the middle points of the diagonals.

9. Prove that, if the sum of the squares on the sides of a quadrilateral is equal to the sum of the squares on the diagonals, the quadrilateral is a parallelogram.

10. *If D is a point in the base BC of a triangle ABC, such that $m \cdot BD = n \cdot DC$, where m and n are any numbers, then*

$$m \cdot AB^2 + n \cdot AC^2 = m \cdot BD^2 + n \cdot CD^2 + (m + n)AB^2.$$

[Proceed as in the second case of Th. 78. Multiply each side of the first equation by m and each side of the second by n and add.]

11. *ABC is a triangle, and a point D is taken in BC produced, such that $m \cdot BD = n \cdot CD$, where m and n are any numbers, show that*

$$m \cdot AB^2 - n \cdot AC^2 = m \cdot BD^2 - n \cdot CD^2 + (m - n)AD^2.$$

[Draw AE perpendicular to BC. Write down expressions for AB^2 and AC^2 from Δs ABD, ACD. Multiply by m and n.]

12. If A and B are fixed points, the locus of a point P which moves, so that $2PA^2 + 3PB^2$ is constant, is a circle.

13. Show that three times the sum of the squares on the sides of a triangle is equal to four times the sum of the squares on the medians of the triangle.

14. If G is the centroid of the triangle ABC, show that

$$GA^2 + GB^2 + GC^2 = \frac{1}{3}(a^2 + b^2 + c^2).$$

15. If G is the centroid of a triangle ABC, show that

$$AB^2 + AC^2 = GB^2 + GC^2 + 4GA^2.$$

16. *If G is the centroid of the triangle ABC and P any point, then*

$$PA^2 + PB^2 + PC^2 = 3PG^2 + GA^2 + GB^2 + GC^2.$$

17. Find a point within a triangle, such that the sum of the squares of its distances from the angular points of the triangle may have its least possible value.

18. *ABCD is any quadrilateral, and O the point of intersection of the straight lines joining the middle points of pairs of opposite sides. If P is any point, then*

$$PA^2 + PB^2 + PC^2 + PD^2 = OA^2 + OB^2 + OC^2 + OD^2 + 4OP^2.$$

19. Find a point within a quadrilateral, such that the sum of the squares of its distances from the angular points of the quadrilateral may have its least possible value.

20. *ABCD* is a quadrilateral and *O* is the point of intersection of the straight lines joining the middle points of pairs of opposite sides. If *a, b, c, d, x, y* are the lengths of the sides and diagonals of *ABCD*, show that

$$OA^2 + OB^2 + OC^2 + OD^2 = \frac{1}{4}(a^2 + b^2 + c^2 + d^2 + x^2 + y^2).$$

[Use Ex. 18; let *P* coincide with *A, B, C, D* in turn.]

XXXIII. RECTANGLES CONTAINED BY SEGMENTS OF A CHORD

THEOREM 79 (Euclid III. 35)

If two chords of a circle meet at a point inside the circle, the rectangle contained by the segments of the one is equal to that contained by the segments of the other.*

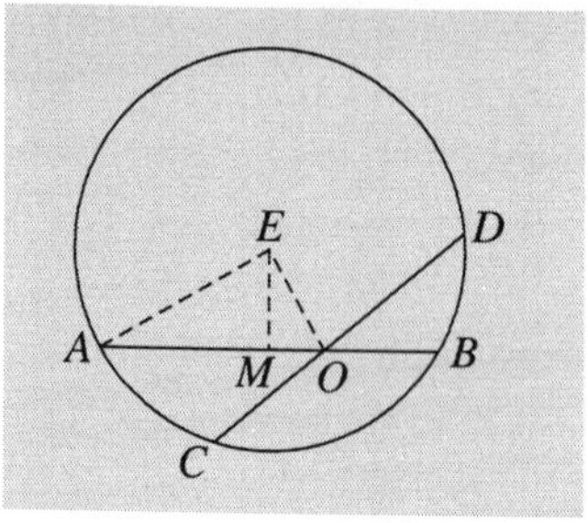

Let *AB*, *CD* be two chords of a circle, meeting at a point *O*, inside the circle.

It is required to prove that

rect. *OA*, *OB* = rect. *OC*, *OD*.

Construction. Let *E* be the centre. Draw *EM* perpendicular to *AB*. Join *EO*, *EA*.

Proof. Because *EM* is the perpendicular from the centre *E* to the chord *AB*, therefore *AM* = *MB*.

Hence, since *AB* is divided equally at *M* and unequally at *O*,

∴ rect. *OA*, *OB* + sq. on *MO* = sq. on *AM*.

To each add the sq. on *EM*;

* *The proof here given is Euclid's. For an easier proof see p. 230.*

∴ rect. OA, OB + sq. on MO + sq. on EM = sq. on AM + sq. on EM

Now the angles at M are right angles;

∴ sq. on MO + sq. on EM = sq. on OE

and sq. on AM + sq. on EM = sq. on AE;

∴ rect. OA, OB + sq. on OE = sq. on AE.

Similarly, it can be shown that

rect. OC, OD+sq. on OE = sq. on CE = sq. on AE;

∴ rect. OA, OB = rect. OC, OD.

THEOREM 80 (Euclid III. 36)

If, from a point outside a circle, a secant and a tangent are drawn, the rectangle contained by the whole secant and the part of it outside the circle is equal to the square on the tangent.*

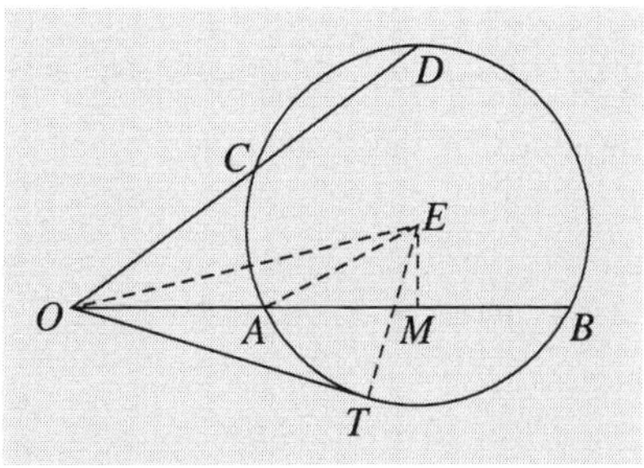

Let O be the point from which a straight line is drawn to cut the circle at A, B, and let OT be a tangent from O to the circle.

It is required to prove that rect. OA, OB = sq. on OT.

Construction. Let E be the centre. Draw EM perpendicular to AB. Join EO, EA, ET.

Proof. Because EM is the perpendicular from the centre E to the chord AB, therefore $AM = MB$.

Hence, since BA is bisected at M and produced to O,

∴ rect. OB, OA + sq. on AM = sq. on OM.

To each add the sq. on EM;

∴ rect. OA, OB + sq. on AM + sq. on EM

= sq. on OM + sq. on EM.

Now, the angle EMA is a right angle;

∴ sq. on AM + sq. on EM = sq. on AE

and sq. on OM + sq. on EM = sq. on OE;

∴ rect. OA, OB + sq. on AE = sq. on OE.

* *The proof here given is Euclid's. For an easier proof see p.* 230.

Again, OT is a tangent and ET a radius;

∴ the angle OTE is a right angle;

∴ sq. on OT + sq. on TE = sq. on OE;

∴ sq. on OT + sq. on TE = rect. OA, OB + sq. on AE.

But $TE = AE$;

∴ rect. OA, OB = sq. on OT.

Cor. If from a point O, outside a circle, two straight lines are drawn, cutting the circle at A, B and C, D respectively, then

rect. OA, OB = rect. OC, OD.

Second Proof of Theorem 79

Proof. In the adjoining figure. The angles ADC, ABC are in the same segment;

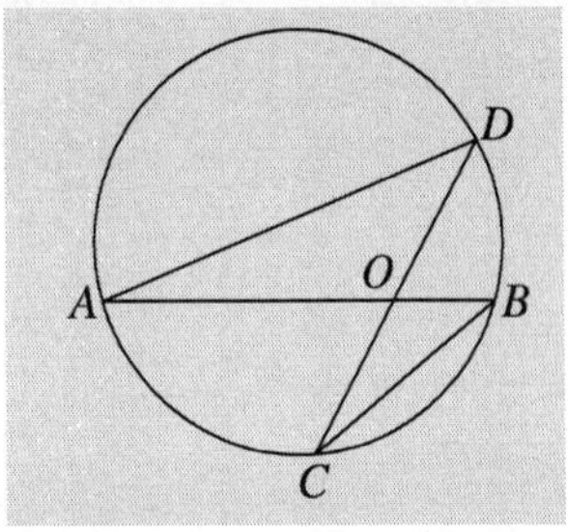

∴ $\angle ADC = \angle ABC$.

Similarly, $\angle BAD = \angle BCD$;

∴ the triangles OBC, ODA are equiangular;

∴ the corresponding sides of these triangles are proportional (*see p.* 127).

$$\therefore \quad \frac{OA}{OC} = \frac{OD}{OB};$$

$$\therefore \quad OA \cdot OB = OC \cdot OD,$$

that is, the rect. OA, OB = the rect. OC, OD.

The student should work out a similar proof for the corollary of Theorem 80, where the straight lines intersect outside the circle.

Second Proof of Theorem 80

Proof. In the adjoining figure, since TO is a tangent and TA a chord of the circle, therefore

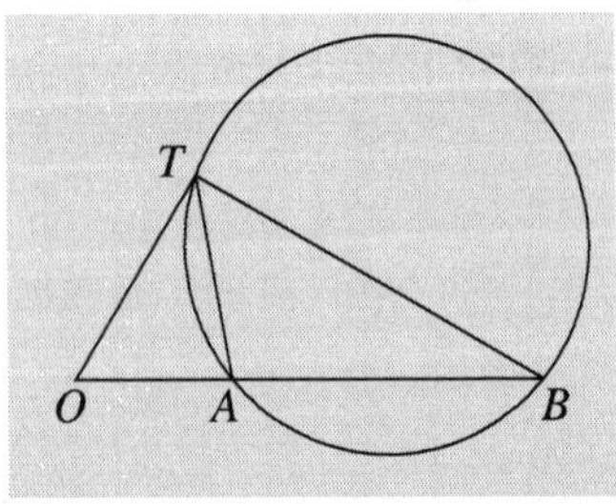

$\angle OTA = \angle ABT$

in the alternate segment.

Hence, in the triangles OTA, OBT,

$\angle OTA = \angle OBT$,

$\angle O$ is common;

∴ the triangles are equiangular;

∴ their corresponding sides are proportional;

$$\therefore \quad \frac{OA}{OT} = \frac{OT}{OB};$$

$$\therefore \quad OA \cdot OB = OT^2;$$

that is, the rect. OA, OB = the sq. on OT.

THEOREM 81

If two finite straight lines intersect, or both being produced intersect, so that the rectangle contained by the segments of the one is equal to that contained by the segments of the other, the extremities of the lines are concyclic.

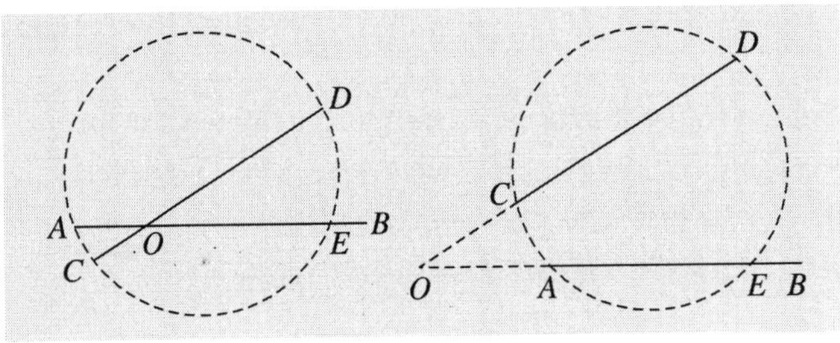

Let the finite straight lines *AB*, *CD*, produced if necessary, intersect at *O*, and let rect. *OA*, *OB* = rect. *CO*, *OD*.

It is required to prove that *A*, *B*, *C*, *D* are concyclic.

Proof. A circle can be drawn through the three points *A*, *C*, *D*.

If this circle does not pass through *B*, let it cut *AB*, or *AB* produced, at *E*.

Then, because the chords *AE*, *CD* meet at *O*,

∴ rect. *OA*, *OE* = rect. *OC*, *OD*.

But rect. *OA*, *OB* = rect. *OC*, *OD* (*given*);

∴ rect. *OA*, *OE* = rect. *OA*, *OB*;

∴ *OE* = *OB*;

∴ *E* coincides with *B*;

∴ the circle through *A*, *C*, *D* also passes through *B*.

THEOREM 82 [Euclid III. 37]

If, from a point outside a circle, a secant is drawn, and also another straight line to meet the circle, and if the rectangle contained by the whole secant and the part of it outside the circle is equal to the square on the other line, then this line touches the circle.

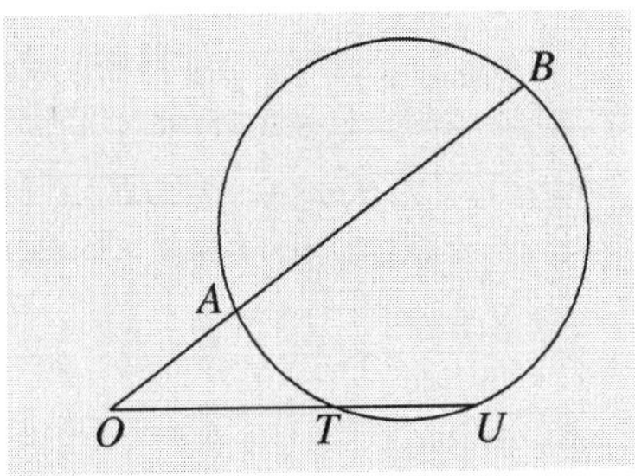

Let O be the point from which one straight line is drawn to cut the circle at A, B and another straight line to meet the circle at T, such that

rect. OA, OB = sq. on OT.

It is required to prove that OT touches the circle.

Proof. If OT does not touch the circle, it will meet the circle at a second point, if produced.

If possible, let OT be produced to meet the circle again at U.

Then, because the chords AB, TU meet at O, when produced,

∴ rect. OA, OB = rect. OT, OU.

But rect. OA, OB = sq. on OT (*given*);

∴ rect. OT, OU = sq. on OT;

∴ $OU = OT$;

∴ U coincides with T;

∴ OT, when produced, does not meet the circle at a second point;

∴ OT touches the circle.

EXERCISE LVII (Theorems 79–82)

1. A point O is at a distance d from the centre of a circle of radius r. Any straight line is drawn through O to cut the circle at A, B. Prove that $OA \cdot OB = r^2 - d^2$ or $OA \cdot OB = a^2 - r^2$, according as O is inside or outside the circle.
2. AB is the common chord of two intersecting circles, and P is any point in AB produced. Prove that the tangents from P to the circles are equal.
3. AB is the common chord of two intersecting circles and P is a point such that the tangents from it to the two circles are equal. Prove that P lies on AB produced.

   [Let PA produced cut the circles again at X, Y. Prove that X, Y coincide.]
4. A, B, C are three points in order in a straight line; a circle is drawn through B and C, and a line AP is drawn to touch the circle at P; find the locus of P for all circles drawn through B and C.
5. AB is diameter of a circle and AP is any chord through A. A straight line perpendicular to AB meets AB at C and AP at Q. Prove that $AP \cdot AQ = AB \cdot AC$.
6. A is a fixed point and P is any point in a fixed straight line LM. In AP, or in AP produced, a point Q is taken, such that $AP \cdot AQ$ is constant. Prove that the locus of Q is a circle.

   [Draw AC perpendicular to LM. Draw QB perpendicular to AQ to meet AC or AC produced at B. The locus is the circle on AB as diameter.]

7. PQ is a chord of a circle whose centre is C and O is any point on PQ. Through O, draw OA perpendicular to OC to meet the circle at A. Prove that $OP \cdot OQ = OA^2$.

8. Through a given point within a circle draw a chord so that the rectangle contained by the whole and one part may be equal to a given square. [Use Ex. 7, and solve by analysis.]

9. ABC, DBC are two intersecting circles, P is a point in the common chord produced, and PA, PD are tangents to the circles; show that the straight line AD cuts off similar segments from the two-circles. [Use Ex. 2.]

10. Given two circles, find a point P on one of them, such that, if the tangent at P meets the other circle at Q, R, the rectangle PQ, PR may be equal to a given square.

11. Find a point D in the base BC of a triangle ABC, such that $AD^2 = BD \cdot DC$. What is the condition that a solution is possible? [Use Ex. XLII. 16 or Ex. XXI(*a*), 11.]

12. AB is a chord of a circle whose centre is C and the tangents at A, B meet in T. Join CT, cutting AB at N. Prove that $CN \cdot NT = AN^2$ and $CN \cdot CT = CA^2$. [Prove that $CATB$ is a cyclic quad.]

13. AB is a chord of a circle whose centre is C and the tangents at A, B meet in T. Join CT cutting AB at N. Through N draw any chord PNQ. Prove that the points C, P, T, Q are concyclic. [Use Ex. 12.]

14. *Given four points A, B, C, D in the same straight line, it is required to find a point O in the same straight line, such that* $OA \cdot OB = OC \cdot OD$.

Prove the following construction,—Draw any circle through A, B and any circle through C, D, cutting the first circle at P, Q. Join PQ and produce PQ to cut AD at O. Then O is the required point.

15. AB is a diameter of a circle, and ACD, BCE two chords meeting inside the circle. Show that the sum of the rectangles $AC \cdot AD$ and $BC \cdot BE$ is equal to the square on the diameter. [Use Ex. 1 and Th. 78]

16. AB, CD are parallel chords of a circle and LM is another chord meeting AB at X and CD at Y. If $AX \cdot XB = CY \cdot YD$, show that the middle point of LM is mid-way between AB and CD. [Use Ex. 1]

17. P is any point in a circle of which AB is a diameter. XPY is a chord parallel to AB, meeting the tangents at A, B in M, N. Prove that $PM \cdot PN = PX \cdot PY + AM^2$. [Draw PT perpendicular to AB, and use Ex. 1.]

18. P is a point in the diameter AB of a circle and PM is the perpendicular from P to the tangent at any point Q on the circle; show that

$$PM \cdot AB = AP \cdot PB + PQ^2.$$

[Draw the tangent parallel to that at Q, and use Ex. 17.]

19. C is the centre of a given circle and PQ is any chord through a fixed point O. The tangents at P, Q meet at T. Prove that T lies on a fixed straight line perpendicular to CO.

[Draw TN perpendicular to CO. Join CT cutting PQ at M. Prove that $CO \cdot CN = CM \cdot CT = CP^2$.]

20. *AB* and *CD* are two chords of a circle intersecting at *E*, and *AC* and *BD* meet in *F*. If circles are described about the triangles *AEC* and *BED*, show that their common chord passes through *F*, and that the angle between their tangents at *E* is equal to *AFB*.

[Let *EF* cut the circle *AEC* at *G* and the circle *BED* at *H*. Prove that *G*, *H* coincide.]

21. *If three circles are such that each intersects the other two, the three common chords meet at a point, or are parallel.*

[Let *AB*, *CD*, *EF* be the common chords. Let *AB*, *CD* meet at *O*. Join *EO*. Let *EO* (produced) cut the circles which pass through *F* at *X*, *Y*. Prove that *X*, *Y* coincide with each other and therefore with *F*.]

22. Any circle is drawn through two given points *A*, *B*, to cut a given circle at *P*, *Q*. Prove that *PQ* meets *AB* at a fixed point.

[Draw some fixed circle through *A*, *B* to cut the given circle at *C*, *D*, and use Ex. 21.]

XXXIV. CONSTRUCTION OF SCALES—PRACTICAL

Representative Fraction. If a plan of a field is drawn to scale, and a length of 6 inches on the plan represents an actual length of 100 yards, we have expressed this shortly by writing

$$6 \text{ in.} = 100 \text{ yd.}$$

The relative sizes of plan and object can also be indicated very conveniently by stating **what fraction any length on the plan is of the actual length represented**.

Thus, in the above case, where 6 in. = 100 yd.,

$$\text{this fraction} = \frac{6}{100 \times 3 \times 12} = \frac{1}{600}.$$

This is called the **Representative Fraction (R.F.)**.

Suppose that the actual distance between two points marked on a map is known to be 3/8 of a mile and that the corresponding distance on the map is 1.36 in., and that it is required to find the R.F. The work is as follows :

$$\text{R.F.} = \frac{1.36}{\frac{3}{8} \times 1760 \times 3 \times 12}$$

$$= \frac{1}{17471} \text{ (nearly).}$$

The R.F. is always expressed as a fraction whose numerator is unity.

In drawing a plan of an object, instead of *calculating* the lengths of the various lines on the plan, we begin by drawing a suitable scale.

The length of any line on the plan can then be found from the scale by *measurement.*

In a plan of a field, a distance of 100 yards is represented by a length of 8 inches on the plan. It is required to draw a scale for the plan showing single yards up to 50.

The work is as follows :

100 yd. is represented by 8 in.;

∴ 50 yd. " " 4 in.

Draw a straight line 4 in. long and divide it into 5 equal parts. Each division thus obtained represents a length of 10 yards, and is called a **Primary Division**. Divide the left-hand primary division into 10 equal parts. Each of these sub-divisions represents 1 yard, and is called a **Secondary Division**.

The scale should be completed and figured as below; the R.F. should be calculated and written above the scale.

Scale of 50 yd. R.F. = $\frac{1}{450}$.

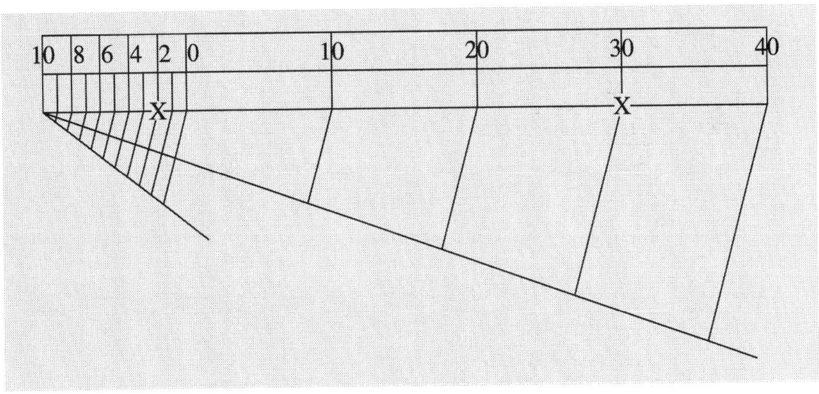

$$\text{R.F.} = \frac{8}{100 \times 36} = \frac{1}{450}.$$

A scale like that in the above figure, in which only one primary division is subdivided, is called a **Plain Scale**,

The distance between the crosses on the scale represents 32 yards.

In the preceding work observe the following points:

(*i*) Distances are set off by making marks with the points of the dividers. It is not necessary to make holes in the paper.

(*ii*) In dividing the line 4 in. long into 5 equal parts, it will be found most convenient to set off along the line used in the construction lengths approximately equal to $\frac{1}{5}$ of 4 in., or 0.8 in.

(*iii*) In dividing the primary division into 10 equal parts, set off along the line used in the construction lengths about equal to, or a little greater than, $\frac{1}{10}$ of the primary division, say about 0.1 in.

(*iv*) In dividing a line into a number of equal parts, it is not necessary to draw the parallels completely as in the given figure. The beginning and end of each should be marked.

(*v*) In constructing a scale, all necessary calculations should be given in full.

EXERCISE LVIII

Drawing to Scale

1. What is the R.F. of the scale in the following cases:

(*i*) 1 in. = 1 foot; (*ii*) 1 in. = 1 yard;

(*iii*) 1 in. = 1 chain; (*iv*) 1 in. = 1 mile;

(*v*) 6 in. = 1 mile; (*vi*) 25 in. = 1 mile?

2. What is the distance between two points on a map, scale 4 in. = 1 mile, whose actual distance apart is 2000 yards?

3. Draw a st. line to represent the length of a flag-pole 35 feet high, scale $\frac{1}{8}$ in. = 1 yard. What is the length of your line? What is the R.F.?

4. If the R.F. = $\frac{1}{64}$, what is the length of a scale which shows distances up to 40 feet?

5. If the R.F. = $\frac{1}{125}$, what number of metres is represented by a length of 5.3 centimetres on the scale?

6. Find the R.F. of a scale on which 13 miles is represented by 1.82 in.

7. In constructing the scale on p. 235 you were instructed to set off a length of 4 in. and divide it into 5 equal parts. It would have been easier to have set off along a line 5 lengths each equal to 0.8 in. Explain why this method would lead to a less accurate result.

8. Using the scale constructed in the preceding work, make a plan of a four-sided field *ABCD*, given the following measurements:

AB = 43 yd., *BC* = 25 yd., *CD* = 16 yd., *DA* = 32 yd., diagonal *AC* = 38 yd., and measure the diagonal *BD*.

9. Draw a straight line 4 in. long. This represents five feet: divide it to show feet and inches, and complete and figure the scale.

10. Draw a plain scale of yards, R.F. = $\frac{1}{655}$, to show 100 yards.

11. Draw a scale of miles and furlongs, R.F. = $\frac{1}{84480}$, showing 5 miles.

12. A distance of 100 yards is represented by $\frac{4}{11}$ in. Draw a scale of furlongs and chains, showing 3 furlongs.

13. Draw a scale to show yards, feet and inches diagonally. R.F. = $\frac{1}{72}$. Show 5 yards.

[To construct this scale, on a line 2.5 in. long, mark a plain scale of yards and feet, showing 5 yards. Draw twelve parallels at equal distances apart and perpendiculars to them through the primary divisions. Join the point 12 in the figure to the secondary division marked 2, and draw parallels to this line through the other secondary divisions. Complete and figure as below. State why you start with a line $2\frac{1}{2}$ in. long.]

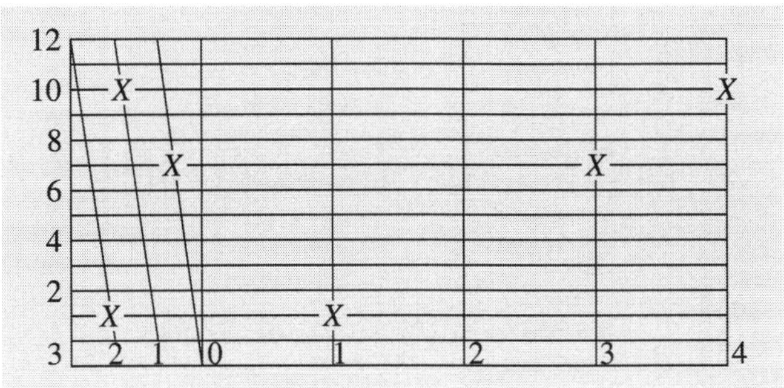

14. Read off the distances in yards, feet and inches between the pairs of crosses on the three horizontal lines in the figure above.

15. Draw a straight line 5 inches long to represent 4 yards. Construct a diagonal scale upon this line to show feet and inches. Complete and figure the scale.

16. Draw a scale for which the R.F. = $\frac{1}{360}$, showing yards and feet diagonally. Show 40 yards.

17. Draw a scale for which the R.F. = $\frac{1}{79200}$ showing miles and furlongs diagonally. Show by two crosses on the scale a distance of 3 miles 5 furlongs.

18. On a Russian map, a distance of 11 versts is represented by 1.42 inches. Draw a scale of English miles for the map, showing 30 miles. [1 verst = 0.661 mile.]

The work is as follows:

$$1 \text{ verst} = 0.661 \text{ mile};$$

$$\therefore \quad 1 \text{ mile} = \frac{1}{0.661} \text{ versts}$$

$$= 1.513 \text{ versts (nearly)};$$

$$\therefore \quad 30 \text{ miles} = 45.39 \text{ versts (nearly)}.$$

Again, 11 versts is represented by 1.42 in.;

$$\therefore \quad 1 \text{ verst} \ldots\ldots\ldots\ldots \frac{1.42 \text{ in.}}{11};$$

$$\therefore \quad 30 \text{ miles} \ldots\ldots\ldots 45.39 \times \frac{1.42 \text{ in.}}{11}$$

$$= 4.126 \times 1.42 \text{ in.}$$

$$= 5.858\ldots \text{ in.}$$

$$= 5.86 \text{ in. (nearly)}.$$

Draw a straight line 5.86 in. long, and divide it into 3 equal parts. Each of these primary divisions represents 10 miles. Divide the left-hand primary division into 10 equal parts. Each of these secondary divisions represents 1 mile.

19. Given that 1 kilometre = 0.621 mile, construct a scale of kilometres for an English map, 10 miles to an inch. Show 50 kilometres.

MISCELLANEOUS EXERCISES

Arranged in Sets for Homework or Revision

PAPER XXI (to Section XXX)

1. *O* is the centre of the circle circumscribing a triangle *ABC*; *D* is any point in *BC*, and the circles circumscribing the triangles *ODC*, *ODB* meet *AC*, *AB* in *E*, *F* respectively. Prove that these circles are equal, and that the triangles *ABC*, *DEF* are equiangular.

2. A line *XYZ* is drawn intersecting the sides *BC*, *CA*, *AB* of a triangle in *X*, *Y*, *Z* respectively. Tangents *AT*, *BT* are drawn to the circle *AYZ*, *BZX*. Prove that the quadrilateral *BCAT* is cyclic, and that *TC* touches the circle *CXY*. [Use Theorem 65]

3. *ABC* is a triangle, right-angled at *A*. *AD* is perpendicular to *BC*. Prove that (*i*) $AD^2 = BD \cdot DC$; (*ii*) $AB^2 = BD \cdot BC$; (*iii*) $AC^2 = CD \cdot CB$.
[Use Theorems 43, 72]

4. Use Ex. 3 to obtain two different constructions for making a square equal in area to a given rectangle.

5. *XY* is a given straight line. *Z* is any point on *XY*. From *Z*, *ZP* is drawn perpendicular to *XY*, such that $ZP^2 = XZ \cdot ZY$. Find the locus of *P*.

6. State, and prove, a geometrical theorem corresponding to the algebraical identity

$$(x + y)^2 + (x - y)^2 = 2(x^2 + y^2).$$

PAPER XXII (to Section XXXI)

1. *ABCD* is a parallelogram; *K* is a point on the diagonal *AC*. If $2AK = KC$, prove that the sum of the complements *BK*, *KD* is equal in area to the parallelogram *KC*.
2. With three given points, not in the same straight line, as centres describe three circles, each of which touches the other two externally. State the lengths of their radii in terms of the lengths of the sides of the triangle whose vertices are the three given points.
3. From the point *T*, external to the circle *QPR*, whose centre is *O*, straight lines *TP*, *TQR* are drawn touching the circle in *P* and cutting it in *Q* and *R* respectively, and *QR* is bisected in *S*. Prove that
   (*i*) the angles *PST*, *POT* are equal;
   (*ii*) if *PS* produced meets the circle again in *U* and if *UV*, drawn parallel to *QR*, meets the circle again in *V*, then *TV* touches the circle.

   [Prove that *TPSO* is cyclic]
4. Prove that the rectangle contained by two parallel chords *AB*, *DC* of a circle *ABCD* is equal to the difference of the squares on *AC* and *BC*.
5. *P*, *Q*, *R* are three points in a straight line. *O* is the middle point of *QR*. A right-angled triangle *ABC* is drawn with its hypotenuse *AC* equal to *OP* and *AB* equal to *OQ*. Prove that $BC^2 = PQ \cdot PR$.

$$[PO = \frac{1}{2}(PQ + PR),\ QO = \frac{1}{2}(PR - PQ)]$$

6. Draw a rectangle, given that the difference between its adjacent sides is 1.5 in., and that its area is 6 sq. in. Measure the sides, and verify your result algebraically.

PAPER XXIII (to Section XXXII)

1. *ABC* is a triangle, and *PBC*, *QCA*, *RAB* are equilateral triangles described externally to the triangle. Prove that *PA*, *QR*, *RC* meet in a point. [Let *AP*, *BQ* meet at *O*. Prove *AOCQ*, *BOCP* cyclic.]
2. *A* is the centre and *BC* a diameter of a given circle. *CD* is a tangent at *C* and equal to *CA*. Join *BD*, and draw *AE* at right angles to *BD*. Join *CE*, and produce it to meet the circumference in *F*. Join *BF*. Prove that $BF = EF$.
3. Two tangents are drawn at the extremities of a diameter *AB* of a circle, and a third tangent meets them in *P* and *Q*. Prove that the rectangle *AP*, *BQ* is equal to the square on the radius of the circle.

   [Show that *PQ* subtends a right angle at the centre.]

4. CD is a chord of a circle; P is any point on a diameter parallel to CD, and Q is one of the extremities of a diameter perpendicular to CD. Prove that $PC^2 + PD^2 = 2PQ^2$.

5. Prove that the locus of the middle point X of a chord PQ of a circle which subtends a right angle at a fixed point O is a circle.

   [If C is the centre and r the radius of the given circle, prove that
   $$OX^2 + CX^2 = r^2]$$

6. The angle ABC of the triangle ABC is 120°; prove that
   $$AC^2 = AB^2 + BC^2 + AB, BC.$$
   Hence calculate the length of AC when AB = 3 in. and BC = 5 in.

   Also if AB is produced to D, so that $CD = CA$, calculate the length of BD.

PAPER XXIV (to Section XXXIII)

1. ABC is a triangle having an acute angle at C. Draw BD at right angles to AC; in DA, or DA produced, take DP equal to DC; also in BA, or BA produced, take BQ equal to BC; join PQ. Prove that the angle PQA is equal to $C + \frac{1}{2}B$.

2. A and B are two points on a circle ADB, whose centre is C. An arc of a circle is described through A, C, B, and any straight line APQ is drawn cutting this arc in P and the circle ADB in Q. Prove that
   $$PB = PQ.$$
   [Note that arc AC = arc CB.]

3. If two chords of a circle intersect at right angles within the circle, prove that the sum of the squares on the four segments is equal to the square on the diameter.

4. AB is a diameter of a circle whose centre is C; on AC as a diameter a second circle is drawn; through P, any point on the circumference of the second circle, a chord QPR of the first circle is drawn. Show that $QP \cdot PR = AP^2$.

5. AB is the diameter of a circle, and CDC' is a chord perpendicular to AB, cutting AB at D. A circle is inscribed in the figure bounded by AD, DC and the arc AC, touching AD in E. Prove that $BE = BC$; hence give a construction for drawing the circle.

   [If the circle touches the arc AC at G, and DC at F, prove that B, F, G are in a straight line.]

6. OAB, OPQ are two straight lines including an angle of 60°. OA is 1.5 in., OB = 3.5 in. Through A and B describe a circle cutting OPQ in P and Q, so that PQ is 1.5 in. in length. Measure the radius.

   [A geometrical construction is demanded, *i.e.* OP must not be calculated. Use Ex. 5 of Paper XXII to prove the following construction: Describe a semicircle on OB as diameter, draw AC perpendicular to AB to meet the semicircle in C; join CB, and along it set off CD equal to half of PQ: then $OD = \frac{1}{2}(OQ + OP)$.]

Paper XXV (to Section XXXIV)

1. Given two points A and B, draw a circle with centre A, such that its diameter is equal to the tangent to it from B.

   [Draw AC, perpendicular to AB, and equal to $\frac{1}{2}AB$. Let BC cut the semicircle on AB as diameter in P. Use the hint given by the construction in Ex. 2 of Paper XXIII.]

2. AB is a chord of a circle whose centre is C and P is any point on the circle. The perpendicular through C to AB meets PA at Q and PB at R. Prove that $CQ \cdot CR = CA^2$. [Prove $\angle CPQ = \angle PRC$; hence show that $CQ \cdot CR = CP^2$.]

3. AB, CD are two chords of a circle which intersect at right angles at the point X. If P, Q are the middle points of AC, BD and O is the centre and r the radius of the circle, prove that
$$PQ^2 = 2r^2 - OX^2.$$

4. AC is a diameter of a given circle; another circle is drawn to touch AC at D and the given circle internally at B. Draw EDF a chord of the given circle, through D, at right angles to AC. Prove that the square on EF is double the rectangle contained by the diameters of the two circles.

5. PCQ is a chord of a circle passing through a fixed point C, and CR is drawn at right angles to PQ to meet the circle on PQ as diameter in R. Find the locus of R.

6. The plan of an estate is roughly in the form of a quadrilateral $ABCD$. The bearings and distances of B, C, D from A are N by 21° E., E. by 42° N., E. by 37° S., and 1.3, 3.7, 2.5 miles respectively. Draw a plan of the estate, on a scale of 1 in. to the mile; find the area of the plan, by reducing the quadrilateral to an equivalent triangle or otherwise, and deduce the area of the estate in acres.

TEST PAPERS

(On Parts I–IV)

I

1. Prove that the area of a parallelogram is equal to the area of a rectangle on the same base and having the same altitude.

   $ABCD$ is a parallelogram, whose diagonals are AC, BD; E is a point in CD; DC is produced to F, so that $CF = DE$, and AE is produced to G, so that $EG = AE$: prove that $BEGF$ is a parallelogram, and is equal in area to $ABCD$.

2. $ABCD$ is a tetrahedron in which $BC = CA = AB = 12$ feet, and $DA = DB = DC = 8$ feet; DP is the perpendicular from D on the plane ABC. Show that $AP = 4\sqrt{3}$ feet, and deduce that $DP = 4$ feet.

3. Prove that the angles in the same segment of a circle are equal.

   OA, *OB*, *OC* are three unequal lines parallel to the sides of an equilateral triangle, and *O*, *A*, *B*, *C* lie on a circle. Prove that *ABC* is an equilateral triangle.

4. Show that if two circles touch each other, the straight line joining their centres must pass through the point of contact.

   Draw a circle of 8 cm. radius, and take a point *P* in its plane distant 5 cm. from the centre. Then show how to draw a circle of radius 6 cm. to touch the first circle and to pass through the point *P*.

5. Two heavy balls each 1.4 inches in diameter lie in a hemispherical bowl of 2 inches internal radius. They touch each other at a point in the vertical radius of the bowl. Enough water is poured in just to cover both balls. Find the depth of the water above the lowest point of the bowl.

6. *ABC* is a triangle, right-angled at *B*, with the side *AB* double of the side *BC*. *CB* is produced to *D*, so that $CD = CA$. From *BA*, *BE* is cut off equal to *BD*. Show that the rectangle contained by *AB*, *AE* is equal to the square on *BE*.

II

1. *PQRS* is a quadrilateral such that the diagonal *QS* bisects the angles at *Q* and *S*. Prove that the diagonals *PR* and *QS* are at right angles.

2. Show that the median of a right-angled triangle which bisects the hypotenuse is equal to half the hypotenuse.

   ABC is a right-angled triangle, *C* being the right angle; *AC* is produced through *C* to *F*, and *BC* through *C* to *G*, so that $CF = CB$ and $CG = CA$. If *X* is the middle point of *FG*, prove that *XC* is at right angles to *AB*.

3. Prove that the opposite angles of any quadrilateral inscribed in a circle are together equal to two right angles.

   ABCD is a quadrilateral inscribed in a circle; *AB*, *DC* produced cut at *I*, and *BC*, *AD* produced cut at *J*; the bisector of the angle *BIC* cuts *BC*, *AD* at *P*, *Q*. Show that $JP = JQ$.

4. A wooden cone, whose slant height is 5 inches, is fixed by its circular base to a flat board. A point in the board which is 8 inches from the nearest point of the base is 12 inches from the vertex; calculate the radius of the base of the cone.

5. *A* is a point outside a circle, *ABC* is a straight line through *A* which cuts the circle at *B* and *C*; *AD* is a tangent to the circle. Prove that the rectangle $AB \cdot AC$ is equal to the square on *AD*.

   Show also that, if *AD* is perpendicular to *ABC*, then $AB^2 + AC^2 + 2AD^2$ is four times the square on the radius of the circle.

6. From the vertices, *B*, *C*, of a triangle *ABC*, perpendiculars are drawn to the bisector of the angle *A*, meeting it in *D* and *E* respectively. If *F* is the middle point of *BC*, prove that the triangle *DEF* is isosceles.

III

1. AOB is an isosceles triangle, and PL, PM are the perpendiculars drawn to the equal sides OA, OB from a point P in the base. Prove that LM is less than the perpendicular drawn from A to OB.
2. Show that the diagonals of a parallelogram bisect each other.

   OA, OB are two given straight lines, P any point in their plane. Show how to draw through P a straight line meeting OA, OB in Q and R so that QR is bisected at P.
3. A point X is taken in the base BC of an isosceles triangle ABC. Prove that $AX^2 = AB^2 - BX \cdot XC$.
4. In a triangle ACB the angle C is a right angle, and $AHKC$, $CLMB$, $ABPQ$, are the squares described on the sides. Prove that

   $$PM^2 + QH^2 = 5KL^2.$$
5. In the quadrilateral $ABCD$ the sum of the angles A, C is equal to two right angles. Prove that the points A, B, C, D lie on a circle.

   From P, a point within the triangle ABC, perpendiculars PL, PM, PN are drawn to BC, CA, AB. Express the magnitude of the angle LNM in terms of the magnitudes of the angles APB, ACB.
6. AO, CO are two radii of a circle at right angles to each other, and from A a line is drawn, cutting CO in X and the circle again in Y; show that the triangle formed by XC produced, XY, and the tangent at Y is isosceles.

IV

1. ABC is an equilateral triangle; BC, CA, AB are produced for equal lengths to A', B', C' respectively. Show that the triangle $A'B'C'$ is also equilateral.
2. Show that parallelograms on the same base and between the same parallels are equal in area.

   $ABCD$, $ABEF$ are two parallelograms in the same plane, with a common side AB. If CE, DF be joined, show that the figure $CDFE$ is a parallelogram, and that its area is either the sum or the difference of the areas of the two given parallelograms.
3. ABC is a triangle with a right angle at A; on the sides of AB, AC remote from BC, squares $ABDE$, $ACFG$ are described. Show that $DF^2 = 2BC^2 + 4AB \cdot AC$.
4. Illustrate and explain by means of a figure the geometrical theorem corresponding to the algebraical identity

   $$a^2 - b^2 = (a + b)(a - b).$$

   In a triangle ABC, AB is greater than AC. D is the foot of the perpendicular to BC from A, and E is the middle point of BC. Prove that $AB^2 = AC^2 + 4BE \cdot ED$.
5. Prove that angles in the same segment of a circle are equal. AB, AC are equal chords of the circle ABC. Parallel straight lines AP, BQ are drawn, cutting the circle again at P and Q. Prove that AQ is parallel to CP.

6. *ACB* is a semi-circle on *AB* as diameter, and *CD* is drawn perpendicular to *AB* to meet it at *D*. A circle touches *AD* at *P*, *DC* at *Q*, and the arc *AC* at *R*. Prove that *RQ* passes through *B*, and that *BP* = *BC*.

V

1. If the angles of a parallelogram are not all equal, prove that the diagonal which joins the acute angles is longer than the other.
2. Show that two triangles on the same base and between the same parallels are equal in area.

   P is a point in the side *AB* of a triangle *ABC*. *PC* is joined, and *BK*, parallel to *PC*, meets *AC* produced in *K*. *AK* is bisected at *Q*, and *PQ* joined. Show that the area of the triangle *APQ* is half that of the triangle *ABC*.
3. Prove that the section of a sphere by a plane is a circle.

   A spherical ball of radius 5.2 in. rests upon three spikes whose points lie in a horizontal plane and are distant 4.2 in. from one another. Find the height of the centre of the ball above the plane.
4. Prove the following construction for finding the direction of the inaccessible point of intersection of two straight lines *AB*, *CD*.

   "Draw any straight line to cut *AB* in *E* and *CD* in *F*. Bisect the angles *AEF*, *CFE* by lines meeting in *G*; and bisect the angles *BEF*, *DFE* by lines meeting in *H*. Then *GH* is the line required."
5. *ABC* is a triangle and points *D*, *E* are taken in *AB*, *AC* respectively, such that $AD = \frac{1}{2}AC$, $AE = \frac{1}{2}AB$; prove that the angles *ABE*, *ACD* are equal.
6. The sides *AB*, *AC* of the triangle *ABC* are equal, and the angle *A* is 90°. *P* is a point in *AC* such that the circles inscribed in the triangles *ABP*, *PBC* touch each other at a point in *BP*. Find the length of *AP* in terms of *BC*.

MISCELLANEOUS RIDERS

It is hoped that these riders, mostly taken from recent Certificate and Matriculation papers, and rendered slightly more difficult by the omission of the theorems on which they depend, and of intermediate steps which give hints, may be of service for final revision.

1. *ABCD*, *AECF* are two parallelograms having the same diagonal *AC*, and *B*, *E*, *D*, *F* are not collinear. Show that *BEDF* is a parallelogram.
2. *ABC* is a triangle right-angled at *A*; *AD* is the perpendicular to *BC*; the bisector of the angle *B* cuts *AD*, *AC* in *E*, *F* respectively. Prove that *AEF* is isosceles.

3. *ABC* is a triangle in which the angle *C* is obtuse; the internal and external bisectors of the angle *A* meet *BC* and *BC* produced in *D* and *E* respectively. If $AD = AE$, prove that

$$C - B = 90°.$$

4. *M* is any point on the base *BC* of a triangle *ABC*; the perpendicular bisectors of *BM*, *BA*, meet in *P*; the perpendicular bisectors of *CM*, *CA*, meet in *Q*. Prove that *PQ* is the perpendicular bisector of *AM*.

5. *ABC* is a triangle, *M* the middle point of *BC*, and *MP*, *MQ* are perpendiculars from *M* to *AB*, *AC* respectively. If $MP = MQ$, prove that the triangle *ABC* is isosceles.

6. Each side of a polygon of *N* sides is produced to meet the side next but one to it, so as to form a star-shaped figure. Prove that the sum of the angles at the points of the star is (2N – 8) right angles.

7. *E* is any point on the side *CD* of a quadrilateral *ABCD*; *DF*, *CF* are parallel to *AE*, *BE* respectively. Prove that the triangle *ABF* is equal in area to the quadrilateral.

8. *ABCD* is any parallelogram; *E* is any point in *BC* produced, and *AE* cuts *CD* in *F*. Prove that the triangles *DFE*, *BCF* are equal in area.

9. *ABC* is a triangle; *AM* is drawn parallel to *BC* and equal to $\frac{1}{2}$*BC*, *M* and *C* being on the same side of *AB*; *E* is the middle point of *AB*, and *EM* cuts *BC* in *F*. Prove that $AC = 4AF$.

10. *ABCD* is a quadrilateral, and *P*, *Q*, *R* are the middle points of the sides *AB*, *BC*, and the diagonal *BD*. *PR*, *QR*, when produced, meet *CD*, *AD* in *E*, *F* respectively. Prove that *EF* passes through the middle point of *RD*.

11. *ABC* is a triangle, and *D* is the middle point of *AB*; the parallel to *BC* through *D* cuts the bisector of the angle *B* in *E*. Prove that *AEB* is a right angle.

12. *A* and *B* are the points of trisection of a given line *PQ*, *ABCD* is any parallelogram in which $BC = 2AB$; *PC* and *DQ*, intersect in *R*. Find the locus of *R*.

13. *A*, *B*, *C*, *D*, *E* are five points in order on the circumference of a circle; *BC*, *CD*, *DE* subtend angles of 30°, 40°, 38° at *A*, and *AB* subtends 35° at *C*. Find the angles of the pentagon *ABCDE*.

14. *ABC* is an isosceles triangle in which $AB = AC$; *D* is a point on *AB* and *E* is a point on *AC* produced, such that $BD = CE$. If *DE* cuts *BC* in *O*, show that *O* is the middle point of *DE*.

15. *ABC* is a triangle, *BQ*, *CR* are the perpendiculars from *B*, *C* to the opposite sides, and *S* is the centre of the circumcircle. Prove that *QR* is perpendicular to *AS*.

16. *AT* is the tangent at a fixed point *A* on a given circle, and *TP* is the other tangent from *T*. Find the locus of the centre of the circle circumscribing the triangle *APT*.

17. *ABD* is a circle whose centre is *C*; the circle circumscribing *ABC* cuts *AD* or *AD* produced, in *E*. Prove that *BDE* is isosceles.

18. *ABC* is a triangle; the interior bisector of *A* meets the exterior bisector of *C* in *D*. Prove that the circumcentre of *ACD* lies on the circumcircle of *ABC*.

19. Two circles, intersecting one another at *A* and *B*, are such that if the tangent at *A* to one circle cuts the other circle in *P*, and the tangent at *A* to the latter cuts the former in *Q*, then *PBQ* is a straight line. Show that *APQ* is a right-angled triangle.

20. *ABC* is an isosceles triangle inscribed in a circle, and $AB = AC$; the bisector of the angle *B* cuts the circle again in *D*, and *AD*, *BC* meet when produced in *E*. Prove that $CE = CA$.

21. *ABC* is a triangle, *D* a point in *BC*; a circle through *B* and *C* cuts *AB*, *AC* again in *E* and *F*; the circle circumscribing *AEF* cuts *AD* in *G*. Prove that *E*, *G*, *B*, *D* lie on a circle, and *F*, *C*, *D*, *G* on another circle.

22. *ABCD* is a cyclic quadrilateral whose diagonals meet in *O*; *OP* is drawn parallel to *BC*. Prove that *OP* touches the circle *AOD*.

23. *ABC* is a triangle; a circle is drawn to touch *AB* at *A*, pass through the middle point of *BC*, and cut *BC* again in *E*. If the circle circumscribing *ACE* cuts *BA* produced in *F*, show that $BA = AF$.

24. *ABC* is a triangle inscribed in a circle; the tangents at *B* and *C* meet in *D*; a parallel through *D* to the tangent at *A* cuts *AB*, *AC* produced in *E* and *F*. Prove that *EBF* and *ECF* are right angles.

25. *ABCD* is a quadrilateral such that the diagonal *AC* bisects the angle *BCD*; and *AD* touches the circle circumscribing *ABC*. Prove that *AB* touches the circle circumscribing *ACD*.

26. *ABCD* is a cyclic quadrilateral; *AD*, *BC*, when produced, meet in *E*; *AB*, *DC*, when produced, meet in *F*. If *B*, *D*, *E*, *F* lie on a circle, prove that *AC* is the diameter of the circle *ABCD*.

27. *ABCD* is a cyclic quadrilateral, and *E* is the middle point of the arc *ADC*. If *AD* is produced to *F*, prove that *ED* bisects the angle *CDF*.

28. *CA*, *CB* are two tangents to a circle; the diameter which passes through *C* cuts *AB* in *D*; *EDF* is any chord through *D*. Prove that *DC* bisects the angle *ECF*.

29. *AT* is the tangent at *A* to a circle whose centre is *O*. *AB* is a chord of the circle which cuts *OT* in *C*, where $TC = TA$. Show that *OB* is perpendicular to *OT*.

30. *AB* is a diameter of a circle, and *D* and *E* are two points on the arc of the same semicircle, such that *DE* is not parallel to *AB*; *AP*, *BQ* are the perpendiculars from *A* and *B* to *DE* produced both ways. Prove that $PD^2 + DQ^2 = PE^2 + EQ^2$.

31. *AOB* is the diameter of a circle perpendicular to a chord *PQ*; *T* is any point on the circumference of the circle, and *PT*, *QT* cut *AOB* in *R* and *S*. Prove that $OR \cdot OS = OP^2$.

32. AB, AE are equal chords of a circle; C and D are points on the arc BE; BE meets AC, AD in F, G. Prove that the difference between the squares on AF and AG is equal to the difference between the rectangles $BF \cdot FG$ and $FG \cdot GE$.

33. Two sides of a parallelogram $ABCD$ are of lengths 9 and 7 units, and the diagonal AC is of length 8 units. Prove that the diagonal BD is double one of the sides of the parallelogram.

34. Two unequal circles intersect in A and B; T, T' are the points of contact of one common tangent. If AB is produced to meet TT' in C, show that $TT'^2 + 4AC^2 = 2AT^2 + 2AT'^2$.

35. ABC is a triangle; BE, CF are the bisectors of the angles B, C; circles are drawn circumscribing ABE, ACF; a parallel to BE through F meets the circle ACF in H; a parallel to CF through E meets the circle ABE in G. If $BE = CF$, prove that (*i*) G, A, H are collinear, (*ii*) the circles are equal, (*iii*) $GE = FH$. Hence, using the properties of similar triangles, show that ABC is isosceles.

36. AB is a given straight line produced to D so that $AD = 4AB$; the circle with A as centre and AB as radius cuts the circle on AD as diameter in E and F. Prove that circles with radius equal to AB and centres E and F, will intersect in the middle point of AB.

Hence, given two points A and B, find the point half-way between them using the compass alone.

37. ABC is a triangle, of which S and O are the circumcentre and the orthocentre respectively. If $AS = AO$, prove that the angle at A is $60°$: and conversely.

The following group of exercises require a knowledge of Parts V and VI

38. ABC, XBY, where each is a circle or a straight line, are inverted with respect to any point O; their inverses are abc, xby. Prove that the angle between the former is equal to the angle between the latter.

39. ABC is a triangle, and O is any point within it; OA_1, OB_1, OC_1 are the perpendiculars from O to BC, CA, AB. This figure is inverted with respect to O, so that a_1, b_1, c_1, are the inverses of A_1, B_1, C_1: prove that a, the foot of the perpendicular from O to b_1c_1, is the inverse of A.

[Show that b_1c_1 is the inverse of the circle drawn through O, B_1, A, C_1; and then use Ex. 38.]

40. Deduce that, to any general homogeneous relation between the rays (OA, OB, OC) and the perpendiculars (OA_1, OB_1, OC_1) of the triangle ABC, there exists a corresponding relation between the reciprocals of the perpendiculars and the reciprocals of the rays.

41. ABC is a triangle and O is any point within it; circles $BOCP$, $COAQ$, $AOBR$ are drawn, and AO, BO, CO, are produced to meet them respectively in P, Q, R. Show that the triangles PBC, AQC, ABR, are similar, the angles of the triangles being the supplements of the angles BOC, COA, AOB.

42. Enunciate and prove the theorem converse to that of Ex. 41. [This is a generalization of Ex. 1]

43. If the ratios of the sides of the triangles, PBC, AQC, ABR, are $p : q : r$, prove that

$$OP = \frac{q}{p} \cdot OB + \frac{r}{p} \cdot OC,$$

with similar equations for OQ, OR.
[Use Ptolemy's theorem, p. 333]

44. If D is the diameter of the circle $BOCP$, and OA_1 is the perpendicular from O to BC, prove that $D \cdot OA_1 = OB \cdot OC$, and

$$\frac{1}{OA_1} \geq \frac{q}{p} \cdot \frac{1}{OC} + \frac{r}{p} \cdot \frac{1}{OB}$$

[Note that $OP \leq D$, and use Ex. 43.]

45. Employ Ex. 44 to prove that

$$\frac{1}{OA_1} + \frac{1}{OB_1} + \frac{1}{OC_1} \geq 2\left[\frac{1}{OA} + \frac{1}{OB} + \frac{1}{OC}\right];$$

and deduce that $OA + OB + OC \geq 2(OA_1 + OB_1 + OC_1)$.*
[Note that $q/r + r/q \geq 2$; for the second part use Ex. 40.]

46. Prove that

$$OB \cdot OC + OC \cdot OA + OA \cdot OB \geq 4(OB_1 \cdot OC_1 + OC_1 \cdot OA_1 + OA_1 \cdot OB_1)$$

with a corresponding relation between the reciprocals; also prove that $OA \cdot OB \cdot OC \geq 8 OA_1 \cdot OB_1 \cdot OC_1$.

47. With the usual notation for a triangle, prove that

(*i*) $s \geq 3\sqrt{3}\, r$, (*ii*) $\Delta \geq 3\sqrt{3}\, r^2$, (*iii*) $a^2 + b^2 + c^2 \geq 36r^2$, and deduce that (*iv*) $OA + OB + OC \geq 6r$.

[Note that $(a + b + c)^3 \geq 27abc$]

48. If $A + B + C = 180°$, prove that $8 \cos A \cos B \cos C \leq 1$. [Take O as the circumcentre of a triangle ABC, and use Ex. 46.]

---

* The second inequality was propounded by *P.* Erdös, and was first proved by L.J. Mordell by the use of projection and trigonometry.

UNIT 5

Similar Figures

XXXV. RATIO AND PROPORTION*

> **NOTE:** In this chapter we shall, as a rule, denote concrete magnitudes by capital letters, and numbers by small letters.

Equality and Inequality of Concrete Magnitudes. In considering how a magnitude or a quantity of any kind is to be measured, at the outset we must define what is meant by saying that "A is equal to B," "A is greater than or less than B," where A and B are quantities of the kind under consideration.

The exact meaning of these statements depends on *the particular kind* of quantity we are considering. For instance, let A and B denote straight lines; if A can be made to coincide with B, we say $A = B$; if A can be made to coincide with a part of B, we say $A < B$ and $B > A$.

Ratio. We assume that a magnitude B can be divided into any number (n) of equal parts. Denoting any one of these parts by C, B is said to contain C (exactly) n times, and C is called the nth part of B. This is expressed shortly by writing $B = nC$ or $C = \frac{1}{n}B$.

If A contains the nth part of B exactly m times, we denote this by writing $A = \frac{m}{n}B$.

DEF. If A and B are magnitudes of the same kind, and it integers m and n exist such that A contains the nth part of B exactly m times $\left(\text{that is, if } A = \frac{m}{n}B\right)$, the magnitudes A and B are said to be **commensurable**, and the ratio of A to B (written $A : B$) is defined as the number m/n.

If no integers m and n exist such that $A = \frac{m}{n}B$, the magnitudes A and B are said to be **incommensurable**.

---

* A "practical" introduction similar to the practical sections in Units I–IV is given in a pamphlet entitled *Ratio, Proportion and Similar Figures,* by J.M. Child (Macmillan, price 1s.).

For convenience, $\frac{m}{n}$ is often written in the form $m : n$, and called *the ratio* of the number m to the number n.

If $A = \frac{m}{n}B$ and C represents the nth part of B, we have $B = nC$ and $A = mC$. If C is taken as the unit of measurement, m and n are the *measures* of A and B. Thus *the ratio of two commensurable magnitudes is equal to the ratio of their measures.*

It can be shown that **the ratio of two incommensurable magnitudes is properly represented by an irrational number.**

In practice the ratio of two incommensurables is represented by some suitable approximate value.

DEFS. A and B are called the terms of the ratio $A : B$. Of these A is called the **antecedent** and B the **consequent**.

The ratio $A : B$ is said to be one of **greater inequality**, one of equality, or one of **less inequality** according as A is greater than, equal to, or less than B.

Proportion. If A, B are magnitudes of the same kind and C, D are magnitudes of the same kind (but not necessarily of the same kind as A and B), A, B, C, D are said to be **proportional** if the ratio of A to B is equal to the ratio of C to D; that is, when

$$A : B = C : D.$$

This is often expressed by saying that *A is to B as C is to D.*

When this is the case, A, B, C, D are called **proportionals**; A and D are called **extremes**, B and C are called **means**, and D is said to be a **fourth proportional** to A, B, C.

If A, B, C are magnitudes of the same kind such that

$$A : B = B : C,$$

then A, B, C are called **three proportionals**, B is called a **mean proportional** to A and C, and C a **third proportional** to A and B. The ratio of A to C is called the **duplicate ratio** of A to B.

If A, B, C, D, ... are magnitudes of the same kind such that

$$A : B = B : C = C : D = \dots ,$$

then A, B, C, D, ... are said to be in **continued proportion**.

If A, B, C, ..., X, Y, Z, ... are magnitudes of the same kind such that

$$A : X = B : Y = C : Z = \dots,$$

then A, B, C, ... are said to be **proportional** to X, Y, Z, ..., which is sometimes expressed by writing

$$A : B : C : \text{etc.} = X : Y : Z : \text{etc.}$$

Theorems on Ratio and Proportion.

I. If a, b, x are positive, then $(a + x) : (b + x) \gtrless a : b$ according as $b \gtrless a$.

Proof. $$\frac{a+x}{b+x} - \frac{a}{b} = \frac{b(a+x)-a(b+x)}{b(b+x)} = \frac{(b-a)x}{b(b+x)},$$

and because x and b are positive, it follows that

$$\frac{a+x}{b+x} \gtrless \frac{a}{b} \text{ according as } b \gtrless a.$$

This theorem is usually stated as follows:

A ratio of greater inequality is diminished and one of less inequality is increased by adding the same positive quantity to each term.

II. If $a : b = c : d$, then $ad = bc$; conversely if $ad = bc$, then $a : b = c : d$.

For if $a/b = c/d$, and we multiply each side by bd, it follows that $ad = bc$.

Conversely if $ad = bc$, and we divide each side by bd, it follows that $a/b = c/d$.

III. If $a : b = b : c$, then $ac = b^2$; conversely if $ac = b^2$, then $a : b = b : c$.

This is a particular case of Theorem II.

IV. If $a : b = b : c$, then $a : c = a^2 : b^2$.

For if $\frac{a}{b} = \frac{b}{c}$, it follows that $\frac{a}{c} = \frac{a}{b} \cdot \frac{b}{c} = \frac{a}{b} \cdot \frac{a}{b} = \frac{a^2}{b^2}$.

Hence $a^2 : b^2$ is the duplicate ratio of a to b.

Observe that if A and B denote concrete quantities, A^2 and B^2 have no meaning, so that the duplicate ratio of A to B cannot be defined as $A^2 : B^2$.

V. If $a : b = c : d$, then $b : a = d : c$. (*Invertendo*)*

For $\frac{b}{c} = 1 \div \left(\frac{a}{b}\right) = 1 \div \left(\frac{c}{d}\right) = \frac{d}{c}$.

VI. If $a : b = c : d$, then $a : c = b : d$ (*Alternando*)

For since $\frac{a}{b} = \frac{c}{d}$,

$$\therefore \quad \frac{a}{b} \cdot \frac{b}{c} = \frac{c}{b} \cdot \frac{b}{c};$$

$$\therefore \quad \frac{a}{c} = \frac{b}{d}.$$

* Theorems V–X are commonly referred to by the names in italics.

VII. If $a : b = c : d$, then $a + b : b = c + d : d$. (*Componendo*)

For $\frac{a}{b} = \frac{c}{d}$;

$$\therefore \qquad \frac{a}{b}+1 = \frac{c}{d}+1;$$

$$\therefore \qquad \frac{a+b}{b} = \frac{c+d}{d}$$

VIII. If $a : b = c : d$, then $a - b : b = c - d : d$. (*Dividendo*)

For $\frac{a}{b} = \frac{c}{d}$;

$$\therefore \qquad \frac{a}{b}-1 = \frac{c}{d}-1;$$

$$\therefore \qquad \frac{c-b}{b} = \frac{c-d}{d}.$$

IX. If $a : b = c : d$, then $a + b : a - b = c + d : c - d$. (*Componendo et Dividendo*)

$$\text{For } \frac{a+b}{a-b} = \frac{\frac{a}{b}+1}{\frac{a}{b}-1} = \frac{\frac{c}{d}+1}{\frac{c}{d}-1} = \frac{c+d}{c-d}.$$

X. If $a : b = x : y$ and $b : c = y : z$, then $a : c = x : z$. (*Ex aequali.*)

For $\frac{a}{b} = \frac{x}{y}$ and $\frac{b}{c} = \frac{y}{z}$;

$$\therefore \qquad \frac{a}{b}\cdot\frac{b}{c} = \frac{x}{y}\cdot\frac{y}{z};$$

$$\therefore \qquad \frac{a}{c} = \frac{x}{z}.$$

XI. If $\frac{x}{a} = \frac{y}{b}$, then each of these fractions is equal to $\frac{lx+my}{la+mb}$, where l and m have any values, positive or negative.

Proof. Denote either of the given fractions by k, so that

$$\frac{x}{a} = \frac{y}{b} = k;$$

$$\therefore \qquad x = ka \text{ and } y = kb;$$

$$\therefore \quad \frac{lx+my}{la+mb} = \frac{lka+mkb}{la+mb} = \frac{k(la+mb)}{la+mb} = k;$$

$$\therefore \quad \frac{lx+my}{la+mb} = \frac{x}{a} = \frac{y}{b}.$$

Important special cases are (*i*) when $l = 1$ and $m = 1$; (*ii*) when $l = 1$ and $m = -1$. These cases may be stated as follows:

If $\frac{\boldsymbol{x}}{\boldsymbol{a}} = \frac{\boldsymbol{y}}{\boldsymbol{b}}$, then each fraction $= \frac{\boldsymbol{x}+\boldsymbol{y}}{\boldsymbol{a}+\boldsymbol{b}} = \frac{\boldsymbol{x}-\boldsymbol{y}}{\boldsymbol{a}-\boldsymbol{b}}$.

In the same way it can be shown that if

$$\frac{x}{a} = \frac{y}{b} = \frac{z}{c} = \ldots ,$$

then each of these fractions $= \frac{lx+my+nz+\ldots}{la+mb+nc+\ldots}$,

where l, m, n, ... are any numbers whatever.

Geometrical Theorems corresponding to some of the algebraical theorems of the preceding article.

If four straight lines (a, b, c, d) are proportionals, the rectangle contained by the extremes (a, d) is equal to the rectangle contained by the means (b, c).

Conversely, if the rectangle contained by one pair of straight lines (a, d) is equal to that contained by another pair (b, c), the four straight lines (a, b, c, d) are proportional. (*Euclid VI.* 16)

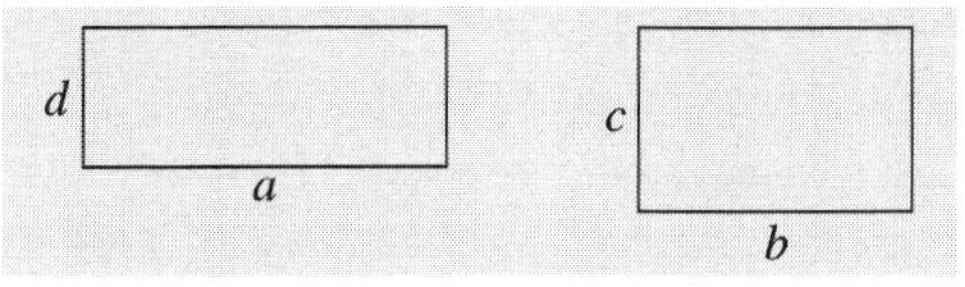

This corresponds to Theorem II of the preceding article.

If three straight lines (a, b, c) are proportionals, the rectangle contained by the extremes (a, c) is equal to the square on the mean (b).

Conversely, if the rectangle contained by two straight lines (a, c) is equal to the square on a third straight line (b), the three straight lines (a, b, c) are proportionals. (*Euclid VI.* 17)

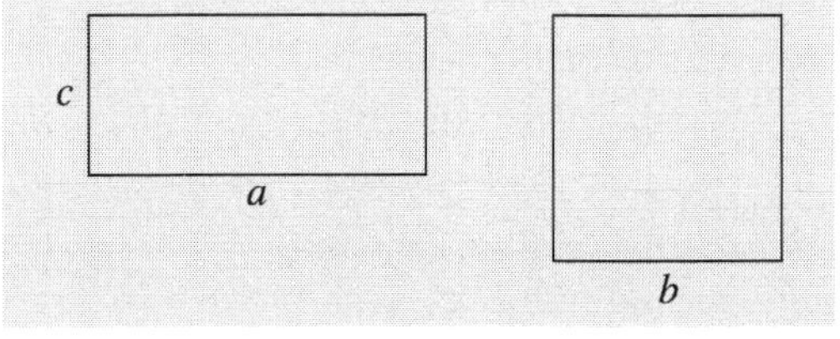

This corresponds to Theorem III of the preceding article.

If three straight lines (a, b, c) *are proportionals, the first is to the third as the square on the first is to the square on the second.*

This corresponds to Theorem IV of the preceding article.

Definitions. If X is a point in the straight line AB, such that

$$AX : XB = p : q,$$

then X is said to divide AB internally in the ratio $p : q$.

If Y is a point in AB produced, such that

$$AY : BY = p : q,$$

then Y is said to divide AB externally in the ratio $p : q$.

It is very important to observe that *there is only one point at which a given line* (AB) *is divided internally in a given ratio.*

For suppose, if possible, that there are two such points X and X'.

Then, by hypothesis,

$$AX : XB = AX' : X'B;$$

$\therefore$ $$AX + XB : XB = AX' + X'B : X'B \text{ (\textit{componendo})};$$

that is, $$AB : XB = AB : X'B;$$

$\therefore$ $$XB = X'B;$$

$\therefore$ X and X' coincide.

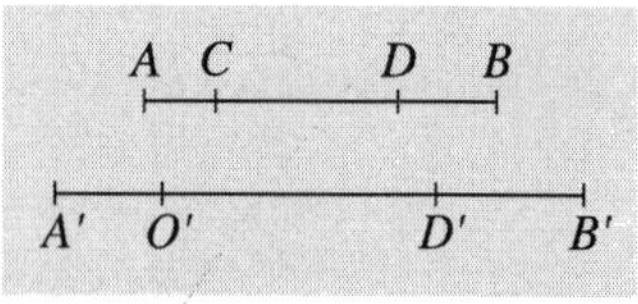

Similarly, by *dividendo*, we can show that *there is only one point at which a given line is divided externally in a given ratio.*

If two straight lines AB and $A'B'$ are divided at C, D and C', D', so that

$$AC : A'C' = CD : C'D' = DB : D'B',$$

then AB is said to be divided similarly to $A'B'$.

EXERCISE LIX (On pages 249-254)

1. In the sides AB, AC of a triangle, points D, E are taken such that

$$AD : DB = AE : EC.$$

Prove that $$AD : AB = AE : AC.$$

2. The sides BA, CA of a triangle are produced, through A, to D and E, so that

$$BD : AD = CE : AE.$$

Prove that $$AD : AB = AE : AC.$$

3. *A* straight line *AB* is divided at *C*, so that

$$AC : CB = p : q.$$

Prove that $AC = \dfrac{p}{p+q} \cdot AB$ and $CB = \dfrac{q}{p+q} \cdot AB$.

4. *A* straight line *AB* is divided externally at *D*, so that

$$AD : BD = p : q \ (p > q).$$

Prove that $AD = \dfrac{p}{p-q} \cdot AB.$ and $BD = \dfrac{q}{p-q} \cdot AB$.

5. *A straight line AB is divided internally at C and externally, in the same ratio, at D, so that*

$$AC : CB = AD : BD.$$

(*i*) *Prove that CD is divided internally and externally in the same ratio at B and A.*

(*ii*) *If O is the middle point of AB, prove that*

$$OB^2 = OC \cdot OD.$$

[Use *Componendo et Dividendo* to show $2OB : 2OC = 2OD : 2OB$.]

(*iii*) *Prove that* $\dfrac{1}{AC} - \dfrac{1}{AB} = \dfrac{1}{AB} - \dfrac{1}{AD}$ or $\dfrac{2}{AB} = \dfrac{1}{AC} + \dfrac{1}{AD}$.

6. *A, C, B, D are points on a straight line and O is the middle point of AB. If* $OB^2 = OC \cdot OD$, and *C*, *D* are on the same side of *O*, *prove that*

$$AC : CB = AD : BD.$$

7. *P* is any point on a circle of which *AB* is a diameter and *PN* is perpendicular to *AB*. If *BA* is produced to meet the tangent at *P* in *T*, prove that $BN : NA = BT : AT$. [Use Ex. 6]

XXXVI. TRIANGLES AND PARALLELOGRAMS

Theorem 83 (Euclid VI. 1)

The ratio of the areas of two triangles of equal altitude is equal to the ratio of their bases.

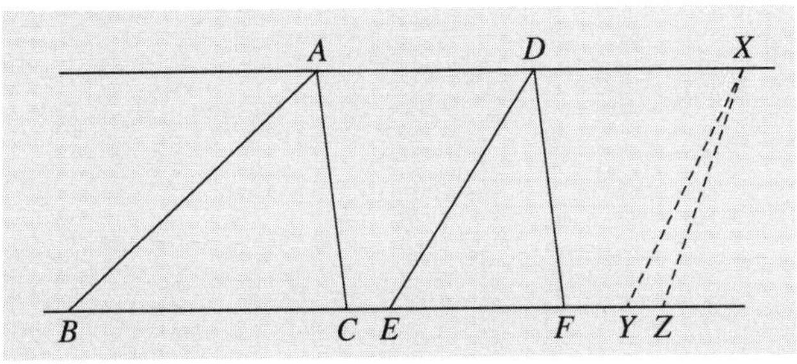

Let *ABC*, *DEF* be two triangles whose bases *BC*, *EF* are commensurable and whose altitudes are equal. It is required to prove that

$$\Delta ABC : \Delta DEF = BC : EF.$$

Proof. Let the bases *BC*, *EF* have a common measure *YZ*, which is contained *p* times in *BC* and *q* times in *EF*, so that

$$BC = pYZ \text{ and } EF = qYZ.$$

Take any triangle *XYZ*, of the same altitude as *ABC* and *DEF*, and place these three triangles so that the bases *BC*, *EF*, *YZ* are in the same straight line and the triangles on the same side of it.

Then, since the altitudes of the triangles are equal,

∴ *A*, *D*, *X* are in a straight line parallel to *BZ*;

and because $BC = pYZ, \quad \therefore \Delta ABC = p\,\Delta XYZ;$

and because $EF = qYZ, \quad \therefore \Delta DEF = q\Delta XYZ;$

$$\therefore \qquad \frac{\Delta ABC}{\Delta DEF} = \frac{p}{q} = \frac{BC}{EF};$$

$$\therefore \qquad \Delta ABC : \Delta DEF = AC : EF.$$

Cor. The ratio of the areas of two parallelograms of equal altitude is equal to the ratio of their bases.

Alternative Proof of Theorem 60

Let *ABC*, *DEF* be two triangles of the same altitude *h*, whose bases are *BC* and *EF*.

Then, the triangles contain $\frac{1}{2}BC \cdot h$ and $\frac{1}{2}EF \cdot h$ units of area respectively;

$$\therefore \qquad \frac{\Delta ABC}{\Delta DEF} = \frac{\frac{1}{2}BC \cdot h}{\frac{1}{2}EF \cdot h} = \frac{BC}{EF};$$

$$\therefore \qquad \Delta ABC : \Delta DEF = BC : EF.$$

Example: *The ratio of the areas of two triangles (ABC, DEF) on equal bases (BC, EF) it equal to the ratio of their altitudes.*

Let *h*, *k* be the altitudes of *ABC*, *DEF*. Then, because *BC* = *EF*, we have

$$\frac{\Delta ABC}{\Delta DEF} = \frac{\frac{1}{2}BC \cdot h}{\frac{1}{2}EF \cdot k} = \frac{h}{k};$$

$$\therefore \qquad \Delta ABC : \Delta DEF = h : k.$$

Alternative proof. Since the bases are equal, let the Δs *ABC*, *DBC* stand on the same base *BC*. Through *C* draw a straight line perpendicular to *BC*; and through *A* and *D* draw *AE*, *DF* parallel to *BC* to meet this perpendicular in *E* and *F*.

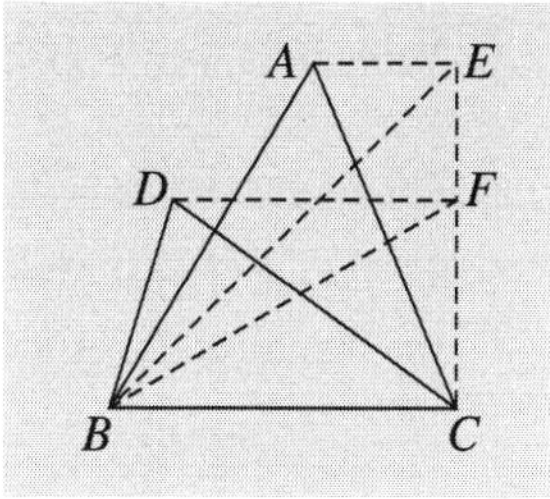

Then *CE*, *CF* are the altitudes of the triangles *ABC*, *DBC*; and

$$\frac{\Delta ABC}{\Delta DBC} = \frac{\Delta EBC}{\Delta FBC} = \frac{EC}{FC} \text{ (by Theorem 60)}$$

Def. The sides of two figures about two of their angles are said to be **reciprocally proportional** when a side of the first is to a side of the second as the remaining side of the second is to the remaining side of the first.

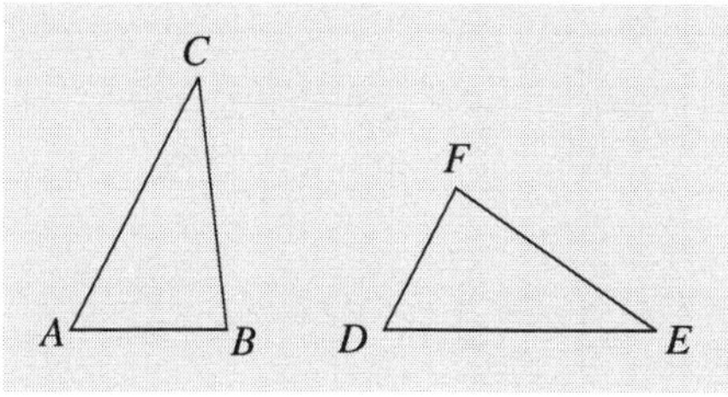

Thus the sides of the triangles *ABC*, *DEF* about the angles *A* and *D* are reciprocally proportional if

$$AB : DE = DF : AC.$$

Theorem 84 (Euclid VI. 14, 15)

(*i*) If two parallelograms (or two triangles) have an angle of the one equal to an angle of the other, and are equal in area, the sides about the equal angles are reciprocally proportional.

(*ii*) Conversely, if two parallelograms (or two triangles) have an angle of the one equal to an angle of the other, and if the sides about the equal angles are reciprocally proportional, the figures are equal in area.

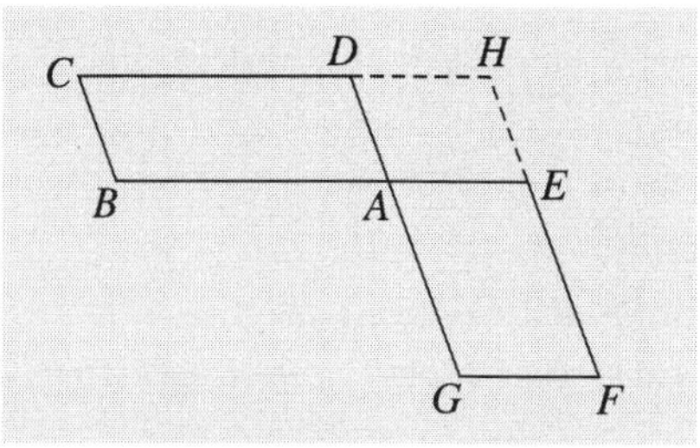

Let *ABCD*, *AEFG* be two parallelograms in which

$$\angle BAD = \angle GAE.$$

(*i*) If the parallelograms are equal in area, it is required to prove that

$$AB : AE = AG : AD.$$

Proof. Place the parallelograms so that AB, AE are in the same straight line. Produce CD, FE to meet at H.

Then, because $\angle BAD = \angle EAG$ (*given*),

$\therefore$ AD, AG are in the same straight line.

Now, the ratio of two parallelograms of equal altitude is equal to the ratio of their bases;

$\therefore$ $\quad \text{▱}^m AC : \text{▱}^m AH = AB : AE,$

and $\quad \text{▱}^m AF : \text{▱}^m AH = AG : AD.$

But $\quad \text{▱}^m AC = \text{▱}^m AF$ (*given*);

$\therefore$ $\quad \text{▱}^m AC : \text{▱}^m AH = \text{▱}^m AF : \text{▱}^m AH;$

$\therefore$ $\quad AB : AE = AG : AD.$

(*ii*) If $AB : AE = AG : AD$, it is required to prove that the parallelograms are equal in area.

As before, $\text{▱}^m AC : \text{▱}^m AH = AB : AE,$

and $\quad \text{▱}^m AF : \text{▱}^m AH = AG : AD.$

But it is given that $\quad AB : AE = AG : AD;$

$\therefore$ $\quad \text{▱}^m AC : \text{▱}^m AH = \text{▱}^m AF : \text{▱}^m AH;$

$\therefore$ $\quad \text{▱}^m AC = \text{▱}^m AF.$

NOTE: Since a parallelogram is bisected by a diagonal, the corresponding theorem for the triangles *BAD*, *EAG* is obviously true.

EXERCISE LX (Theorems 83, 84)

1. The four triangles into which a quadrilateral is divided by its diagonals are proportionals.
2. If the diagonals of a quadrilateral $ABCD$ meet at E, show that

   $$\Delta ABC : \Delta ADC = BE : ED.$$

   [Use Theorem 83 and *Componendo*]
3. If X is any point in the base BC of a triangle ABC and O is any point in AX, prove that

   $$\Delta AOB : \Delta AOC = BX : XC.$$

4. X is a point in the side BC of the triangle ABC and O is a point in AX. If $BX : XC = 2 : 3$ and $XO : OA = 3 : 4$, prove that

(*i*) $\Delta\, OBC = \frac{3}{7}\Delta\, ABC$.

(*ii*) $\Delta\, OBX = \frac{2}{5}\Delta\, OBC$ and $\Delta\, OBX = \frac{3}{4}\Delta\, OAB$.

(*iii*) $\Delta\, OBC : \Delta\, OAB = 15 : 8$.

(iv) $\Delta\, OBC : \Delta\, OCA = 5 : 4$.

(*v*) Produce BO, CO to meet CA, AB at Y, Z, and find the values of the ratios $CY : YA$ and $AZ : ZB$. [Use Ex. 3.]

5. In the triangle ABC straight lines AD, BE, CF are drawn through a common point P to meet the sides BC, CA, AB in D, E, F respectively; show that

$$\frac{PD}{AD}+\frac{PE}{BE}+\frac{PF}{CF} = 1 \text{ and } \frac{AP}{AD}+\frac{BP}{BE}+\frac{CP}{CF} = 2.$$

6. $OACB$ is a parallelogram and P is any point on the diagonal AB. Draw MPH parallel to OA to meet OB, AC at M, H. Draw NPK parallel to OB to meet OA, BC at N, K. Prove that $PM \cdot PN = PH \cdot PK$.

7. In Ex. 6, if $OA = a$, $OB = b$, $MP = x$, $NP = y$, show that

$$\frac{x}{a}+\frac{y}{b} = 1.$$

[Use the result of Ex. 6.]

XXXVII. RATIO OF THE SEGMENTS OF LINES AND ARCS OF CIRCLES

Theorem 85 (Euclid VI. 2)

(*i*) If a straight line is drawn parallel to one side of a triangle, it divides the other two sides proportionally.

(*ii*) Conversely, if a straight line divides two sides of a triangle proportionally, it is parallel to the third side.

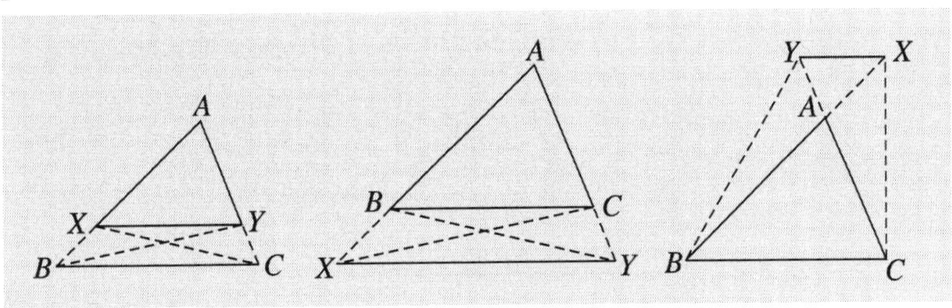

Let ABC be a triangle, and let a straight line XY cut the sides AB, AC (or these sides produced) at X, Y respectively.

(*i*) If XY is parallel to BC, it is required to prove that

$$AX : XB = AY : YC.$$

Construction. Join BY, CX.

Proof. The ratio of the areas of two triangles of equal altitude is equal to the ratio of their bases;

$\therefore$ $\Delta AXY : \Delta BXY = AX : XB,$

and $\Delta AXY : \Delta CXY = AY : YC.$

Now XY is parallel to BC (*given*);

$\therefore$ $\Delta BXY = \Delta CXY;$

$\therefore$ $\Delta AXY : \Delta BXY = \Delta AXY : \Delta CXY;$

$\therefore$ $AX : XB = AY : YC.$

(*ii*) Conversely, if $AX : XB = AY : YC$, it is required to prove that XY is parallel to BC.

Proof. As before,

$\Delta AXY : \Delta BXY = AX : XB,$

and $\Delta AXY : \Delta CXY = AY : YC.$

But it is given that $AX : XB = AY : YC;$

$\therefore$ $\Delta AXY : \Delta BXY = \Delta AXY : \Delta CXY;$

$\therefore$ $\Delta BXY = \Delta CXY.$

And the triangles BXY, CXY are on the same base XY and on the same side of it;

$\therefore$ XY is parallel to BC.

Cor. 1. Since $AX : XB = AY : YC,$

By componendo, or dividendo, it follows that

$$AB : AX = AC : AY.$$

Hence, the segments of AXB are proportional to the corresponding segments of AYC.

Cor. 2. The intercepts made by three parallel straight lines (AB, CD, EF) on any two straight lines (PT, $P'T'$) which cut them are proportional.

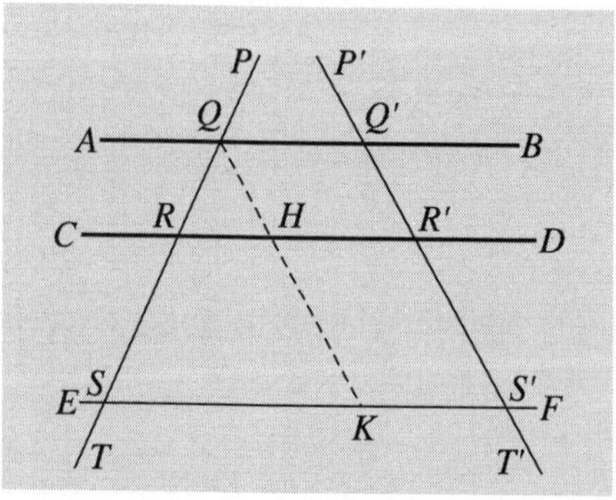

That is, in the figure,

$$QR : RS = Q'R' : R'S',$$
$$QR : QS = Q'R' : Q'S',$$
$$RS : QS = R'S' : Q'S'.$$

Prove by drawing *QHK* parallel to *Q'S'* to cut *CD* at *H* and *EF* at *K*.

Conversely, if the intercepts made on two straight lines by three straight lines which cut them are proportional, and two of these lines are parallel to one another, the third straight line is parallel to the other two.

CONSTRUCTION 30

Divide a given straight line internally, or externally, in a given ratio.

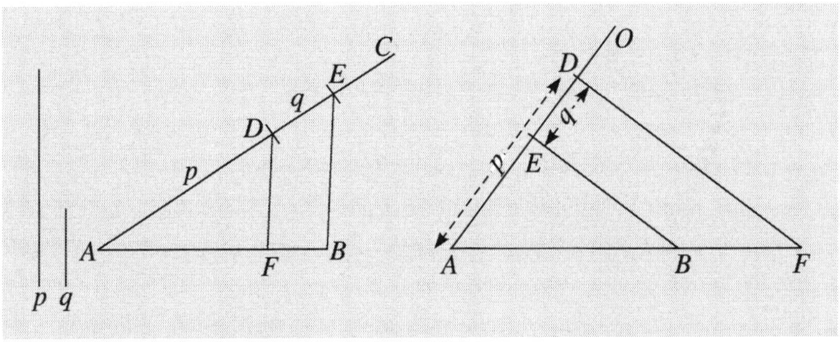

Fig. (a) Fig. (b)

Let *p*, *q* be two straight lines which are in the given ratio, and let *AB* be the given straight line.

It is required to divide *AB* internally, or externally, in the ratio $p : q$.

Construction. Draw a straight line *AC*, of any length, making any angle with *AB*.

From *AC* cut off *AD* equal to *p*. From *DC* or *DA*, according as it is desired to divide *AB* internally or externally, cut off *DE* equal to *q*. Join *EB*, and draw *DF* parallel to *EB* to meet *AB* or *AB* produced at *F*.

Then *F* is the required point of division.

Proof. Because *FD* is parallel to the side *BE* of the triangle *ABE*,

$\therefore$ $$AF : FB = AD : DE.$$

But $$AD : DE = p : q;$$

$\therefore$ $$AF : FB = p : q.$$

EXERCISE LXI (On pages 259–261)

1. $ABCD$ is a trapezium with AB parallel to CD, and the diagonals AC, BD meet at E. Prove that

$$AE : EC = BE : ED.$$

2. $ABCD$ is a quadrilateral and P is any point in AB. Draw PQ parallel to BC to meet AC at Q. Draw QR parallel to AD to meet CD at R. Prove that

$$DR : RC = AP : PB.$$

3. $ABCD$ is a quadrilateral; in AB, CD points P, Q, are taken so that $AP : PB = DQ : QC$; in AD any point O is taken, and PM, QN are drawn parallel to AD, meeting BO, CO in M, N respectively; show that MN is parallel to BC.

4. $ABCD$ is a trapezium, with AB parallel to CD ; AC, BD meet at E, and AD, BC are produced to meet at F. Join EF, and prove that

   (*i*) $\Delta ADE = \Delta BCE$,

   (*ii*) $\Delta DEF = \Delta CEF$,

   (*iii*) hence show that EF bisects CD and AB.

5. Given a straight line AB and a straight line CD parallel to AB, give a construction for bisecting AB, using the ruler only. [Use Ex. 4]

6. P is any point in a given straight line and A is a fixed point. In AP, or in AP produced, take a point Q so that $AQ : AP$ is equal to a given ratio. Prove that the locus of Q is a straight line.

7. A is a given point on a circle and BC is a given chord. Through A draw a straight line AXY to meet BC at X and the circle again at Y, such that $XY = 2AX$. [Use Ex. 6]

8. ABC is a triangle, and MN is a variable straight line parallel to AC, cutting the sides BC, BA in M, N respectively. Find the locus of the intersection of AM and CN. [Use Ex. 4]

9. $ABCD$ is a parallelogram and E any point in AB; any parallel to AB cuts AD, ED, EC, BC in H, F, G, K respectively. Show that the parallelogram $AHKB$ is double either of the triangles EDG or ECF.

10. P is any point in the base BC of a triangle ABC. Draw PL parallel to BA to meet CA at L and PM parallel to CA to meet AB at M. Prove that the triangle PLM is a mean proportional to the triangle: BMP, PLC.

11. $ABCD$ is a quadrilateral. Through A, B draw parallels to meet CD at L, M, such that $DL : CM$ is a given ratio. [Divide AB, CD at P, Q, so that $AP : PB = DQ : QC$ = given ratio; draw parallels to PQ through A and B.]

12. *Given a straight line AB, divided in any manner at X and Y. Divide another given straight line CD* similarly *to AB*. [Use Ex. 2]

13. Any straight line is drawn on a sheet of exercise paper ruled with equidistant parallel straight lines: using a pair of set-squares only, divide the given straight line into three segments in the ratio 2 : 3 : 4.

14. *Construct a triangle of given perimeter, whose sides are in the ratio* 2 : 3 : 4.

[Draw *HK* equal to the given perimeter, and divide it at *P* and *Q* in the ratio 2 : 3 : 4. Construct a triangle with sides equal to *HP*, *PQ*, *QR*.]

15. *O is a given point and AB, AC two given straight lines, of indefinite length. Through O draw a straight line so that the segments intercepted between O and the given lines may be in the ratio* 2 : 3.

Analysis. Let *XY* be the line required. Draw *OD* parallel to *AB* to meet *AC* in *D*; then

$$\frac{AD}{DY} = \frac{XO}{OY} = \frac{2}{3}.$$

Hence, supply the construction and proof.

16. Draw any triangle *ABC*. Bisect *AB* at *F*. Divide *BC* at *D* so that *BD* : *DC* = 2 : 3. Join *AD*, *CF*, meeting at *Y*. Prove that $CY = \frac{3}{4}CF$.

[Draw *FL* parallel to *AD* to meet *BC* at *L*.]

17. Draw any triangle *ABC*. Bisect *AB* at *F*. Divide *CA* at *E* so that *CE* : *EA* = 1 : 2. Join *BE*, *CF*, meeting at *X*. Prove that *CX* = *XF*.

[Draw *FM* parallel to *BE* to meet *CA* at *M*.]

18. Draw any triangle *ABC*. Divide *BC* at *D* so that *BD* : *DC* = 2 : 3. Divide *CA* at *E* so that *CE* : *EA* = 1 : 2. Join *AD*, *BE*, meeting at *Z*. Prove that *EZ* = *ZB*.

[Draw *EN* parallel to *AD* to meet *BC* at *N*.]

19. Draw any triangle *ABC*. Divide *AB* at *F* so that *BF* : *FA* = 2 : 1. Bisect *AC* at *E*. Join *FE* and produce it to meet *BC* produced in *D*. Prove that *BC* = *CD*.

[Draw *CH* parallel to *EF* to meet *AB* in *H*.]

20. Draw any triangle *ABC*. Divide *AC* at *E* so that *AE* : *EC* = 3 : 2. Divide *BC* at *D* so that *BD* : *DC* = 1 : 2. Join *ED* and produce it to meet *AB* produced in *F*. Prove that *AB* = 2*BF*.

[Draw *BH* parallel to *DE* to meet *AC* in *H*.]

Ceva's Theorem*

If the straight lines which join the vertices of a triangle ABC to any point O cut the opposite sides, or these sides produced, at X, Y, Z respectively, then

$$\frac{BX}{XC} \cdot \frac{CY}{YA} \cdot \frac{AZ}{ZB} = 1.$$

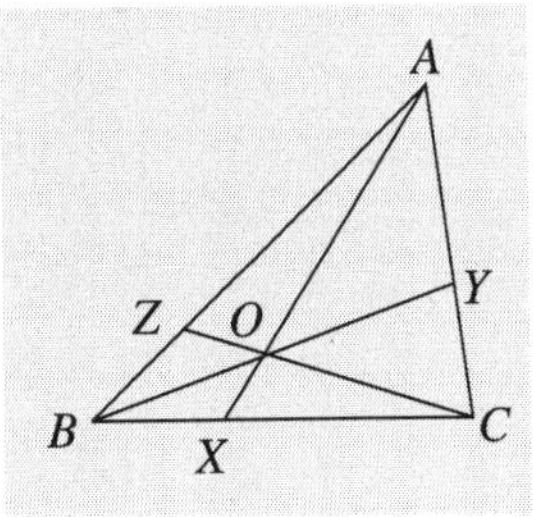

[By Ex. LX. 3, $\frac{BX}{XC} = \frac{\Delta AOB}{\Delta AOC}$, with similar values for *CY* : *YA* and *AZ* : *ZB*.]

*Pronounced Ká'-vâ and Më-nê-la'-us.

Menelaus' Theorem*

If a straight line cuts the sides BC, CA, AB of a triangle ABC, or the sides produced, at X, Y, Z respectively, then

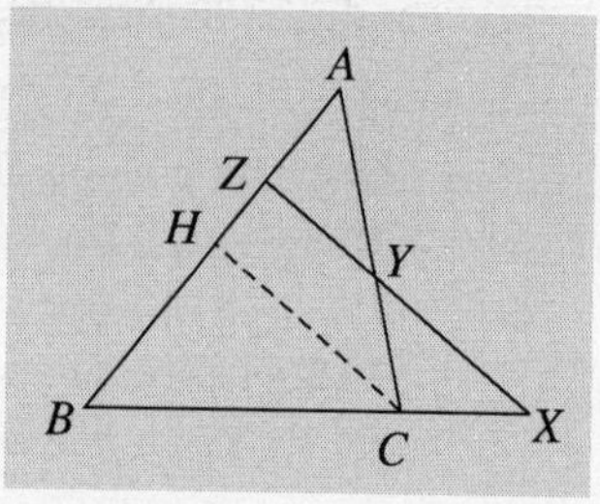

$$\frac{BX}{CX} \cdot \frac{CY}{YA} \cdot \frac{AZ}{ZB} = 1.$$

[Draw CH parallel to YZ to meet AB or AB produced in H. Prove $BX : CX = BZ : HZ$ and $CY : YA = HZ : ZA$.]

EXERCISE LXII

1. Points X, Y are taken in the sides BC, CA of a triangle, such that $BX = \frac{2}{5}BC$ and $CY = \frac{3}{7}CA$. If AX, BY meet at O and CO is produced to meet AB at Z, prove that $AZ = 2ZB$. [Use Ceva's Theorem.]
2. Points Y, Z are taken in the sides CA, AB of a triangle ABC, such that $AY = \frac{1}{3}AC$, $AZ = \frac{1}{4}AB$. If YZ, BC are produced to meet at X, show that $CX = 2BC$. [Use Menelaus' Theorem]
3. In Ex. 16 of the preceding exercise, apply Menelaus' Theorem to the triangle BCF, to prove $CY = \frac{3}{4}CF$.
4. Obtain similar proofs for Exs. 17–20.
5. Use Ceva's Theorem, to prove that the medians of a triangle meet in a point.
6. The inscribed circle of a triangle ABC touches the sides BC, CA, AB in D, E, F respectively; prove that AD, BE, CF meet in a point.
7. *The straight lines which join the vertices of a triangle ABC to any point O cut the opposite sides, or these sides produced at X, Y, Z respectively, such that* $BX : XC = m : n$, $CY : YA = p : q$, $AZ : ZB = r : s$. *Prove that* $AO : OX = \frac{r}{s} + \frac{q}{p} : 1$; *and obtain similar expressions for the ratios* $BO : OY$, $CO : OZ$.

*Pronounced Ká'-vâ and Më-nê-la'-us.

THEOREM 86 (Euclid VI. 3 and *A*)

(*i*) If the vertical angle of a triangle is bisected, internally or externally, by a straight line which cuts the base, or the base produced, it divides the base, internally or externally, in the ratio of the other sides of the triangle.

(*ii*) Conversely, if a straight line through the vertex of a triangle divides the base, internally or externally, in the ratio of the other sides, it bisects the vertical angle, internally or externally.

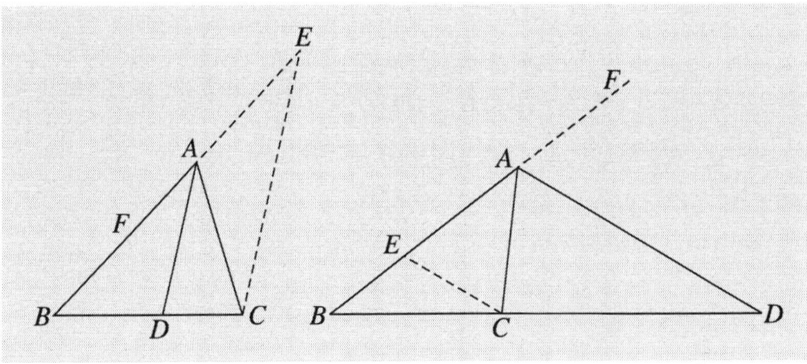

Fig. (a) Fig. (b)

(*i*) Let *AD* bisect the angle *A* of the triangle *ABC*, either internally as in Fig. (*a*), or externally as in Fig. (*b*), and cut *BC* or *BC* produced at *D*.

It is required to prove that

$$BD : DC = BA : AC.$$

Construction. In Fig. (*a*), let *F* be any point in *BA*, and in Fig. (*b*), let *F* be any point in *BA* produced. Draw *CE* parallel to *DA* to cut *BA* or *BA* produced at *E*.

Proof. Because *AD* is parallel to *CE*,

$\therefore$ $$\angle DAC = \angle ACE$$

and $$\angle FAD = \angle AEC;$$

and $$\angle FAD = \angle DAC \ (given);$$

$\therefore$ $$\angle ACE = \angle AEC;$$

$\therefore$ $$AE = AC.$$

Again, because *AD* is parallel to *EC*, a side of the triangle *BEC*,

$\therefore$ $BD : DC = BA : AE.$

But $$AE = AC \ (proved)$$

$\therefore$ $$BD : DC = BA : AC.$$

(*ii*) Let *D* be the point in *BC*, or in *BC* produced, such that

$$BD : DC = BA : AC.$$

It is required to prove that *AD* bisects the angle *A*, internally or externally.

Construction. The same as before.

Proof. Because AD is parallel to EC, a side of the triangle BEC,

$\therefore \quad BD : DC = BA : AE.$

But $\quad BD : DC = BA : AC$ (*given*);

$\therefore \quad BA : AE = BA : AC;$

$\therefore \quad AE = AC;$

$\therefore \quad \angle AEC = \angle ACE.$

But, because AD is parallel to EC,

$\therefore \quad \angle AEC = \angle FAD$

and $\quad \angle ACE = \angle DAC;$

$\therefore \quad \angle FAD = \angle DAC;$

$\therefore$ AD bisects the angle A, internally or externally.

EXAMPLE 1: *Given the base of a triangle and the ratio of its sides, prove that the vertex lies on a fixed circle.*

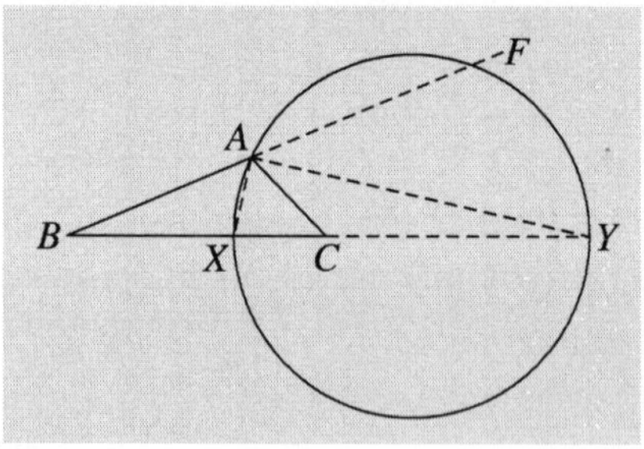

Let ABC be a triangle on a given base BC, and let $AB : AC = p : q$, where $p : q$ is the given ratio.

It is required to prove that A lies on a fixed circle.

Construction. Produce BA to F. Bisect the angles BAC, CAF by Straight lines meeting BC, or BC produced, at X and Y.

Proof. Because AX, AY are the bisectors of the interior and exterior angles at A of the triangle ABC,

$\therefore \quad BX : XC = BA : AC,$

and $\quad BY : YC = BA : AC.$

But $\quad BA : AC = p : q;$

$\therefore$ X, Y are the points at which BC is divided internally and externally in the ratio $p : q$;

$\therefore$ X and Y are fixed points.

Also, because AX, AY are the bisectors of the angles BAC, CAF,

$\therefore$ $\angle XAY$ is a right angle;

$\therefore$ A is on the circle on XY as diameter, and this is a fixed circle.

EXAMPLE 2: *If a straight line BC is divided internally at X and externally at Y in the ratio p : q, and if A is any point on the circle on XY as diameter, then*

$$AB : AC = p : q.$$

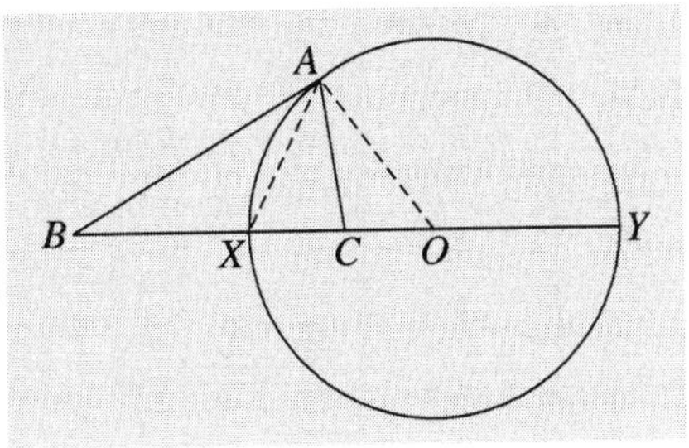

Construction. Let O be the middle point of XY. Join AX, OA

Proof. Because $\quad BX : XC = BY : CY$ (Hyp),

$\therefore \quad XC : CY = BX : BY$;

that is, XY is divided internally and externally in the same ratio at C and B. And because O is the middle point of XY,

$\therefore \quad OX^2 = OB \cdot OC$ (Ex. LIX. 5 (ii));

$\therefore \quad OA^2 = OB \cdot OC$;

$\therefore$ the circle ABC touches OA;

$\therefore \quad \angle OAC = \angle ABC$ in the alternate segment;

Now, $\quad \angle BAX = \angle OXA - \angle ABC$;

$\therefore \quad \angle BAX = \angle OAX - \angle OAC = \angle CAX$;

$\therefore$ AX bisects $\angle BAC$;

$\therefore \quad AB : AC = BX : XC = p : q.$

The last two examples prove the following important theorem: The locus of a point which moves so that the ratio of its distances from two fixed points is constant is a circle.*

THEOREM 87 (Euclid VI. 33)

In equal circles, angles at the centres have the same ratio as the arcs on which they stand; so also have the sectors.

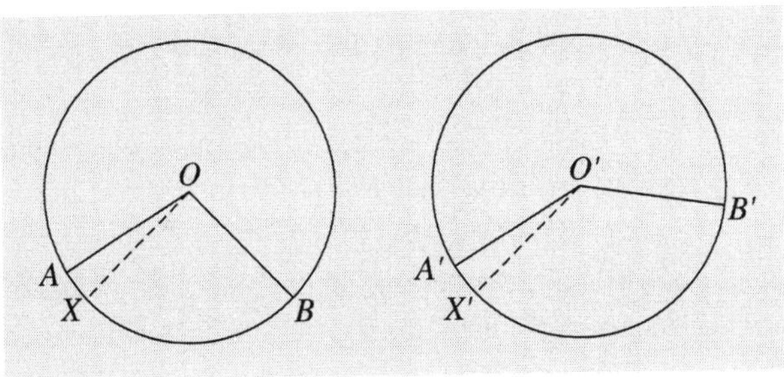

* *This is called the* Circle of Apollonius.

Let AB, $A'B'$ be arcs of equal circles of which O, O' are the centres. It is required to prove that

$$\frac{\angle AOB}{\angle A'O'B'} = \frac{\text{arc } AB}{\text{arc } A'B'} = \frac{\text{sector } AOB}{\text{sector } A'O'B'}.$$

Construction. Let the arcs AB, $A'B'$ be commensurable, and from them cut off arcs AX, $A'X'$ respectively, each equal to a common measure of the arcs AB, $A'B'$, so that

$$\text{arc } AB = p.\ \text{arc } AX$$

and $$\text{arc } A'B' = q.\ \text{arc } A'X',$$

where p and q are whole numbers. Join OX, $O'X'$.

Proof. Since, equal arcs subtend equal angles at the centres,

∴ $$\angle AOB = p.\ \angle AOX$$

and $$\angle A'O'B' = q.\ \angle A'O'X'.$$

Also, since the sectors which stand on equal arcs are equal,

∴ $$\text{sector } AOB = p.\ \text{sector } AOX,$$

$$\text{sector } A'O'B' = q.\ \text{sector } A'O'X'.$$

Now, $$\angle AOX = \angle A'O'X',$$

and $$\text{sector } AOX = \text{sector } A'O'X'\ ;$$

∴ $$\frac{\angle AOB}{\angle A'O'B} = \frac{\text{arc } AB}{\text{arc } A'B'} = \frac{\text{sector } AOB}{\text{sector } A'O'B'}.$$

Cor. In equal circles, angles at the circumferences have the same ratio as the arcs on which they stand, for each is half the corresponding angle at the centre.

EXERCISE LXIII(a) (Theorem 86)

1. ABC is a triangle with the angle A a right angle ; AX, AY meet BC, or BC produced, in X, Y, and are equally inclined to AB. Show that
$$BX : BY = XC : CY.$$
2. ABC is a triangle, in which $AB > AC$. The bisector of the angle A meets BC in D, and O is the middle point of BC. Show that
$$OD : OB = AB - AC : AB + AC.$$
3. The angle C of a triangle ABC is bisected by a line which cuts AB in F. The angle B is bisected by a line which cuts CF in I. Prove that
$$AF : FI = AC : CI.$$
4. $PQRS$ is a quadrilateral; show that if the bisectors of the angles P and R meet in the diagonal QS, then the bisectors of the angles Q and S will meet in PR.

5. ABC is a triangle, D the middle point of the side BC; the straight line DE which bisects the angle ADB meets AB in E, and the straight line DF which bisects the angle ADC meets AC in F. Prove that EF is parallel to BC. [Use Th. 85]

6. The bisector of the angle BAC of a triangle ABC meets the side BC at D. The circle described about the triangle BAD meets CA again at E, and the circle described about the triangle CAD meets BA again at F. Show that BF is equal to CE. [Use Th. 80]

7. A and B are fixed points. Another point, P, moves so that PA is always equal to $3PB$. Prove that P lies on a circle whose diameter is equal to $\frac{3}{4}AB$. [See Example 1, p. 266]

8. A and B are given points and CD a given straight line; find, when possible, a point P in CD such that $PA = 3PB$.

9. A, B, C are three given points in the same straight line. Prove that the points, at which AB and BC subtend equal angles, lie on a fixed circle.

10. Given four collinear points A, B, C, D, find a point at which AB, BC, CD subtend equal angles.

11. Draw a triangle ABC, in which $BC = 2$ inches, $3AB = 2AC$ and the area of ABC is 1 square inch. Show that there are two solutions, and measure AB in each case.

12. Draw a straight line 2 inches long, and on it, as base, construct the triangle of greatest possible area with its sides in the ratio 3 : 4. Explain the construction, and calculate (without measurement) the area of the triangle.

13. Construct a parallelogram whose sides are given and whose diagonals are in the ratio of 1 : 2. Explain the construction.

14. A and B are the centres of two given circles and CD is any straight line parallel to AB, meeting the first circle at C and the second at D. If P is the point of intersection of AC, BD, prove that P lies on a fixed circle.

EXERCISE LXIII(b)

1. The bisector of the angle A of the triangle ABC meets BC at X. If $BC = a$, $CA = b$, $AB = c$, prove that

$$BX = \frac{ac}{b+c} \text{ and } CX = \frac{ab}{b+c}.$$

2. The bisector of the exterior angle at A of the triangle ABC meets BC produced at X'. If $AB > AC$, prove that

$$BX' = \frac{ac}{c-b} \text{ and } CX' = \frac{ab}{c-b}.$$

3. The bisectors of the interior and exterior angles at A of the triangle ABC meet BC at X and X'. If $AB > AC$. prove that

$$XX' = \frac{2abc}{c^2 - b^2}. \text{ [Use Exs. 1, 2]}$$

4. ABC is a triangle in which $AB > AC$. If L is the middle point of BC, AP is perpendicular to BC and the bisectors of the interior and exterior angles at A meet BC in X, X', prove that

(*i*) $LX = \frac{a}{2} \cdot \frac{c-b}{c+b}$;

(*ii*) $LX' = \frac{a}{2} \cdot \frac{c+b}{c-b}$;

(*iii*) $LP = \frac{c^2 - b^2}{2a}$.

(*iv*) $LX \cdot LP$ is equal to the square on the tangent from L either to the inscribed circle or to the escribed circle opposite A.

(*v*) $LP \cdot LX'$ is equal to the square on the tangent from L to either of the escribed circles opposite B and C respectively.

[For (*iii*) see Ex. LII 16, and for (*iv*), Ex. XLVI(*a*) 9.]

5. The bisector of the angle A of the triangle ABC meets BC at X. Prove that $b \cdot BX = c \cdot CX$. Hence show that

$$bc(b+c) = b \cdot BX^2 + c \cdot CX^2 + (b + c)AX^2.$$

[See Ex. LVI. 10]

6. The bisector of the angle A of the triangle ABC meets BC at X. Prove that

$$AX^2 = bc\left\{1 - \frac{a^2}{(b+c)^2}\right\}.$$

[Use Exs. 1, 5]

7. The bisector of the exterior angle at A of the triangle ABC meets BC produced at X'. Prove that $b.\ BX' = c.CX'$. Hence show that, if $c > b$,

(*i*) $bc(c - b) = b \cdot BX'^2 - c \cdot CX'^2 - (c - b)AX'^2$.

(*ii*) $AX'^2 = bc\left\{\frac{a^2}{(c-b)^2} - 1\right\}$.

[See Ex. LVI, II]

8. Use the values for AX^2 and AX'^2 found in Exs. 6, 7 to verify the value found for XX' in Ex. 3.

9. If I is the centre of the circle inscribed in the triangle ABC and AI cuts BC at X, show that

$$AI = \frac{b+c}{a+b+c} \cdot AX,$$

[BI, CI bisect $\angle$s B, C ; $\therefore AI : IX = AB : BX = AB : CX$.]

10. If I is the centre of the circle inscribed in the triangle ABC, prove that

$$AI^2 = bc \cdot \frac{s-a}{s},$$

where $2s = a + b + c$. [Use Ex. 6]

11. If I_1 is the centre of the escribed circle of the triangle ABC opposite A, and AI_1 cuts BC at X, show that

$$AI_1 = \frac{b+c}{b+c-a} \cdot AX.$$

[BI_1, CI_1 bisect the exterior angles at B, C;
$\therefore\ AI_1 : XI_1 = AB : BX = AC : CX$.]

12. If I_1 is the centre of the escribed circle of the triangle ABC opposite A, show that

$$AI_1^2 = bc \cdot \frac{s}{s-a}. \quad \text{[Use Ex. 6]}$$

XXXVIII. SIMILAR TRIANGLES

THEOREM 88 (Euclid VI. 4)

If two triangles are equiangular, their corresponding sides are proportional.*

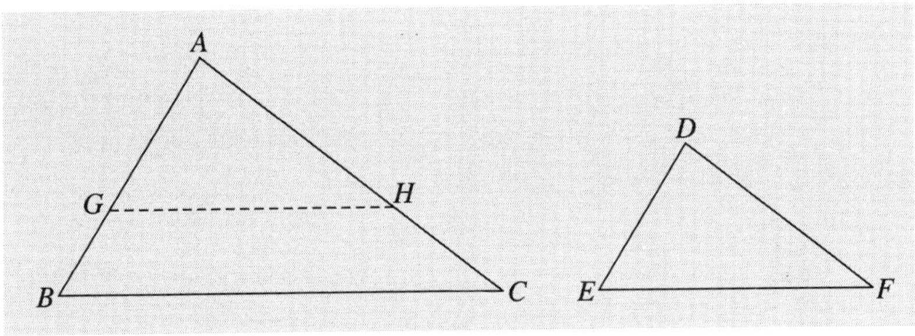

Let ABC, DEF be two triangles in which the angles A, B, C are equal to the angles D, E, F respectively.

It is required to prove that

$$AB : DE = AC : DF = BC : EF.$$

Construction. Along AB, AC, set off AG, AH, equal to DE, DF respectively. Join GH.

Proof. In the triangles AGH, DEF,

$$\begin{cases} AG = DE \text{ (construction)}, \\ AH = DF \text{ (construction)}, \end{cases}$$

* When it is said that the sides a, b, c of one triangle are proportional to the sides a', b', c' of another triangle, the meaning is that

$$a : a' = b : b' = c : c'.$$

$$\angle A = \angle D \ (given);$$

∴ The triangles are congruent;

∴ $$\angle AGH = \angle E.$$

But $$\angle B = \angle E \ (given);$$

∴ $$\angle AGH = \angle B,$$

and these are corresponding angles;

∴ GH is parallel to BC;

∴ $$AB : AG = AC : AH;$$

∴ $$AB : DE = AC : DF.$$

Similarly, it can be proved that

$$AC : DF = BC : EF.$$

DEFINITIONS. Two rectilineal figures are said to be equiangular when the angles of one, taken in order, are equal to the angles of the other, taken in order.

Two rectilineal figures are said to be similar when they are equiangular and the sides about every pair of equal angles are proportional.

In Theorem 88 it has been shown that if, in the triangles ABC, DEF, $\angle A = \angle D$, $\angle B = \angle E$, and, consequently, $\angle C = \angle F$,

then $$AB : DE = AC : DF = BC : EF;$$

that is, the sides about every pair of equal angles are proportional.

Hence, *if two triangles are equiangular, they are also similar*, but the corresponding statement is not true for figures of more than three sides.

The sides of similar triangles which are opposite equal angles are called corresponding sides.

Observe that the two sides (such as AB, DE) which are the terms of any of the ratios in the above statement of proportion are corresponding sides.

In working with similar triangles, *it is important to arrange the letters in order*. Thus, if ABC, DEF are similar triangles, the order of letters should be such that $\angle A = \angle D$, $\angle B = \angle E$, $\angle C = \angle F$. If this is so, the relations between the sides can be written down without looking at the figure.

It is obvious that two similar triangles can be placed so that their corresponding sides are parallel. If this can be done *without turning one of the triangles over*, the triangles are said to be directly similar. Otherwise, the triangles are said to be inversely similar.

In the figure of Theorem 89, the triangles ABC, DEF are directly similar and the triangles ABC, GEF are inversely similar.

A triangle which varies in such a way as to be always directly similar to a given triangle is said to be of **given species**.

THEOREM 89 (Euclid VI. 5)

If the three sides of one triangle are proportional to the three sides of another triangle, the two triangles are equiangular.

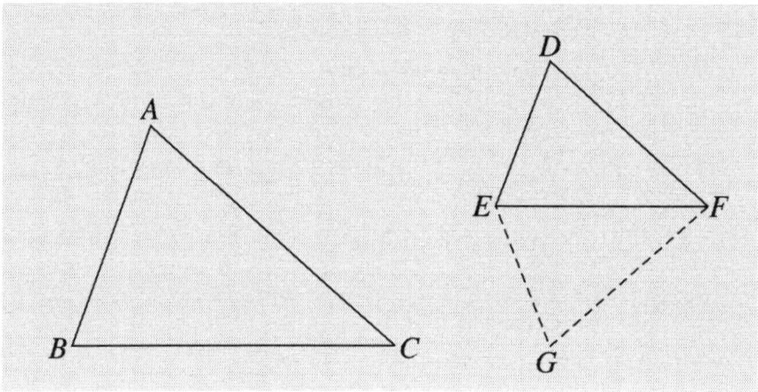

Let ABC, DEF be two triangles in which

$$AB : DE = AC : DF = BC : EF.$$

It is required to prove that ABC, DEF are equiangular.

Construction. On the side of EF opposite to D, make $\angle FEG$ equal to $\angle B$ and $\angle EFG$ equal to $\angle C$.

Proof. Because $\angle FEG = \angle B$ and $\angle EFG = \angle C$,

$\therefore$ the triangles ABC, GEF are equiangular;

$\therefore$ $AB : GE = BC : EF.$

But $AB : DE = BC : EF$ (*given*);

$\therefore$ $AB : GE = AB : DE$;

$\therefore$ $GE = DE.$

Similarly $GF = DF.$

Hence, in the triangles GEF, DEF,

$$\begin{cases} GE = DE \ (\textit{proved}), \\ GF = DF \ (\textit{proved}), \\ EF \text{ is common.} \end{cases}$$

$\therefore$ The triangles GEF, DEF are congruent;

$\therefore$ $\angle GEF = \angle DEF$

and $\angle GFE = \angle DFE.$

But $\angle GEF = \angle B$

and $\angle GFE = \angle C$ (*construction*);

$\therefore$ $\angle B = DEF,$

$\angle C = \angle DFE,$

and consequently $\angle A = \angle EDF.$

EXERCISE LXIV (Theorems 88, 89)

1. *ABCD* is a parallelogram. *CD* is bisected at *E* and *BE* meets *AC* at *X*. Prove that $AX = \frac{2}{3}AC$. [Consider Δs *AXB*, *CXE*]

2. *ABCD* is a parallelogram. *AB* is divided at *X* and *CD* at *Y*, so that $AX : XB = 3 : 2$ and $CY : YD = 4 : 1$. If *XY* meets. *AC* at *Z*, prove that $AZ = \frac{3}{7}AC$.

3. *ABCD* is a trapezium with *AB* parallel to *CD*, and $AB = 2CD$. If *AC*, *BD* meet at *E*, prove that $AE = \frac{2}{3}AC$. Also, if *AD*, *BC* are produced to meet at *F*, prove that $DF = AD$.

4. *OA*, *OB*, *OC* are three given straight lines and *P* is any point in *OC*. If *PM*, *PN* are perpendicular to *OA*, *OB*, show that the ratio $PM : PN$ is constant. [Consider the angles of the triangle *PMN*.]

5. *OX, OY, OZ are three given straight lines; AB, CD are two straight lines parallel to one another, cutting OX, OY, OZ in P, Q, R and P', Q', R' respectively. Prove that* $PQ : QR = P'Q' : Q'R'$.

6. *AB* and *CD* are the parallel sides of a trapezium *ABDC*, whose diagonals *AD*, *BC* cut in *E*. Through *E* a straight line *LEM* is drawn parallel to *AB* to cut *AC* in *L* and *BD* in *M*. Show that $LE = EM$.

7. *P* is any point on a circle of which *C* is the centre and *AB* is a diameter. *PN* is perpendicular to *AB* and the tangent at *P* meets *BA* produced at *T*. Show that the triangles *CNP*, *CPT* are equiangular, and that

$$CN : CP = CP : CT.$$

8. The tangent at *P* to a circle, whose centre is *C*, meets two parallel tangents in *Q*, *Q'* ; show that $PQ \cdot PQ' = CP^2$.

9. *ABC* is a triangle, in which $AB = AC$; *AD* is the perpendicular to *BC*; *R* is the radius of the circle *ABC*. Prove that $2R \cdot AD = AB^2$.

10. *ABC*, *DEF* are equiangular triangles and *AM*, *DN* are the perpendiculars from *A*, *D* to the corresponding sides *BC*, *EF*. Prove that

$$AM : DN = BC : EF.$$

11. *ABC*, *A'B'C'* are equiangular triangles and *BC*, *B'C'* are corresponding sides. If *R*, *R'* are the radii of the circumcircles and *r*, *r'* the radii of the inscribed circles of the triangles, prove that

$$R : R' = BC : B'C' \text{ and } r : r' = BC : B'C'.$$

12. Two circles of radii *r*, *r'* intersect at *AB*. Any straight line through *A* meets the circles again at *C*, *D*. Prove that $BC : BD = r : r'$. [Let *O*, *O'* be the centres. Prove Δs *BOC*, *BO'D* equiangular.]

13. AB, CD are chords of a circle and AD, BC meet at P. If PM, PN are the perpendiculars from P to AB, CD, prove that

$$PM : PN = AB : CD.$$

14. $ABCD$ is a parallelogram and AXY is any straight line through A, meeting BC at X and DC produced at Y. Prove that $BX \cdot DY$ is constant.

15. $ABCD$ is a parallelogram and any straight line is drawn through A to meet BD at X, BC at Y and DC produced at Z. Prove that

$$AX : XZ = AY : AZ.$$

[Consider Δs ABX, ZDX and Δs ABY, ZDA.]

16. Through the vertex A of the triangle ABC, a straight line DAE is drawn parallel to BC, such that $DA = AE$. CD meets AB at P and BE meets AC at Q. Prove that PQ is parallel to BC.

17. In a triangle ABC, BC is bisected at D, AD is bisected at E and BE is produced to meet AC at F. Prove that $3AF = AC$ and $4EF = BF$.

[Draw DK parallel to BF to meet AC at K, or use Menelaus' Theorem.]

18. In a triangle ABC, AB is divided at F, so that $AF : FB = 2 : 3$. FC is divided at H, so that $FH : HC = 3 : 5$. AH is produced to meet BC at D. Prove that

$$BD : DC = 3 : 2 \text{ and } AH : HD = 5 : 3.$$

[Draw FL parallel to AD to meet BC at L, or use Menelaus' Theorem.]

19. One of the sides containing the right angle of a right-angled triangle is double the other, and circles are described on these sides as diameters. Show that the common chord of these circles is equal to two-fifths of the hypotenuse.

20. Given four points A, B, C, D in a straight line. It is required to find a point O in the same straight line, such that

$$OA : OB = OC : OD.$$

Prove the following construction,—On AC as base, draw any triangle AXC. Draw BY, DY parallel to AX, CX respectively, meeting at Y. Join XY, meeting AD, or AD produced, at O. Then O is the required point.

21. *Given four points A, B, C, D in a straight line. It is required to find a point O in the same straight line, such that*

$$OA \cdot OB = OC \cdot OD.$$

Prove the following construction,—On AD as base, draw any triangle AXD. Draw BY, CY parallel to DX, AX respectively, meeting at Y. Join XY, meeting AD at O. Then O is the required point. [Compare Ex. LVII, 14]

22. $ABCD$, $AECF$ are two parallelograms and EF is parallel to AD. If AF, DE meet at G and BF, CE meet at H, prove that GH is parallel to AB.

THEOREM 90 (Euclid VI. 6)

If two triangles have an angle of the one equal to an angle of the other, and the sides about these equal angles proportional, the triangles are equiangular.

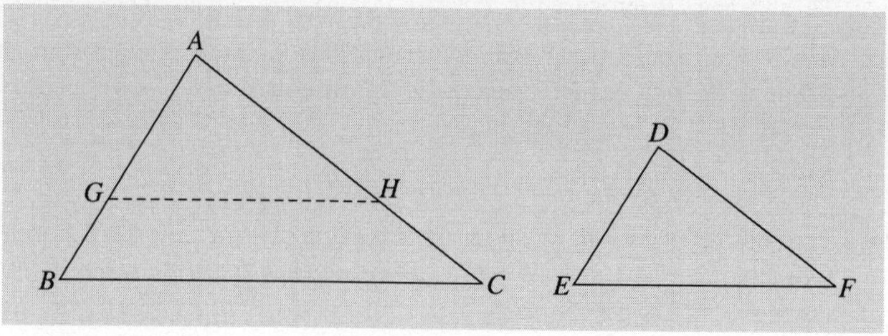

Let *ABC*, *DEF* be two triangles, in which $\angle A = \angle D$

and $$AB : DE = AC : DF.$$

It is required to prove that the triangles are equiangular.

Construction. Along *AB*, *AC* set off *AG*, *AH* equal to $DE \cdot DE$ respectively. Join *GH*.

Proof. In the triangles *AGH*, *DEF*,

$$\begin{cases} AG = DE \ (construction), \\ AH = DF \ (construction) \\ \text{and} \quad \angle A = \angle D \ (given). \end{cases}$$

∴ The triangles are congruent;

∴ $$\angle AGH = \angle E.$$

Now, by hypothesis,

$$AB : DE = AC : DF;$$

∴ $$AB : AG = AC : AH;$$

∴ *GH* is parallel to *BC*;

∴ $\angle AGH$ = the corresponding $\angle B$.

But $$\angle AGH = \angle E \ (proved);$$

∴ $$\angle B = \angle E;$$

also $$\angle A = \angle D \ (given);$$

∴ $\angle C = \angle F$, and the triangles *ABC*, *DEF* are equiangular.

EXAMPLE 1: *A common tangent to two circles divides the straight line joining the centres, externally or internally, in the ratio of the radii.*

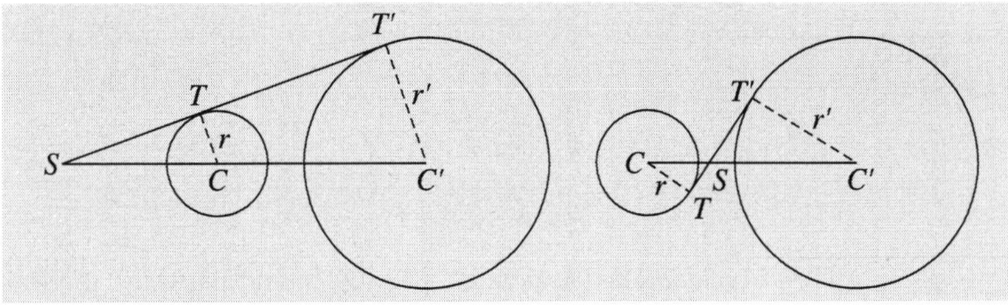

Let TT' be a common tangent to two circles whose centres are C, C' and radii r, r'. Let TT' meet CC', or CC' produced, at S.

Because TT' touches both circles,

$\therefore$ $\angle$ s CTS, $C'T'S$ are right angles;

$\therefore$ Δs CTS, $C'T'S$ are equiangular;

$\therefore$ $$SC : SC' = CT : C'T' = r : r'.$$

EXAMPLE 2: *P is any joint on a circle whose centre is C, and S is a fixed point. If SP it divided externally or internally at P' in a constant ratio, the locus of P' is a circle.*

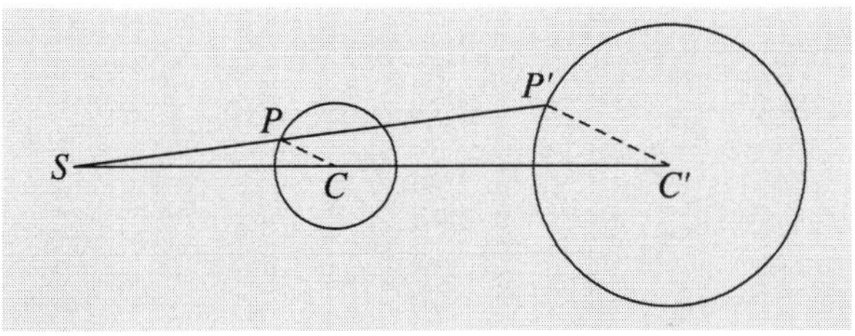

Let $p : q$ be the given ratio, so that $SP : SP' = p : q$.

Along SC set off SC', so that $SC : SC' = p : q$ (Construction 29).

Join $C'P'$.

In the triangles SCP, $SC'P'$,

$$\begin{cases} \text{the angle } S \text{ is common,} \\ \text{and } SC : SC' = SP : SP'. \end{cases}$$

$\therefore$ The triangles are equiangular;

$\therefore$ $$CP : C'P' = SP : SP' = p : q;$$

$\therefore$ $C'P'$ is of constant length,

and C' is a fixed point;

$\therefore$ the locus of P' is a fixed circle.

THEOREM 91 (Euclid VI. 7)

If two triangles have an angle of the one equal to an angle of the other, and the sides about two other angles proportional, the third angles are equal or supplementary.

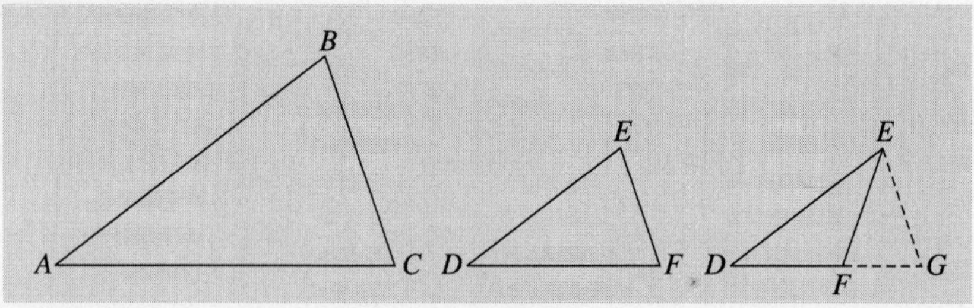

Let ABC, DEF be two triangles, in which $\angle A = \angle D$, and the sides about the angles B and E proportional, so that

$$AB : DE = BC : EF.$$

It is required to prove that the angles C and F are equal or supplementary.

Proof. If $\angle C$ is not equal to $\angle F$, draw EG, on the same side of ED as F, making $\angle DEG$ equal to $\angle B$. Let EG meet DF, or DF produced, at G.

Because $\angle A = \angle D$ (*given*)

and $\angle B = \angle DEG$ (*construction*),

$\therefore$ $\angle C = \angle DGE$,

and the triangles ABC, DEG are equiangular);

$\therefore$ $AB : DE = BC : EG$.

But $AB : DE = BC : EF$ (*given*);

$\therefore$ $BC : EG = BC : EF$;

$\therefore$ $EG = EF$;

$\therefore$ $\angle EGF = \angle EFG$

$\therefore$ $\angle C = \angle EFG$.

But the angles EFG and EFD are supplementary;

$\therefore$ the angles C and EFD are supplementary.

Theorem 92 (Euclid VI. 8)

In a right-angled triangle, if a perpendicular is drawn from the right angle to the hypotenuse, the triangles on each side of the perpendicular are similar to the whole triangle, and to one another.

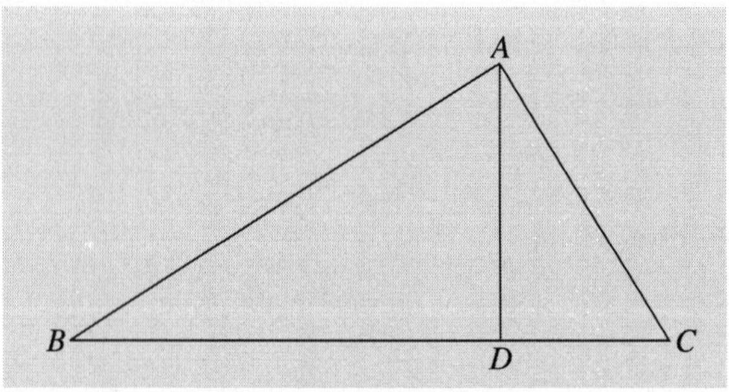

Let ABC be a triangle, in which $\angle A$ is a right angle, and let AD be perpendicular to BC.

It is required to prove that the triangles DBA, DAC are similar to the triangle ABC, and to one another.

Proof. In the triangles DBA, ABC,

$\angle$s BDA, BAC are right angles, and $\angle B$ is common.

Hence, the triangles DBA, ABC are equiangular, and therefore similar.

In the same way, it can be shown that the triangles DAC, ABC are similar;

$\therefore$ also the triangles DBA, DAC are similar.

COR. DA is a mean proportional to DB and DC.

For since the triangles DBA, DAC are similar,

$$\therefore \qquad DB : DA = DA : DC.$$

BA is a mean proportional to BD and BC.

For since the triangles DBA, ABC are similar,

$$\therefore \qquad DB : AB = AB : CB$$

CA is a mean proportional to CD and CB.

For since the triangles DAC, ABC are similar,

$$\therefore \qquad CD : CA = CA : CB.$$

EXERCISE LXV (Theorems 90–92)

1. ABC, $A'B'C'$ are equiangular triangles and BC, $B'C'$ are corresponding sides. If X, X' are the middle points of BC, $B'C'$, prove that $AX : A'X' = BC : B'C'$.
2. ABC, $A'B'C'$ are equiangular triangles, in which $\angle A = \angle A'$ and $\angle B = \angle B'$. If BC, $B'C'$ are divided in the same ratio at X, X', so that $BX : XC = B'X' : X'C'$, prove that the triangles ABX, $A'B'X'$ are similar.
3. In the triangle ABC take any point P. Find a point Q outside the triangle, such that $\angle CBQ = \angle ABP$ and $\angle BCQ = \angle BAP$. Prove that the triangles PBQ, ABC are equiangular.
4. Two obtuse-angled triangles have an acute angle of the one equal to an angle of the other, and the sides about the other acute angles proportional; prove that the triangles are equiangular.
5. OA, OB are given straight lines ; PM, $P'M'$ are perpendiculars from points P, P' to OA, and PN, $P'N'$ are perpendiculars to OB. If $PM : PN = P'M' : P'N'$, prove that O, P, P' are in the same straight line.

   [Use the triangles PMN, $P'M'N'$ to show that $\angle PMN = \angle P'M'N'$, and consider the cyclic quadrilaterals $OMPN$, $O'M'P'N'$.]

6. *PM, PN are the perpendiculars from a point P to two given straight lines OA, OB. If P moves so that PM : PN is constant, prove that the locus of P is a straight line through O.*
7. Draw any triangle *ABC*, and construct the locus of a point *P* which moves so that the ratio of the perpendiculars from *P* to *AB*, *AC* are in the ratio 2 : 3.
8. Draw any triangle *ABC*, and find a point *P* within it such that the perpendiculars from *P* to *BC*, *CA*, *AB* are in the ratio 2 : 3 : 4.
9. *ABC* is a triangle with the angle *A* a right angle, and *AD* is perpendicular to *BC*. Prove that $BD : DC = AB^2 : AC^2$.
10. *AB* is the diameter of a semicircle. Through, *A* draw any straight lines *AP*, *AQ* to meet the circumference again at *P*, *Q*. Draw *PM*, *QN* perpendicular to *AB*. Prove that $AM : AN = AP^2 : AQ^2$.
11. From the vertex *A* of the triangle *ABC*, perpendiculars *AL*, *AM* are drawn to the bisectors of the exterior angles of the triangle at *B* and *C*. Prove that *LM* is parallel to *BC*. [Let *I* be the centre of the inscribed circle. Prove that $AL : AM = IB : IC$, etc.]
12. *A* is a fixed point and *P* is any point on a given circle. From *AP* cut off $AQ = \frac{2}{3} AP$. Construct the locus of *Q*.
13. Given two circles and a point *A*, through *A* draw a straight line *APQ* to meet one of the circles at *P* and the other at *Q*, so that $PQ = 2AP$.
14. Three concurrent straight lines *OA*, *OB*, *OC* are cut by two transversals at *X*, *Y*, *Z* and *X'*, *Y'*, *Z'*. If $XY : YZ = X'Y' : Y'Z'$, show that the transversals are parallel. [If *X'Y'Z'* is not parallel to *XYZ*, draw *X'MN* parallel to *XYZ*, cutting *OB*, *OC* at *M*, *N*, etc.]
15. *ABC* is a triangle, *P* is the foot of the perpendicular from *A* to *BC*, *X* is the point where the bisector of the angle *A* cuts *BC*, *D* the point of contact with *BC* of the inscribed circle, D_1 the point of contact with *BC* of the escribed circle which touches *AB*, *AC* produced:
    (*i*) Prove that DD_1 is divided internally at *X* and externally at *P* in the same ratio. [Use Example 1, p. 277]
    (*ii*) Hence, show that, if *L* is the middle point of *BC*,
    then $$LX \cdot LP = LD^2 \text{ and } LP \cdot DX = LD \cdot DP.$$
16. *O is any point within a triangle ABC and O' is another point within the triangle, such that*
$$\angle BAO' = \angle CAO \text{ and } \angle ABO' = \angle CBO.$$
Join CO, CO', and prove that $\angle ACO' = \angle BCO$.
[Let *p*, *q*, *r* be the perpendiculars from *O* and *p'*, *q'*, *r'* those from *O'* to *BC*, *CA*, *AB*. Adapt the method of Ex. 5, to prove that $q : r = r' : q'$ and $r : p = p' : r'$. Hence show that $p : q = q' : p'$, and consequently $\angle ACO' = \angle BCO$.]

EXERCISE LXVI

Two important Loci

1. *A is a given point and P is any point on a given straight line BC. On AP draw a triangle APQ, directly similar to a given triangle DEF. Find the locus of Q.*

[Draw AM perpendicular to BC. On AM, draw $\Delta\,AMN$ directly similar to $\Delta\,DEF$. Join QN. We shall prove that $\angle\,ANQ$, is a right angle.

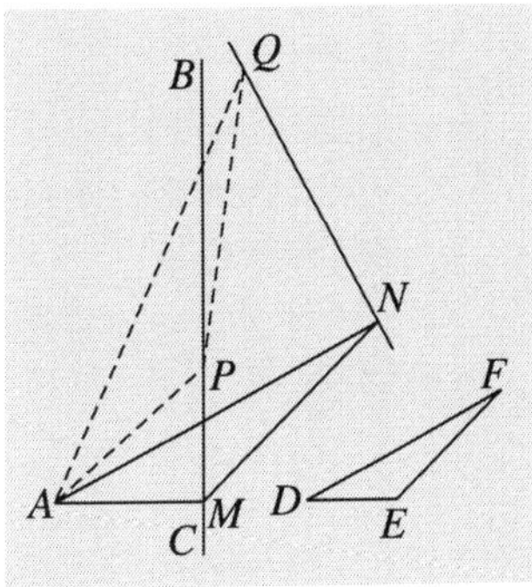

For Δs APQ, AMN are equiangular;

$\therefore$ $AQ : AN = AP : AM$

and $\angle\,QAP = \angle\,NAM$;

$\therefore$ $\angle\,QAN = \angle\,PAM$.

Hence Δs QAN, PAM have the sides about the equal angles QAN, PAM proportional;

$\therefore$ Δs QAN, PAM are equiangular;

$\therefore$ $\angle\,ANQ = \angle\,AMP$ = a right angle;

$\therefore$ the locus is the straight line through N perpendicular to AN.]

2. A is a given point and X is any point on a given straight line BC. On AX draw a square $AXYZ$. Construct the locus of Y.

3. ABC is a given triangle. Draw a square $XYZW$, with the vertex X at a given point in AB, Y on BC and Z on AC. [Use the last example to find the locus of Z if Y is supposed to move along BC.]

4. ABC is a given triangle. Draw an equilateral triangle XYZ with the vertex X at a given point in AB, Y on BC and Z on AC.

5. Explain how to inscribe, in a triangle ABC, a triangle equiangular to a triangle DEF, with one vertex at a given point in AB. Show that, in general, there are two solutions.

6. A is a given point and BC is a chord of a given circle. Through A draw two straight lines AP, AQ, at right angles, to meet BC at P and the circle at Q, such that $AQ = 2AP$. Show that, in general, there are two solutions or no solution.

7. Construct an equilateral triangle with its angular points on three given parallel straight lines.

[Take any point on one of the lines as one vertex, and use Example 1.]

8. *A is a given point and P is any point on a given circle with centre C. On AP draw a triangle APQ, directly similar to a given triangle XYZ. Find the locus of Q.*

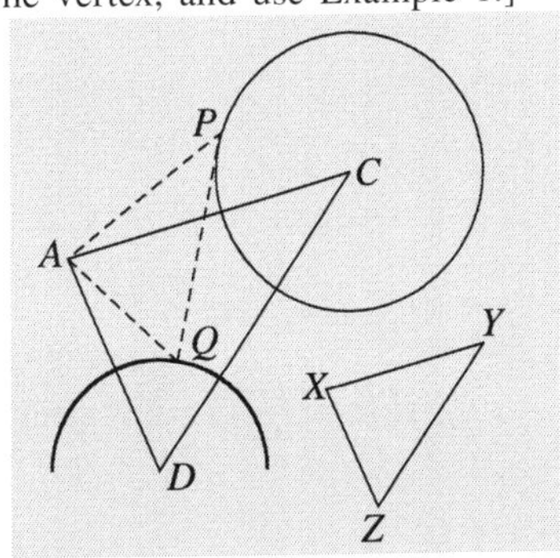

[On AC draw $\Delta\,ACD$, directly similar to XYZ. Then D is a fixed point and DQ is of constant length; hence the locus is the circle with centre D and radius DQ.

Proof. Join PC, QD.

As in Ex. 1, Δs AQD, APC can be shown to be similar;

$$\therefore \quad DQ : CP = AQ : AP = XZ : XY;$$

$\therefore$ DQ is of constant length.]

9. Through a given point A draw two straight lines at right angles to meet a given circle at P and Q respectively, such that

$$AP = 2AQ.$$

10. A and B are fixed points on the circumference of a given circle and P is any other point on the circumference ; if AP is produced to Q, so that $PQ : PB$ is a given ratio, find the locus of Q.

11. Given three concentric circles, construct triangle, similar to a given triangle, with one vertex on each of the given circles.

XXXIX. THIRD, FOURTH AND MEAN PROPORTIONALS

Construction 31

Find a fourth proportional to three given straight lines.

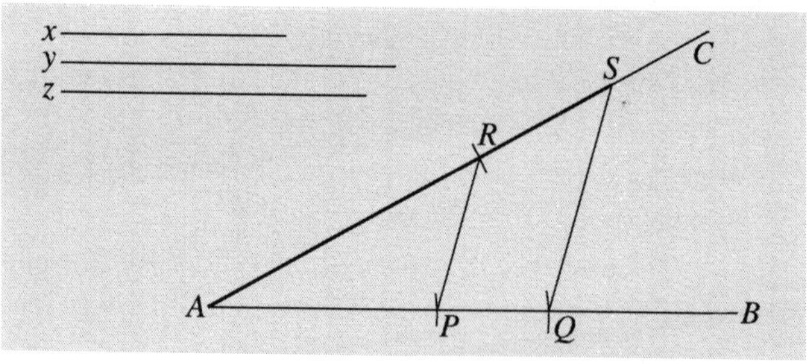

Let x, y, z be the given straight lines. It is required to find a fourth proportional to x, y, z.

Construction. Draw two straight lines AB, AC, making any convenient angle. Along AB set off AP and AQ equal to x and y respectively. Along AC set off AR equal to z. Join PR. Draw QS parallel to PR to meet AC in S.

Then AS is the required fourth proportional.

Proof. Because PR is parallel to QS,

$$\therefore \quad \angle APR = \text{the corresponding } \angle AQS;$$

$\therefore$ the triangles APR, AQS are equiangular;

$$\therefore \quad AP : AQ = AR : AS;$$

that is, $$x : y = z : AS;$$

$\therefore$ AS is the fourth proportional to x, y, z,

EXERCISE LXVII

1. Show graphically that $\frac{6 \times 3}{2} = 9$.

2. Find graphically the value of $\frac{4.3 \times 2.7}{5.9}$.

3. Find graphically the value of 2.7×5.9.

4. Construct a line of length $\frac{1}{1.67}$ inches. [Find a third proportional to 1.67, 1]

5. *Given a straight line x units long, construct one of length x^3 units.*
[Draw two straight lines *AB*, *AC*, making any convenient angle. Along *AB* set off *AD* of unit length, and along *AC* set off *AE* *x* units in length. With centre *A* and radius *AE* draw an arc of a circle to cut *AB* in *F*. Draw *FG*, parallel to *DE*, to cut *AC* in *G*. Then *AG* is x^2 units in length. Complete the construction and supply proof.]

6. Find graphically, to two places of decimals, the values of (*i*) $(1.4)^3$, (*ii*) $(1.32)^4$.

7. Find graphically the values of $\sqrt{2}$, $\sqrt{3}$, $\sqrt{5}$, correct to two places of decimals.
[Find a mean proportional to 1 and 2, etc., using any suitable unit.]

8. Find graphically the value of $\sqrt{31}$.
[Finding a mean proportional to 1 and 31 involves the use of too long a line. Write $31 = \frac{31}{5} \times 5$, and find a mean proportional to 6.2 and 5.]

9. Find graphically $\sqrt[4]{3}$ and $\left(\sqrt{2}\right)^3$.

10. *Divide a straight line AB in the ratio $\sqrt{2} : \sqrt{3}$.*
Draw a straight line *AX*, making any angle with *AB*.
Along *AX* set off *AC* = 2 units, *CD* = 3 units. On *AD* draw a semicircle. Draw *CE* perpendicular to *AD* to meet the circumference at *E*. Along *CX* set off *CF* equal to *CE*. Join *BF*. Draw *CG* parallel to *FB* to meet *AB* at *G*. Then *G* is the point required.

Proof. $$\frac{AG^2}{GB^2} = \frac{AC^2}{CF^2} = \frac{AC^2}{CE^2} = \frac{AC^2}{AC \cdot CD} = \frac{AC}{CD} = \frac{2}{3};$$

$$\therefore \quad AG : GB = \sqrt{2} : \sqrt{3}.$$

11. Divide a straight line internally at *X* and externally at *Y*, so that
$$AX^2 : XB^2 = AY^2 : YB^2 = 2 : 5$$

XL. AREAS OF SIMILAR TRIANGLES

THEOREM 93 (Euclid VI. 19)

The ratio of the areas of two similar triangles is equal to the ratio of the squares on corresponding sides.

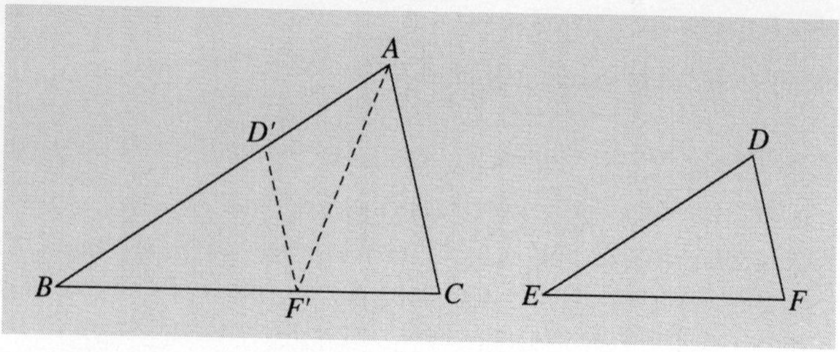

Let *ABC*, *DEF* be two similar triangles, having $\angle B$ equal to $\angle E$ and $\angle C$ equal to $\angle F$. It is required to prove that

$$\Delta ABC : \Delta DEF = BC^2 : EF^2.$$

Construction. Along *BA*, or *BA* produced, cut off $BD' = ED$, and along *BC*, or *BC* produced, cut off $BF' = EF$; join $D'F'$, AF'.

Proof. In the Δs $D'BF'$, *DEF*,

$$BD' = ED,\ BF' = EF \text{ and } \angle D'BF' = \angle DEF;$$

$\therefore$ Δs are congruent and equal in area;

$$\therefore \qquad \frac{\Delta ABC}{\Delta DEF} = \frac{\Delta ABC}{\Delta BD'F'} = \frac{\Delta ABC}{\Delta ABF'} \cdot \frac{\Delta ABF'}{\Delta BD'F'}.$$

$$\text{Now} \qquad \frac{\Delta ABC}{\Delta ABF'} = \frac{BC}{BF'} = \frac{BC}{EF}$$

$$\text{and} \qquad \frac{\Delta ABF'}{\Delta BD'F'} = \frac{AB}{BD'} = \frac{AB}{DE} = \frac{BC}{EF};$$

$$\therefore \qquad \frac{\Delta ABC}{\Delta DEF} = \frac{BC}{EF} \cdot \frac{BC}{EF} = \frac{BC^2}{EF^2}.$$

EXERCISE LXVIII (Theorem 93)

1. If *ABC*, *DEF* are two triangles, in which $\angle A = \angle D$, prove that

$$\Delta ABC : \Delta DEF = AB \cdot AC : DE \cdot DF.$$

[Draw *BH*, *EK* perpendicular to *AC*, *DF*.]

2. If *ABCD*, *EFGH* are equiangular parallelograms, prove that

$$\text{▱}^m\, ABCD : \text{▱}^m\, EFGH = AB \cdot AD : EF \cdot EH.$$

3. If *ABC*, *XYZ* are two triangles, in which $AB : AC = XY : XZ$ and the angles *A* and *X* are supplementary, prove that

$$\Delta\, ABC : \Delta\, XYZ = AB^2 : XY^2.$$

4. *AT*, *BT* are tangents to a circle at any points *A*, *B* and *O* is the centre. Show that $\Delta\, ATB : \Delta\, OAB = AT^2 : AO^2$.

5. *ABC* is a triangle, in which $\angle A$ is a right angle. Equilateral triangles *BXC*, *CYA*, *AZB* are drawn on the sides, and *AD* is perpendicular to *BC*. Prove that

$$\Delta\, CYA = \Delta\, CDX \text{ and } \Delta\, AZB = \Delta\, BDX.$$

6. From Ex. 5, deduce a construction for an equilateral triangle of area equal to the sum of the areas of two given equilateral triangles.

7. *C* is the centre of a circle, *AB* is a diameter and *P* is any point on the circumference. *PN* is perpendicular to *AB*, the tangent at *P* meets *BA* produced at *T* and the tangents at *P* and *A* meet at *Q*. Prove that

   (i) $\Delta\, TAQ : \Delta\, TPC = TA : TB.$

   (ii) $\Delta\, PQA : \Delta\, PCA = AN : NB.$

   [For (*ii*) show that Δs *QPC*, *APB* are equiangular.]

8. Take any triangle *ABC*. Draw *AD*, *BE* perpendicular to *BC*, *CA*. Draw *EG* perpendicular to *BC*. Join *DE*. Prove that

$$\Delta\, DEC : \Delta\, ABC = CG : CB.$$

   [Show that $\Delta\, DEC : \Delta\, ABC = CE^2 : CB^2$.]

9. Take any acute-angled triangle *ABC*. Draw *AD*, *BE*, *CF* perpendicular to *BC*, *CA*, *AB*. Draw *EG*, *FH* perpendicular to *BC*. Join *DE*, *DF*. Prove that *DE*, *DF* divide the triangle *ABC* into parts which are in the ratio *CG* : *GH* : *HB*. [Use *E*x. 8]

Problems Connected with Areas

1. *Divide the triangle ABC into three equal parts by straight lines parallel to BC.*

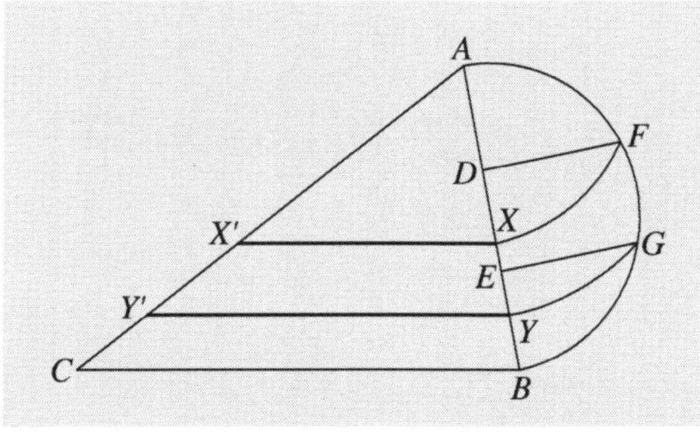

Analysis. Suppose the construction effected, and that XX', YY' are the lines required.

Then $\Delta AXX' : \Delta AYY' : \Delta ABC = 1 : 2 : 3;$

and because these are similar triangles,

$$\therefore \quad \Delta AXX' : \Delta YY' : \Delta ABC = AX^2 : AY^2 : AB^2;$$

$$\therefore \quad AX : AY : AB = 1 : \sqrt{2} : \sqrt{3}.$$

Hence the following construction:

Trisect AB in D and E. Draw the semicircle $AFGB$. Draw DF and EG perpendicular to AB to meet the semicircle in F, G. With centre A and radii AF, AG, draw arcs cutting AB in X, Y, etc.

Proof.
$$\begin{aligned}\Delta AXX' : \Delta YY' : \Delta ABC &= AX^2 : AY^2 : AB^2 \\ &= AF^2 : AG^2 : AB^2 \\ &= AD\,.\,AB : AE\,.\,AB : AB^2 \\ &= AD : AE : AB \\ &= 1 : 2 : 3.\end{aligned}$$

2. Assuming that the area of a circle varies as the square of the radius, show how to effect the following construction: Draw two circles, concentric with a given circle, and dividing it into three parts of equal area.

3. Divide a parallelogram $ABCD$ into three equal parts by straight lines parallel to the diagonal AC.

[Draw a line parallel to AC cutting AB in X and BC in Y so that $\Delta BXY = \frac{2}{3}\Delta BAC$. Divide the ΔDAC in a similar manner by the line RS. XY and RS are the lines required.]

4. *Draw an equilateral triangle equal in area to a given triangle ABC*

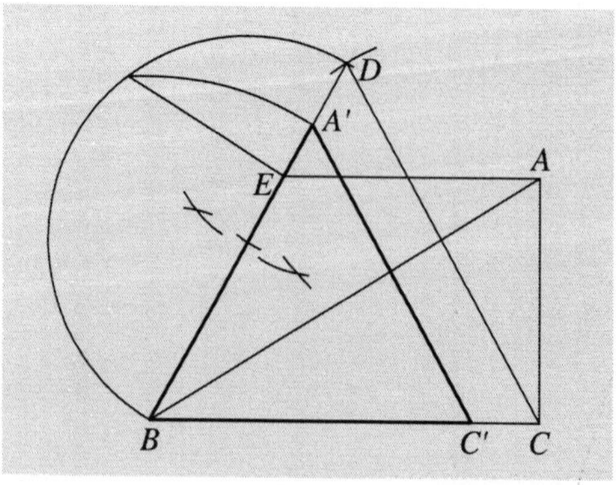

On one side BC of ΔABC draw an equilateral ΔDBC.

Draw AE parallel to CB to meet DB in E. Along BD set off BA', a mean proportional to BE, BD. Draw $A'C'$ parallel to DC to meet BC in C'.

Then $A'BC'$ is the required triangle.

Proof.
$$\frac{\Delta A'BC'}{\Delta DBC} = \frac{BA'^2}{BD^2} = \frac{BE \cdot BD}{BD^2}$$

$$= \frac{BE}{BD} = \frac{\Delta BEC}{\Delta BDC} = \frac{\Delta ABC}{\Delta DBC};$$

$$\therefore \qquad \Delta A'BC' = \Delta ABC;$$

also $\Delta A'BC'$ is equilateral.

By the method of the last example, we can make a triangle similar to a given triangle ABC and equal in area to another given triangle XYZ.

5. Construct an equilateral triangle of area equal to 3 square inches.

6. Draw ΔABC, in which $BC = 1.6$ in., $CA = 1.2$ in., $AB = 2$ in. Construct a triangle equal in area to ΔABC and with its sides in the ratio 3 : 4 : 5. Find the area of this triangle by taking suitable measurements, and verify that it is equal to that of ABC.

7. *Bisect the triangle ABC by a straight line perpendicular to BC.*

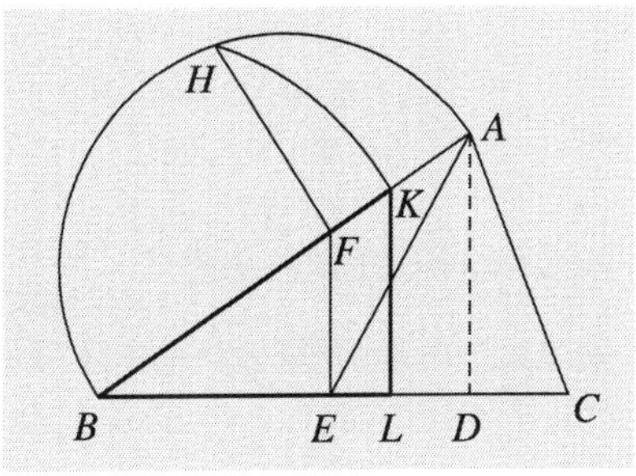

Analysis, Suppose that KL is the line required, and that

$$\Delta BKL = \frac{1}{2}\Delta ABC.$$

Bisect BC at E: draw EF perpendicular to BC to meet BA at F: draw AD perpendicular to BC.

Then by hypothesis $\qquad \Delta BKL = \Delta BAE,$

and
$$\frac{\Delta BAE}{\Delta BAD} = \frac{BE}{BD} = \frac{BF}{BA};$$

also
$$\frac{\Delta BKL}{\Delta BAD} = \frac{BK^2}{BA^2},$$

$$\therefore \qquad \frac{BK^2}{BA^2} = \frac{BF}{BA}.$$

$\therefore BK^2 = BF \cdot BA$, and the construction is that indicated in the figure. Complete the construction and supply proof.

8. Give a construction similar to that in Ex. 7 for bisecting a triangle by a straight line parallel to any given straight line.

9. *Draw a parallelogram, given the area, one angle, and the ratio of adjacent sides.*

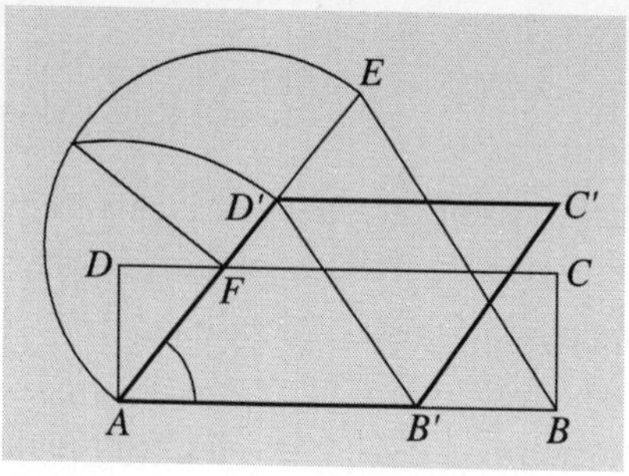

Draw a rectangle $ABCD$, equal to the given area. Draw ΔABE, having $\angle BAE$ equal to the given angle of the parallelogram, and $AB : AE$ equal to the given ratio. Let AE cut CD, or CD produced, in F. Complete as in the figure. Then $AB'C'D'$ is the required parallelogram. Supply proof.

10. Draw a parallelogram of 3 sq. in. area, with one angle equal to 60°, and the ratio of adjacent sides 2 : 3. Measure the sides, and verify by calculation.

XLI. TRIGONOMETRICAL FORMULAE

1. *If I is the centre of the inscribed circle of the triangle ABC and I_1 the centre of the escribed circle which touches AB, AC produced, then*

$$AI \cdot AI_1 = AB \cdot AC.$$

Proof. For A, I, I_1 are in the same straight line; and

$$\angle BAI_1 = \frac{1}{2}\angle BAC = \angle IAC.$$

Also, since BI and BI_1 are the internal and external bisectors of $\angle B$,

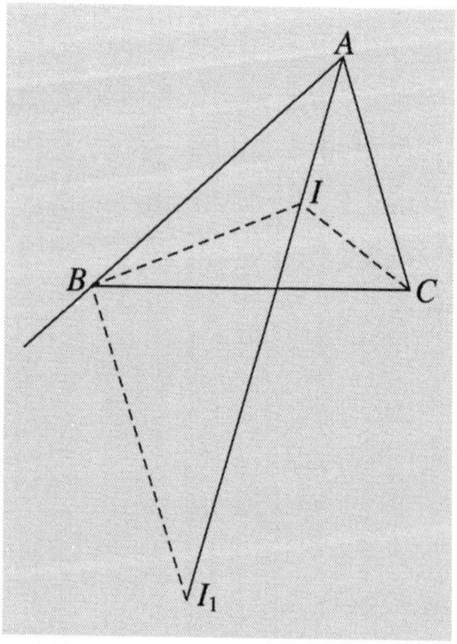

$$\therefore \quad I_1BA = 90° + \frac{1}{2}B$$

$$= 180° - \frac{1}{2}(A + C)$$

$$= \angle CIA;$$

$\therefore$ Δs BAI_1, IAC are equiangluar;

$$\therefore \quad BA : AI_1 = IA : AC;$$

$$\therefore \quad AI \cdot AI_1 = AB \cdot AC.$$

2. *If I be the centre of the inscribed circle of the triangle ABC, I_1 that of the escribed circle, which touches AB and AC produced, r, r_1 their respective radii, and s the semi-perivteter, then*

$$rr_1 = (s - b)(s - c).$$

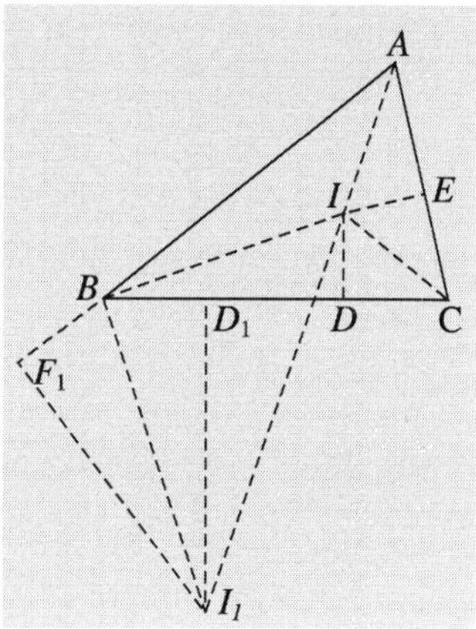

Proof. Let these circles touch BC in D, D_1 respectively,

Then because BI and BI_1 arc the internal and external bisectors of $\angle B$,

$\therefore$ $\angle IBD = 90° - I_1BD_1 = \angle BI_1D_1$,

and $\angle$s at D, D_1 are rt. $\angle$s;

$\therefore$ Δs IBD, BI_1D_1 are equiangular;

$$ID : BD = BD_1 : I_1D_1;$$

$\therefore$ $ID \cdot I_1D_1 = BD \cdot BD_1$;

$\therefore$ $rr_1 = (s - b)(s - c)$. (*See p.* 174)

Similarly, $rr_2 = (s - c)(s - a)$

and $rr_3 = (s - a)(s - b)$

3. *If Δ is the area of the triangle ABC, then*

$$\Delta = \sqrt{\{s(s-a)(s-b)(s-c)\}},$$

where $2s = a + b + c$.

We have $r = \dfrac{\Delta}{s}$ and $r_1 = \dfrac{\Delta}{s-a}$. (*See p.* 173)

Also $rr_1 = (s - b)(s - c)$; (*Ex.* 2.)

$\therefore$ $$\frac{\Delta^2}{s(s-a)} = (s - b)(s - c);$$

$\therefore$ $$\Delta^2 = s(s - a)(s - b)(s - c);$$

$\therefore$ $$\Delta = \sqrt{\{s(s-a)(s-b)(s-c)\}}.$$

4. *For the triangle ABC, find expressions for sin $\frac{A}{2}$, cos $\frac{A}{2}$, tan $\frac{A}{2}$, in terms of the sides.*

In the figure of Example 2, let the inscribed circle touch BC, CA, at D, E, and let the escribed circle opposite A touch BC at D_1 and AB produced at F_1.

Then $$\sin\frac{A}{2} = \frac{EI}{AI} = \frac{r}{AI}$$

and $$\sin\frac{A}{2} = \frac{F_1I_1}{AI_1} = \frac{r_1}{AI_1}.$$

$$\sin^2\frac{A}{2} = \frac{rr_1}{AI \cdot AI_1} = \frac{(s-b)(s-c)}{bc};$$

∴ $$\sin\frac{A}{2} = \sqrt{\frac{(s-b)(s-c)}{bc}}.$$

Again, $$\cos\frac{A}{2} = \frac{AE}{AI} = \frac{AF_1}{AI_1};$$

∴ $$\cos^2\frac{A}{2} = \frac{AE \cdot AF_1}{AI \cdot AI_1} = \frac{(s-a)\cdot s}{bc};$$

∴ $$\cos\frac{A}{2} = \sqrt{\frac{s(s-a)}{bc}}.$$

Hence, $$\tan\frac{A}{2} = \sqrt{\frac{(s-b)(s-c)}{s(s-a)}}.$$

The trigonometrical ratios of $\frac{B}{2}$ and $\frac{C}{2}$ have similar values.

These formulae are of extreme importance, and afford a ready method of finding the angles of a triangle when the three sides are given, as they are adapted to logarithmic computation.

5. *Find the sine, cosine and tangent of half the greatest angle of the triangles whose sides are as below, giving the results as surds (or rational quantities):*

(*i*) $a = 15,\ b = 13,\ c = 4$;

(*ii*) $a = 5,\ b = 29,\ c = 30$;

(*iii*) $a = 65,\ b = 80,\ c = 17$.

6. *For the triangle ACB, if R is the radius of the circum-circle, and A and B are the acute angles, then*

(*i*) $\dfrac{a}{\sin A} = \dfrac{b}{\sin B} = 2R$; (*ii*) $R = \dfrac{abc}{4\Delta}$.

(*i*) Draw the diameter *BD* of the circle *ABC*.

Join *CD*.

Then BCD is a semicircle;

$\therefore \angle BCD$ is a right angle;

$$\therefore \qquad \sin BDC = \frac{CD}{DB} = \frac{a}{2R}.$$

Also, because the angles BAC, BDC are in the same segment,

$$\therefore \qquad \angle BAC = \angle BDC;$$

$$\therefore \qquad \sin A = \frac{a}{2R},$$

$$\therefore \qquad \frac{a}{\sin A} = 2R.$$

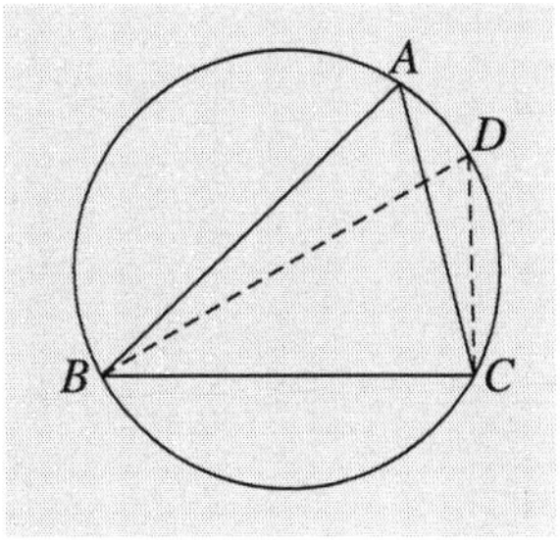

Similarly, it can be shown that $\dfrac{b}{\sin B} = 2R$.

(*ii*) We have (*see* p. 135) $\Delta = \dfrac{1}{2} bc \sin A$;

$$\therefore \qquad R = \frac{a}{2\sin A} = \frac{abc}{2bc\sin A} = \frac{abc}{4\Delta}.$$

7. If ABC is any triangle, prove that

(*i*) $\dfrac{abc}{2\Delta} = \dfrac{a}{\sin A} = \dfrac{b}{\sin B} = \dfrac{c}{\sin C}$, when C is an acute angle;

(*ii*) $\dfrac{abc}{2\Delta} = \dfrac{a}{\sin A} = \dfrac{b}{\sin B} = \dfrac{c}{\sin(180° - C)}$, when C is an obtuse angle;

(*iii*) $\dfrac{abc}{2\Delta} = \dfrac{a}{\sin A} = \dfrac{b}{\sin B} = c$, when C is a right angle.

8. *In the triangle ABC, if $\angle C > \angle B$, prove that*

$$\frac{\tan\frac{1}{2}(C+B)}{\tan\frac{1}{2}(C-B)} = \frac{c+b}{c-b}.$$

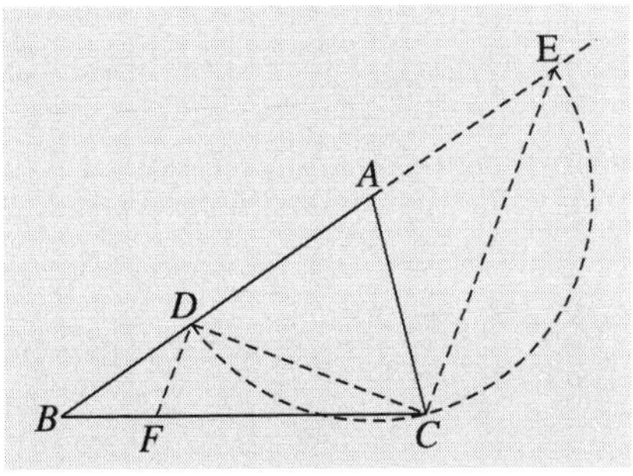

With centre A and radius AC describe a circle, cutting BA and BA produced in D and E respectively; join CE and CD, and draw DF parallel to EC to meet BC in F.

Then $$BD = BA - AD = c - b,$$
and $$BE = BA + AE = c + b.$$

Also, $\angle ECD$, an $\angle$ in a semicircle, is a rt. $\angle$.

$\therefore$ $\angle CDF$ is a rt $\angle$;

and since A is the centre of circle ECD,

$$\therefore \quad \angle ADC = \frac{1}{2} \angle EAC$$
$$= \frac{1}{2}(B + C);$$
$$\therefore \quad \angle BCD = \angle ADC - \angle DBC$$
$$= \frac{1}{2}(B + C) - B$$
$$= \frac{1}{2}(C - B).$$

Again, because Δs BEC, BDF are equiangular,

$$\therefore \quad \frac{EC}{DF} = \frac{BE}{BD} = \frac{c+b}{c-b};$$

$$\therefore \quad \frac{\tan\frac{1}{2}(C+B)}{\tan\frac{1}{2}(C-B)} = \frac{EC}{CD} \div \frac{DF}{CD}$$
$$= \frac{EC}{DF}$$
$$= \frac{c+b}{c-b}.$$

This formula is most useful in doing examples on the solution of triangles where two sides and the included angle *are given.*

EXERCISE LXIX

Harder Examples on Similar Triangles

1. *If the straight lines which join the vertices of a triangle ABC to any point O cut the opposite sides, or these sides produced, at X, Y, Z respectively, then*

$$\frac{BX}{XC} \cdot \frac{CY}{YA} \cdot \frac{AZ}{ZB} = 1. \text{ (Ceva's Theorem*)}$$

*See also p. 263.

[Draw $B'AC'$ parallel to BC to meet BO, CO (produced) at B', C'. Prove that

$$\frac{BX}{XC} = \frac{AB'}{AC'}; \quad \frac{CY}{YA} = \frac{BC}{AB'}; \quad \frac{AZ}{ZB} = \frac{AC'}{BC}.$$

2. *If a straight line cuts the sides BC, CA, AB of a triangle ABC, or these sides produced at X, Y, Z respectively, then*

$$\frac{BX}{CX} \cdot \frac{CY}{YA} \cdot \frac{AZ}{ZB} = 1. \text{ (Theorem of Menelaus*)}$$

[Draw AL, BM, CN perp. to XYZ. Prove that

$$\frac{BX}{CX} = \frac{BM}{CN}; \quad \frac{CY}{YA} = \frac{CN}{AL}; \quad \frac{AZ}{ZB} = \frac{AL}{BM}.$$

3. *A is a fixed point and P is any point on a given straight line BC. In AP, or in AP produced, a point Q is taken, such that $AP \cdot AQ = k^2$, where k is constant. Prove that the locus of Q, is a circle,*

[Draw AM perpendicular to BC. Along AM set off AD, such that $AM \cdot AD = k^2$. The locus of Q is the circle on AD as diameter. Supply proof.]

4. *S is a fixed point and P is any point on a given circle. In SP, or in SP produced, a point Q is taken, such that $SP \cdot SQ = k^2$, where k is constant. Prove that the locus of Q is a circle except when S is on the circumference, in which case it is a straight line.*

[Let SP cut the given circle again at P'. Draw QC' parallel to $P'C$ to meet SC, or SC produced at C'. Prove that C' is a fixed point and that $C'Q$ is of constant length. For the exceptional case, see Example 3.]

5. The perpendicular bisector of AC, one of the equal sides of an isosceles triangle, meets the base AB produced in D. Show that AC is a mean proportional to AB, AD.

[Show that CA touches the circle CBD.]

6. ABC is a triangle and AD is the perpendicular from A to $BC \cdot AD$ is produced to meet the circumcircle of ABC at E, and AF is drawn parallel to CE to meet BC at F. Show that $AF \cdot BE = AB \cdot AC$. [Consider Δs AFC, BAE.]

7. The inscribed circle of a triangle ABC touches BC in D, and an escribed circle touches BC in E and also touches the other two sides produced; let F be the point of the inscribed circle furthest from BC; show that E, F, A lie on a straight line. [Show $AI : AI_1 = r : r_1$.]

8. D and E are points on the side BC of the triangle ABC, such that $BC \cdot BD = BA^2$ and $CB \cdot CE = CA^2$. Prove that

$$AD^2 = AE^2 = BD \cdot EC.$$

[Prove $\angle BAD = \angle C$ and $\angle CAE = \angle B$.]

9. On the sides AB and AC of a triangle ABC points F and E are taken, so that $CE : EA = AF : FB = 2 : 1$. Find the ratio of the areas of the triangles ABC and AEF. Also prove that if L and M are the feet of the perpendiculars let fall on BC from E and F, then $EL : FM = 2 : 1$.

---

*See also page 264

10. A, B are two points on a circle, and the tangents at A, B meet in T. If P is any point on the circle and PL, PM, PN are the perpendiculars from P to AT, BT, AB respectively, show that $PL \cdot PM = PN^2$.

[Consider Δs PLN, PNM.]

11. ABC is a triangle, in which $CA = CB$. If PL, PM, PN are the perpendiculars from a point P to BC, CA, AB respectively, and if $PL \cdot PM = PN^2$, show that the locus of P is the circle which touches CA, CB at A and B.

[Show Δs PLN, PNM equiangular.]

12. $ABCD$ is a cyclic quadrilateral and P is any point on the circumscribing circle. If PL, PM, PN, PR are the perpendiculars to AB, BC, CD, DA, prove that $PL \cdot PN = PM \cdot PR$.

13. ABC is a triangle, in which the angle A is a right angle and AD is the perpendicular from A to BC. If I, I' are the centres of the circles inscribed in the triangles ABD, ACD, prove that the triangles DII', ABC are equiangular. [Show that $ID : I'D = AB : AC$]

14. $ABCD$ is a quadrilateral; AF drawn parallel to BC meets BD, or BD produced, in F, and BE drawn parallel to AD meets AC, or AC produced, in E; show that EF is parallel to CD.

15. $ABCD$ is a trapezium, AC, BD meet in E and AD, BC meet in F. Show that if EF meets AB, CD in P and Q, then

$$EP : EQ = FP : FQ.$$

16. Through a given point O draw any three straight lines OAD, OBE, OCF to meet two parallel straight lines in A, B, C; D, E, F respectively. Let AE, BD meet in X and CE, BF meet in Y. Show that XY is parallel to AC or DF.

17. Given two parallel straight lines AB, DE and a point X. Using the ruler only, draw a parallel to AB through X. [Make the figure of Ex. 16.)

18. $ABCD$ is a parallelogram and LM a straight line. AB, AD, produced if necessary, cut LM at P and Q. If PD meets BC at X and QB meets CD at Y (the lines being produced if necessary), show that XY is parallel to LM. [Produce BC to meet LM at T.]

19. Given a parallelogram $ABCD$, a straight line LM and a point P. Using the ruler only, draw a parallel to LM through P. [Use Exs. 18, 17.]

20. ACB is a straight line : circles are described on AC, CB as diameters. AP is a chord of the first circle which touches the second at T. BQ is a chord of the second circle which touches the first at T'. Prove that

$$AP \cdot BQ = 4PT \cdot QT'.$$

21. If a straight line PQ cuts the equal sides AB, AC of an isosceles triangle at P, Q respectively, so that $4BP \cdot CQ = BC^2$, show that AB, AC, PQ always touch a fixed circle. [Bisect BC at M.]

22. A, B, C, D are four points in a plane, and O is the middle point of AB. Prove that if $OC \cdot OD = OA^2$, and OC, OD are equally inclined to OA and on opposite sides of it, then A, B, C, D are concyclic.

23. From a point P on a circle the line PRR' is drawn perpendicular to a diameter ACA' and AR, $A'R'$ are drawn parallel to the tangent at P. Show that, C being the centre, the triangles PCR, $PR'C$ are equiangular.

[Prove that CR bisects $\angle PCA$.]

24. PR, SQ are parallel chords of a circle, such that PQ, RS meet at a point X within the circle; if the tangents at P and S meet at T, prove that TX is parallel to PR. [Let PT, QS produced meet at U; prove Δs UTS, SXP are equiangular.]

25. *OA, OB, OC, OD are four given straight lines. Any straight line meets them at P, Q, R, S respectively. Through Q draw a parallel to OD, meeting OA at X and OC at Y. Prove that*

$$PQ \cdot RS : QR \cdot SP = QX \cdot QY.$$

Hence show that the ratio $PQ \cdot RS : QR \cdot SP$ is constant for all transversals.

26. *$ABCD$ is a quadrilateral; AB and DC are produced to meet at E; BC and AD are produced to meet at F. AC and BD are produced to meet EF at X and Y. Prove that EF is divided internally and externally in the same ratio at X and Y.*

[Apply Ceva's Theorem to ΔAEF with the point C, and apply Menelaus' Theorem to ΔAEF with, the transversal YDB.]

27. ABC is any triangle, and points D, E, F are taken in BC, CA, AB respectively, such that $3BD = 2DC$, $2CE = EA$, $AF = FB$.

If XYZ is the triangle formed by AD, BE, CF, prove that

$$\Delta XYZ = \frac{1}{24} \Delta ABC.$$

[Let AD meet CF at Y and BE at Z. Use Ex. LXVIII, I to find the values of $\Delta XYZ : \Delta XFB$ and $\Delta XFB : \Delta CFB$.]

XLII. SIMILAR POLYGONS

In what follows, the word **polygon** will be used to mean **any rectilineal figure**.

Def. Two polygons are said to be **similar** when they are equiangular and the sides about every pair of equal angles are proportional.

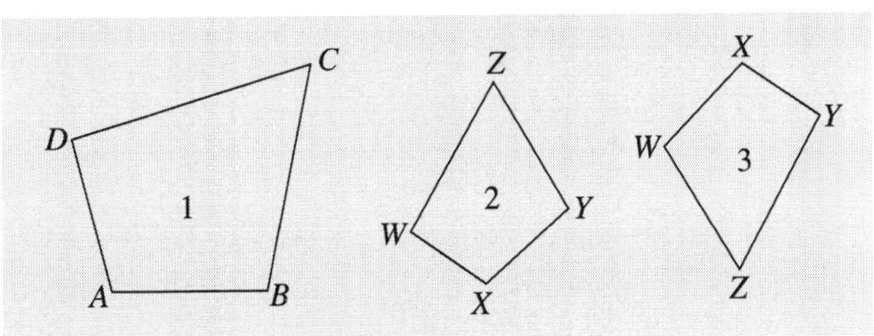

For example, let *ABCD*, *XYZW* be two quadrilaterals, in which the angles *A*, *B*, *C*, *D* are equal to the angles *X*, *Y*, *Z*, *W* respectively. In order that *ABCD*, *XYZW* may be similar, we must have

$$DA : WX = AB : XY, \qquad AB : XY = BC : YZ,$$
$$BC : YZ = CD : ZW, \qquad CD : ZW = DA : WX.$$

The vertices *A*, *B*, *C*, *D* are said to correspond to *X*, *Y*, *Z*, *W* respectively, and the sides *AB*, *BC*, etc., are said to correspond to the sides *XY*, *YZ*, etc.

If two polygons are similar, since they are equiangular, they can be placed so that corresponding sides are parallel. When placed in this way, they are said to be similarly situated: they are also called homothetic figures.

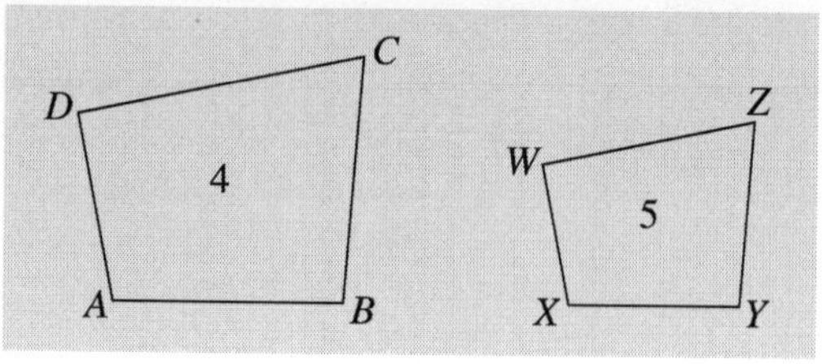

Thus the figures 4 and 5 are similar and similarly situated, or homothetic, figures.

In Theorem 94, it will be proved that if two polygons are similar and similarly situated, the straight lines joining corresponding vertices pass through a point. This point is called the **homothetic centre** of the **figures**.

If two similar polygons can be placed so that their corresponding sides are parallel, *without turning one of them over*, the polygons are said to be **directly similar**. But if, in order to bring them into this position, one of the polygons has to be turned over, they are said to be **inversely similar**.

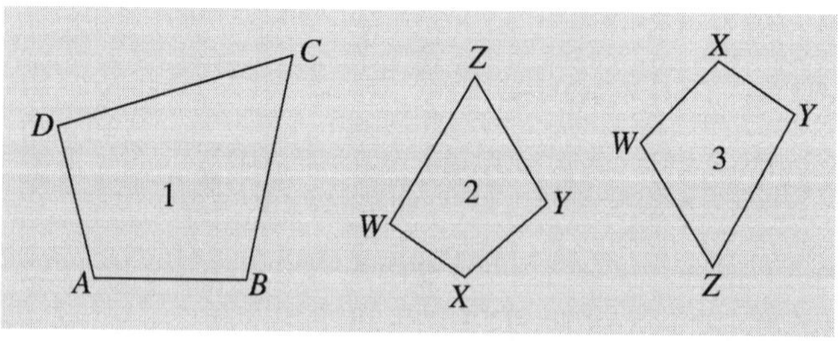

Thus, figures 1 and 2 are directly similar and figures 1 and 3 are inversely similar.

Theorem 94

If the straight lines joining a point to the vertices of a given polygon are divided, all internally or all externally, in the same ratio, the points of division are the vertices of a polygon which is similar to the given polygon.

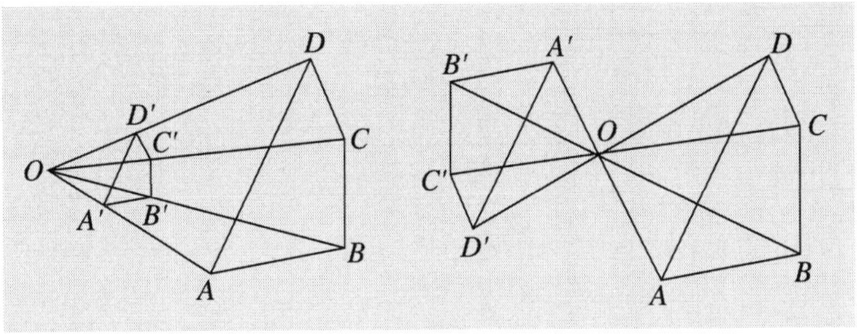

Let $ABCD$ be the given polygon and O the given point. Let OA, OB, OC, OD be divided, all internally or all externally, in the same ratio at A', B', C', D'.

It is required to prove that the polygon $A'B'C'D'$ is similar to the polygon $ABCD$.

Proof. Because $OA' : A'A = OB' : B'B$ (*given*),

$\therefore$ $$OA' : OA = OB' : OB.$$

$\therefore$ The sides of the triangles OAB, $OA'B'$ about the angles at O are proportional, and the angles of the triangles at O are either common, or vertically opposite, and are therefore equal.

$\therefore$ The triangles OAB, $OA'B'$ are similar and $\angle OBA = \angle OB'A'$.

In the same way, it can be shown that the triangles OBC, $OB'C'$ are similar and $\angle OBC = \angle OB'C'$;

$\therefore$ $$\angle ABC = \angle A'B'C'.$$

In the same way, it can be shown that the angles at C, D, A of $ABCD$ are equal to the angles at C', D', A' of $A'B'C'D'$.

$\therefore$ The polygons $ABCD$, $A'B'C'D'$ are equiangular.

Again, because the triangles OAB, $OA'B'$ are similar (*proved*),

$\therefore$ $$AB : A'B' = OB : OB';$$

and because the triangles OBC, $OB'C'$ are similar,

$\therefore$ $$BC : B'C' = OB : OB';$$

$\therefore$ $$AB : A'B' = BC : B'C';$$

that is, the sides of $ABCD$, $A'B'C'D'$ about the equal angles ABC, $A'B'C'$ are proportional.

In the same way it can be shown that the sides of $ABCD$, $A'B'C'D'$ about any pair of equal angles are proportional.

But it has been shown that the polygons are equiangular;

$\therefore$ the polygons are similar.

THEOREM 95

If two similar polygons are similarly placed, the straight lines joining corresponding vertices meet in a point.

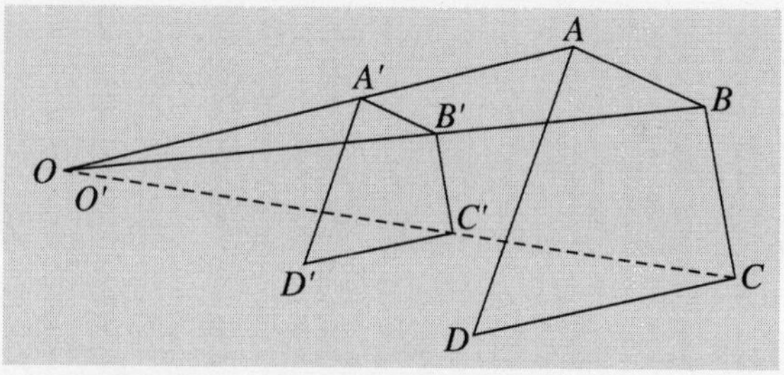

Let $ABCD$, $A'B'C'D'$ be similar polygons, placed so that the sides AB, BC, etc., are parallel to the sides $A'B'$, $B'C'$, etc.

It is required to prove that the straight lines AA', BB', CC', DD' meet at a point.

Proof. Let AA', BB' meet at O. If CC' does not pass through O, let it cut BB' at some point O'.

Then, because AB is parallel to $A'B'$ (*given*),

$\therefore$ $$\angle OAB = \angle OA'B' \text{ and } \angle OBA = \angle OB'A';$$

$\therefore$ the triangles OAB, $OA'B'$ are equiangular,

$$OB : OB' = AB : A'B'.$$

Similarly, the triangles $O'BC$, $O'B'C'$ are equiangular

and $$O'B : O'B' = BC : B'C'.$$

Now the polygons $ABCD$, $A'B'C'D'$ are similar (*given*);

$\therefore$ $$AB : A'B' = BC : B'C';$$

$\therefore$ $$OB : OB' = O'B : O'B',$$

Hence the points O, O' divide BB', both externally or both internally, in the same ratio;

$\therefore$ O' coincides with O;

that is, CC' passes through O.

Similarly, it can be shown that DD' passes through O.

THEOREM 96 (Euclid VI. 20 (*I*))

If a polygon is divided into triangles by lines joining its vertices to a given point, any similar polygon can be divided into similar triangles, by lines joining its vertices to a corresponding point.

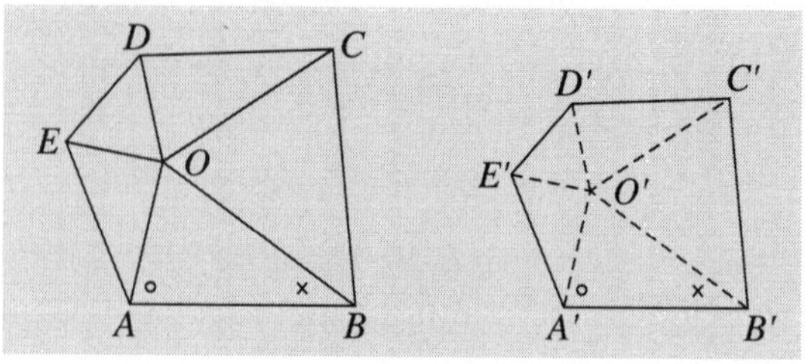

Let the polygon *ABCDE* be divided into the triangles *OAB*, *OBC*, etc., by joining its vertices to a given point *O*, and let *A'B'C'D'E'* be a similar polygon.

It is required to prove that *A'B'C'D'E'* can be divided into triangles similar to the triangles *OAB*, *OBC*, etc., by joining its vertices to some point *O'*.

Construction. Let *AB*, *A'B'*; *BC*, *B'C'*, etc., be corresponding pairs of sides. Make the angles *B'A'O'*, *A'B'O'* equal to the angles *BAO*, *ABO* respectively. Join *O'C'*, *O'D'*, *O'E'*.

Proof. The triangles *OAB*, *O'A'B'* are equiangular (*construction*);

$\therefore \qquad AB : A'B' = OB : O'B';$

and because the polygons are similar,

$\therefore \qquad AB : A'B' = BC : B'C';$

$\therefore \qquad OB : O'B' = BC : B'C'.$

Again, because the polygons are similar,

$\therefore \qquad \angle ABC = \angle A'B'C'.$

But $\qquad \angle ABO = \angle A'B'O'$ (*const.*);

$\therefore \qquad \angle OBC = \angle O'B'C'.$

Hence, in the triangles *OBC*, *O'B'C'*, the angles *OBC*, *O'B'C'* are equal, and the sides about these angles are proportional;

∴ the triangles *OBC*, *O'B'C'* are similar.

In the same way, it can be shown that the pairs of triangles *OCD*, *O'C'D'*; *ODE*, *O'D'E'*; *OEA*, *O'E'A'* are similar.

CONSTRUCTION 34 (Euclid VI. 18)

On a given straight line *A'B'*, draw a polygon, similar to a given polygon *ABCDE*, so that *A'B'* and *AB* may be corresponding sides.

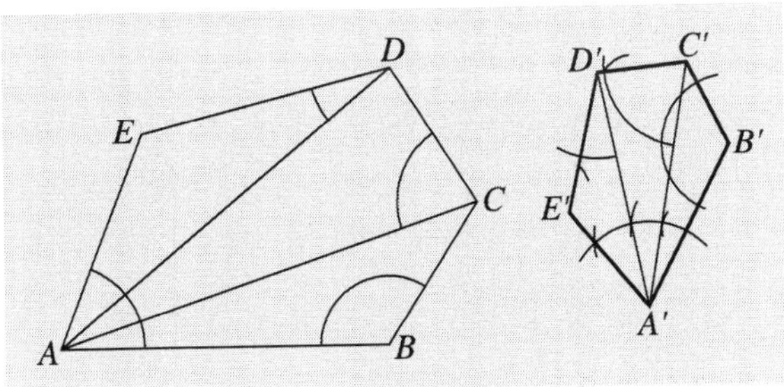

Construction. Join *AC*, *AD*. Make the angles *B'A'C'*, *A'B'C'* equal to *BAC*, *ABC* respectively.

Make the angles *C'A'D'*, *A'C'D'* equal to *CAD*, *ACD* respectively. Make the angles *D'A'E'*, *A'D'E'* equal to *DAE*, *ADE* respectively.

Then $A'B'C'D'E'$ is the required polygon.

Proof. By construction, the polygons $ABCDE$, $A'B'C'D'E'$ are equiangular.

Also, because the triangles ABC, $A'B'C'$ are equiangular,

$\therefore$ $AB : A'B' = BC : B'C'$

and $BC : B'C' = CA : C'A'$;

and, because the triangles ACD, $A'C'D'$ are equiangular,

$\therefore$ $CA : C'A' = CD : C'D'$.

Hence also $BC : B'C' = CD : C'D'$.

Similarly, $CD : C'D' = DE : D'E'$

and $DE : D'E' = EA : E'A'$.

Hence also $EA : E'A' = AB : A'B'$.

Therefore the sides of the polygons $ABCDE$, $A'B'C'D'E'$ about any pair of equal angles are proportionals, and it has been shown that they are equiangular;

$\therefore$ the polygons are similar.

EXAMPLE 1: *Draw a polygon similar to a given polygon ABCDE, so that the side corresponding to a given side AB may be of given length.*

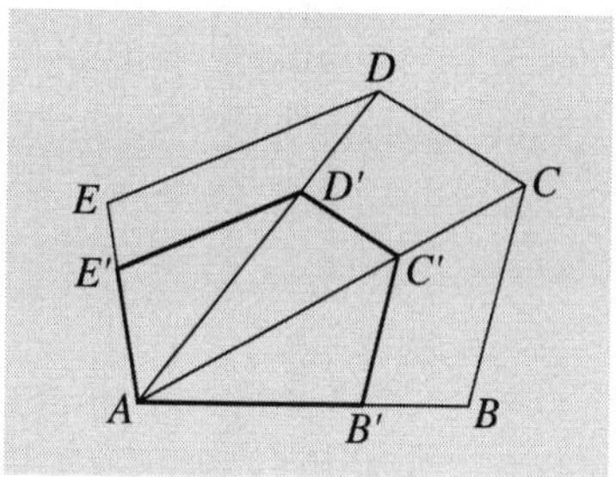

Take a point B' in AB, or in AB produced, so that AB' is equal to the given length. Join AC, AD. Draw $B'C'$, $C'D'$, $D'E'$ parallel to BC, CD, DE, to meet AC, AD, AE, or these lines produced, respectively, in C', D', E'.

Show that $AB'C'D'E$ is the polygon required.

EXAMPLE 2: *Given two straight lines AB, CD which would meet, if produced, at some point off the sheet of paper. It is required to draw, through a given point X, a straight line which, if produced, would pass through the intersection of AB, CD.*

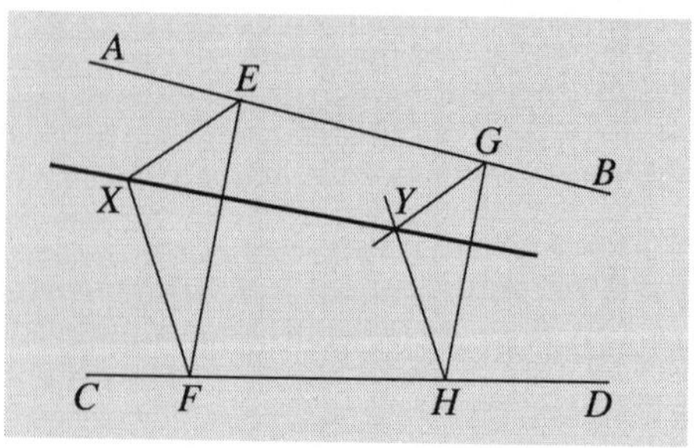

Construction. Take any convenient points *E* and *F*, in *AB*, *CD* respectively. Join *EF*, *EX*, *FX*. Draw any straight line *GH* parallel to *EF*, to cut *AB* in *G* and *CD* in *H*. Draw *GY*, *HY* parallel to *EX*, *FX* respectively. Join *XY*.

This is the required line.

Proof. The Δs *EFX*, *GHY* are similar and similarly placed;

∴ the straight lines joining the corresponding vertices are concurrent;

∴ *XY* passes through the intersection of *AB* and *CD*.

EXERCISE LXX

Similar Polygons

1. Draw any quadrilateral *ABCD*. Draw a straight line parallel to *CD*, meeting *BC* at *X* and *DA* at *Y*. Prove that the quadrilaterals *ABCD*, *ABXY* are equiangular but are not similar.

2. If the quadrilaterals *ABCD*, *A'B'C'D'* are both similar to the quadrilateral *XYZW*, prove that *ABCD* is similar to *A'B'C'D'*.

3. The parallelograms *ABCD*, *AXYZ* have a common angle at *A*, and the point *Y* is on the diagonal *AC*. Prove that the parallelograms are similar.

4. If the parallelograms *ABCD*, *AXYZ* are similar and similarly placed, show that *A*, *Y*, *C* are in the same straight line.

5. *ABC*, *XYZ* are triangles, such that *BC*, *CA*, *AB* are respectively parallel to *YZ*, *ZX*, *XY*. Prove that the straight lines *AX*, *BY*, *CZ* meet at a point.

6. If two polygons are similar, prove that the triangle formed by joining any three vertices of one polygon is similar to the triangle formed by joining the corresponding vertices of the other.

7. Explain how to apply Theorem 93 to reduce or enlarge the scale of a map.

8. On a given straight line *A'C'* draw a quadrilateral similar to a given quadrilateral *ABCD*, so that *A'C'*, *AC* may be corresponding lines.

9. Draw a polygon similar to a given polygon *ABCDE*, so that the diagonal corresponding to *AC* may be of given length.

10. Draw any quadrilateral *ABCD*. Join *AC*, *BD*, meeting at *O*. Without producing *AB*, *CD*, draw the straight line through *O* which would, if produced, pass through the intersection of *AB* and *CD*.

11. Draw a quadrilateral *A'B'C'D'* similar and similarly placed to a given quadrilateral *ABCD*, so that the sides *A'B'*, *B'C'*, *C'D'* pass through given points *P*, *Q*, *R* respectively.

12. *AB, A'B' are given straight lines. It is required to find a point S, such that the triangles SAB, SA'B' are directly similar.*

Prove the following construction: Let AB, A'B', produced if necessary, meet at O. Draw the circles OAA'. OBB', meeting again at S. Then S is the required point.

13. *In Ex. 12, M and M′ are points which divide AB, A′B in the same ratio, so that AM : MB = A′M′ : M′B′. Prove that*

$$SM : SM' = SA : SA' \text{ and that } \angle MSM' = \angle ASA'$$

On this account, S is called the centre of similitude of the line segments AB, $A'B'$.

14. *In Ex. 12, P is any vertex of a polygon described on AB and P′ is the corresponding vertex of the directly similar polygon described on A′B′. Prove that SP : SP′ = SA : SA′ and that ∠ PSP′ = ∠ ASA′.*

On this account, S is called the centre of similitude of the polygons. *If one of the polygons is rotated about S through the angle ASA′, the polygons will then be similarly placed, and S will be the homothetic centre.*

15. OA, OB are two given straight lines. It is required to find a point S, such that the triangles SAO, SOB are directly similar. Prove the following construction: Draw the circle through A to touch OB at O. Draw the circle through B to touch OA at O. Let the circles meet again at S. Then S is the required point.

16. A variable straight line cuts the sides AB, AC of a triangle ABC at M, N, so that $AM : MB = CN : NA$. Prove that MN subtends a constant angle at a certain fixed point, and find this point. [Use Ex. 15]

17. Given two equilateral triangles ABC, $A'B'C'$, both lettered counter-clockwise. Find a point S, such that

$$SA : SA' = SB : SB' = SC : SC' \text{ and } \angle ASA' = \angle BSB' = \angle CSC'.$$

18. ABC, LMN are triangles, such that BC, CA, AB are respectively parallel, to MN, NL, LM; O is the point where AL, BM, CN meet [see Ex. 5]; E, F are the centres, R, R' the radii of the circum-circles of the triangles ABC, LMN. Prove that

(*i*) $OE : OF = OA : OL = AB : LM = R : R'$. [The Δs AEB, LFM are homothetic.]

(*ii*) If a line through O cuts the circles in P, Q and R, S respectively, prove that

EP, EQ are parallel respectively to FR, FS or FS, FR :

and if EP is parallel to FR, then $OP : OR = OQ : OS = R : R'$. [Use Theorem 91]

> **NOTE:** O is called a centre of similitude of the circles.

19. L, M, N are the middle points of the sides BC, CA, AB of a triangle ABC; S is the circumcentre, O the orthocentre, G the centroid, J the centre of the circle through L, M, N. Prove that

(*i*) G is a centre of similitude of the circles ABC, LMN.

(*ii*) The radius of the circle ABC is twice, that of circle LMN.

(*iii*) S, G, J, O are in a straight line.

[S, J are the circumcentres, O, S the orthocentres of the Ds ABC, LMN; *i.e.* pairs of corresponding points.]

(*iv*) J bisects OS. [$SG : GJ = 2 : 1 = OG : GS = OS : JS$.]

(*v*) O is the other centre of similitude of the circles.

[$OS : OJ = 2; 1$, and S and J are corresponding points.]

XLIII. INSCRIBED AND CIRCUMSCRIBED FIGURES

> **NOTE:** || stands for 'parallel to' or 'is parallel to.'
> ⊥ " 'perpendicular to 'or is perpendicular to.'

One rectilineal figure is said to be inscribed in another, when as many as possible of the vertices of the first lie on sides of the second.

One rectilineal figure is said to be described (or circumscribed) about another, when as many as possible of the sides of the first pass through vertices of the second.

EXAMPLE 1: *In a given triangle ABC, inscribe a triangle, similar and similarly placed to a given triangle XYZ.*

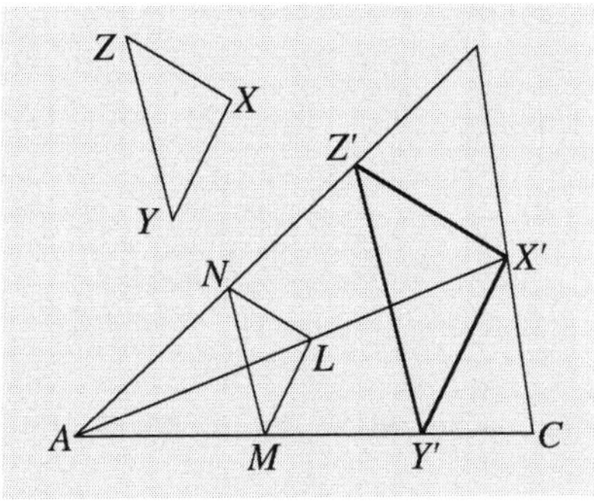

Draw any straight line *MN* || *YZ*, to cut *AC*, *AB* in *M*, *N* respectively.

Draw *ML* || *YX* and *NL* || *ZX*. Join *AL*, and let *AL*, produced if necessary, meet *BC* in *X'*. Draw *X'T'* || *XY* and *X'Z'* || *XZ*, to meet *AC* in *Y'* and *AB* in *Z'*.

Show that *X'Y'Z'* is the triangle required.

EXAMPLE 2: *In a given triangle ABC, inscribe a quadrilateral X'Y'Z'W' similar to a given quadrilateral XYZW, so that the side X'Y', corresponding to XY, may be in BC, and the vertices Z', W', in AC, AB respectively.*

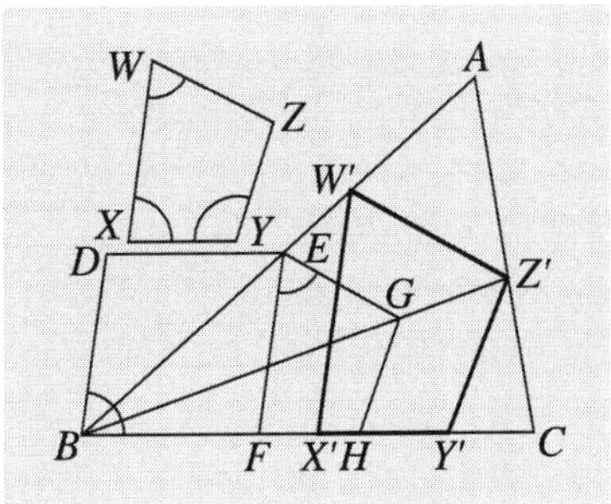

Make ∠ *CBD* = ∠ *X*. Set off *BD* equal to *XW*. Draw *DE* || *BC* to meet *AB* in *E*. Draw *EF* || *DB* to meet *BC* in *F*. On *EF* draw *EFHG* congruent with *WXYZ*. Join *BG*, and let *BG* (produced if necessary) meet *AC* in *Z'*. Draw *Z'Y'*, *Z'W'*, *W'X'* || *GH*, *GE*, *EF* to meet *BC*, *AB*, *BC* respectively in *Y'*, *W'*, *X'*.

Show that *X'Y'Z'W'* is the required quadrilateral.

EXAMPLE 3: *In a given triangle ABC, inscribe a square with two of its vertices in BC.*

Proceed as indicated in the figure. Explain the construction, and give a proof.

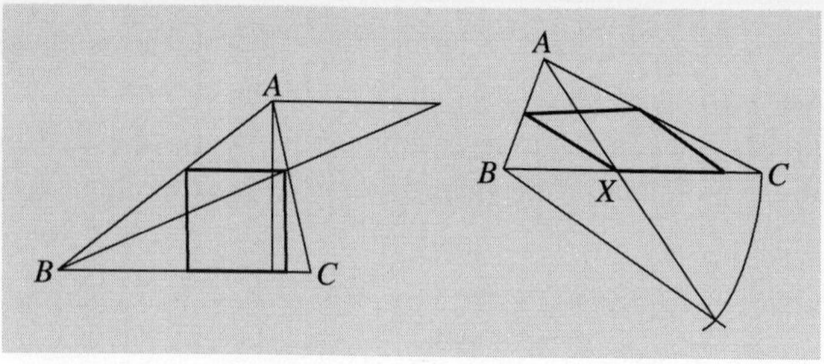

EXAMPLE 4: *In a given triangle ABC, inscribe a rhombus with one side along BC, and an extremity of this side at a given point X in BC.*

Explain the construction indicated in the figure, and give a proof.

EXERCISE LXXI

Numerical

1. Draw a triangle *ABC*, where *BC* = 2.5 in., *CA* = 3.0 in., *AB* = 3.5 in. Inscribe in it an equilateral triangle, with one of its sides perpendicular to *AB*. Measure a side.
2. Draw the triangle of Ex. 1 and, in it, inscribe a rectangle *XYZW* with the side *XY* in *BC* and *XY* = 2*YZ*. Measure *XY*.
3. Draw the triangle of Ex. 1 and, in it, inscribe a parallelogram *XYZW* with *X* at the middle point of *BC*, *XY* along *BC*, and $XY = \frac{1}{2}YZ$. Measure *XY*.
4. Draw an equilateral triangle *ABC* of 3 in. side and, in it, inscribe a right-angled isosceles triangle with its hypotenuse perpendicular to *BC*. Measure the hypotenuse.
5. Draw the triangle *ABC*, given *BC* = 3 in., *CA* = 2.5 in., *AB* = 2 in. In it, inscribe a triangle *PQR* with *P*, *Q*, *R* on *BC*, *CA*, *AB* respectively, and *PQ*, *QR*, *RP* respectively perpendicular to *BC*, *CA*, *AB*. Measure *PQ*.
6. Draw the triangle *ABC*, given *BC* = 3 in., *CA* = 2.5 in., *AB* = 2 in., and, in it, inscribe a square with one side along *BC*. Measure a side of the square.
7. Draw an equilateral triangle of 2 in. side and, in it, inscribe a rhombus with one angle equal to 70°. Measure a side of the rhombus.
8. Draw an equilateral triangle of 2 in. side and, in it, inscribe a rhombus with one of its sides along a side of the triangle, and an extremity of this side at a point of trisection of the side of the triangle. Measure a side of the rhombus.

Example 5: *Describe a square about a given quadrilateral ABCD.*

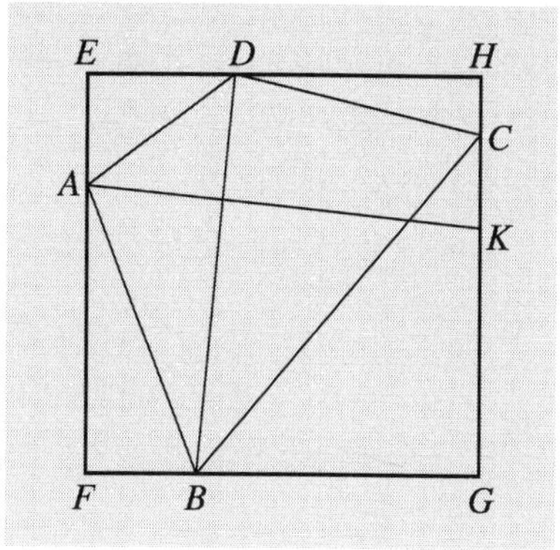

Analysis. Suppose $EFGH$ to be the square required. Join BD, and draw $AK \perp DB$ to meet GH in K.

Prove that $$AK = BD.$$

Hence the following construction:

Join BD, draw $AK \perp DB$, making $AK = DB$, then K is a point in the side of the square required, which passes through C.

Complete the construction and proof.

Example 6: *About a given quadrilateral ABCD describe a quadrilateral similar to a given quadrilateral XYZW.*

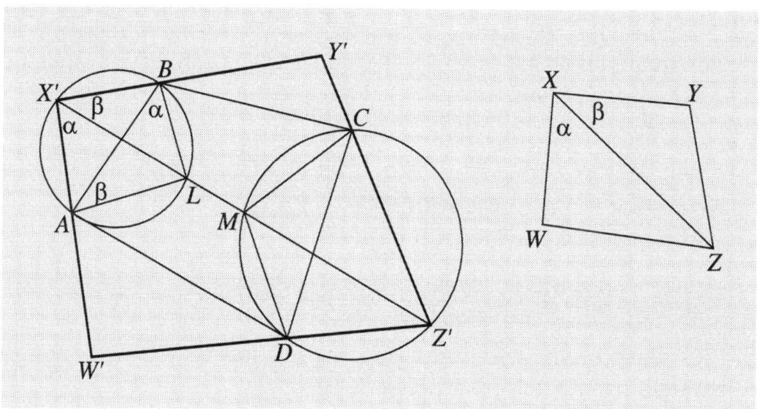

Analysis. Let $X'Y'Z'W'$ be the required quadrilateral. Join $X'Z$, XZ, and draw the circle $X'AB$, cutting $X'Z'$ in L.

Then, since $$\angle LAB = \angle LX'B = \angle ZXY$$

and $$\angle LBA = \angle LX'A = \angle ZXW,$$

$\therefore$ the point L on $X'Z'$ can be found.

A second point M on $X'Z'$ can be found in a similar manner.

Hence the following construction:

Draw two lines *AL*, *BL*, so that $\angle ABL = \angle WXZ$ and $\angle BAL = \angle YXZ$. Find *M* in a similar manner: describe circles round *ABL*, *CDM*, and join and produce *LM* to meet the circumferences again in *X'* and *Z'*.

Complete the construction and apply proof.

EXAMPLE 7: *Inscribe a square in a given quadrilateral ABCD.*

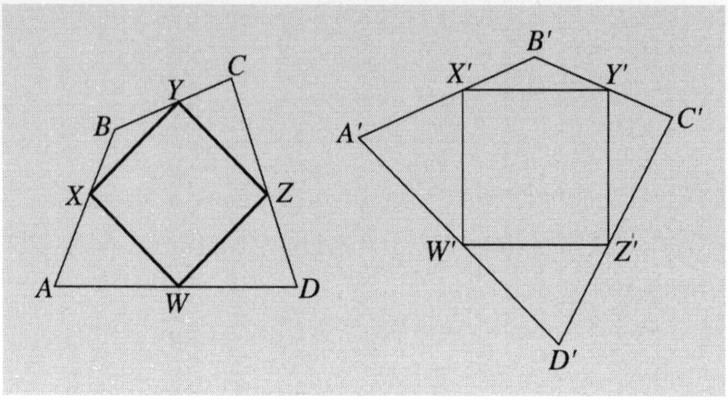

Draw any square *X'Y'Z'W'*, and about it describe a quadrilateral *A'B'C'D'* similar to *ABCD*, so that *X'* is in *A'B'*, etc.

Divide *AB*, *BC*, etc., in *X*, *Y*, etc., in the ratios *A'X'* : *X'B'*, *B'Y'* : *Y'C'*, etc.

Then *XYZW* is the square required.

Supply proof by showing Δs *BXY*, *B'X'Y'*, etc., similar.

The method employed in the last example is very useful. We first make a figure of the required shape (*i.e.* one similar to the required figure), and finally a figure of the proper size.

EXERCISE LXXII

1. Apply a construction similar to Ex. 5, p. 307, to describe a rhombus about a given quadrilateral, one angle of the rhombus being equal to 60°.
2. Describe a square *ABCD*, given the vertex *A* and that the sides *BC*, *CD* pass through fixed points *X*, *Y* respectively.

   [Draw $AZ \perp AX$, making $AZ = AX$; *Z* is another point on *BC*.]
3. Draw a square, such that three of its sides pass each through a given point and its fourth side is parallel to a given straight line.
4. Draw an equilateral triangle with a side of given length and its vertices on three given concurrent lines *OA*, *OB*, *OC*.

   [First use Ex. LXVI. 1 to construct an equilateral triangle having one vertex at any chosen point on *OA*, and the other vertices on *B* and *C* respectively. *O* is the homothetic centre of this triangle and the triangle required.]

5. Construct an equilateral triangle with its vertices on three given concurrent lines *OA*, *OB*, *OC*, so that one of its sides may pass through a given point *X*.

[Proceed as in Ex. 4.]

6. Draw a square with a side of given length, so that three of its vertices may be each on one of three given concurrent straight lines.

7. *Inscribe a square in a given parallelogram ABCD.*

[It is best to proceed thus,—Let *AC*, *BD* meet at *O*. Then *O* is the centre of the square. Find points *X* in *AB* and *Y* in *BC*, such that $OX = OY$ and $\angle XOY$ is a right angle. See Ex. LXVI. 1.]

8. *Inscribe a regular pentagon in a square ABCD, so that one vertex of the pentagon may be on each side of the square.*

[Draw a regular pentagon *BEFGD* on *BD* so that *C* is within the pentagon. Join *AE*, cutting *BC* in *E'*. Draw *E'F'*, *F'G'*, *G'D'*, *D'B'* ∥ *EF*, *FG*, *GD*, *DB* to meet *AF*, *AG*, *AD*, *AB* in *F'*, *G'*, *D'*, *B'*. Join *B'E'*. Then *B'E'F'G'D'* is a regular pentagon.]

9. Draw a regular hexagon, whose sides are each 1 in. Inscribe in it a regular heptagon with one vertex on each side of the hexagon. Measure a side of the heptagon.

10. *Inscribe a square in a given regular figure* (*say a heptagon ABCDEFG*).

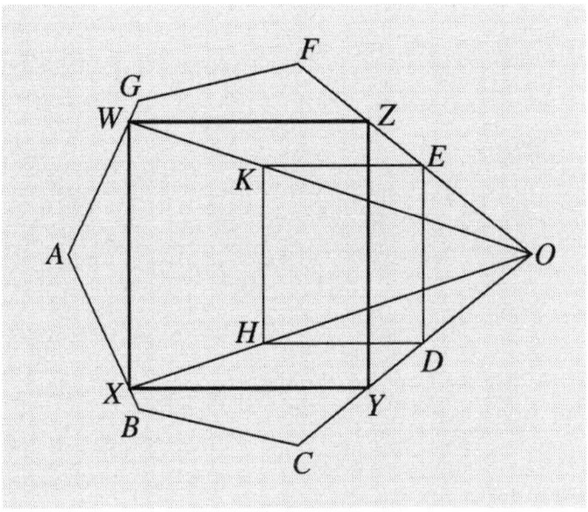

Analysis. Produce *CD*, *FE* to meet in *O*. Join *AO*. Then the heptagon is symmetrical with regard to *AO*, and it is clearly possible to draw an inscribed square, which will also be symmetrical with regard to *AO*.

Hence the following construction: Draw the square *DHKE*. Join *OH*, *OK*, and produce them to cut sides of the heptagon in *X*, *W* respectively. Draw the square *XYZW*.

This is the square required. Supply proof.

11. Inscribe squares in the following regular figures: (*i*) a hexagon, (*ii*) a pentagon, (*iii*) a decagon, the sides of each being 1 in. Measure the side of the square in each case.

12. The construction in Ex. 10 fails for a polygon with more than 12 sides. Modify it for a regular polygon of 15 sides, each 1 in. Measure the side of the inscribed square.

13. *Inscribe a regular octagon in an equilateral triangle ABC, so that one side of the octagon may be along a side of the triangle, and two other vertices on sides of the triangle.*

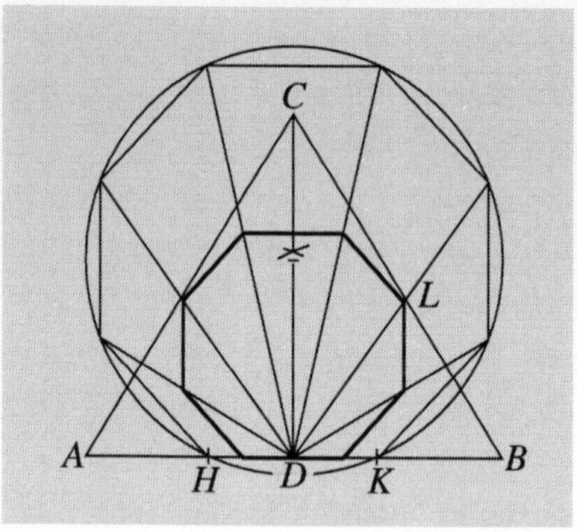

Bisect *AB* at *D*. Along *DA*, *DB* set off any two equal lengths *DA*, *DK*. On *HK* draw a regular octagon, and proceed as indicated in the figure.

Complete the construction and proof, noting that *L* is the vertex, which is first found.

14. *In a regular figure, say an equilateral triangle, inscribe a regular figure of double the number of sides.*

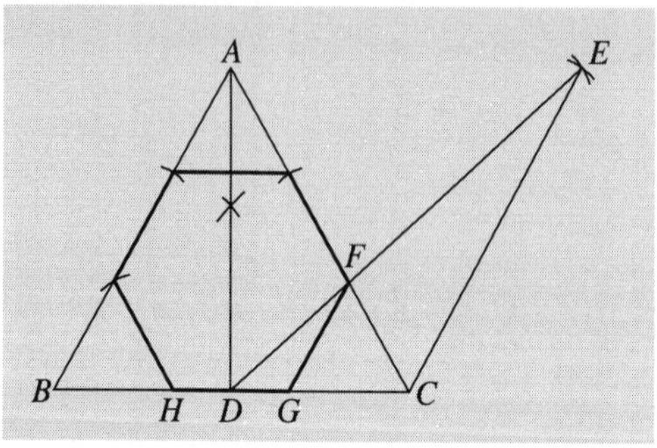

Bisect *BC* in *D*. Draw *CE*, the side adjacent to *BC*, of the regular hexagon on *BC*. Join *DE*, cutting *CA* in *F*. Draw *FG* || *EC* to meet *BC* in *G*.

Then *FG* is a side of a regular hexagon, which, with that on *BC*, has *D* as homothetic centre.

Complete the construction and proof.

15. In a square of 3 in side, inscribe a regular octagon, and measure the length of one side.

CONSTRUCTION 35

Draw a rectilineal figure similar to one given rectilineal figure *ABCD*, and equal in area to another, *XYZ*.

Construct squares equal in area to *ABCD*, *XYZ*. (This part of the construction is not shown.) Along any straight line through *A*, set off lengths *AE*, *AF* respectively, equal to sides of these squares. Join *EB*, *AC*. Draw *FB'*, *B'C'*, *C'D'* ∥ *EB*, *BC*, *CD* to meet *AB*, *AC*, *AD* in *B'*, *C'*, *D'* respectively.

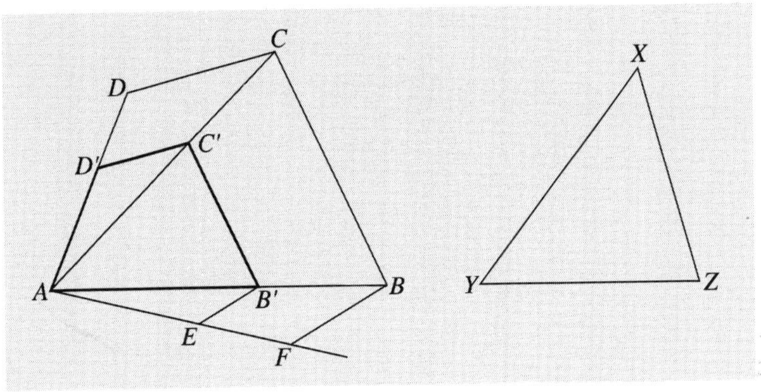

Then *AB'C'D'* is the required figure.

Proof.
$$\frac{ABCD}{XYZ} = \frac{AE^2}{AF^2} = \frac{AB^2}{AB'^2} = \frac{ABCD}{AB'C'D'}$$

$\therefore$
$$AB'C'D' = XYZ;$$

and by construction *AB'C'D'* is similar to *ABCD*.

A modified construction can be used with advantage in particular cases, as in the following:

Draw a regular pentagon of 4 square inches area.

Construct a right-angled Δ *ABC*, of area $\frac{1}{5}$ of 4 sq. in., with the right angle at *C* (making *BC* = 2 in., *CA* = $\frac{4}{5}$ in.). On *BC* describe an isosceles Δ *DBC*, similar to a triangle formed by joining the centre of a regular pentagon to two consecutive vertices. Draw *AE* ∥ *CB* to meet *DB* in *E*. Along *BD* set off *BO*, a mean proportional to *BE*, *BD*. Draw *OC'* ∥ *DC* to meet *BC* in *C'*. Then *O* is the centre, and *BC'* a side of the required pentagon. Supply proof as in Ex. 4, p. 289.

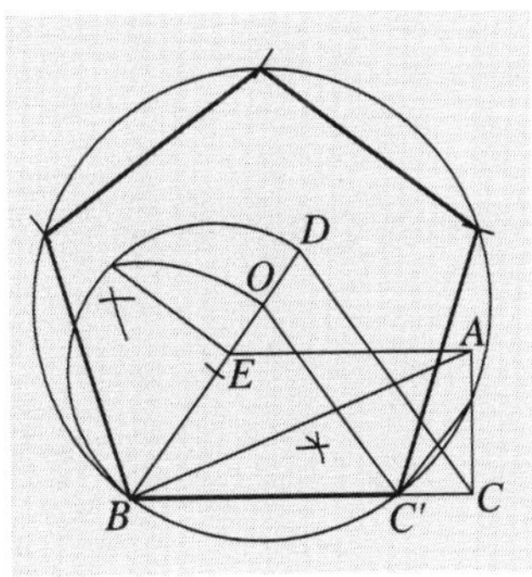

> **NOTE:** The figure, being drawn half-size linear, gives a pentagon whose area. is 1 sq. in., *i.e.* a quarter of the area required.

XLV. MEDIAL SECTION AND THE REGULAR PENTAGON

CONSTRUCTION 36 (Euclid II. 11)

Divide a given straight line into two parts, such that the rectangle contained by the whole line and one part may be equal to the square on the other part.

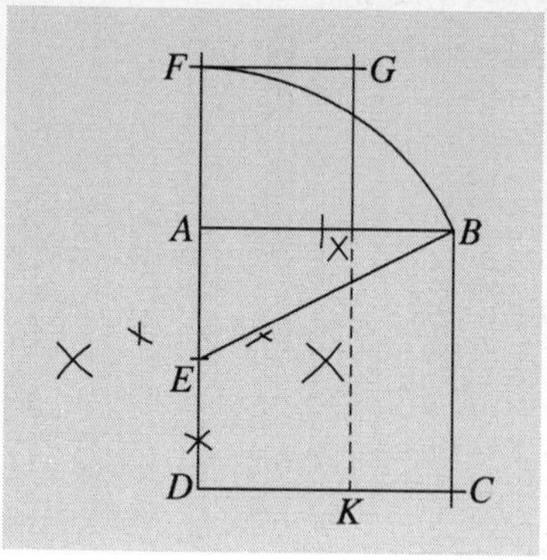

Let *AB* be the given straight line. It is required to divide *AB* at *X*, so that the rect. *AB*, *BX* may be equal to the sq. on *AX*.

Construction. On *AB* draw the square *ABCD*. Bisect *AD* at *E*. Join *EB*. Along *EA* produced set off *EF* equal to *EB*. On *AF* draw the square *AFGX*. Then *X* is the required point of division.

Proof. Produce *GX* to meet *CD* at *K*.

Because *DA* is bisected at *E* and produced to *F*,

∴ rect. *DF*, *FA* + sq. on *EA* = sq. on *EF*.

But *EF* = *EB* (*construction*);

∴ rect. *DF*, *FA* + sq. on *EA* = sq. on *EB*.

Now ∠ *EAB* is a right angle;

∴ sq. on *EB* = sq. on *EA* + sq. on *AB*;

∴ rect. *DF*, *FA* + sq. on *EA* = sq. on *EA* + sq. on *AB*.

From each of these equals take the sq. on *EA*;

∴ rect. *DF*, *FA* = sq. on *AB*.

Now, by construction, *AF* = *FG*;

∴ the rect. *DF*, *FA* = the fig. *DFGK*;

∴ the fig. *ABCD* = the fig. *DFGK*.

From each take the fig. *AXKD*;

∴ the fig. *XBCK* = the fig. *AFGX*;

that is, the rect. *AB*, *BX* = the sq. on *AX*.

DEF. A straight line is said to be divided in **medial section** when the rectangle contained by the whole line and one part is equal to the square on the other part.

In practice, when we wish to divide a straight line AB in medial section, we draw the figure of Construction 36 in the modified form, shown below. We begin by bisecting AB at L and proceed as indicated.

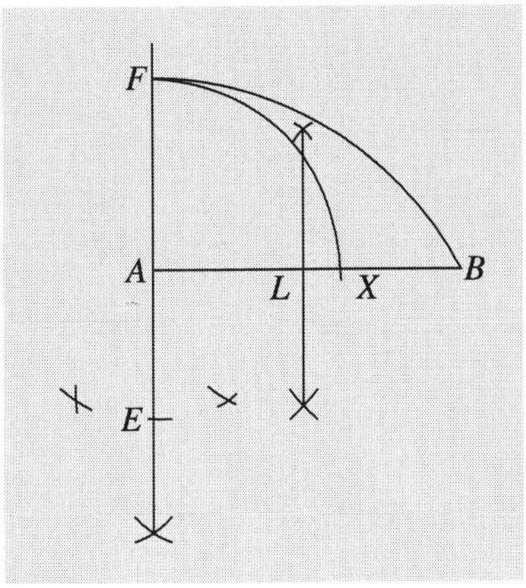

EXAMPLE 1: *Solve algebraically the problem of dividing a given straight line AB at X, so that*

$$AB \cdot BX = AX^2.$$

Let $AB = a$, $AX = x$;

$\therefore$ $$BX = a - x;$$

$\therefore$ $$a(a - x) = x^2;$$

$\therefore$ $$x^2 + ax = a^2; \qquad \text{...}(i)$$

$\therefore$ $$x^2 + ax + \left(\frac{1}{2}a\right)^2 = a^2 + \frac{1}{4}a^2;$$

$\therefore$ $$\left(x + \frac{1}{2}a\right)^2 = \frac{5a^2}{4};$$

$\therefore$ $$x + \frac{1}{2}a = \pm\frac{1}{2}\sqrt{5} \cdot a;$$

$\therefore$ $$x = \frac{1}{2}\left(\pm\sqrt{5} - 1\right) \cdot a.$$

The negative value of x is inapplicable;

$\therefore$ $$x = \frac{1}{2}\left(\sqrt{5} - 1\right) \cdot a;$$

that is, $$AX = \frac{1}{2}\left(\sqrt{5} - 1\right) \cdot AB;$$

EXAMPLE 2: *Interpret geometrically the negative root of equation (i) of Ex. 1.*

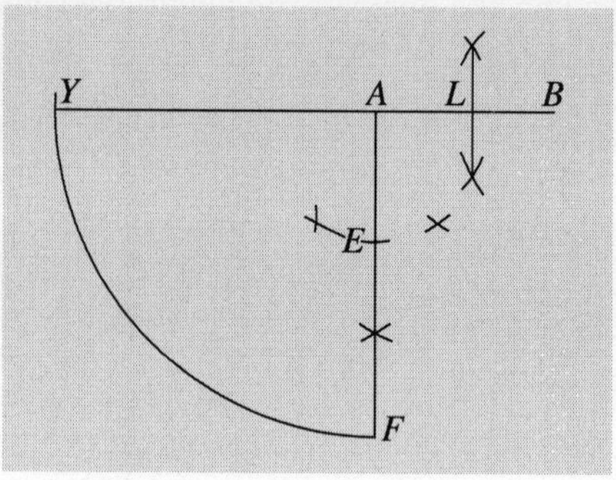

The negative root is $-\frac{1}{2}\left(\sqrt{5}+1\right)a$.

Denote this by $-y$, so that

$$y = \frac{1}{2}\left(\sqrt{5}+1\right)a.$$

The value $(-y)$ of x satisfies the equation

$$x^2 + ax = a^2;$$

$$\therefore \quad (-y)^2 + a(-y) = a^2;$$

$$\therefore \quad y^2 = a(a + y).$$

Along BA, *produced through* A, set off $AY = y = \frac{1}{2}\left(\sqrt{5}+1\right)a$. [This length can be constructed geometrically, as in the figure.]

Then $\quad BA = a,\ AY = y;$

$\therefore \quad BY = a + y.$

But $\quad y^2 = a(a + y);$

$\therefore \quad AY^2 = BA \cdot BY.$

Hence, a point Y has been found in BA produced, such that $BA \cdot BY = AY^2$. This point Y is said to divide ***BA* externally** in medial section.

EXAMPLE 3: *The construction most often, used in Practical Geometry for dividing a line in medial section is that shown in the figure on the right.*

Prove geometrically or algebraically that the construction is correct.

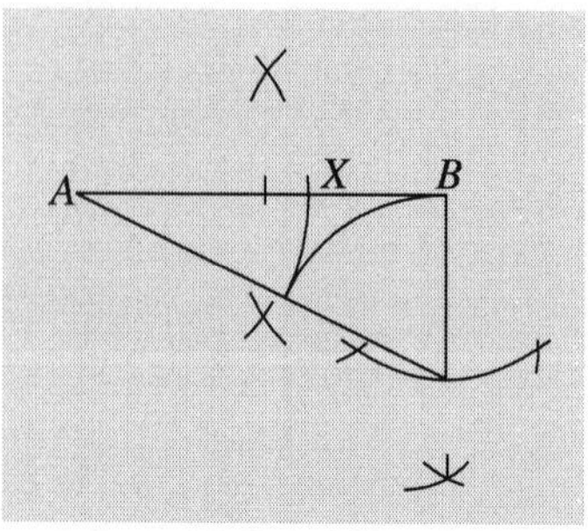

CONSTRUCTION 37 (Euclid IV. 10)

Construct an isosceles triangle with each of the angles at the base twice the third angle.

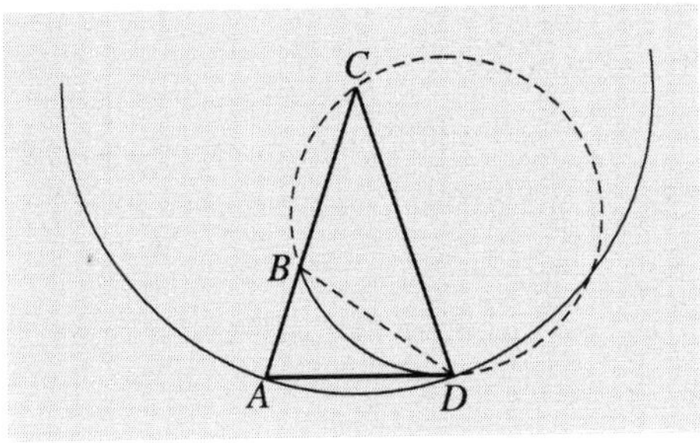

Construction. Draw any straight line *AC*.

Divide *AC* at *B*, so that the rect. *AC*, *AB* = the sq. on *BC*. With centre *C* and radius *CA*, draw a circle. In the circle place a chord *AD* equal to *BC*. Join *CD*.

Then *CAD* is an isosceles triangle with each of its angles at *A* and *D* twice the angle at *C*.

Proof. Draw a circle through *B*, *C*, *D*.

Then, because the rect. *AC*, *AB* = the sq. on *BC* (*construction*);

and $AD = BC$ (*construction*),

∴ rect. *AC*, *AB* = sq. on *AD*;

∴ *AD* touches the circle *BCD*;

∴ $\angle ADB = \angle BCD$, in the alternate segment.

To each add $\angle BDC$;

∴ $\angle ADC = \angle BCD + \angle BDC$.

Now, the side *CB* of the triangle *CBD* is produced to *A*;

∴ the ext. $\angle ABD = \angle BCD + \angle BDC$;

∴ $\angle ABD = \angle ADC$.

Again, by construction, $CA = CD$;

∴ $\angle DAC = \angle ADC$;

∴ $\angle ABD = \angle DAC$;

that is, $\angle ABD = \angle BAD$;

∴ $AD = BD$.

But, by construction, $AD = BC$;

∴ $BD = BC$;

∴ $\angle BDC = \angle BCD$.

Again, $\angle ADB = \angle BCD$ (*proved*);

∴ $\angle ADC = 2\angle BCD$;

∴ Also $\angle DAC = 2\angle BCD$.

The lines which are actually necessary for the construction of a triangle CAD with the angles A and D each twice the angle C are shown in the accompanying figure.

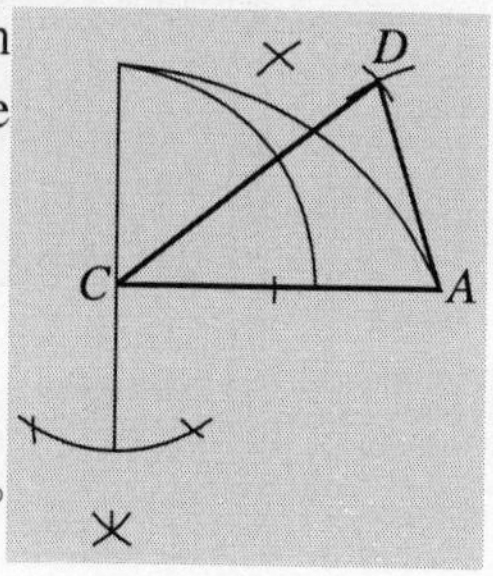

Since $$A = D = 2C$$
and $$C + A + D = 180°;$$
$\therefore$ $$C = 36° \text{ and } A = D = 72°.$$

Hence, *this figure gives a construction for angles of* 36° *and* 72°.

EXAMPLE 1: *Construct an angle of 18° and find its sine in a surd form.*

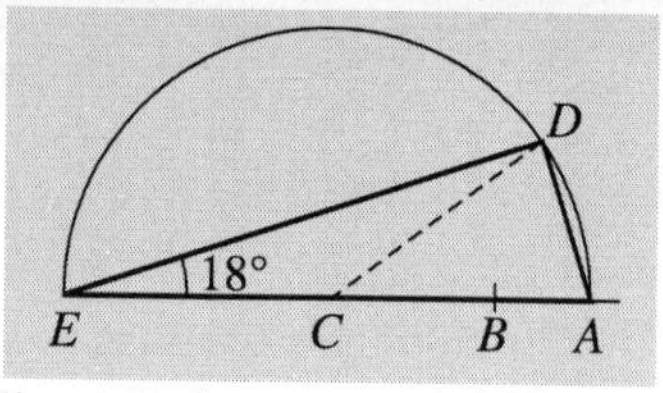

With centre C and radius equal to any convenient unit, draw a circle.

Let ECA be a diameter.

Divide CA at B, so that

$$CA \cdot BA = CB^2.$$

In the circle place a chord AD equal to CB. Join DE.

Then $$\angle AED = 18°.$$

Again, since $$CA = 1,$$

$\therefore$ by Ex. 1, $$CB = \frac{1}{2}\left(\sqrt{5} - 1\right);$$

$\therefore$ $$AD = \frac{1}{2}\left(\sqrt{5} - 1\right) \text{ and } EA = 2.$$

Now, EDA is a semicircle;

$\therefore \angle EDA$ is a rt. $\angle$;

$\therefore$ $$\sin AED = \frac{AD}{EA} = \frac{\sqrt{5} - 1}{4};$$

$\therefore$ $$\sin 18° = \frac{\sqrt{5} - 1}{4}.$$

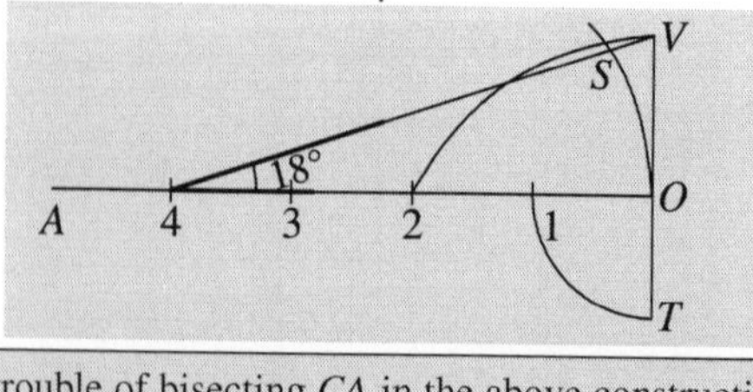

NOTE: To save the trouble of bisecting CA in the above construction, we can begin by setting off four equal lengths along a straight line OA, and proceed as in the figure above.

EXERCISE LXXIII

Medial Section

1. In the figure of Construction 36, if $AB = a$, prove that

(*i*) $EB = \frac{1}{2}\sqrt{5} \cdot a$, $AX = \frac{1}{2}\left(\sqrt{5}-1\right)a$, $XB = \frac{1}{2}\left(3-\sqrt{5}\right) \cdot a$.

(*ii*) Hence show that $\frac{1}{3}\text{AB} < XB < \frac{1}{2}AB$.

(*iii*) Show that the middle point of AX is equidistant from B and G.

(*iv*) If CB, FG are produced to meet at L, show that the triangles DAX, XGL are equiangular. Hence show that D, X, L are in the same straight line.

2. AB is divided at C, so that $AB \cdot CB = AC^2$. If BA is produced to D, so that $AD = AC$, show that $DA \cdot DB = AB^2$. [Let $AB = a$, $AC = x$, etc.]

3. Explain how to produce a given straight line AB to C, so that $AC \cdot BC = AB^2$. [See Ex. 2.]

4. AB is divided at C, so that $AB \cdot CB = AC^2$, and from CA a part CD is cut off equal to CB. Prove that $CA \cdot DA = CD^2$.

5. If AB is divided at C, so that $AB \cdot CB = AC^2$, prove that

$$AB^2 + CB^2 = 3AC^2.$$

6. AB is divided at C, so that $AB \cdot CB = AC^2$. If BA is produced to D, so that $BD = 3BC$, prove that $BC^2 = BA \cdot AD$.

7. If BA is a given straight line, solve algebraically the problem of finding a point Y in BA produced, such that $BA \cdot BY = AY^2$.

8. From the algebraical solution, deduce a geometrical solution of the problem in Ex. 7, and give a geometrical proof, similar to that of Construction 36.

9. In the figure of Construction 37, prove that

(*i*) $\angle ACD = 36°$, $\angle CBD = 108°$.

(*ii*) DAB is an isosceles triangle with the angles at A and B each twice the angle at D.

(*iii*) BCD is an isosceles triangle with the vertical angle equal to three times each of the base angles.

(*iv*) AD is a side of a regular decagon inscribed in the circle with centre C.

(*v*) BC is a side of a regular pentagon inscribed in the circle BCD.

(*vi*) If the two circles meet again at E, show that

$$\angle CDE = \angle CED = \angle ABD.$$

(*vii*) Hence show that $DE = AD$.

10. In the figure of Construction 37, show that the circumcentre of the triangle ABD is the point where the bisector of the angle ACD cuts the arc BD.

Theorem 100

(*i*) If the radius of a circle is divided in medial section, the greater part is equal to a side of a regular decagon inscribed in the circle.

(*ii*) The sum of the squares on the radius and on a side of the decagon is equal to the square on the side of a regular pentagon inscribed in the circle.

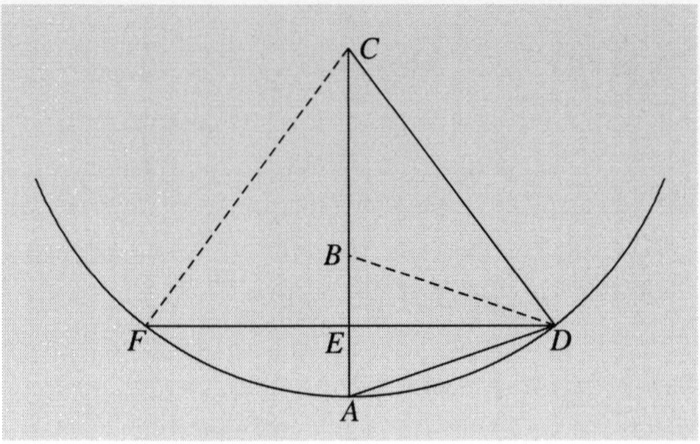

Let CA be a radius of a circle with centre C, and let CA be divided at B, so that $AC \cdot AB = BC^2$.

(*i*) It is required to prove that BC is equal to a side of a regular decagon inscribed in the circle.

Construction. Place a chord AD in the circle equal to BC. Join AD, CD.

Proof. As in the proof of Construction 37,

$$\angle CAD = \angle CDA = 2\angle ACD;$$

∴ the sum of the angles of the triangle $CAD = 5\angle ACD$.

But this sum = 2 right angles;

$$\therefore \quad \angle ACD = \frac{1}{5} \text{ of 2 right angles};$$

$$= \frac{1}{10} \text{ of 4 right angles};$$

∴ AD is a side of a regular decagon inscribed in the circle,

and (as in Construction 37) $AD = BC$;

∴ BC is equal to a side of a regular inscribed decagon.

(*ii*) It is required to prove that the sum of the squares on CA, AD is equal to the square on a side of a regular inscribed pentagon.

Construction. Draw a chord DF of the circle to cut CA at right angles at E. Join BD, CF.

Proof. Because CE is perpendicular to the base DF of the isosceles triangle CDF,

$$\therefore \quad ED = EF$$

and $\angle ECD = \angle ECF$;

$\therefore$ $\angle DCF = 2\angle ACD$

$= \frac{1}{5}$ of 4 right angles;

$\therefore$ *DF* is a side of a regular inscribed pentagon.

Again, because *AC* is divided at *B*,

$\therefore$ $AC^2 + AB^2 = 2AC \cdot AB + BC^2$;

and $AC \cdot AB = BC^2$,

$\therefore$ $AC^2 + AB^2 = 3BC^2 = 3AD^2$.

To each add the square on *AD*,

$\therefore$ $AC^2 + AD^2 + AB^2 = 4AD^2 = 4AE^2 + 4ED^2$.

Now *DE* is perpendicular to the base *AB* of the isosceles triangle *DAB*,

$\therefore$ $AE = EB$

and $AB^2 = 4AE^2$;

$\therefore$ $AC^2 + AD^2 = 4ED^2 = FD^2$.

CONSTRUCTION 38

Inscribe (*i*) a regular pentagon, (*ii*) a regular decagon in a given circle.

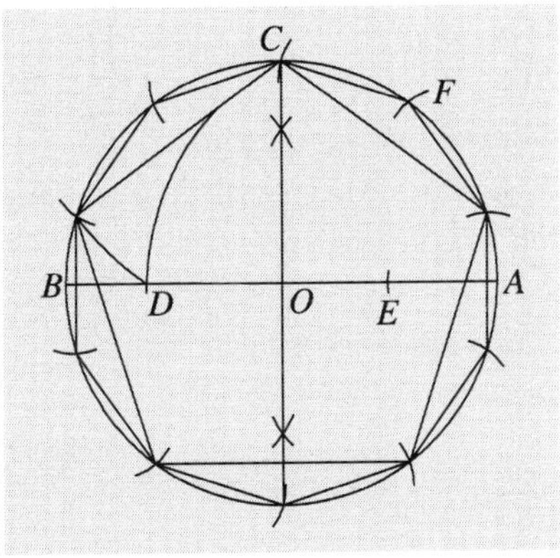

Let *O* be the centre of the given circle. Draw a diameter *AOB* and a radius *OC* perpendicular to *OA*. Bisect *OA* at *E*. With centre *E* and radius *BC*, draw an arc cutting *OB* in *D*.

Then *OD*, *CD* are respectively equal to sides of the required decagon and pentagon, and the figures may be constructed by placing chords equal to these lines round the circle.

Proof. Because *OB* is divided in medial section at *D*,

$\therefore$ the greater segment is equal to the side of a regular inscribed decagon.

Again, the sum of the squares on the radius and on the side of an inscribed decagon is equal to the square on the side of an inscribed pentagon.

∴ sq. on the side of pentagon = sq. on OC + sq. on OD

= sq. on CD;

∴ side of pentagon = CD.

EXERCISE LXXIV

1. In the figure of Construction 37, let the circles meet again at E. Prove that AE is a side of a regular pentagon inscribed in the circle with centre C.
2. In the figure of Construction 37, if DB is produced to meet the circle with centre C at F, prove that AF is a side of a regular pentagon inscribed in this circle.
3. It is required to draw on a given straight line AB an isosceles triangle DAB with the angles A and B each twice the angle D. Prove the following construction,—Produce AB to C, so that $AB \cdot AC = BC^2$ (by Ex. 2, p. 317). With centres A and B and radii each equal to BC, draw arcs cutting at D. Join AD, BD. Then DAB is the required triangle. [The proof is obvious from Construction 37.]
4. From Ex. 3, deduce a construction for drawing a regular pentagon on a given straight line AB.
5. *Prove the following construction for drawing a regular pentagon on a given straight line AB*:

   Draw CDF bisecting AB at right angles.

   Set off $CD = AB$. Join BD. Produce BD to E, making DE equal to AC. With centre B and radius BE, draw an arc cutting CF in F. With centres A, F, B, and radius equal to AB, draw arcs cutting in G, H.

   Show that $AGFHB$ is the required pentagon.
6. Describe a regular decagon on a given line.
7. In the figure of Theorem 100, show that $CE = \frac{1}{4}\left(\sqrt{5}+1\right)\cdot CA$. Hence show that cos, $36° = \frac{1}{4}\left(\sqrt{5}+1\right)$.
8. If DAB is an isosceles triangle, in which $\angle A = \angle B = 2\angle D$ and $AB = a$, prove that

   (*i*) $AD = \frac{1}{2}(\sqrt{5}+1)\cdot a$;

   (*ii*) Area of $\Delta\, DAB = \frac{1}{4}\sqrt{\left(5+2\sqrt{5}\right)}\,a^2$.

9. If $ABCDE$ is a regular pentagon and $AB = a$, prove that

(*i*) $AD = \frac{1}{2}\left(\sqrt{5}+1\right)a$;

(*ii*) If AB, DC are produced to meet at F, prove that $BF = AD$;

(*iii*) If the perpendicular from B to AB meets CD at H, prove that

$$CH = \frac{1}{2}\left(3-\sqrt{5}\right)a.$$

10. If r is the radius of the circle inscribed in a regular pentagon and R the radius of the circumscribing circle, prove that $r = \frac{1}{4}\left(\sqrt{5}+1\right)R$.

11. If R is the radius of the circle described about a regular decagon of side a, prove that $R = \frac{1}{2}\left(\sqrt{5}+1\right)a$.

MISCELLANEOUS EXERCISES

Arranged in Sets for Homework or Revision

Paper XXVI (to Section XXXVI)

1. Two equal circles, having their centres at A and B, touch at C. A point D, in AB produced, is the centre of a third circle passing through C. Draw a common tangent (other than that at C) to the circles whose centres are A and D, and let P and Q be the points of contact; draw the line CQ, cutting the circle with centre B in M, and produce it to meet in N the tangent to this circle at the point E, which is diametrically opposite to C. Show that $EN = PQ$ and $CM = QN$.

[Prove that Δs EMN, PCQ are congruent.]

2. AB is a diameter of a circle and CD a parallel chord; if P is any point in AB, show that the sum of the squares on PC, PD is equal to the sum of the squares on PA, PB.

3. If P, Q are points in the sides AB, AC of a triangle ABC, prove that

$$\Delta\, APQ : \Delta\, ABC = AP \cdot AQ : AB \cdot AC.$$

4. If ABC, XYZ are triangles in which the angles A and X are either equal or supplementary, prove that

$$\Delta\, ABC : \Delta\, XYZ = AB \cdot AC : XY \cdot XZ.$$ [Use Ex. 3.]

5. A, B, C, D are four given points and BA, DC produced meet in O. Find a point E on AB produced, such that $OA : OE = OA \cdot OC : OB \cdot OD$.

6. ABC is a triangle; BE and CF are the perpendiculars from B, C to the opposite sides. Find the ratio of the triangle AFE to the triangle ABC, (*i*) when $A = 30°$, (*ii*) $A = 45°$, (*iii*) $A = 60°$.

PAPER XXVII (to Section XXXVII)

1. With three given points (not in the same straight line) as centres, describe three circles each of which shall touch the other two externally.
2. *ABCD*, *AB'C'D'* are squares with a common vertex *A* (both lettered counter clock-wise). Prove that *BB'*, *CC' DD'* meet in a point.

   [This point is the second point of intersection of the circles *ABCD*, *AB'C'D'*.]
3. *ABC* is a triangle, *E* the middle point of *AB*, *D* the point of trisection of *CA* nearer to *A*; *BD*, *CE* cut in *O*. Show that *DO* is a quarter of *DB*.

   [Draw *EF*, parallel to *BD*, to meet *AC* in *F*.]
4. In the side *AB* of the triangle *ABC* a point *D* is taken, and in *BC* produced a point *E*, such that $CE = AD$; *DE* is drawn, cutting *AC* in *F*. Prove that $DF : FE = BC : AB$.

   [Use Menelaus' Theorem.]
5. *A* and *B* are given points within a circle. Find a point *P* on the circle such that if *AM*, *BN* are the perpendiculars from *A* and *B* to the tangent at *P*, $PM : PN$ may be equal to a given ratio.
6. Draw two circles, radii 2.5 in. and 1.5 in., with their centres 3 in. apart. Through one point of intersection draw a straight line so that the chord intercepted on it by the larger circle is twice the chord intercepted on it by the smaller circle. Measure the greater chord.

PAPER XXVIII (to Section XXXVIII)

1. If, in the triangles *ABC*, *XYZ*, the angles *A* and *X* are equal, and the angles *B* and *Y* are supplementary, prove that the sides about the angles *C* and *Z* are proportional.
2. A circle is drawn to touch the circumcircle of a triangle *ABC* at the point *A*. Prove that the tangents drawn to it from the points *B*, *C* are in the ratio of the sides *BA*, *CA*.

   [Produce *BA*, *CA* to meet the circle in *B'*, *C'*, and prove that *B'C'* is parallel to *BC*.]
3. *P* and *Q* are two points in the sides *AB*, *CD* respectively of a quadrilateral *ABCD*, such that $AP : PB = CQ : QD$. Prove that, if *QA*, *QB*, *PC*, *PD* are drawn, the sum of the triangles *QAB*, *PCD* is equal to the area of the quadrilateral.
4. A circle *PAR* is touched internally by a circle *QAS*. The tangent to the latter at *Q* cuts the former circle in *P*, *R*. Join *AP*, *AR*, cutting the circle *QAS* in *S* and *T*. Prove that $PQ : QR = PS : RT$.

   [Prove that *AQ*, bisects $\angle SAT$ and apply Ex. 1 to Δs *QSP*, *QTR*.]
5. *MPN*, *M'P'N'* are two tangents to a given circle *PQP'*; *AM*, *BN*, *AM'*, *BN'* are the perpendiculars to them respectively from two given points *A* and *B*. Prove that, if $MP : PN = M'P' : P'N'$, then either *PP'* is a diameter of the circle, or *A*, *B* lie on a diameter.

6. ABC is a triangle : BC is 3.5 in., the ratio of $AB : AC$ is equal to the ratio 5 : 2, and its area is 2.45 sq. in. Construct the triangle and measure AB. [Use the Circle of Apollonius.]

PAPER XXIX (to Section XXXVIII)

1. The arc AB of the circle $ABCDE$ is equal to the arc BC, the chords AE, ED, DC are drawn, and DC is produced to meet the tangent at B in P. Show that the angles AED and DPB are supplementary.
2. C is the middle point of AB, a chord of a circle whose centre is O; a point P is taken on the circumference, whose distance PD from AB is equal to AC; M is the middle point of PD, and CF is drawn parallel to OM to meet PD in F; show that $CF = FP$.
3. $ABA'B'$ is a rectangle inscribed in a circle. AC is a chord equal to AB, meeting $A'B'$, $A'B$, produced if necessary, in F, E respectively. Prove that $AF : AE = CF : CA$.
4. CD is perpendicular to AB, a diameter of the circle ABC, meeting it in D. The tangents at A and B meet the tangent at C in E and F respectively, Show that AF and EB intersect at the middle point of CD.
   [Produce BC, AE to meet at K; show $AE = EK$.]
5. P is any point in the diameter AB of a given circle. Bisect AP at D and BP at E. Describe a circle on DE as diameter. Through P, draw any line PQR, cutting this circle in Q and the given circle in R. Prove that $PQ = QR$, and that the tangent at Q, to the inner circle is parallel to the tangent at R to the outer circle.
   [Take $QR' = PQ$; prove that $\angle AR'B$ is a right angle.]
6. Draw a circle of 2.5 in. radius. Describe a square, of which one side lies along a tangent to the circle and the opposite side is a chord of the circle. Measure the side of the square.

PAPER XXX (to Section XL)

1. ABC is a triangle and two circles are drawn, one touching AB at B and the other AC at C. If the circles intersect at P, Q, prove that the difference between the angles BPC, BQC is equal to the angle A.
2. AB and CD are chords of a circle whose centre is O, intercepted on lines at right angles to one another, which meet at a point P outside the circle. Prove that $AB^2 + CD^2 + 4OP^2 = 8AO^2$.
3. A and B are the centres of two circles which do not intersect, and P is the point of intersection of a direct and a transverse common tangent to the circles. Prove that, if PN is the perpendicular from P to AB, then PN is a mean proportional between AN and NB.

4. *A* straight line *POQ* is drawn through *O*, one of the common points of two given circles, meeting the circles in *P*, *Q*, and *R* divides *PQ* in a given ratio. Find the locus of *R*.

[If *O′* is the other point of intersection of the circles, prove that Δ *PO′Q* is of fixed species, and that ∠ *ORO′* is constant.]

5. *ABC* is a triangle: points *L*, *M*; *P*, *Q*; *R*, *S* are taken on the sides *BC*, *CA*, *AB* respectively, so that the middle segment of each side is half each of the other two segments of that side. Prove that the area of the hexagon *LMPQRS* is thirteen-twenty-fifths of the area of the triangle.

6. *OX*, *OY* are two straight lines inclined at an angle of 40°. *OP* makes an angle of 25° with *OX* and 65° with *OY*, and is 2 in. in length. From *P*, draw two equal lines *PQ*, *PR* to *OX*, *OY* respectively, so that the contained angle *RPQ*, is 75°. Measure *QR*.

Paper XXXI (to Section XLI)*

1. *A*, *B*, *C*, *D* are four collinear points; two circles, one through *A*, *B* and the other through *C*, *D*, are drawn to touch one another in *P*. Find the locus of *P*.

2. On the sides *AB*, *AC* of the triangle *ABC*, squares *ABHK*, *ACGF* are described externally to the triangle; if *BF*, *CK* meet in *X*, prove that *AX* passes through the centre of the square on *BC*.

[*Prove.* ∠ *BXK* a right angle, and that *AX* bisects the angle *BXC*.]

3. A point is taken within a parallelogram, such that the line joining it to any vertex subtends equal angles at the two adjacent vertices. Prove that the line joining the point to any vertex will subtend equal angles at adjacent vertices;

[Draw parallels, through the given point to the sides of the parallelogram.]

4. *ABCD* is a quadrilateral inscribed in a circle and *P* any point on the circles: *X*, *Y*, *Z*, *W* are the feet of the perpendiculars on *AB*, *BC*, *CD*, *DA* respectively. Prove that $PX \cdot PZ = PY \cdot PW$.

[*Prove.* Δs *PYZ*, *PWX* equiangular.]

5. Points *D*, *E*, *F* are taken in the sides *BC*, *CA*, *AB* of a triangle *ABC*, so that $BD = CE = AF = x$. Through *D*, *E*, *F*, parallels are drawn to the sides *CA*, *AB*, *BC* respectively, to form a triangle *PQR*. Prove that, if the lengths of *BC*, *CA*, *AB* are *a*, *b*, *c*,

$$\Delta PQR : \Delta ABC = \left[2 - \left(\frac{x}{a} + \frac{x}{b} + \frac{x}{c}\right)\right]^2 : 1.$$

[Through *B* draw a parallel to *PR* to meet *QR* produced, and prove *PE*; $c = a - x : a$ and $c - x - QE : c = x : b$.]

* Some of the questions in Papers XXXI–XXXV are considerably harder than those in Papers XXVI–XXX and may require further hints from the teacher.

6. ABC is a triangle, I the centre of the inscribed circle, I' the centre of the escribed circle opposite to A. $II_1 = 1\frac{1}{4}$ in., $AB = 2$, $AC = 1\frac{1}{8}$ in. Draw the triangle and calculate the length of BC. Verify by measurement.

[$AI \cdot AI_1 = AB \cdot AC = \frac{9}{4}$ and $II_1 = AI_1 - AI = 1\frac{1}{4}$: hence find AI, AI_1. Also $s - a : s = AI : AI_1$; and $b + c = 2s - a = 3\frac{1}{8}$: hence find a.]

PAPER XXXII (to Section XLII)

1. ABC is an equilateral triangle, and two circles are described, one with A as centre and AC as radius, the other on AB as diameter: P is a point in AC, such that a circle can be described with centre P to touch both circles, the first internally and the second externally. Prove that AP is four-fifths of AC. [If M is the middle point of AB, prove that $MP^2 = AP^2 + PB^2 - AP \cdot PB$, and solve algebraically.]

2. Prove that the side of a regular polygon of twelve sides inscribed in a circle is a mean proportional between the radius of the circle and the difference between the diameter of the circle and a side of an inscribed equilateral triangle.

3. AB, the diameter of a circle, is trisected at C, D. PCQ, PDR are two chords of the circle and QR, AB, produced, meet in N. Prove that $PC^2 : PD^2 = NR : NQ$. [Prove $PC \cdot CQ = PD \cdot DR$, and draw RT parallel to QC to meet AB in T.]

4. $ABCD$, $AEFG$ are two similar and similarly situated parallelograms, A, E, B being in one straight line. Prove that BG, ED meet on the diagonal AC. [Let BG, ED cut the diagonal AC in O, O'; prove $AO : OC = AO' : O'C$.]

5. From two given points A, B, draw two equal straight lines AP, BQ to a given straight line PQ, so that AP, BQ, may be inclined to one another at a given angle.

[Suppose the construction effected: complete the parallelogram $APQT$: then $\Delta\, BQT$ is given in species.]

6. OA, OB are given straight lines and X is a given point. Draw a straight line perpendicular to OA, cutting OA, OB at P, Q, such that $XP = PQ$.

Let $\angle BOA = 40°$, $\angle BOX = 65°$ (both angles measured in the same sense) and $OX = 2.5$ in. Through any point R on OA, draw RS perpendicular to OA, cutting OB in S. Let the circle with centre R and radius RS cut OX in Y; draw XP parallel to YR to meet OA in P.

Prove that P is the point required, showing that there are two solutions. Measure the smaller value of PQ.

Paper XXXIII (to Section XLIII)

1. P is any point on the circumference of a circumcircle of a triangle ABC whose centre is S. PD and PE are the feet of the perpendiculars from P to BC and CA respectively. Show that the locus of the centre of the circle circumscribing the triangle PDE is the circle on SC as diameter.
2. AB is a chord at right angles to a diameter CD of a given circle, and AB is nearer to C than to D. It is required to draw through C a chord CQ cutting AB in P, so that PQ may be of given length.

   Prove the correctness of the following construction: Along AD mark off AX equal $\frac{1}{2}PQ$: with centre X and radius XA, draw an arc cutting CX in Y : with centre C and radius CY, describe a circle cutting AB in P.

   [Prove that $CX^2 = CP \cdot CQ + AX^2$.]
3. A, B, C, D are four points along a straight line. On AB and CD as bases, equiangular triangles XAB, YCD are constructed. Prove that XY meets AD produced in a fixed point.
4. $ABCD$ is a parallelogram and EF is a straight line parallel to AB, cutting AD at E and BC at F. Find a point P in EF such that $\angle ABP = \angle ADP$.

   [Use Paper XXVI, 2 to obtain $DP : PB$.]
5. OA, OB, OC are three concurrent straight lines and X is a point in OA. Through X draw a straight line to cut OB, OC at Y, Z, such that $XY : YZ = p : q$, where p, q are given numbers.
6. Draw a parallelogram with one angle 60°, and the adjacent sides equal to 2.5 in, and 3 in. In it inscribe a square. Measure a diagonal of the square.

Paper XXXIV (to Section XLIV)

1. C, D are given points on the same side of a given straight line AB. Find a point P in AB such that the angle APC may be equal to twice the angle BPD.

   [Draw DN perpendicular to AB, and produce it to E, so that $NE = DN$. With centre E and radius EN, draw a circle. Draw a tangent to this circle from C, etc.]
2. Through a given point A, within a given circle, draw a chord PAQ such that $PA : AQ$, may be equal to a given ratio.
3. $ABCD$ is a quadrilateral inscribed in a circle : AB and DC are produced to meet at E : BC and AD are produced to meet at F. Prove that the circumcircles of the triangles ACE, DCF meet on EF.

   Hence show that if t, t' are the tangents to the circle $ABCD$ from E and F, $EF^2 = t^2 + t'^2$.

Note: *EF* is called the third diagonal of the quadrilateral *ABCD*.

4. EF is a diameter of a circle and t, t' are the tangents from E and F to another circle. If $EF^2 = t^2 + t'^2$, prove that the circles cut one another at right angles. Hence show that *the circle on the third diagonal of a cyclic quadrilateral as diameter cuts the circle circumscribing the quadrilateral at right angles.*

5. Three circles have two common points O and O': a variable straight line through O cuts them in P, Q, R. Prove that the circumscribing circle of the triangle formed by the tangents at P, Q, R passes through O'.

   [Prove that, if the tangents at P, Q, meet at S, and the tangents at Q, R meet at T, the quadrilaterals $O'QSP$, $O'QRT$ are cyclic.]

6. $ABCD$ is a quadrilateral; BH, DK are drawn perpendicular to AC. Prove that the rectangle AC, HK is equal to the difference between $AB^2 + CD^2$ and $BC^2 + AD^2$.

PAPER XXXV (to Section XLV)

1. *A is a given point on a circle and P is any point on the circumference. In AP, or in AP produced, a point P' is taken such that* $AP \cdot AP' = k^2$, *where k is constant. Prove that the locus of P' is a straight line.**

   [Draw the diameter AB. Draw $P'N$ perpendicular to AB, meeting it in N. Show that N is a fixed point.]

2. *A is a given point, within or without a given circle, and P is any point on the circumference. In AP, or in AP produced, a point P' is taken such that* $AP \cdot AP' = K^2$, *where k is constant. Prove that the locus of P' is a circle.**

   [Let C be the centre of the circle and let AP cut the circle again at Q. Draw $P'C'$ parallel to QC to meet AC, or AC produced, at C. Prove that C' is a fixed point and that $C'P'$ is of constant length.]

3. P is a given point on the diameter AB of a circle of which O is the centre; through P any chord RPS of the circle is drawn; the tangents at R, S meet the tangent at B in U, V. Prove that the rectangle BU, BV is constant.

   [Prove that the triangles ARS, OVU are similar: draw a chord ANT, cutting RS at right angles in N, and show that

   $$BU \cdot BV : OB^2 = RN \cdot NS : AN^2 = BP : PA.]$$

4. If, in the figure of Construction 37, the circle BCD cuts the circle with centre C again at E and cuts AE again at H, prove that

   (*i*) $AH \cdot AE$ = sq. on a side of a regular decagon inscribed in the circle with centre C.

   (*ii*) $EH \cdot EA$ = sq. on a side of a regular inscribed pentagon.

   (*iii*) Hence give an alternative proof of Theorem 100 (*ii*). [For (*ii*) show that EC touches the circle AHC, and, for (*iii*), use Ex. LXXIII 9.]

* **NOTE:** The locus of P' is called the inverse of the given circle; A is called the **centre** and k the **constant of inversion**.

5. *ABCD* is a quadrilateral; *BH*, *DK* are the perpendiculars from *B* and *D* to the diagonal $AC \cdot DK$ is produced to meet a straight line through *B* parallel to *AC* in *L*. If the lengths of the sides of the quadrilateral and its area are given, prove that the shape of the triangle *BDL* can be determined.

[Use Ex. 6, Paper XXXIV.]

6. Construct the quadrilateral *ABCD*, given *AB* = 1 in., *BC* = 2.5 in., *CD* = 2 in., *DA* = 1.5 in., and that its area is 2.65 sq. in. Measure the diagonal *AC*.

[Use the preceding exercise : fix *B* and *C*, then locus of *D* is a circle, centre *C* : again, since *BLD* is of given species, whilst *B* is fixed, and *D* moves on a circle, therefore *L* moves on a circle. Now show that, if the parallelogram *BRCA* is completed, $BL \cdot BR = AC \cdot HK$; hence locus of *R* is a circle, by Ex. 2. But *R* lies also on a circle whose centre is *C*, since *CR* = *AB*. Hence there are two positions of *R*.]

UNIT 6

Miscellaneous Propositions

XLVI. TRIANGLES AND QUADRILATERALS

THEOREM 101 (Euclid VI. B)

If the vertical angle of a triangle is bisected by a straight line which cuts the base, the rectangle contained by the sides of the triangle is equal to the rectangle contained by the segments of the base, together with the square on the line which bisects the angle.

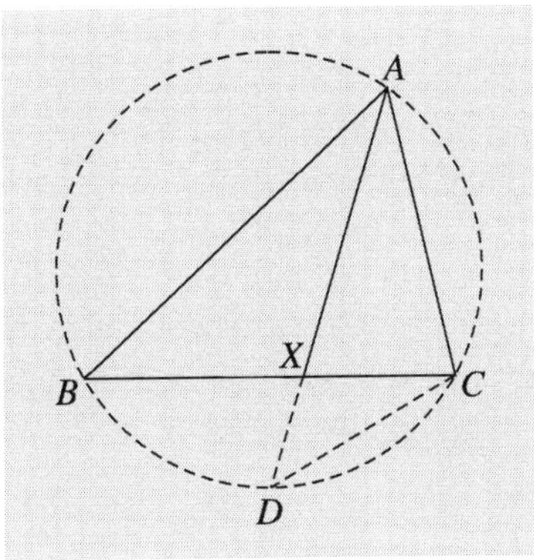

Let the bisector of the angle A of the triangle ABC meet BC at X. It is required to prove that

$$AB \cdot AC = BX \cdot XC + AX^2.$$

Construction. Draw the circle through A, B, C, and produce AX to meet the circle again at D. Join CD.

Proof. The angles ABX, ADC are in the segment $ABDC$;

$\therefore$ $\angle ABX = \angle ADC$

But, by hypothesis, $\angle XAB = \angle CAD$;

$\therefore$ the triangles ABX, ADC are equiangular;

$\therefore$ $AB : AD = AX : AC$;

$\therefore$ $AB \cdot AC = AD \cdot AX$.

Also, because AD is divided at X,

$\therefore$ $$AD \cdot AX = AX \cdot XD + AX^2.$$

Now, AD, BC are chords of the circle meeting at X;

$\therefore$ $$AX \cdot XD = BX \cdot XC;$$

$\therefore$ $$AD \cdot AX = BX \cdot XC + AX^2;$$

$\therefore$ $$AB \cdot AC = BX \cdot XC + AX^2.$$

Theorem 102 (Euclid VI. C)

If from the vertical angle of a triangle a perpendicular is drawn to the base, the rectangle contained by the sides of the triangle is equal to the rectangle contained by the perpendicular and the diameter of the circumcircle.

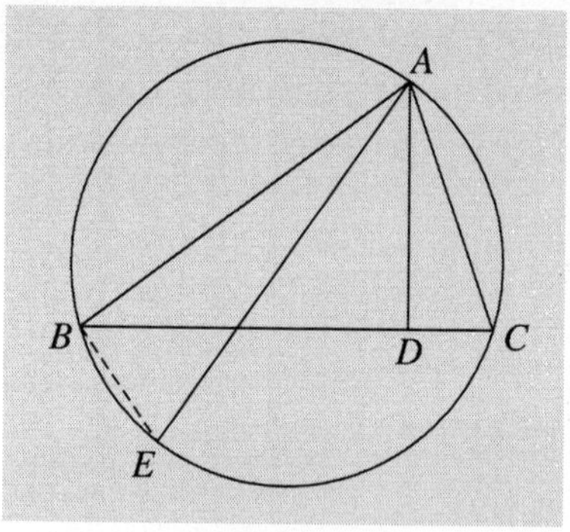

Let ABC be a triangle, let AD be the perpendicular from A to BC, and let AE be the diameter of the circumcircle through A. It is required to prove that

$$AB \cdot AC = AD \cdot AE$$

Construction. Join BE.

Proof. Because ABE is a semicircle,

$\therefore$ $\angle ABE$ is a right angle;

$\therefore$ $$\angle ABE = \angle ADC.$$

Again, the angles AEB, ACD are in the segment $ACEB$;

$\therefore$ $$\angle AEB = \angle ACD;$$

$\therefore$ the triangles ABE, ADC are equiangular;

$\therefore$ $$AB : AD = AE : AC;$$

$\therefore$ $$AB \cdot AC = AD \cdot AE.$$

Cor. If R is the radius of the circumcircle and Δ the area of the triangle ABC, then

$$R = \frac{abc}{4\Delta}.$$

For $$2R \cdot AD = AB \cdot AC \text{ and } BC \cdot AD = 2\Delta;$$

∴ (by division) $$\frac{2R}{BC} = \frac{AB \cdot AC}{2\Delta};$$

∴ $$R = \frac{BC \cdot AB \cdot AC}{4\Delta} = \frac{abc}{4\Delta}.$$

THEOREM 103 (Euclid VI. D)

The rectangle contained by the diagonals of a cyclic quadrilateral is equal to the sum of the rectangles contained by pairs of opposite sides.

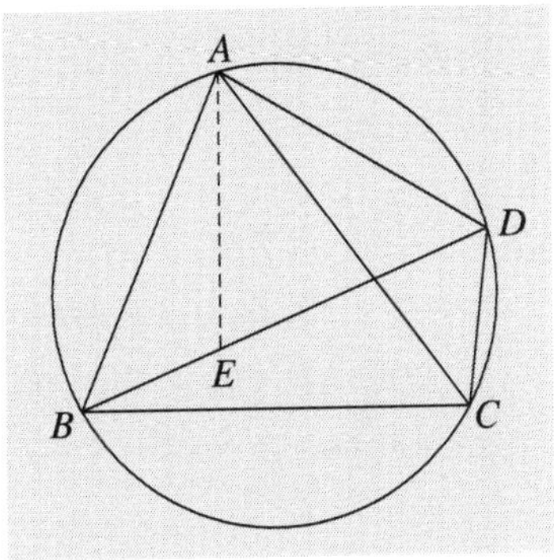

Let *ABCD* be a quadrilateral inscribed in a circle. It is required to prove that

$$AC \cdot BD = BC \cdot AD + AB \cdot CD.$$

Construction. Draw *AE*, cutting *BD* at *E*, so that

$$\angle BAE = \angle CAD.$$

Proof. Because $\angle BAE = \angle CAD$ (*construction*),

and $\angle ABE = \angle ACD$, being in the segment *ABCD*,

∴ the triangles *ABE*, *ACD* are equiangular;

∴ $$AB : AC = BE : CD;$$

∴ $$AB \cdot CD = AC \cdot BE.$$

Again, $$\angle BAE = \angle CAD;$$

∴ also, $$\angle BAC = \angle EAD,$$

and $\angle ACB = \angle ADE$, being in the segment *ADCB*;

∴ the triangles *ABC*, *AED* are equiangular;

∴ $$BC : ED = AC : AD;$$

∴ $$BC \cdot AD = AC \cdot ED.$$

Now $$AC \cdot BD = AC \cdot BE + AC \cdot ED;$$

∴ $$AC \cdot BD = AB \cdot CD + BC \cdot AD.$$

EXAMPLE: *If α, β are acute angles, prove that*

$$\sin(\alpha + \beta) = \sin\alpha\cos\beta + \cos\alpha\sin\beta.$$

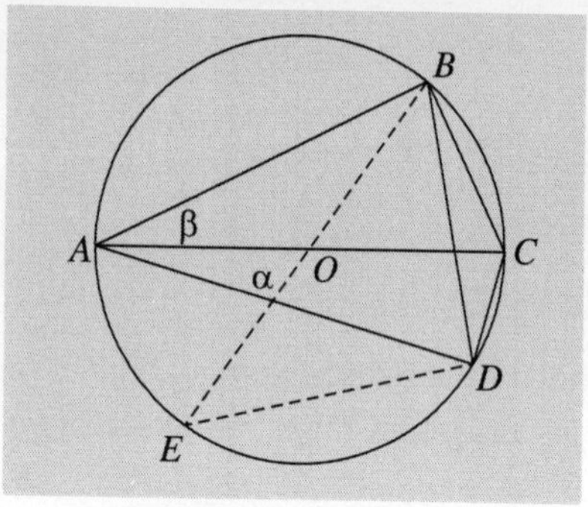

Draw a circle *ABCD* with diameter *AC* of unit length.

Make $\angle CAD = \alpha,\ \angle CAB = \beta.$

Draw the diameter *BE*.

Then $\sin(\alpha + \beta) = \sin BED = BD.$

Also $\sin\alpha = CD,\ \cos\alpha = AD,\ \sin\beta = BC,\ \cos\beta = AB,$

But $AC \cdot BD = AB \cdot CD + BC \cdot AD;$

∴ $\sin(\alpha + \beta) = \sin\alpha\cos\beta + \cos\alpha\sin\beta.$

EXERCISE LXXV

1. By means of Theorem 100, show that, if the bisector of the angle *A* of the triangle *ABC* meets *BC* at *X*, then

$$AX^2 = bc\left\{1 - \frac{a^2}{(b+c)^2}\right\}.$$

2. If the bisector of the exterior angle at *A* of the triangle *ABC* meets *BC* produced at *Y*, prove that $AB \cdot AC = YB \cdot YC - AY^2$.

   [Produce *YA* to cut the circle *ABC* at *D*. Prove Δs *ABY*, *ADC* equiangular.]

3. By means of Ex. 2, show that if $c > b$, then

$$AY^2 = bc\left\{\frac{a^2}{(c-b)^2} - 1\right\}.$$

4. If *AB*, *A'B'* are chords of two circles which subtend equal angles at the circumferences, and if *D*, *D'* are the diameters of the circles, prove that $AB : A'B' = D : D'$.

5. If *X* is any point on the circumcircle of the triangle *ABC* and *XM*, *XN* are perpendicular to *CA*, *AB* respectively, prove that $MN : BC = XA : D$, where *D* is the diameter of the circle *ABC*.

[Prove Δs *XNM*, *XBC* equiangular, and apply Theorem 102 to Δ*AXC*.]

6. If from any point on the circumference of a circle straight lines are drawn to the vertices of an inscribed equilateral triangle, one of these lines is equal to the sum of the other two.

7. If the radius of a circle increases indefinitely, the circumference tends to become a straight line. From Ptolemy's Theorem, we may therefore infer that if *A*, *B*, *C*, *D* are four points, in order, on a straight line, then $AC \cdot BD = BC \cdot AD + AB \cdot CD$.

Verify this, using theorems on segments of a straight line.

8. *P* is any point on an arc *AB* of a circle. Prove that the sum of the chords *PA*, *PB* is greatest when $PA = PB$.

[Let the perpendicular bisector of *AB* cut the remaining part of the circumference at *C*. Apply Ptolemy's Theorem to the quad. *APBC*.]

9. *P* is any point on an arc *AB* of a circle. Prove that the rectangle contained by the chords *PA*, *PB* is greatest when $PA = PB$.

[Apply Theorem 102 to the triangle *PAB*.]

10. Through two given points *P*, *Q* of a circle draw two parallel chords *PX*, *QY*, so that the rectangle contained by them may be equal to a given rectangle.

11. If *R* is the radius of the circumcircle of the triangle *ABC*, and *BD*, *CD* are perpendicular to *AB*, *AC*, prove that

$$AC \cdot BD + AB \cdot CD = 2R \cdot BC.$$

12. If *ABC*, *A'B'C'* are two triangles, in which the angles *B*, *B'* and also the angles *C*, *C'* are complementary, prove that

$$Ra' = R'a = \frac{1}{2}(bc' + b'c),$$

where *a*, *b*, *c*, *a'*, *b'*, *c'* are the sides and *R*, *R'* the radii of the circumcircles of *ABC*, *A'B'C'*. [Draw *BD*, *CD* perpendicular to *AB*, *AC*, and use Ex. II.]

13. *If ABCD is a quadrilateral which is not cyclic, prove that*

$$BC \cdot AD + AB \cdot CD > AC \cdot BD.$$

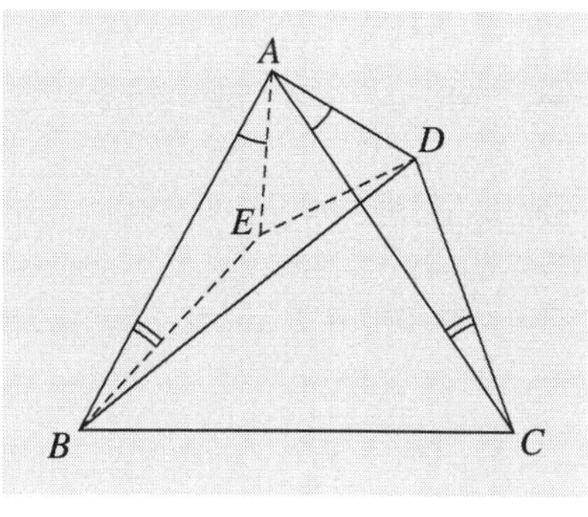

[Make $\angle BAE = \angle CAD$ and $\angle ABE = \angle ACD$. Join *ED*. Prove that $AB \cdot CD = AC \cdot BE$, also that Δs *ABC*, *AED* are similar, and

$$\therefore \quad BC \cdot AD = AC \cdot ED$$

Next show that *B*, *E*, *D* are not in the same straight line.]

14. Explain why the following statement is incomplete and complete it,—If A, B, C, D are four points, such that

$$AC \cdot BD = BC \cdot AD + AB \cdot CD,$$

then A, B, C, D lie on the circumference of a circle.

15. By applying Ptolemy's Theorem to the quadrilateral whose angular points are four consecutive vertices of a regular pentagon inscribed in a circle, find the value of sin 18° in a surd form.

16. If A, B, C, D are four consecutive vertices of a regular heptagon, prove that $AD^2 = AC(AB + AC)$.

17. From the vertex A of a triangle ABC, AD and AE are drawn to the base, making the angle BAD equal to the angle CAE; prove that

$$BD \cdot BE : CE \cdot CD = AB^2 : AC^2.$$

[Produce AD, AE to meet the circle ABC at X, Y. Prove that $BD \cdot BE : AB^2 = CX \cdot CY : AX \cdot AY$.]

XLVII. THE ORTHOCENTRE AND THE PEDAL TRIANGLE

In Theorem 37 it has been shown that

The perpendiculars from the vertices of a triangle to the opposite sides are concurrent. The following is a more direct proof of this theorem.

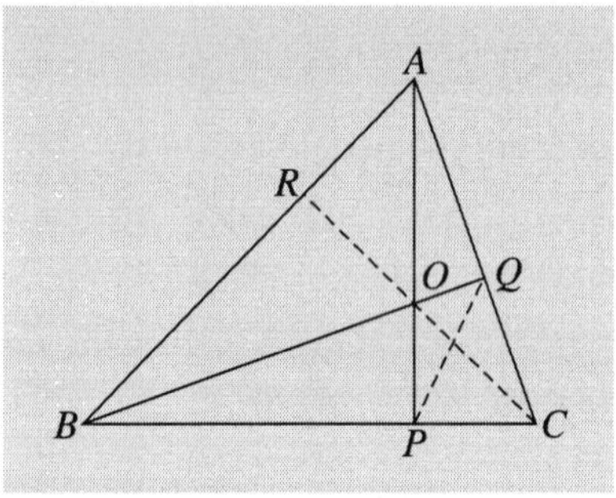

Let ABC be a triangle, and let AP, BQ be the perpendiculars from A, B to BC, CA respectively, meeting at O. Join CO, and produce it to meet AB at R.

We shall prove that CR is perpendicular to AB.

Construction. Join PQ.

Proof. Because the angles APB, AQB are right angles,

$\therefore$ $ABPQ$ is a cyclic quadrilateral;

$$\therefore \qquad \angle CQP = \angle ABC.$$

Also, because the angles OPC, OQC are right angles,

$\therefore$ $CPOQ$, is a cyclic quadrilateral;

$$\therefore \qquad \angle CQP = \angle COP;$$

$$\therefore \qquad \angle COP = \angle ABC;$$

∴ $BPOR$ is a cyclic quadrilateral;

∴ $\angle ORB + \angle OPB = 2$ right ∠s. But $\angle OPB$ is a right angle;

∴ $\angle ORB$ is a right angle;

∴ CR is perpendicular to AB.

Def. The point of concurrence of the perpendiculars from the vertices of a triangle to the opposite sides is called the **orthocentre** of the triangle, and the triangle whose vertices are the feet of these perpendiculars is called the **pedal** triangle.

Theorem 104

If O is the orthocentre of the triangle ABC and AO is produced to meet BC at P and the circumcircle of the triangle at A', then

$$OP = PA'.$$

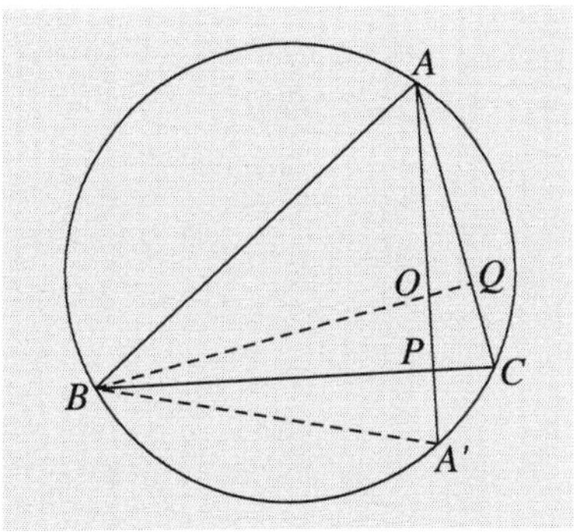

Construction. Join BA', BO. Produce BO to meet CA at Q.

Proof. Because AP, BQ are perpendicular to BC, CA,

∴ the quadrilateral $POQC$ is cyclic;

∴ $\angle BOP = \angle ACB$.

Again, the angles ACB, $AA'B$ are in the segment $ACA'B$;

∴ $\angle ACB = \angle AA'B$;

∴ $\angle BOP = \angle BA'P$.

In the triangles BOP, $BA'P$,

$\angle BPO = \angle BPA'$ (*right angles*),
$\angle BPO = \angle BA'P$ (*proved*),
and BP is common;

∴ the triangles are congruent;

∴ $OP = PA'$.

THEOREM 105

If O is the orthocentre and PQR the pedal triangle of the triangle ABC, then A, B, C, O are the centres of the four circles which can be drawn to touch the sides, or the sides produced, of the triangle PQR.

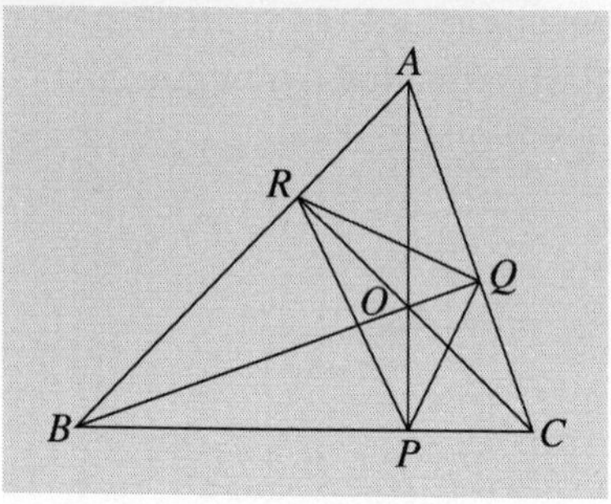

Proof. Because AP, BQ, CR are perpendicular to BC, CA, AB,

$\therefore$ the quadrilaterals $OPCQ$, $BRQC$, $OPBR$ are cyclic:

$$\therefore \quad \angle OPQ = \angle OCQ = \angle OBR = \angle OPR;$$

$\therefore$ AP, BC are the bisectors of the interior and exterior angle at P of the triangle PQR.

Similarly, BQ, CA and CR, AB are the bisectors of the angles at Q and R of the triangle PQR;

$\therefore$ A, B, C, O are the centres of the circles touching the sides.

THEOREM 106

If S is the centre and R the radius of the circumcircle of the triangle ABC, and I the centre and r the radius of the circle inscribed in the triangle, then

$$SI^2 = R^2 - 2Rr.$$

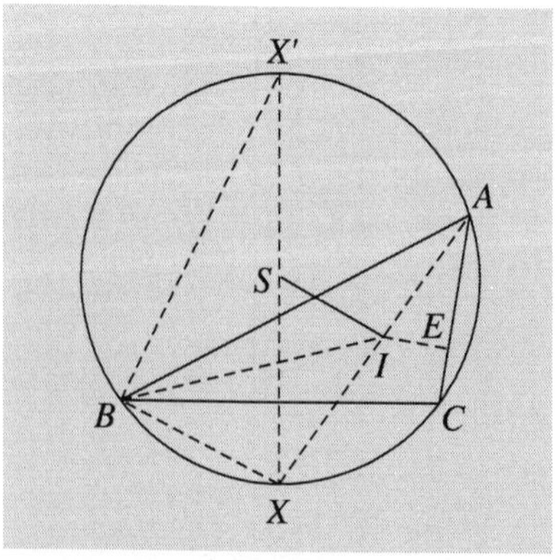

Construction. Join AI, BI.

Draw IE perpendicular to CA.

Produce AI to meet the circumcircle at X. Draw the diameter XSX'. Join BX, BX'.

Proof. From the triangle AIB,

the ext. $\angle BIX = \angle IAB + \angle IBA$;

but $\angle IAB = \angle XAC = \angle XBC$,

and $\angle IBA = \angle IBC$;

$\therefore$ $\angle BIX = \angle XBC + \angle IBC = \angle IBX$;

$\therefore$ $XI = XB$.

Now, because XIA is a chord of the circumcircle,

$\therefore$ $R^2 - SI^2 = XI \cdot IA = XB \cdot IA$.

Also, the angles $XX'B$, XAB are in the segment $XAX'B$;

$\therefore$ $\angle XX'B = \angle XAB = \angle IAE$,

and the angles XBX', IEA are right angles;

$\therefore$ the triangles XBX', IEA are equiangular;

$\therefore$ $XB : IE = XX' : IA$;

$\therefore$ $XB \cdot IA = XX' \cdot IE = 2Rr$;

$\therefore$ $R^2 - SI^2 = 2Rr$,

or $SI^2 = R^2 - 2Rr$.

EXERCISE LXXVI

Orthocentre and Pedal Triangle

1. If O is the orthocentre of the triangle ABC, show that the four points A, B, C, O are such that anyone of them is the orthocentre of the triangle of which the remaining three points are the vertices.
2. If O is the orthocentre of the triangle ABC, show that the circumcircles of the triangles OBC, OCA, OAB, ABC are equal.
3. If O is the orthocentre and S the circumcentre of the triangle ABC, and if A', B', C' are the circumcentres of the triangles OBC, OCA, OAB, show that O is the circumcentre and S the orthocentre of the triangle A', B', C'. [Use Ex. 2.]
4. If AP, BQ, CR are perpendicular to BC, CA, AB, the sides of the triangle ABC, and O is the orthocentre, show that
   (*i*) $OA \cdot OP = OB \cdot OQ = OC \cdot OR$.
   (*ii*) $AP \cdot OP = BP \cdot PC$.
5. Prove that the angles of the pedal triangle of the triangle ABC are the supplements of $2A$, $2B$, $2C$ respectively.
6. Prove that the angles of the triangle formed by joining the centres of the escribed circles of the triangle ABC are the complements of $\frac{1}{2}A$, $\frac{1}{2}B$, $\frac{1}{2}C$ respectively.

7. The triangle whose vertices are the centres of the escribed circles of a triangle is similar to that formed by joining the points of contact of the inscribed circle.

8. Construct a triangle, having given the circumcircle, one vertex and the orthocentre.

9. Given the base and the magnitude of the vertical angle of a triangle, prove that the locus of the orthocentre is a circle.

10. *If S is the circumcentre and O the orthocentre of a triangle ABC, in which AB > AC, prove that (i) ∠ SAO = C – B; (ii) the angle SAO is bisected by the bisector of the angle A.*

11. From the last example, deduce a construction for a triangle, having given the base, the vertical angle and the difference of the base angles.

12. Construct a triangle, having given the circumcircle, the orthocentre and the difference of two angles. [Use Ex. 9.]

13. *O* is the orthocentre and *S* the circumcentre of the triangle *ABC*. If *AO*, *AS* are produced to meet the circumcircle at *X*, *U* respectively, prove that

(*i*) *UX* is parallel to *BC*.

(*ii*) *BUCO* is a parallelogram.

(*iii*) If *L*, *H* are the middle points of *BC*, *AO*, then *LH* is parallel to *SA* and equal to *SA*. Also *AO* = 2*SL*.

(*iv*) *AL* divides *SO* internally in the ratio 1 : 2.

(*v*) If *G* is the centroid of the triangle *ABC*, *G* divides *SO* internally in the ratio 1 : 2.

14. Construct the triangle *ABC*, having given the radius of the circumcircle, the vertex *A* and the lengths of the lines joining *A* to the middle point of *BC* and to the orthocentre.

15. A circle of constant magnitude passes through a fixed point *A* and intersects two fixed straight lines *AB*, *AC* in *B* and *C*. Prove that the locus of the orthocentre of the triangle *ABC* is a circle.

16. *ABCD* is a quadrilateral inscribed in a circle, *E* and *F* the orthocentres of the triangles *ABC*, *ABD* respectively; prove that *CDFE* is a parallelogram. [Use Ex. 13 (*iii*).]

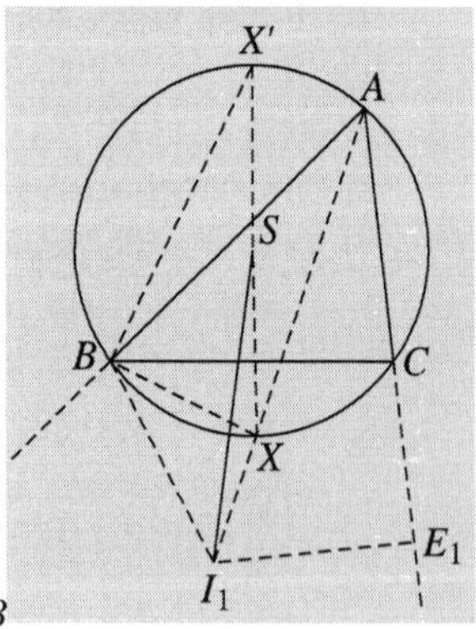

17. *If S is the centre of the circumcircle of the triangle ABC and I_1 the centre and r_1 the radius of the escribed circle opposite to the angle A, prove that*

$$SI_1^2 = R^2 + 2Rr_1.$$

[Let AI_1 cut the circumcircle at *X*. Draw the diameter *XSX'*. Draw I_1E_1 perpendicular to *AC*. Prove that

$$I_1A \cdot I_1X = SI_1^2 - R^2,\ XI_1 = XB$$

and the triangles *XBX'*, I_1E_1A are similar. Cf. Theorem 106.]

18. *If two circles are such that a triangle can be inscribed in one of the circles and circumscribed to the other, prove that an unlimited number of triangles can be so described.*

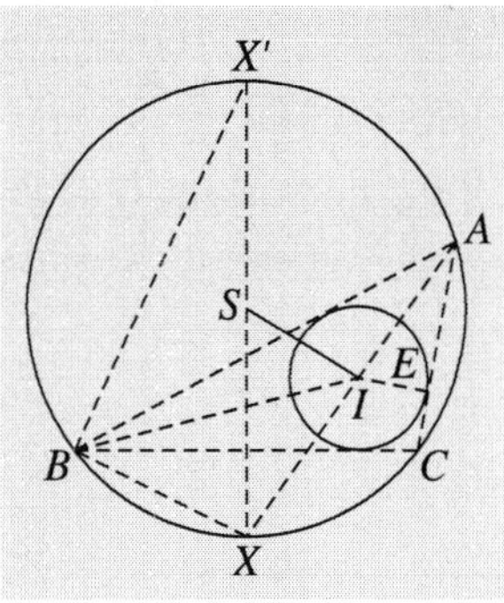

[Let I, S be the centres and r, R the radii of the inner and outer circles. From any point A on the outer circle draw tangents AF, AE to the inner, meeting the outer again at B, C. Join BC. Prove that BC touches the inner circle.

Let AI produced meet the outer circle in X.

Draw the diameter XSX'.

By Theorem 106, $SI^2 = R^2 - 2Rr$;

$$\therefore \quad AI \cdot IX = XX' \cdot IE.$$

Next prove Δs XBX', IEA similar and $AI \cdot BX = XX' \cdot IE$;

$$\therefore \quad BX = IX. \text{ Now prove } \angle IBC = \angle IBA.]$$

19. ABC is a triangle, AP the perpendicular to BC, L the middle point of BC. I, I_1 the centres of the inscribed circle and of the escribed circle opposite A, D, D_1 the points of contact of these circles with BC, X the point of intersection of AII_1 and BC. Prove that

(*i*) I_1I is divided internally and externally in the same ratio at X and A.

(*ii*) D_1D is divided internally and externally in the same ratio at X and P.

(*iii*) Hence show that $LX \cdot LP$ is equal to the square on the tangent from L either to the inscribed circle or to the escribed circle opposite A. [Cf. Ex. LXIII(*b*)]

(*iv*) $LP \,.\, DX = LD \,.\, DP$.

20. *Show that each of the six lines joining the centres of the four circles touching the sides, or sides produced of a triangle, are bisected by the circumcircle.*

[If AII_1 cut the circumcircle in X, $XI = XB = XI_1$.]

21. *The circle circumscribing the pedal triangle of a triangle ABC passes through the middle points of BC, CA, AB.*

[Use Theorem 105 and Ex. 20.]

22. *If S is the circumcentre of a triangle ABC, and O is the orthocentre, R the radius of the circumcircle, and r' the radius of the circle inscribed in the pedal triangle, prove that*

$$SO^2 = R^2 - 4Rr'.$$

[If J is the centre, and ρ the radius, of the circle circumscribing the pedal triangle, then $JO^2 = \rho^2 - 2\rho r'$.

But, by Ex. 21, J is the centre of the circle through the middle points of the sides of the triangle ABC; therefore, by Ex, LXX, 10,

$$JO = \frac{1}{2} SO \text{ and } \rho = \frac{1}{2} R.]$$

XLVIII. THE SIMSON LINE

THEOREM 107

The feet of the perpendiculars on the sides of a triangle, from any point on the circumcircle, are collinear.

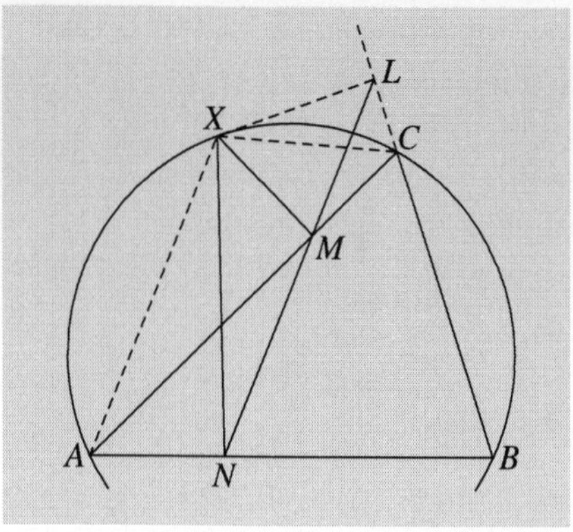

Let X be any point on the circumcircle of the triangle ABC, and let XM, XN be perpendicular to CA, AB respectively. Join MN, cutting BC or BC produced at L. Join XL.

It is required to prove that XL is perpendicular to BC.

Construction. Join XA, XC.

Proof. Because the angles XMA, XNA are right angles,

$\therefore$ $XMNA$ is a cyclic quadrilateral;

$\therefore$ $$\angle XML = \angle XAB.$$

Also, because $AXCB$ is a cyclic quadrilateral,

$\therefore$ $$\angle XCL = \angle XAB;$$

$\therefore$ $$\angle XCL = \angle XML;$$

$\therefore$ $XLCM$ is a cyclic quadrilateral;

$\therefore$ $$\angle XLC + \angle XMC = 2 \text{ right angles}.$$

But $\angle XMC$ is a right angle;

$\therefore$ XL is perpendicular to BC

The line LMN is called the **Simson** *or* **Pedal line** *of the point X.*

EXERCISE LXXVII

1. If the feet of the perpendiculars from a point to the sides of a triangle are collinear, the point is on the circumcircle.
2. In the figure of Theorem 107, the angle which the Simson line of X makes with AC is the complement of the angle which the arc BX subtends at any point on the circumference.

3. The angle between the Simson lines of any two points X, Y, on the circumcircle of a triangle, is equal to the angle subtended by the arc XY at any point on the circumference of the circle. [Use Ex. 2.]

4. ABC is a triangle and D is any point. The straight lines DB, DC cut the circumcircle of ABC again at X, Y. Prove that the angle between the Simson lines of X and Y is equal to the difference between the angles BDC and BAC.

5. Find a point X on the circumcircle of the triangle ABC, such that its Simson line may be parallel to a given straight line.

6. In the figure of Theorem 107, let X, A, C be fixed points, and let B move along the circumference and tend to coincidence with C. Show that we may infer the following theorem, and give an independent proof,—If A, B, X are any points on a circle and XL, XM are perpendicular to the tangent at B and AB respectively, then LM touches the circle on AX as diameter.

7. Find a point such that the feet of the four perpendiculars from it to the sides of a given quadrilateral may be collinear.

8. *In the figure of Theorem* 107, *if R is the radius of the circle ABC, prove that*

$$2R \cdot MN = BC \cdot AX;\ 2R \cdot NL = CA \cdot BX;\ 2R \cdot LM = AB \cdot CX.$$

Hence deduce Ptolemy's Theorem for the cyclic quadrilateral $AXCB$.

9. *If $ABCD$ is a quadrilateral which is not cyclic, by a method similar to that indicated in Ex. 8, prove that*

$$BC \cdot AD + AB \cdot CD > AC \cdot BD.$$

10. In the figure of Theorem 107, if $LM = MN$, prove that

$$XA : XC = AB : BC.$$

Hence, show how to find a point X, on the circumcircle of the triangle ABC, of which the Simson line is divided into two equal segments by the sides of the triangle.

11. In the figure of Theorem 107, if $LM : MN = p : q$, where p, q are given numbers, explain how to find the position of X.

12. In the figure of Theorem 107, if XL', XM', XN' are drawn, meeting BC, CA, AB at L', M', N', and making the angles LXL', MXM', NXN' equal and measured in the same sense, prove that L', M', N' are collinear and that $L'M' : M'N' = LM : MN$.

13. If XX' is any diameter of the circumcircle of the triangle ABC, prove that the Simson lines of X and X' meet at right angles at a point on the nine-point circle.

[For the first part, use Ex. 3. Let LMN, $L'M'N'$ be the Simson lines of X, X', meeting at Q. Let D, E, F be the middle points of BC, CA, AB. Prove that E, F bisect MM', NN'. Consider the right-angled triangles MQM', NQN', and prove that $\angle$s EQF, EDF are equal or supplementary.]

14. O is the orthocentre of the triangle ABC and X is any point on the circumcircle. XM, XN are perpendicular to CA, AB respectively, and CO (produced) cuts the circumcircle again at C'. Prove that MN and XC' are equally inclined to CC'. [Use the cyclic quads. $AXNM$, $AXCC'$.]

15. If, in the last example, XC' and AB (produced if necessary) meet at Q, prove that MN is parallel to OQ and bisects XQ. [Use the right-angled triangle XMQ.]

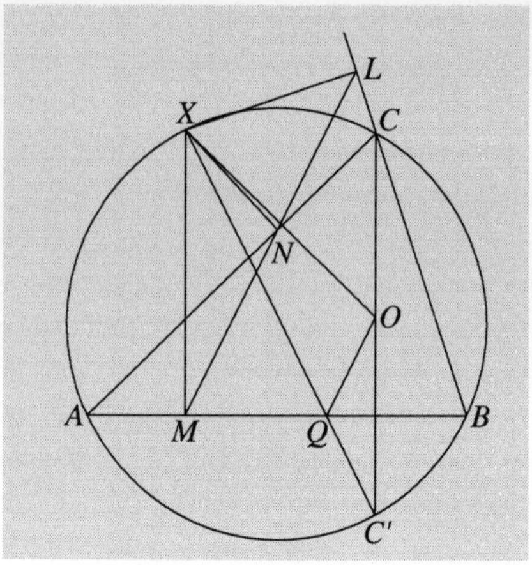

16. By means of the last two examples, show that:—*The Simson line of any point X on the circumcircle of a triangle bisects the straight line joining X to the orthocentre of the triangle.*

17. P is a point on the circumcircle of the triangle ABC and the perpendiculars from P to BC, CA, AB meet the circle again in X, Y, Z. Show that the perpendiculars from A, B, C to the sides of XYZ meet in a point Q on the circle, such that the pedal lines of P and Q with respect to ABC and XYZ respectively are coincident.

18. Let X be any point on the circumcircle of a triangle ABC, and let the perpendiculars from A, B, C to the opposite sides cut the circle again in A', B', C' ; then, if XA', XB', XC' cut BC, CA, AB in U, V, W. Prove that U, V, W lie on a straight line through the orthocentre parallel to the pedal line of X.

XLIX. THE NINE-POINT CIRCLE

Theorem 108

The circle through the middle points of the sides of a triangle passes through

(*i*) the feet of the perpendiculars from the vertices of the triangle on the opposite sides;

(*ii*) the middle points of the lines joining the orthocentre to the vertices.

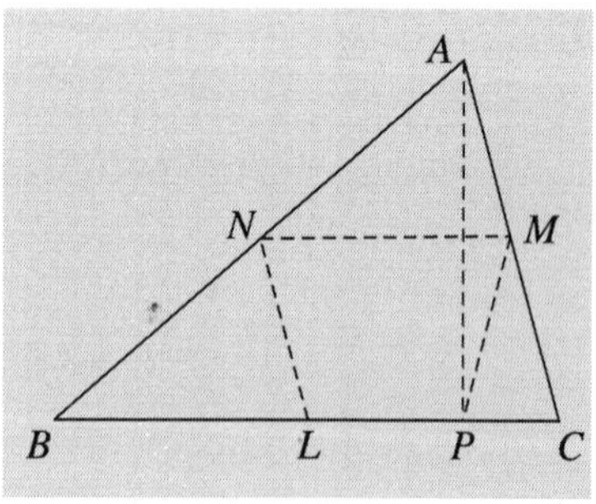

Let P, Q, R be the feet of the perpendiculars from the vertices A, B, C of a triangle to the opposite sides and L, M, N the middle points of BC, CA, AB.

It is required to prove that the circle *LMN* passes through *P*, *Q*, *R*.

Construction. Join *LN*, *MN*, *MP*.

Proof. Because *M*, *N* are the middle points of *CA*, *AB*,

$\therefore$ *MN* is parallel to *BC*.

Similarly, *LN* is parallel to *CA*;

$\therefore$ *CMNL* is a parallelogram;

$\therefore \qquad \angle LNM = \angle C.$

Also, because $\angle APC$ is a right angle and *CA* is bisected at *M*,

$\therefore \qquad MP = MC;$

$\therefore \qquad \angle MPC = \angle C;$

$\therefore \qquad \angle MPC = \angle LNM;$

$\therefore$ the circle *LMN* passes through *P*.

Similarly, it can be shown that this circle passes through *Q*, *R*.

(*ii*) Let *O* be the orthocentre of the triangle *ABC* and *U*, *V*, *W* the middle points of *OA*, *OB*, *OC*.

It is required to prove that the circle *LMN* passes through *U*, *V*, *W*.

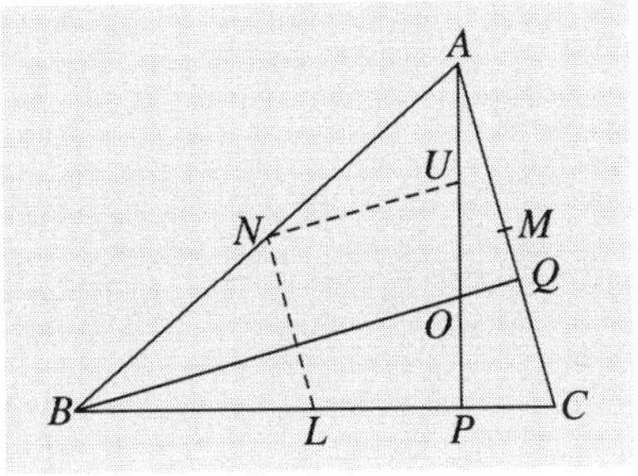

Construction. Join *LN*, *NU*.

Proof. Because *N*, *U* are the middle points *AB*, *AO*,

$\therefore$ *NU* is parallel to *BQ*.

Similarly, it can be shown that *LN* is parallel to *AC*. But *BQ* is perpendicular to *AC*,

$\therefore$ *NU* is perpendicular to *LN*.

Hence, in the quadrilateral *LNUP*, $\angle$s *P*, *N* are right angles;

$\therefore$ *U* lies on the circle through *L*, *N*, *P*, *i.e.* the circle *LMN* passes through *U*.

Similarly, it can be shown that the circle *LMN* passes through *V*, *W*.

Cor. *LU*, *MV*, *NW* are **diameters of the circle *LMN***.

On account of the properties proved in Theorem 108, the circle *LMN* is called the **Nine-point Circle** of the triangle *ABC* (abbrev. N.P.⊙).

It is also occasionally referred to as the **Medioscribed Circle**.

THEOREM 109

(*i*) The centre of the nine-point circle of any triangle is the middle point of the line joining the circumcentre and orthocentre of the triangle;

(*ii*) The radius of the nine-point circle of a triangle is equal to half that of the circumcircle.

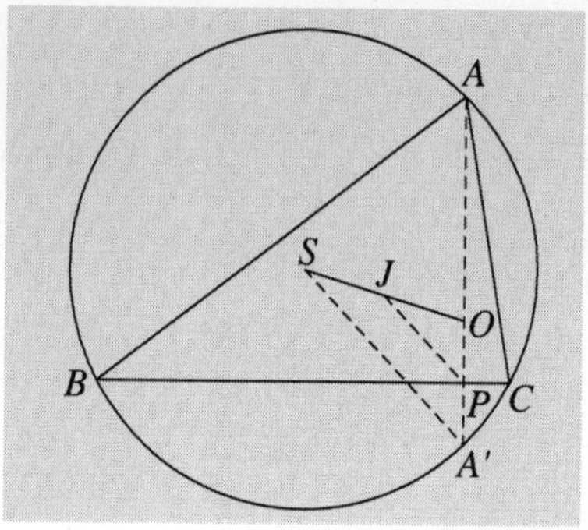

Let O, S be the orthocentre and circumcentre of the triangle ABC; let AP, BQ, CR be the perpendiculars from A, B, C to the sides BC, CA, AB. Let J be the middle point of SO.

It is required to prove that J is the centre of the circle PQR and that $JP = \frac{1}{2}R$.

Construction. Produce AP to meet the circumcircle in A' join JP, SA'.

Proof. Because J, P are the middle points of OS, OA',

$$\therefore \qquad JP = \frac{1}{2}SA' = \frac{1}{2}R.$$

Similarly, it can be shown that JQ, JR are each equal to $\frac{1}{2}R$.

$\therefore$ J is the centre of the circle PQR, which is the nine-point circle; and its radius is equal to half that of the circumcircle.

THEOREM 110

The centroid of a triangle is a point of trisection of the line joining the circumcentre and the orthocentre.

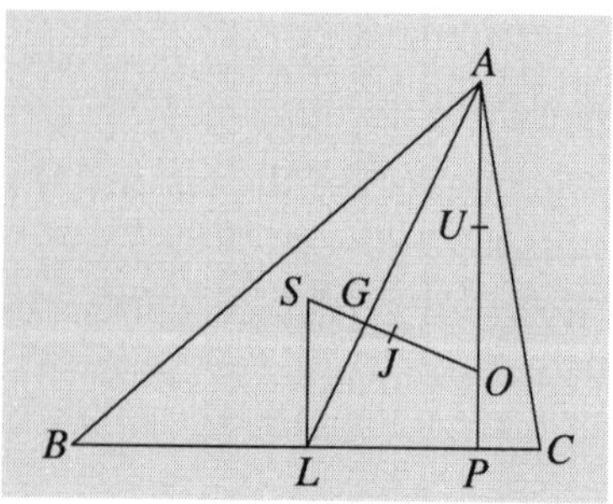

Let S, O be the circumcentre and orthocentre of the triangle ABC ; let AL cut SO in G.

It will be proved that G is the centroid of ΔABC, and that $OG = 2GS$.

Proof. Let U, J be the middle points of OA, SO, respectively. Then J is the centre and LU is a diameter of the nine-point circle,

$\therefore$ LU bisects SO.

Now, SL is parallel to AO,

$\therefore$ $$OU = SL \text{ and } OA = 2SL$$

Again, the triangles OAG, SLG are equiangular,

$\therefore$ $$AG : LG = OG : SG = OA : SL$$

But $$OA = 2SL,$$

$\therefore$ $$AG = 2LG \text{ and } OG = 2SG,$$

$\therefore$ G is the centroid of ΔABC, and is a point of trisection of SO.

NOTE: An alternative proof of Theorems 108, 109 is given in Ex. LXXVIII. 12.

EXERCISE LXXVIII [Theorems 108–110]

1. If O is the orthocentre of the triangle ABC, show that the four triangles OBC, OCA, OAB, OBC have the same nine-point circle.
2. If I, I_1, I_2, I_3 are the centres of the inscribed and escribed circles of the triangle ABC, prove that the circumcircle of ABC is the nine-point circle of each of the triangles II_2I_3, II_3I_1, II_1I_2, $I_1I_2I_3$.
3. If O is the orthocentre of a triangle ABC, in which $AB > AC$ and L, U, M are the middle points of BC, AO, AC, prove that
$$\angle LUP = \angle LMP = C - B.$$
4. The nine-point circle of the triangle ABC cuts the side BC at an angle equal to the difference between the angles B and C.

   [This follows from Ex. 3, for LU is a diameter.]
5. Construct a triangle, given the nine-point circle, the orthocentre and the difference between two angles. [Use Ex. 3.]
6. Given the base of a triangle and the magnitude of the vertical angle, prove that the nine-point circle touches a fixed circle.
7. Lines are drawn through the middle points of the sides of a triangle ABC perpendicular to the bisectors of the opposite angles. Prove that the triangle $A'B'C'$ formed by these lines has the same nine-point circle as the original triangle.

   [If L, M, N are the middle points of BC, CA, AB, prove that LMN is the pedal triangle of $A'B'C'$.]

8. B and C are points on two fixed straight lines AX, AY, such that $AB + AC$ is constant. The circle ABC meets the bisector of the angle XAY at a second point Q. Prove that Q is a fixed point.

[Along BX set off BC' equal to AC. Prove $\angle BC'Q = \frac{1}{2}A$.]

9. B and C are points on two fixed straight lines AX, AY such that $AB + AC$ is constant. Prove that the locus of each of the following points is a straight line perpendicular to the bisector of the angle XAY, —(1) the middle point of BC, (2) the centroid of ABC, (3) the circumcentre of ABC, (4) the centre of the nine-point circle of ABC. [For (3) use Ex. 8.]

10. The nine-point circle of an isosceles triangle, which has each angle at the base double that at the vertex, intercepts parts of the equal sides, such that a regular pentagon can be inscribed in the circle and have these parts as two of its sides.

11. If perpendiculars are drawn from the orthocentre of a triangle ABC to the bisectors of the angle A, their feet are collinear with the middle point of BC and the nine-point centre.

12. ABC is a triangle; AP, BQ, CR are the perpendiculars from A, B, C to the opposite sides of the triangle; AP, BQ, CR when produced cut the circumcircle in A', B', C'; O is the orthocentre, S the circumcentre; U, V, W are the middle points of OA, OB, OC ; L, M, N the middle points of BC, CA, AB; and J the middle point of SO.

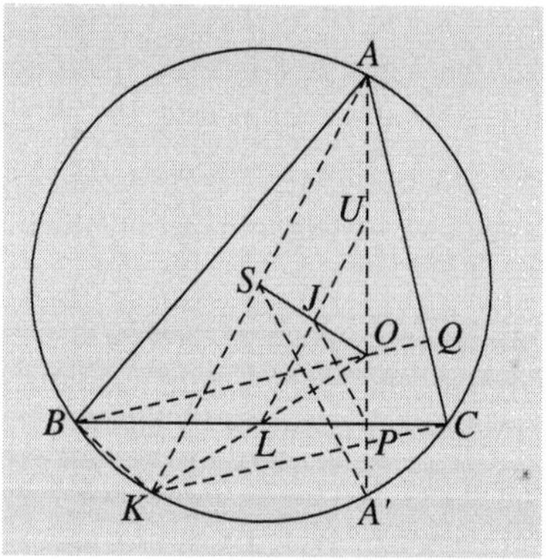

(*i*) Show that $JU = JV = JW = \frac{1}{2}R$. [Join SA and prove $JU = \frac{1}{2}SA$.]

(*ii*) Show that $JP = JQ = JR = \frac{1}{2}R$. [Join SA', and prove $JP = \frac{1}{2}SA'$.]

(*iii*) Show that $JL = JM = JN = \frac{1}{2}R$.

[Draw diameter ASK and prove that $BKCO$ is a parallelogram: hence $KL = LO$ and $JL = \frac{1}{2}SK$.]

NOTE: The student should draw a separate figure for each part of the proof.

13. *The nine-point circle of a triangle touches the inscribed circle.*

(Feuertach's Theorem)

Let ABC be the triangle ($AB > AC$), LM the middle points of BC, AC, AX the bisector of $\angle A$, meeting BC at X, D the point of contact of BC with the inscribed circle, AP the perpendicular from A to BC, O the orthocentre, I, J the centres of the inscribed and nine-point circles. Draw KJK' a diameter of the nine-point circle perpendicular to BC, K being the extremity outside the triangle; join KD and produce it to meet the circle again at T. Let r, ρ be the radii of the inscribed and nine-point circles.

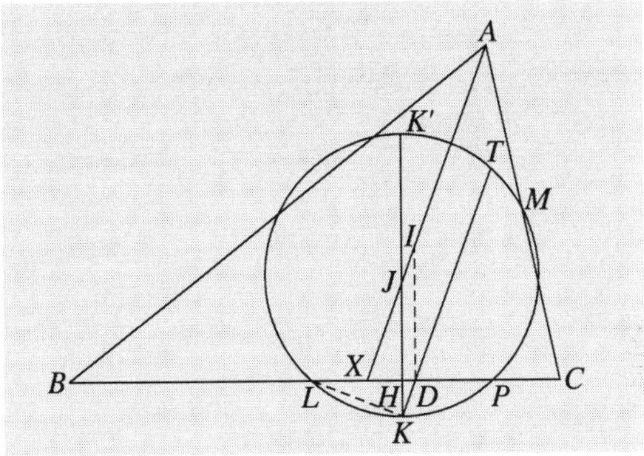

It will be proved that the inscribed circle touches the nine-point circle at T.

Construction. Join LK, ID, and let JK cut BC in H.

Proof. Since $\quad \text{arc } LK = \text{arc } KP,$

$$\therefore \quad \angle KLP = \frac{1}{2}\angle LMP = \frac{1}{2}(C - B). \text{ (Ex. LXXVIII, 3)}$$

But $\quad \angle XID = \angle XAP$

$$= \frac{1}{2}(C - B);$$

$$\therefore \quad \angle KLP = \angle XID.$$

$\therefore$ Δs XID, KLH are equiangular.

$$\therefore \quad DX : DI = HK : HL$$

It follows that $\quad 2r \cdot HK = 2HL \cdot DX = LP \cdot DX.$

But $\quad LP \cdot DX = LD \cdot DP.$ [Ex. LXXVI, 19 (*iv*)]

$$\therefore \quad 2r \cdot HK = LD \cdot DP = KD \cdot DT.$$

Also, since the points D, H, T, K' are concyclic,

$$\therefore \quad KD \cdot KT = KH \cdot KK' = 2\rho \cdot KH;$$

$$\therefore \quad r : \rho = DT : KT.$$

Hence, the triangles IDT, JKT are equiangular;

$\therefore$ T, I, J are collinear,

and $\quad DI : IT = JK : JT,$

$$\therefore \quad DI = IT.$$

It follows that T is on the inscribed circle and, since T, I, J are collinear, the two circles have the same common tangent, and therefore touch at T.

14. *The nine-point circle of a triangle touches each of the escribed circles.* [Let *ABC* be the triangle, D_1 the point of contact with *BC* of the escribed circle which touches *AB*, *AC* produced. Join KD_1 and produce it to cut the nine-point circle at T_1. Then T_1 is the point at which this escribed circle touches the nine-point circle. The proof is practically the same as in Ex. 13.

15. *If J is the centre of the nine-point circle of the triangle ABC and I, I_1 the centres of the inscribed circle and the escribed circle opposite A, show that*

$$IJ = \frac{1}{2}R - r; \quad I_1J = \frac{1}{2}R + r_1.$$

16. If two circles are such that a triangle can be inscribed in one circle and circumscribed to the other, it has been shown that an unlimited number of triangles can be so described. Prove that

(*i*) The centres of the escribed circles of these triangles lie on a fixed circle.

(*ii*) The centres of the nine-point circles of the triangles lie on a fixed circle.

(*iii*) The orthocentres lie on a fixed circle.

L. RADICAL AXIS OF TWO CIRCLES

The circle of which C is the centre and r the radius is often denoted by (C, r).

1. *If (C, r), (C′, r′) are two given circles, and P is a point such that the tangents PT, PT′ to the two circles are equal, then P lies on a fixed straight line, perpendicular to CC′, called the* **Radical Axis** *of the circles.*

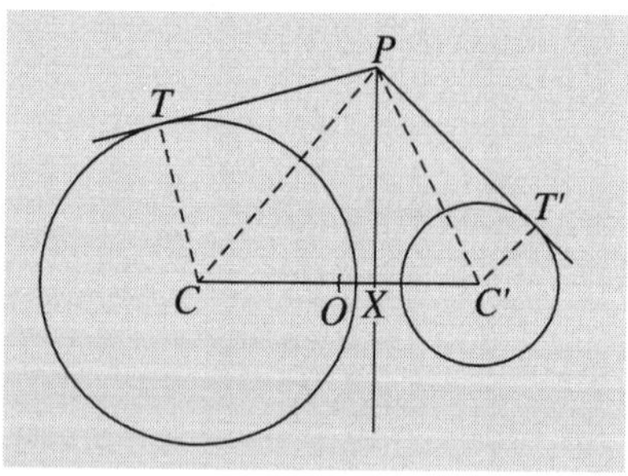

Proof. Let $r > r'$. Draw *PX* perpendicular to *CC′*. Bisect *CC′* at *O*.

Because $\angle$s *PXC*, *PTC* are rt. $\angle$s,

$$\therefore \quad CX^2 + PX^2 = CP^2 = CT^2 + PT^2;$$

$$\therefore \quad CX^2 + PX^2 = r^2 + PT^2.$$

Similarly, it can be shown that

$$C'X^2 + PX^2 = r'^2 + PT'^2.$$

Now, by hypothesis, $\quad PT = PT'$.

Hence, by subtraction,

$$CX^2 - C'X^2 = r^2 - r'^2.$$

But O is the middle point of CC';

$$\therefore \quad CX - C'X = 2OX;$$

$$\therefore \quad CX^2 - C'X^2 = (CX - C'X)(CX + C'X)$$
$$= 2OX \cdot CC';$$

$$\therefore \quad 2OX \cdot CC' = r^2 - r'^2;$$

$\therefore$ X is a fixed point and P lies on a fixed straight line perpendicular to CC'.

2. If the circles intersect at A, B, the radical axis passes through A, B.

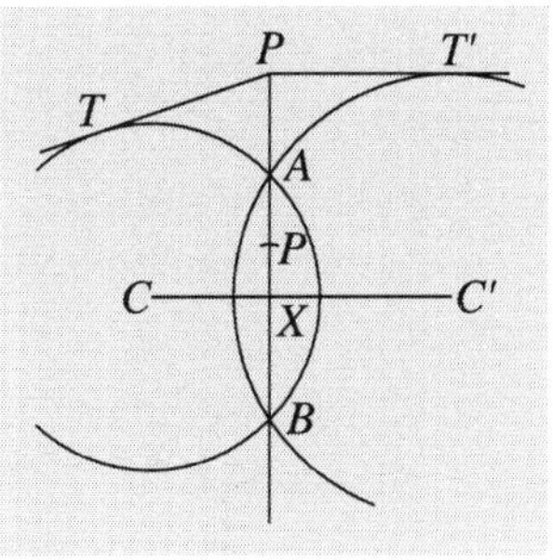

If P is any point on the radical axis, either within or without the circles, since PX is perpendicular to CC', we have

$$CP^2 - C'P^2 = CX^2 - C'X^2$$
$$= r^2 - r'^2;$$

$$\therefore \quad CP^2 - r^2 = C'P^2 - r'^2.$$

The radical axis of the circles (C, r), (C', r') *is therefore defined as the locus of a point* P*, such that*

$$CP^2 - r^2 = C'P^2 - r'^2.$$

3. Prove the following construction for the radical axis of the circles (C, r), (C', r'):

Draw any circle to cut the given circles at A, B *and* A', B'. *Join* AB, $A'B'$, *meeting in* P. *Draw* PX *perpendicular to* CC'. *Then* PX *is the radical axis.*

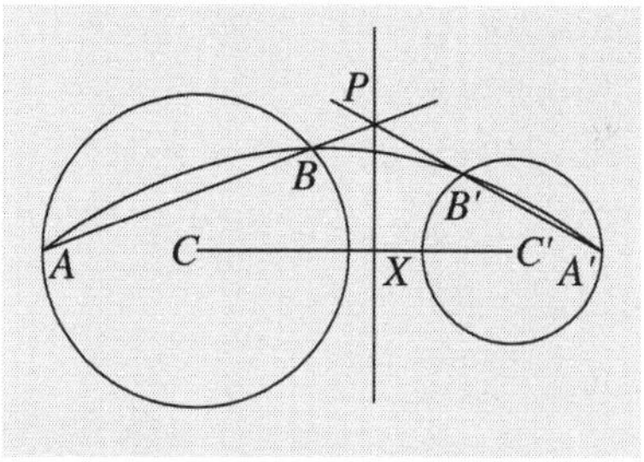

[For $PA \cdot PB = PA' \cdot PB'$; $\therefore$ the tangents from P to the given circles are equal; $\therefore$ P is on the radical axis.]

4. If two circles touch, show that the radical axis is the tangent at the point of contact.

5. If T, T' are the points of contact of a common tangent to two circles, show that the radical axis bisects TT'.

6. Show that the middle points of the four common tangents which can be drawn to two circles, each of which is outside the other, lie on a straight line.

7. *For any three circles, the radical axes of the first and second, second and third, third and first, meet in a point called the* **Radical Centre of the three circles**.

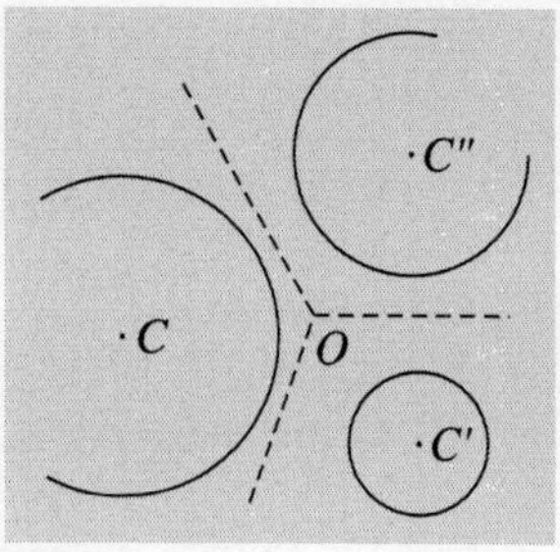

Let (C, r), (C', r'), (C'', r'') be the circles, and let the radical axis of the first and second and second and third circles meet in O.

Then, by definition,

$$CO^2 - r^2 = C'O^2 - r'^2$$

and $$C'O^2 - r'^2 = C''O^2 - r''^2;$$

$\therefore$ $$CO^2 - r^2 = C''O^2 - r''^2;$$

$\therefore$ O is on the radical axis of the first and third circles.

8. Find a point O, such that the tangents from it to three given circles are equal.

9. If three circles are such that each intersects the other two, prove that the three chords common to pairs of the circles meet in a point. [This is a particular case of § 7.]

10. Given two points A, B and a circle C, if any circle is drawn through A, B to cut the circle C at P, Q, prove that PQ meets AB in a fixed point.

[Draw a fixed circle through A, B to cut the circle C in D, E. Let DE meet AB at O. Then (by § 7) PQ passes through the fixed point O.]

11. AA', BB', CC' are the perpendiculars from the vertices of a triangle to the opposite sides and X, Y, Z are the points of intersection of BC, $B'C'$; CA, $C'A'$; AB, $A'B'$. Prove that X, Y, Z are in a straight line which is the radical axis of the circumcircle and the nine-point circle of the triangle ABC.

[X is on the radical axis, for $XB \cdot XC = XB' \cdot XC'$.]

12. *If PT, PT' are tangents from any point P to two circles (C, r), (C', r') and PM is the perpendicular from P to the radical axis of the circles, then*

$$PT^2 - PT'^2 = 2CC' . PM,$$

it being assumed that $PT > PT'$.

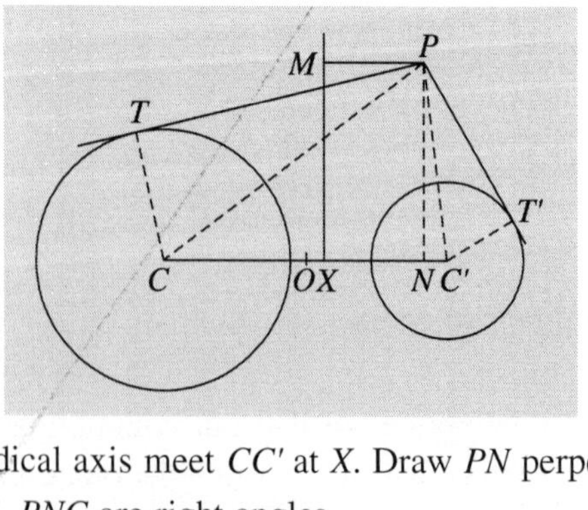

Proof. Let the radical axis meet CC' at X. Draw PN perpendicular to CC'.

Because $\angle$s PTC, PNC are right angles,

$\therefore$ $$PT^2 + CT^2 = CP^2 = CN^2 + PN^2;$$

$\therefore$ $$PT^2 + r^2 = CN^2 + PN^2.$$

Similarly,

$$PT'^2 + r'^2 = C'N^2 + PN^2;$$

$$\therefore \quad PT^2 - PT'^2 + r^2 + r'^2 = CN^2 - C'N^2.$$

Also, since MX is the radical axis,

$$r^2 - r'^2 = CX^2 - C'X^2 = 2OX \cdot CC' ;$$

and

$$CN^2 - C'N^2 = (CN - C'N)(CN + C'N)$$

$$= 2ON \cdot CC';$$

$$\therefore \quad PT^2 - PT'^2 + 2OX \cdot CC' = 2ON \cdot CC';$$

$$\therefore \quad PT^2 - PT'^2 = 2XN \cdot CC' = 2CC' \cdot PM.$$

13. As a particular case of (12), let P be any point on the circle (C', r'). In this case PT' vanishes, and we have

$$PT^2 = 2CC' \cdot PM.$$

Hence, *if the square on the tangent from a point P to a given circle varies as its distance from a given line, the locus of P is a circle.*

14. DEFINITIONS: The angle of intersection of two circles, at a point where they cut, is the angle between their tangents at this point.

Since the angle between two straight lines is equal to the angle between two perpendiculars to these lines, the angle at which two circles cut is equal to that between the radii drawn to a point of intersection of the circles.

Circles which cut at right angles are said to cut **orthogonally**.

15. The **cosine** of the angle at which two circles cut is $\pm (r^2 + r'^2 - d^2)/2rr'$, where r, r' are the radii and d is the distance between the centres.

16. If the circles (C, r), (C', r') cut orthogonally, then

$$CC'^2 = r^2 + r'^2.$$

17. If a point P is taken on the radical axis of two circles, and PT is a tangent to one of them, prove that the circle whose centre is P and radius PT cuts each of the circles orthogonally.

18. Prove that the locus of the centre of a circle which cuts two given circles orthogonally is the radical axis of the given circles.

19. Draw a circle to cut three given circles orthogonally.

20. *Any circle which cuts two given non-intersecting circles (C, r), (C', r') orthogonally, passes through two fixed points in CC',*

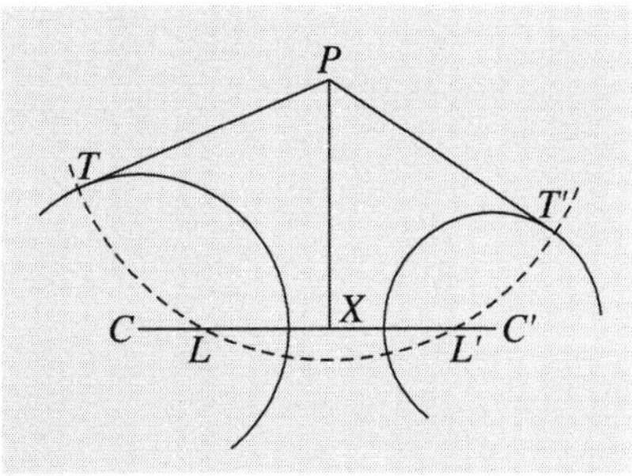

Let P be the centre of a circle which cuts the given circles orthogonally and which meets CC' at L, L'.

Since PT, PT' touch the given circles and $PT = PT'$

$\therefore$ P is on the radical axis PX. Since CT touches the circle TLL',

$$\therefore \qquad CL \cdot CL' = CT^2 = r^2.$$

Now X bisects LL';

$$\therefore \qquad XC^2 - XL^2 = CL \cdot CL' = r^2;$$

$$\therefore \qquad XL^2 = XC^2 - r^2;$$

$\therefore$ L, L' are fixed points.

21. In the figure of (§ 20), show that $C'L \cdot C'L' = r'^2$.

22. If the circle (C, r) meets CC' at A, B, prove that AB is divided internally and externally in the same ratio L, L'.

LI. CENTRES OF SIMILITUDE

1. If C, C' are the centres and r, r' the radii of two circles, the points S, S' which divide CC' externally and internally in the ratio r: r' are called the **external and internal centres of similitude** of the circles. Thus, in the figure, we have

$$SC : SC' = CS' : S'C' = r : r'.$$

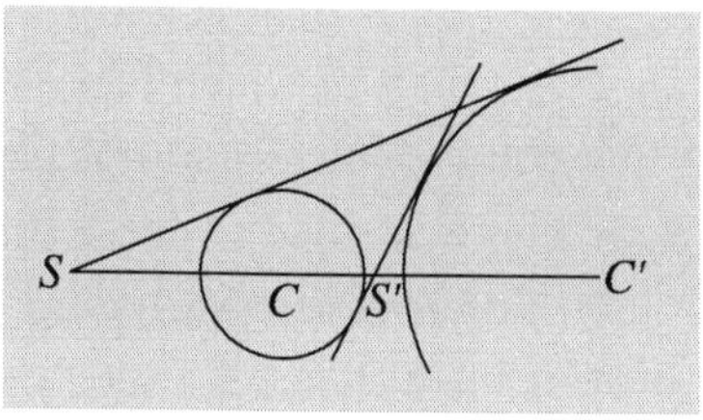

It has been proved that *any common tangent to the circles meets CC' in a centre of similitude* (Ex. 1, p. 277).

2. *If S is a centre of similitude of two circles, of which C, C' are the centres and r, r' the radii, and if any straight line through S cuts the circles at P, Q; P', Q', then*

$$SP : SP' = r : r' \text{ and } SQ : SQ' = r : r'.$$

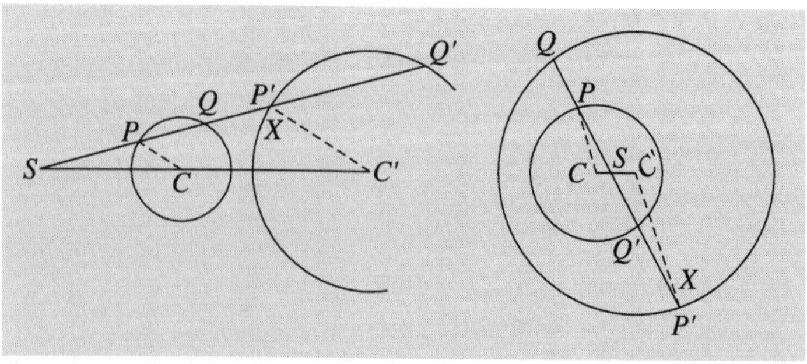

Construction. Join CP. Draw $C'X$ parallel to CP to meet SP (*produced*) at X.

Proof. Because $C'X$ is parallel to CP,

∴ The triangles SCP, $SC'X$ are equiangular ;

∴ $CP : C'X = SC : SC'$.

But $SC : SC' = CP : C'P'$ (*given*);

∴ $CP : C'X = CP : C'P'$;

∴ $C'X = C'P'$.

∴ X coincides with either P' or Q'. Taking P' as the point which coincides with X, it follows that the triangles SCP, $SC'P'$ are equiangular;

∴ $SP : SP' = CP : C'P' = r : r'$.

Similarly, $SQ : SQ' = r : r'$.

3. If two circles touch externally, the point of contact is the internal centre of similitude; if they touch internally, the point of contact is the external centre of similitude.

4. If A is a point at which two circles intersect one another, the bisectors of the angles between the tangents at A pass through the centres of similitude.

5. If the bisector of the angle A of the triangle ABC meets BC at X, then A and X are the centres of similitude of the inscribed circle and the escribed circle opposite A.

6. If the bisector of the exterior angle at A of the triangle ABC meets BC produced at X', then A and X' are the centres of similitude of the escribed circles opposite B and C.

7. In each figure of § 2 on the preceding page, prove that

(*i*) $PQ : P'Q' = r : r'$.

(*ii*) $SP \cdot SQ' = SP' \cdot SQ$ = constant for all lines through S.

(*iii*) The tangents at P, P' are parallel; so are those at Q, Q'.

(*iv*) The tangents at P, Q' meet on the radical axis of the circles; so do those at Q, P'.

(*v*) If the tangents at P, Q meet in A and the tangents at P', Q' meet at A', then S, A, A' are collinear.

8. In the left-hand figure of § 2 on the preceding page 352, if STT' is a common tangent, prove that

(*i*) $SP \cdot SQ' = ST \cdot ST' = SQ \cdot SP'$.

(*ii*) The quadrilaterals $PTT'Q'$, $QTT'P'$ are cyclic.

(*iii*) PT, $Q'T'$ meet on the radical axis ; so also do QT, $P'T'$.

9. The centroid and the orthocentre of a triangle are the centres of similitude of the circumcircle and the nine-point circle.

10. Every straight line drawn from either the centroid or the orthocentre of a triangle to the circumference of the circumcircle is bisected by the nine-point circle.

11. If C, C' are the centres and CP, $C'P'$ are parallel radii of two circles, show that PP' passes through a centre of similitude of the circles. Explain how to decide whether this is the external or internal centre of similitude.

[If P, P' are on the same side of CC', it is the external centre of similitude.]

12. Prove that the straight line joining the points of contact of two parallel tangents to two circles passes through a centre of similitude. Explain how to decide whether this is the external or internal centre of similitude.

[If the points of contact are on the same side of the line of centres, it is the external centre of similitude.]

13. AB, $A'B$ are parts of the same straight line, with the same middle point S. On AB, $A'B'$ similar segments of circles are described, on the same side of the line. Prove that S is a centre of similitude of the circles. [Use Ex. 12.]

14. Draw an equilateral triangle ABC. Draw a segment of a circle with its base along BC, such that its arc contains three-quarters of the circumference, and touches AB, AC.

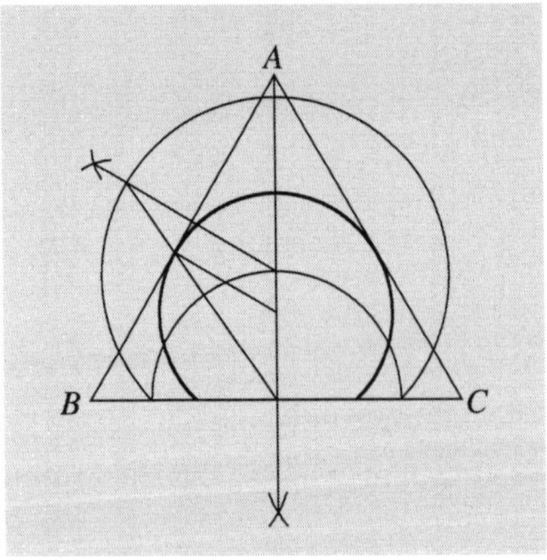

[The construction depends on Ex. 13, and is indicated in the figure. Explain and supply proof.]

15. If a circle is drawn to touch two given circles C, C' at P, P', prove that PP' passes through a centre of similitude of the given circles. [Let PP' cut the circle C again at Q. Prove that the tangents at Q, P' are parallel.]

16. If from any point A, on the radical axis of two circles, tangents AP, AQ are drawn to the circles, prove that PQ passes through a centre of similitude.

[Let PQ, cut the second circle again at Q'. Prove that the tangents at P, Q' are parallel.]

17. A straight line, through a centre of similitude S of two circles, cuts the circles at P, Q and P', Q'. Another straight line through S cuts the circles at X, Y and X', Y'. Prove that

(*i*) PX, $P'X'$ are parallel; so also are QY, $Q'Y'$.

(*ii*) The quadrilaterals $PXY'Q'$, $QYX'P'$ are cyclic.

(*iii*) PX, $Q'Y'$ meet on the radical axis; so also do QY, $P'X'$.

18. *Prove that the locus of a point such that the tangents from it to two circles are in the ratio of the radii is the circle described on the line joining the centres of similitude as diameter.**

[If PT, $P'T'$ are tangents to the circles (C, r), (C', r'), such that $PT : PT' = r : r'$, prove that $PC : PC' = r : r'$.]

19. Prove that the locus of a point at which two given circles subtend equal angles is the circle described on the line joining the centres of similitude as diameter.

20. Find a point at which the escribed circles of a triangle ABC subtend equal angles. Show that there are, in general, two such points.

[Let the bisectors of the exterior angles at A, B meet BC, CA at X', Y'. The circles on AX', BY' as diameters meet at the required points. See Exs. 6, 19.]

21. C, C' are two given circles and PT, QT are tangents to C. Two tangents $P'T'$, $Q'T'$ are drawn to C', parallel to PT, QT respectively, and such that $P'T'$, $Q'T'$ are either both in the same sense as PT, QT or both in the opposite sense to PT, QT. Show that TT' passes through a centre of similitude of the circles.

Explain how to select the parallel tangents in order that TT' may pass through the internal centre of similitude S'.

[If P, Q, P', Q' are the points of contact, PP', QQ' must pass through S'.]

22. *It is required to draw a circle to touch two given straight lines and a given circle C.*

(*i*) Explain how Ex. 21 leads to the following construction for the point of contact with C,—Draw a properly selected pair of tangents (PT, QT) to the circle C, parallel to OA, OB and meeting at T. Join TO. Then TO passes through the point where the required circle touches C.

(*ii*) Explain how to find the centre of the required circle.

(*iii*) Taking the case in which the circle C is within the angle AOB, draw the two circles which touch OA, OB and touch the circle C externally.

[PT, QT must be drawn on the opposite sides of C to OA, OB respectively. Explain why. See Exs. 3, 21.]

---

* *This circle is called the* **Circle of Similitude** *of the given circles.*

(*iv*) Taking the case in which the circle C is within the angle AOB, draw the two circles which touch OA, OB and touch the circle C internally.

[PT, QT must be drawn on the sides of C nearest to OA, OB respectively. Explain why.]

(*v*) Taking the case in which O is within the circle C, draw the circle which touches OA, OB (neither being produced backwards through O) and the circle C internally.

23. Show that the number of circles which can be drawn to touch two given straight lines and a given circle, is determined by the number of intersection, of the circle and the straight lines.

State the number of circles, when

(*i*) both lines cut the circle;

(*ii*) one line cuts, the other touches the circle;

(*iii*) one line only meets the circle;

(*iv*) both lines touch the circle;

(*v*) one line touches the circle, the other does not meet it;

(*vi*) neither line meets the circle.

Application to Questions of Geometrical Drawing

1. *Draw an equilateral* ΔABC, *and in it inscribe three arcs of circles, each containing* $\frac{3}{4}$ *of the circumference, touching two sides of ABC, and meeting two by two in P, Q, R, the vertices of an equilateral triangle.**

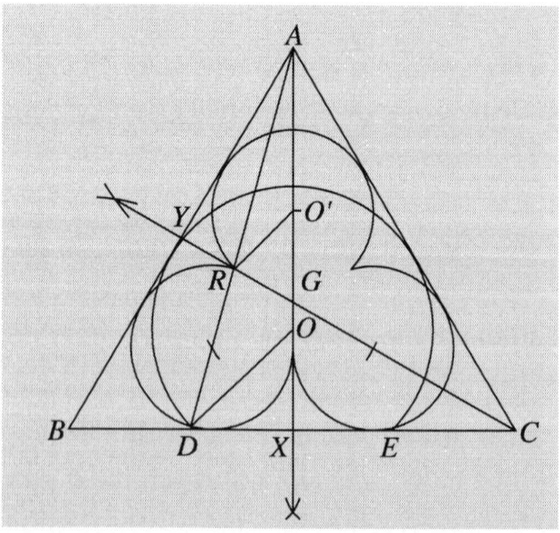

Draw AX, $CY \perp BC$, BA. Draw a segment of a circle, with base DE along BC, whose arc contains $\frac{3}{4}$ of the circumference and touches AB, AC.

* Such arcs are mis-named *continuous* in some books on geometrical drawing.

∵ AB, AC are common tangents to the required arc QR and the arc DE.

∴ A is a centre of similitude;

∴ R lies in the straight line AD.

Also, by symmetry, R lies in YC and can therefore be found.

Let O be centre of ⊙ DE.

Draw $RO' \parallel DO$ to meet AX in O'.

Then ∵ A is a centre of similitude,

∴ O' is the centre of ⊙ QR.

The centres of the circles RP, PQ can now be found by setting off distances from G, along GB, GC, equal to GO'.

2. *Draw six equal and 'continuous' semi-circumferences to touch a given circle.*

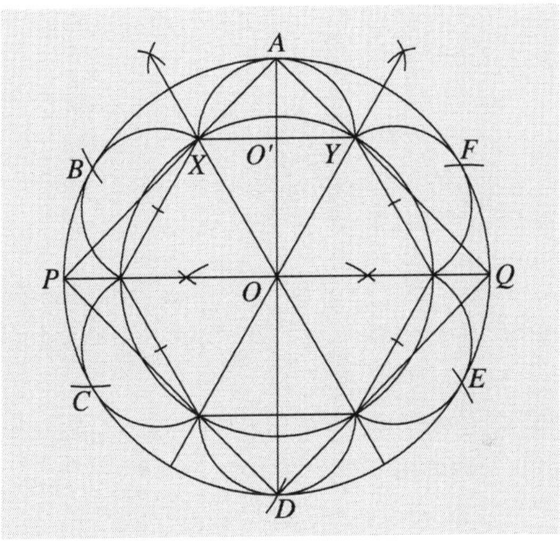

Find A, B, C, D, E, F, the vertices of a regular inscribed hexagon. Draw the diameter $POQ \perp AOD$. Then, A is a centre of similitude of the given circle and the required circle, touching at A; therefore the extremities X, Y of the diameter of the required circle lie on AP, AQ.

Also, by symmetry, X lies on the bisector of $\angle AOB$. Thus, X can be found.

Similarly, Y lies on the bisector of $\angle AOF$, and hence can be found by drawing a circle, with centre O and radius OX. The centre O' of the circle touching at A is the intersection of XY and AO. The other centres can be stepped out as in the figure above.

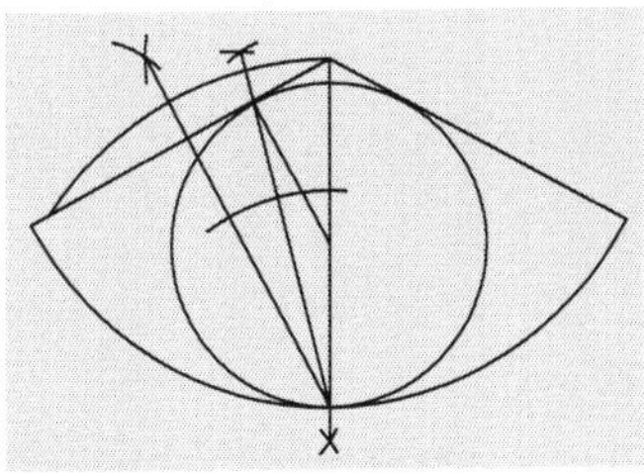

3. *Inscribe a circle in a given sector of a circle.*

4. Draw a circle to touch a given circle externally, and also to touch two given radii of the given circle.

5. Draw four equal circles, each touching a given circle internally, and each of the four touching two of the others.

6. Draw four equal circles, each touching a given circle externally, and each of the four touching two of the others.

7. *Within a given regular pentagon ABCDE, draw 10 equal circles, such that each side of the pentagon touches three of them, and each circle touches the two adjacent circles.*

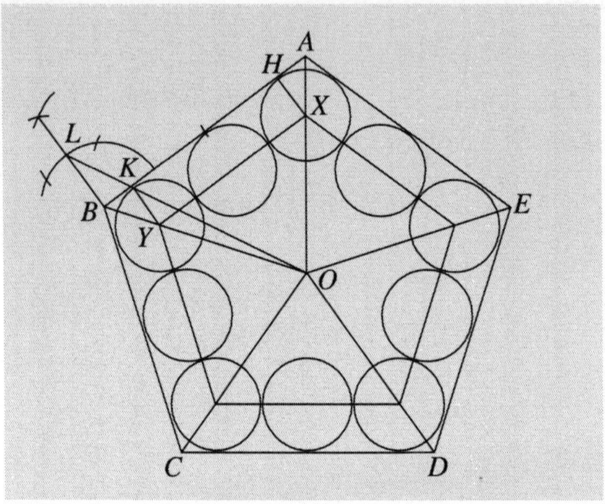

Analysis. Let O be the centre of $ABCDE$. Then because OA, OB... bisect the angles A, B..., therefore the centres X, Y,... of the circles which touch two sides are on OA, OB ....

Again, if XH, YK are perpendicular to AB, then $XY = 4YK$.

Therefore XK is a rectangle inscribed in ΔOAB, with its sides in the ratio 4 : 1; hence the following:

Construction. Draw $BL \perp BA$, making $BL = \frac{1}{4} BA$. Join OL, cutting AB in K. Draw $KY \perp AB$, cutting OB in Y. Draw $YX \parallel AB$. Then YX contains the centres of the required circles which touch AB. The other lines of centres may be found by drawing parallels to the sides of the pentagon.

EXERCISE LXXIX

(*In the following Numerical Examples measure the radii of the constructed arcs.*)

1. Draw a regular hexagon of 1.5 in. side. Draw three equal and continuous circular arcs, each two-thirds of the circumference, these arcs touching alternate sides of the hexagon.

2. Draw a square of 2 in. side. Draw four equal and continuous circular arcs, each arc touching two sides of the square, and each arc being a semi-circumference.
3. In an equilateral triangle of 3 in. side, draw three equal and continuous semi-circular arcs, each touching one side of the triangle.
4. Draw a square of 2 in. side. Draw two equal and continuous arcs, each two-thirds of the circumference, to touch opposite sides of the square at their middle points.
5. Draw a circle of 1.5 in. radius. Draw four equal and continuous arcs of circles, each three-quarters of the circumference, to touch the given circle.
6. Draw a circle of radius 1.5 in. Draw six equal circles, each touching this circle (*i*) internally, (*ii*) externally, each of the six touching two of the others.
7. Draw an equilateral triangle of 3 in. side, and within it draw three equal circles, each touching the other two, and two sides of the triangle.
8. Draw a regular hexagon of 1.5 in. side. Within this, draw 12 equal circles, such that each side of the hexagon touches three of them, and each circle touches the two adjacent circles.

LII. CONSTRUCTION OF CIRCLES TO SATISFY GIVEN CONDITIONS

1. *Draw a circle through two given points A, B to touch a given straight line CD.*

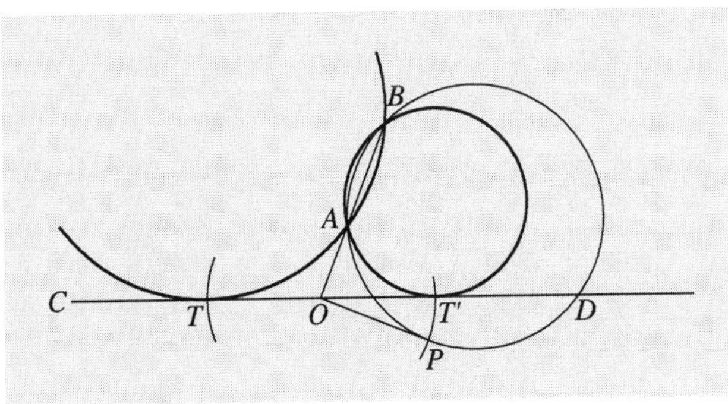

Analysis. Let AB produced cut CD in O. Draw any circle through A and B, and let OP be a tangent to it.

Then $OP^2 = OA \cdot OB$ = a constant for all circles through A and B.

Hence the following construction:

Join *AB*. Produce *AB* to cut *CD* in *O*. Draw any circle through *A* and *B*. Draw a tangent *OP* to this circle. Set off *OT*, *OT′* along *OC*, *OD* respectively, each equal to *OP*.

The circles *ABT*, *ABT′* are the circles required.

Supply proof.

2. The above construction fails when *AB* || *CD*. Give a construction for this case.

3. Draw a circle, with its centre on a given straight line, to pass through a given point and touch a given straight line.

4. *Draw a circle to touch two given straight lines OA, OB, and pass through a given point C.*

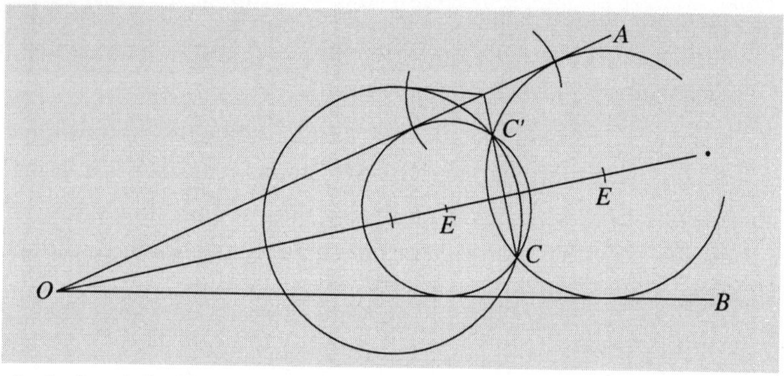

Analysis. Let *E* be the centre of the required circle. Then *OE* bisects $\angle AOB$. Also, if *C′* is the image of *C* in this bisector, *C′* is on the circle.

Hence, a circle through *C* and *C′*, touching *OA* (or *OB*), is the circle required.

Supply the construction and proof.

5. *Draw a circle to touch a given circle (centre C and radius r) and also to touch two given straight lines OA, OB.*

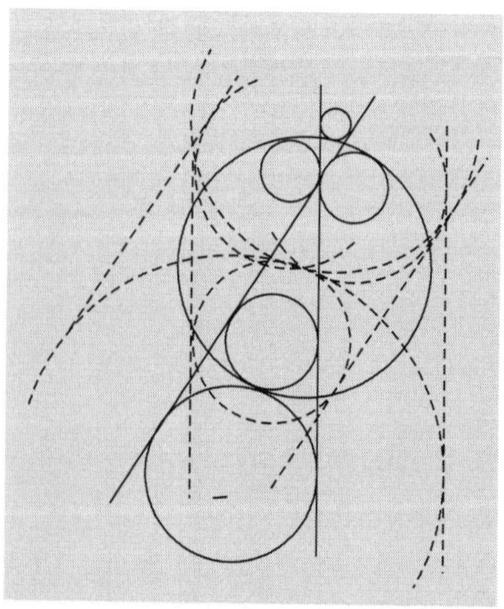

This question has been considered in Ex. 2, p. 357, or we may proceed as follows:

The circle is concentric with one passing through *C*, and touching two straight lines parallel to *OA*, *OB*, and at a distance *r* from them. There are eight circles in the most general case satisfying the given conditions, of which five are shown in the figure.

Supply the construction for the three which are not shown and give a proof.

6. Draw any triangle *ABC*. Bisect *BC*, *CA*, *AB* at *D*, *E*, *F*. Draw the circle *DEF*, and construct the eight circles which touch the circle *DEF* and *AB*, *AC* (or these lines produced). Verify that *BC* (produced if necessary) touches four of these circles.

7. *Draw a circle through two given points A, B to touch a given circle.*

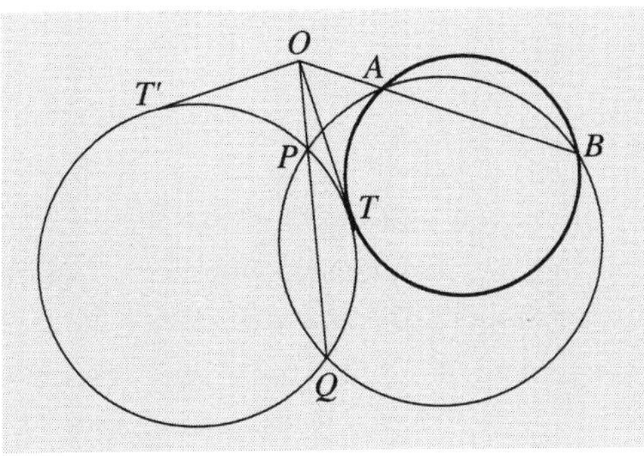

Analysis. Suppose a circle through *A*, *B* to touch the given circle at *T*. Let any circle be drawn through *A*, *B* to cut the given circle at *P*, *Q*.

The radical axis of the three circles shown in the figure, taken in pairs, are *AB*, *PQ* and the tangent at *T*.

Therefore these three lines meet in a point *O*. (See § 7, p. 350.)

Hence the following construction :

Through *A*, *B* draw any circle to cut the given circle at *P*, *Q*. Join *AB*, *PQ*, meeting at *O*. From *O* draw tangents *OT*, *OT'* to the given circle.

Describe circles about the triangles *ABT*, *ABT'*. These satisfy the given conditions.

Supply proof.

8. Draw a circle, with its centre on a given straight line, to pass through a given point and touch a given circle.

9. Find a point in a given straight line, such that its distance from a given point is equal to its distance from the circumference of a given circle (measured along a radius).

10. Find a point in a given straight line, such that (*i*) the sum or (*ii*) the differences of its distance from two given points is equal to a given straight line.

11. *Draw a circle to pass through a given point A, to touch a given straight line BC and a given circle.*

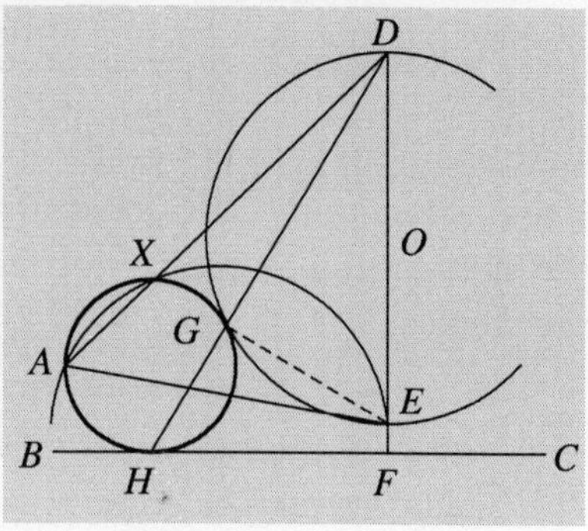

Analysis. Suppose a circle to touch the given circle at G and BC at H. Then, since G is a centre of similitude of the two circles, HG will cut the given circle at a point D, where the tangent is parallel to BC, *i.e.* at one end of the diameter DE, perpendicular to BC.

Draw this diameter, meeting BC at F. Join DA, cutting the circle AGH at X.

Now, $\angle$s DGE, EFH are right angles;

$\therefore$ $GEFH$ is a cyclic quadrilateral;

$\therefore$ $$DE \cdot DF = DG \cdot DH = DA \cdot DX;$$

$\therefore$ $AFEX$ is a cyclic quadrilateral.

Hence the following:

Construction. Draw the diameter DE of the given circle, perpendicular to BC, meeting BC at F. Join A to one end D of the diameter DE. Draw the circle AFE, meeting AD again at X. Through A, X draw a circle to touch the given circle. This circle will also touch BC and will satisfy the given conditions.

Proof. Let the circle through A, X touch the given circle at G. Let DG produced cut BC at H : join GE.

Then because the angles at G, F are right angles,

$\therefore$ $GEFH$ is a cyclic quadrilateral;

$\therefore$ $$DE \cdot DF = DG \cdot DH.$$

But, by construction, $XEFA$ is a cyclic quadrilateral;

$\therefore$ $$DE \cdot DF = DX \cdot DA;$$

$\therefore$ $$DX \cdot DA = DG \cdot DH;$$

$\therefore$ the circle AXG passes through H.

Again, the tangents at D, G, H make equal angles with DH;

$\therefore$ the tangents at D, H are parallel;

$\therefore$ BC is the tangent at H.

NOTE: *Two circles can be drawn through. A, X to touch the given circle. Also two more circles can be drawn to fulfil the conditions, by joining AE. Thus, in general, there are four solutions.*

12. *Draw a circle to touch two given circles and a given straight line AB.*

Let C_1, r_1 ; C_2, r_2 be the centres and radii of the given circles, and let $r_1 > r_2$. Let a circle be drawn (1) to pass through C_2, (2) to touch a straight line parallel to AB and distant r_2 from it, and (3) to touch a circle whose centre is C_1 and radius either $r_1 - r_2$ or $r_1 + r_2$.

In general, eight such circles can be drawn; and each of these is concentric with a circle which satisfies the given conditions.

There are consequently, in general, eight solutions.

13. *Draw a circle through a given point A to touch two given circles C, C′.*

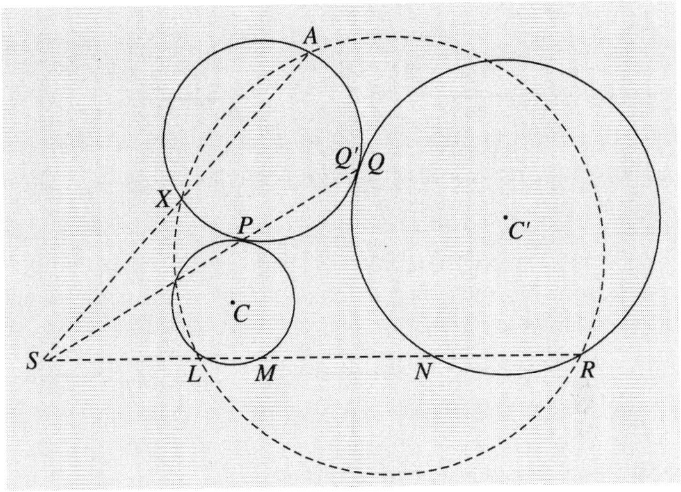

Analysis. Let the required circle touch the two circles C, C', in P, Q respectively. Then the common tangents at P, Q are equally inclined to PQ; hence, if PQ meets the circle C again in P', the tangents at Q and P' will be parallel; therefore PQ passes through S, a centre of similitude of the circles C, C'. (See § 7 (v), p. 353.)

Draw any straight line through S, to cut C, C' in L, M, N, R. Let SA cut the required circle again in X.

Then $SX \cdot SA = SP \cdot SQ = SL \cdot SR$, by §7 ($ii$), p. 353;

∴ $LRAX$ is cyclic.

Hence the following :

Construction. Find a centre of similitude, S, of C, C'. Draw any straight line through S, to cut the circles C, C' in L, M, N, R. Draw the circle through A, L, R, cutting SA in X. Through A, X draw a circle to touch the circle C in P.

Then this circle will also touch the circle C'.

Proof. Let SP cut ⊙ C' in Q, and ⊙ AXP in Q'.

Then, since S is a centre of similitude of C, C',

∴ $$SP \cdot SQ = SL \cdot SR$$
$$= SX \cdot SA.$$

Also, since $XAQ'P$ is cyclic,

∴ $$SX \cdot SA = SP \cdot SQ';$$

$$\therefore \qquad SP \cdot SQ = SP \cdot SQ';$$

hence, Q and Q' coincide.

Again, since S is a centre of similitude of C, C', the tangents at P and Q to the circles C, C' are equally inclined to PQ ; but the tangents at P, Q to ⊙ $XPQA$ are also equally inclined to PQ ; hence, since the circles C and $XPQA$ have a common tangent at P,

∴ the circles C' and $XPQA$ have a common tangent at Q, *i.e.* these circles touch one another at Q.

Two circles can be drawn through A, X to touch C, each of which also touches C'. Starting with the other centre of similitude, two more circles can be found to satisfy the conditions.

Thus, in general, four circles can be drawn to satisfy the conditions.

14. Deduce, from the preceding, a construction for a circle which touches three given circles; and show that, in general, there are eight solutions.

15. Identify the eight circles in the case where the three given circles are the three escribed circles of a triangle.

LIII. HARMONIC DIVISION OF A LINE

1. A straight line is said to be **divided harmonically** when it is divided internally and externally in the same ratio, and each point of division is called the **harmonic conjugate** of the other.

2. If AA' is divided harmonically at B, B', then BB' is divided harmonically at A, A'.

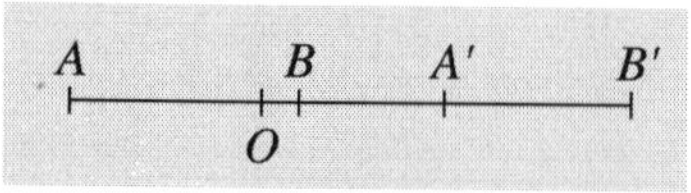

For, by hypothesis,

$$AB : BA' = AB' : A'B';$$

$$\therefore \qquad AB : AB' = BA' : A'B'.$$

3. A set of collinear points A, B, C,... is called the **range** (ABC...). The straight line on which the points lie is the axis of the range.

If A, A' are harmonic conjugates to B, B', the points A, A', B, B' are said to form the **harmonic range** (AA', BB').

4. *If* (AA', BB') *is a harmonic range and* O *is the middle point of* AA', *then*

$$OA^2 = OA'^2 = OB \cdot OB'.$$

For it is given that $\quad AB : BA' = AB' : A'B'.$

∴ By componendo and dividendo,

$$AB + BA' : AB - BA' = AB' + A'B' : AB' - A'B';$$

$$\therefore \qquad 2OA' : 2OB = 2OB' : 2OA';$$

$$\therefore \qquad OA'^2 = OB \cdot OB'.$$

5. *Conversely, if O is the middle point of AA', and points B, B' are taken in the straight line AA', on the same side of O, such that* $OA^2 = OB \cdot OB'$, *then B, B' are harmonic conjugates of A, A'.*

For the steps in the proof of § 4 can be taken in the reverse order.

6. If (*AA', BB'*) is a harmonic range, *A* being outside *BB'*, then

$$\frac{2}{AA'} = \frac{1}{AB} + \frac{1}{AB'}.*$$

In the figure of § 2, we have to prove that

$$\frac{1}{AB} - \frac{1}{AA'} = \frac{1}{AA'} - \frac{1}{AB'};$$

that is, $$\frac{BA'}{AB \cdot AA'} = \frac{A'B'}{AA' \cdot AB'}, \text{ or } \frac{BA'}{AB} = \frac{A'B'}{AB'},$$

which follows from the hypothesis.

7. Take any three points *A*, *B*, *B'* in order, on a straight line. Bisect *BB'* at *O*. Draw a circle on *BB'* as diameter and a tangent *AT* to this circle. Draw *TA'* perpendicular to *BB'* and prove that (*i*) the range (*AA', BB'*) is harmonic; (*ii*) *AO*, *AT*, *AA'* are the arithmetic, geometric and harmonic means of *AB*, *AB'*.

8. Explain how to construct lines which are the arithmetic, geometric and harmonic means of two given straight lines *x*, *y*.

If these are denoted by *a*, *g*, *h*, prove geometrically that $a > g > h$ and that $ah = g^2$.

9. *AA', BB' an two non-overlapping segments of a straight line. Find points X, X' which divide both AA' and BB' harmonically.*

[Let *O* be the middle point of *XX'*, then $OX^2 = OA \cdot OA' = OB \cdot OB'$, and the point *O* can be found by Ex. LXXVII, 14.]

10. *Let (AA', BB') be a harmonic range. Take any point O and join OA, OA', OB, OB'. Through B draw a parallel to meet OB' to meet OA, OA' at X, X'. Prove that XB = BX'.*

Proof. By similar triangles,

$$XB : OB' = AB : AB'$$

and $$BX' : OB' = BA' : A'B'.$$

But (*AA', BB'*) is a harmonic range;

$$\therefore \quad AB : AB' = BA' : A'B';$$

$$\therefore \quad XB : OB' = BX' : OB';$$

$$\therefore \quad XB = BX'.$$

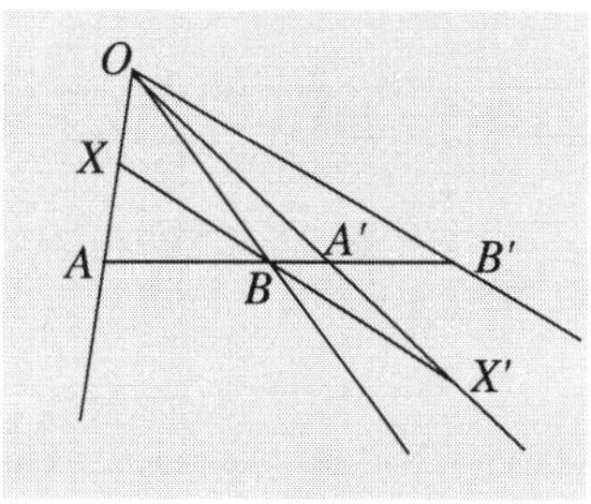

* *Hence the term* harmonic *range.*

11. Conversely, if in the figure of § 10, $XB = BX'$, prove that (AA', BB') is a harmonic range.

12. *Given three collinear points A, A', B. Find B', the harmonic conjugate of B with regard to A, A'.*

[Through B draw any straight line XBX', making $XB = BX'$. Join AX, $A'X'$, meeting at O. Draw OB' parallel to XX', meeting AA' at B'.]

13. A set of concurrent straight lines OA, OB, OC, ...is called the **pencil** $O(ABC...)$; O is the **vertex** and OA, OB, OC, ...are the rays of the pencil.

14. (AA', BB') is a harmonic range and O is any point. If any transversal meets OA, OA', OB, OB' at P, P', Q, Q', then (PP', QQ') is a harmonic range.

[Through B, Q draw parallels to OB', meeting OA, OA' at X, X' and Y, Y'. Prove that $XB : BX' = YQ : QY'$ and use §§ 10 and 11.]

15. From § 13 it follows that *if a pencil of four rays divides any transversal harmonically, it divides every transversal harmonically.*

A pencil of this kind is called a **harmonic pencil**.

LIV. POLE AND POLAR

Theorem 111

If through a given point any straight line is drawn to cut a circle, the tangents at the points of contact meet on a fixed straight line.

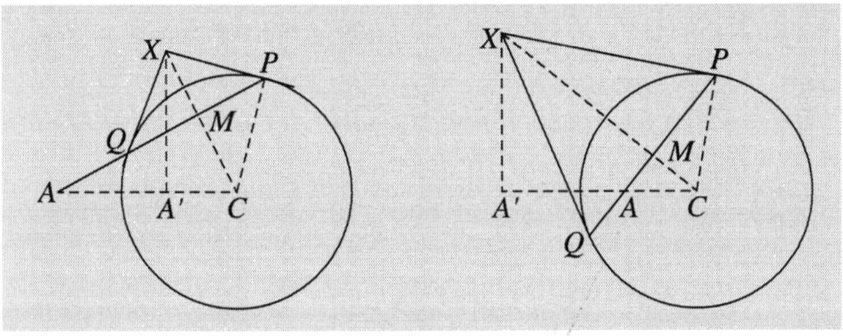

Let any straight line, through the given point A, meet the circle, of which C is the centre and r the radius, at P and Q. Let the tangents at P, Q meet at X. It is required to prove that X lies on a fixed straight line.

Construction. Join CA. Draw XA' perpendicular to CA, produced if necessary. Join CX, meeting PQ at M. Join CP.

Proof. In the right-angled triangle CPX, because PM is perpendicular to the hypotenuse,

∴ the triangles CPM, CXP are similar;

∴ $$CM : CP = CP : CX;$$

∴ $$CM \cdot CX = CP^2 = r^2.$$

Also, because the angles at M and A' are right angles,

∴ a circle can be described through the points M, X, A', A;

∴ $$CM \cdot CX = CA \cdot CA';$$

∴ $$CA \cdot CA' = r^2;$$

∴ A' is a fixed point;

∴ X lies on a fixed straight line, namely, the line through A' perpendicular to CA'.

THEOREM 112

If tangents are drawn to a circle from any point on a given straight line, the chord of contact of these tangents passes through a fixed point.

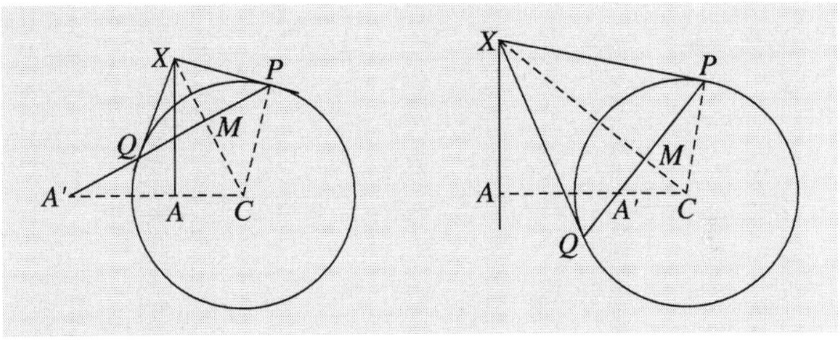

Let X be any point on the given straight line, and let XP, XQ be the tangents from X to a circle, of which C is the centre and r the radius. It is required to prove that the chord of contact PQ passes through a fixed point.

Construction. Draw CA perpendicular to the given line, meeting PQ at A'. Join CX, meeting PQ at M. Join CP.

Proof. This is identically the same as that of the last theorem, with the omission of the last two lines.

1. DEFINITIONS. If C is the centre and r the radius of a circle, and A, A' are points collinear with C and on the same side of it, such that $CA \cdot CA' = r^2$, either of the points A, A' is called the **inverse** of the other with regard to the circle.

The **polar** of a point (A) with regard to a circle is the perpendicular ($A'X$) through the inverse point (A') to the straight line joining it to the centre.

The **pole** of a straight line (AX) with regard to a circle is the inverse point (A'), of the foot of the perpendicular (A) from the centre to the line.

2. The fundamental properties of *pole and polar* are as follows:

(*i*) If the point A is outside the circle, its polar passes through the points of contact (T, T') of the tangents from A to the circle.

[For the triangles CAT, CTA' are similar;

∴ $$CA \cdot CA' = r^2.]$$

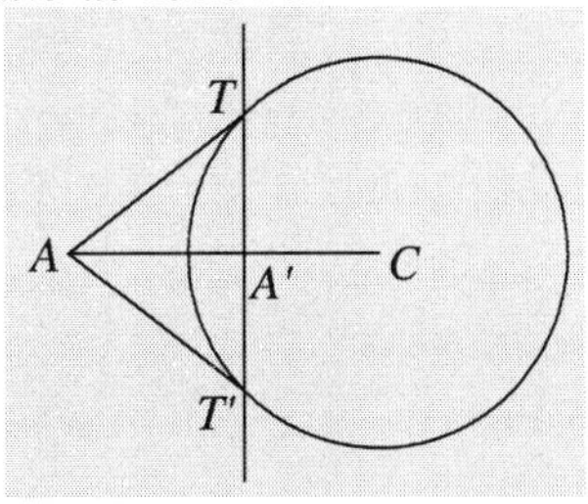

(*ii*) If a straight line cuts a circle at T, T', its pole is the intersection of the tangents at T, T'.

(*iii*) If through a point A any straight line is drawn to cut a circle, the tangents at the points of contact meet on the polar of A. (Theorem 111)

(*iv*) If tangents are drawn to a circle from any point on a straight line AX, the chord of contact of the tangents passes through the pole of AX. (Theorem 112)

(*v*) The property stated in the next theorem.

THEOREM 113

If the polar of a point A with regard to a circle passes through a point B, then the polar of B passes through A.

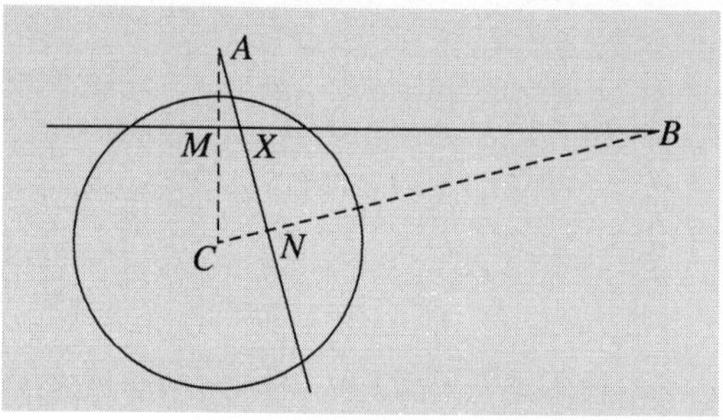

Let C be the centre and r the radius of a circle. Join CA, CB. Draw AN, BM perpendicular to CB, CA respectively.

Proof. The polar of A is perpendicular to CA and, by hypothesis, it passes through B;

$\therefore$ BM is the polar of A.

Hence, by definition,

$$CM \cdot CA = r^2.$$

Again, the angles at M, N are right angles;

$\therefore$ a circle can be drawn through the points A, B, N, M,

$$\therefore \qquad CM \cdot CA = CN \cdot CB;$$

$$\therefore \qquad CN \cdot CB = r^2;$$

$\therefore$ AN is the polar of B;

$\therefore$ the polar of B passes through A.

3. The pole of a straight line with regard to a circle is the point of intersection of the polars of any two points on the line.

4. The polar of a point with regard to a circle is the straight line joining the poles of any two straight lines through the point.

5. The straight line joining a pair of inverse points (A, A') is divided harmonically at the points (B, B'), where it meets the circle.

6. If any diameter BB' of a circle is divided harmonically at AA', then A, A' are inverse points with regard to the circle.

7. If A, A' are inverse points with regard to a circle, any circle through A, A' cuts the given circle orthogonally.

8. Any circle which cuts a given circle orthogonally divides any diameter of the given circle harmonically.

9. Any circle which passes through a given point A and cuts a given circle orthogonally, passes through another fixed point A'.

[A' is the inverse of A for the given circle.]

10. Draw a circle "through two given points (A, B) to cut a given circle (C) orthogonally. [Use § 9 to find another point on the circle.]

11. Draw a circle through a given point to cut two given circles orthogonally. [Use § 9 to find two other points on the circle.]

12. (*i*) If ABC is an obtuse-angled triangle and AA', BB', CC' are the perpendiculars from A, B, C to the opposite sides, meeting at P, show that

$$PA \cdot PA' = PB \cdot PB' = PC \cdot PC'.$$

(*ii*) If $PA \cdot PA' = k^2$, show that each vertex of the triangle ABC is the pole of the opposite side with regard to the circle with centre P and radius k.

This circle is called the **Polar circle** *of the triangle, and the triangle is said to be* **Self-conjugate** *with regard to the circle.*

13. *If A, B are any two points, and if AX is the perpendicular from A to the polar of B with regard to a circle of which C is the centre, and BY is the perpendicular from B to the polar of A, then*

$$AX : BY = CA : CB \text{ (Salmon's Theorem)}$$

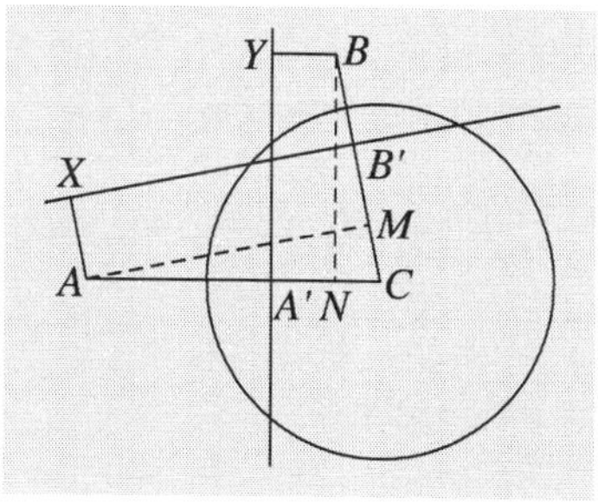

Construction. Let CA, CB cut the polars of A, B at A', B'. Draw AM, BN perpendicular to CB, CA.

Proof. Then $\quad AX \cdot CB = MB' \cdot CB$

$$= CB' \cdot CB - CM \cdot CB$$

$$= r^2 - CM \cdot CB.$$

Similarly, $\quad BY \cdot CA = r^2 - CN \cdot CA.$

But $ANMB$ is a cyclic quadrilateral;

$\therefore \qquad CN \cdot CA = CM \cdot CB;$

$\therefore \qquad AX \cdot CB = BY \cdot CA$

or $\qquad AX : BY = CA : CB.$

14. In § 13, prove that

$$2AX \cdot CB = 2BY \cdot CA$$
$$= \pm(CA^2 + CB^2 - AB^2 - 2r^2).$$

15. *If A, A' are inverse points with regard to a given circle, and P is any point on the circle, the ratio $PA : PA'$ is constant.*

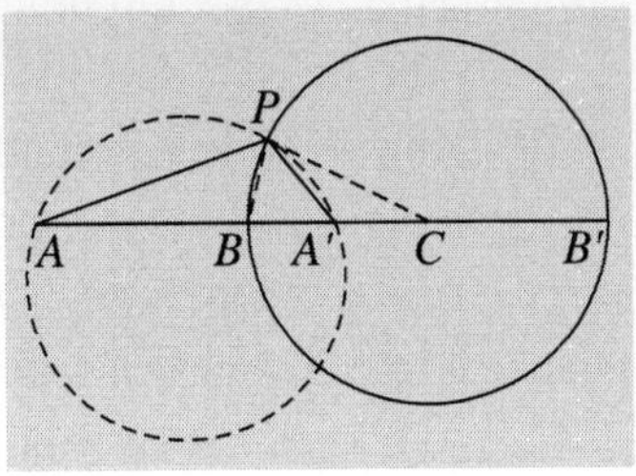

For if C is the.centre and AA' meets the circle at B, B',

$$CA \cdot CA' = CB^2 = CP^2;$$

$\therefore$ CP touches the circle APA';

$\therefore \qquad \angle CPA' = \angle CAP;$

$\therefore \qquad \angle BPA = \angle CBP - \angle CAP$

$\qquad = \angle CPB - \angle CPA'$

$\qquad = \angle A'PB;$

$\therefore \qquad PA : PA' = AB : BA' =$ constant.

16. *Any straight line through a given point which cuts a circle is divided harmonically by the circle, the point and the polar of the point.*

Let A be the point, $A'PQ$ the straight line, meeting a circle with centre C at P, Q. Let the polar of A meet CA, PQ at $A'R$.

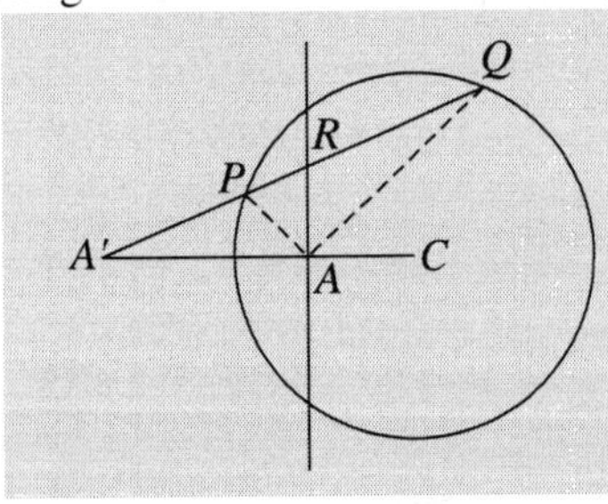

Then A, A' are inverse points, and since P, Q are points on the circle.

$\therefore$ By §15, $PA : PA' = QA : QA'$,

$\therefore \qquad PA' : QA' = PA : QA;$

$\therefore$ $A'A$, $A'R$ are the bisectors of $\angle PA'Q$;

$\therefore$ (PQ, AR) is a harmonic range.

17. Through a given point A, outside a circle, any straight line APQ is drawn to cut the circle at P, Q. In APQ, a point R is taken such that

$$\frac{1}{AR} = \frac{1}{AP} + \frac{1}{AQ}.$$

Find the locus of R.

18. Through a fixed point A, outside a circle with centre C, two straight lines are drawn, equally inclined to CA, and meeting the concave and convex arcs respectively at P, Q. Show that PQ passes through a fixed point.

19. Through a point A, any straight line is drawn to meet a given circle at P, Q. If A' is the inverse of A, show that PA'. QA' is constant.

20. AT, AT' are the tangents from a point A to a circle with centre C. Any straight line is drawn through A to cut the circle at P, Q. If CA meets TT' at A', prove that

(*i*) The triangles PTQ, $PA'T'$, $T'A'Q$ are similar.

(*ii*) $PT : TQ = PT' : T'Q$.

(*iii*) $PT \cdot QT' = QT \cdot PT' = \frac{1}{2} PQ \cdot TT'$.

LV. THE 'POINT O' THEOREM

THEOREM 114

If ABC is a given triangle and X, Y, Z are points in the sides BC, CA, AB, or in the sides produced, such that the triangle XYZ is directly similar to a given triangle, the circles AYZ, BZX, CXY meet in a fixed point O.

Let the circles AYZ, BZX meet at O, and consider the case in which O is within the triangle ABC and X, Y, Z are in BC, CA, AB respectively.

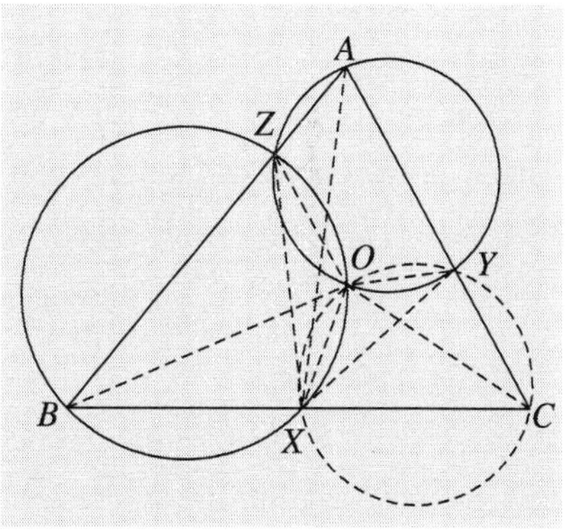

Construction. Join OX, OY, OZ.

Proof. (*i*) It is required to prove that the circle CXY passes through O.

Because $AYOZ$ is a cyclic quadrilateral,

$\therefore \quad \angle OYC = \angle OZA$;

and, because $BZOX$ is a cyclic quadrilateral,

$\therefore \quad \angle OZA = \angle OXB$;

$\therefore \quad \angle OXB = \angle OYC$;

$\therefore$ The circle CXY passes through O.

(*ii*) It is required to prove that O is a fixed point.

Construction. Join AX, OB, OC, XZ, XY.

Proof. Then $\quad \angle BOC = \angle BOX + \angle COX$.

Now $\quad \angle BOX = \angle BZX = \angle ZAX + \angle ZXA$,

and $\quad \angle COX = \angle CYX = \angle YAX + \angle YXA$;

$\therefore \quad \angle BOC = A + X$.

Similarly, $\quad \angle COA = B + Y$ and $\angle AOB = C + Z$.

Hence O is the point of intersection of arcs of segments drawn on BC, CA, AB, containing angles equal to $A + X$, $B + Y$, $C + Z$ respectively, and is therefore a fixed point.

Any triangle which is *directly similar* to a given triangle is said to be of given species.

1. The proof given in Theorem 114, with slight alterations, applies to any other particular case. The only changes which have to be made are that occasionally a sign has to be changed and an angle replaced by its supplement.

The values of the angles *BOC*, *COA*, *AOB* for the different particular cases are given below in §§ 8-11.

2. Take the case in which *O is within the triangle* and *X*, *Y*, *Z* are in *BC*, *CA*, *AB*, *produced.* Show that, as before,

$$\angle BOC = A + X, \quad \angle COA = B + Y, \quad \angle AOB = C + Z.$$

3. In the figure of Theorem 114, prove that the triangles *OYZ*, *OZX*, *OXY* are of fixed species, and that *O* is invariably connected with the triangle *XYZ*.

4. The locus of any point (except *O*) which is invariably connected with the triangle *XYZ*, for example, its centroid *G*, is a straight line.

[For the triangle *GOX* is of fixed species and *X* lies on *BC*.]

5. If *OL*, *OM*, *ON* are perpendicular to *BC*, *CA*, *AB*, prove that

$$\angle XOL = \angle YOM = \angle ZON,$$

and that the triangles *XYZ*, *LMN* are similar.

6. If *k* is a chord of the circumcircle of the triangle *ABC* which subtends at the circumference an angle equal to *OXB*, prove that

$$\frac{YZ}{BC} = \frac{OA}{k}, \quad \frac{ZX}{CA} = \frac{OB}{k}, \quad \frac{XY}{AB} = \frac{OC}{k}.$$

Hence show that

$$YZ : ZX : XY = OA \cdot BC : OB \cdot CA : OC \cdot AB.$$

7. Suppose that, in the figure of Theorem 114, the point *Y* approaches *A* and tends to *A* as a limiting position. In the limit, the circle *AYOZ* touches *AC* at *A*.

Hence we infer the following construction:

For a given triangle ABC and a triangle XYZ of given species, to find the point O, draw a triangle XAZ of the given species, with X in BC and Z in AB, or in these lines produced. Draw the circle through A, Z to touch AC. Draw the circle ZBX. These circles meet at O.

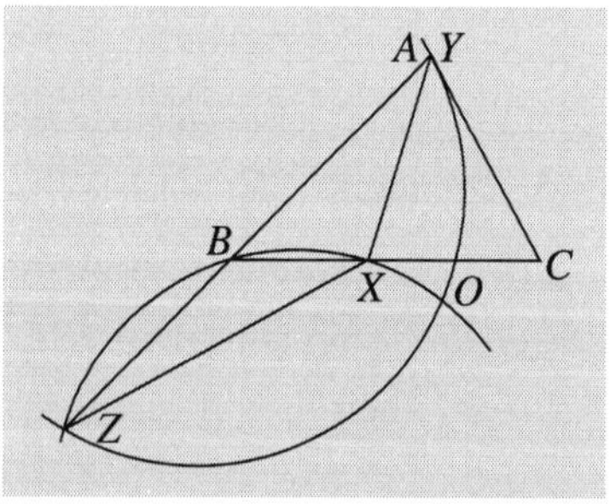

8. If, as in the figure of § 7, *O* is outside the triangle *ABC* and within the angle *A*, and if, in moving round the triangles in the same sense, the vertices *A*, *B*, *C* ; *X*, *Y*, *Z* occur in this order, prove that

$$\angle BOC = 360 - A - X, \ \angle COA = B + Y, \ \angle AOB = C + Z.$$

[This follows easily from the figure of § 7, remembering that *Y* coincides with *A*.]

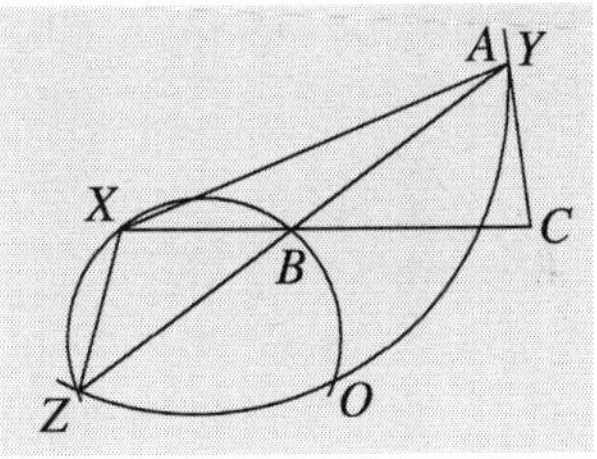

9. If, as in the figure, O is outside the triangle ABC, and if, in moving round the triangles ABC, XYZ, in the same sense, the vertices occur in the order A, B, C ; X, Z, Y, prove that

$$\angle BOC = X - A,$$
$$\angle COA = B - Y,$$
$$\angle AOB = C - Z;$$

[For $COXY$, *i.e.* $COXA$, is cyclic ; $\angle COA = \angle CXA = B - Y$.
Also $\angle AOB = \angle AOX - \angle BOX = \angle ACX - \angle BZX = C - Z$.]

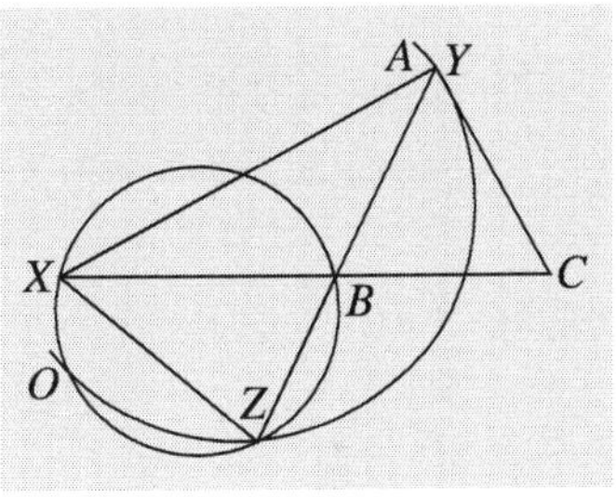

10. If, as in the figure, O is within the angle formed by producing AB, CB through B, it can be shown that in moving round the triangles ABC, XYZ, in the same sense, the vertices occur in the order A, B, C ; X, Y, Z. In this case, prove that

$$\angle COA = B - Y,$$
$$\angle AOB = Z - C,$$
$$\angle BOC = X - A.$$

11. Summary. All the distinct cases of the Point O Theorem have now been considered. In general, two sets of triangles, equiangular to a given triangle XYZ, can be drawn with their vertices on the sides, or on the sides produced, of a given triangle ABC.

For one set, the sides BC, CA, AB subtend angles equal to $A + X$, $B + Y$, $C + Z$ at O, unless one of these angles, as $A + X$, is greater than 180°. In this case the angles subtended by the sides are $360 - A - X$, $B + Y$, $C + Z$.

For the other set, BC, CA, AB subtend at O angles equal to

$$\pm(A - X),\ \pm(B - Y),\ \pm(C - Z),$$

where a set is taken, with two upper signs and one lower, or one upper sign and two lower, that makes the angles all positive.

12. In a given triangle ABC inscribe a triangle XYZ similar to a given triangle, with its vertex X at a given point in BC.

13. Draw a triangle XYZ congruent with a given triangle DEF and with its vertices X, Y, Z on the sides BC, CA, AB of a given triangle ABC, or on the sides produced. Show that, in general, there are two solutions. [Use § 3 to find OX.]

14. In a given triangle ABC, inscribe the least triangle similar to a given triangle.

15. The sides BC, CA, AB of a variable triangle ABC, or the sides produced, pass through fixed points X, Y, Z. If ABC is always directly similar to a given triangle, prove that

(*i*) The circles AYZ, BZX, CXY are fixed and have a common point O.

(*ii*) The triangles *BOC*, *COA*, *AOB* are of fixed species.

(*iii*) The locus of any point invariably connected with the triangle *ABC*, for example, its orthocentre, is a circle.

16. About a given triangle *XYZ*, describe the greatest triangle *ABC* similar to a given triangle *DEF*. [Find *O*, then *BC*, *CA*, *AB* are perpendicular to *OX*, *OY*, *OZ*.]

17. *In the figure of Theorem* 114, *consider various cases in which XYZ is similar to ABC.*

(*i*) Let $X = A$, $Y = B$, $Z = C$. Show that *O* is the circumcentre of *ABC* and the orthocentre of *XYZ*.

(*ii*) Let $X = B, Y = C, Z = A$. Show that *BOC*, *COA*, *AOB* are the supplements of *C*, *A*, *B* respectively, and that

$$\angle OBC = \angle OCA = \angle OAB.$$

(*iii*) Let $X = C, Y = A, Z = B$. Show that *BOC*, *COA*, *AOB* are the supplements of *B*, *C*, *A*, respectively and that

$$\angle OCB = \angle OAC = \angle OBA.$$

18. If a straight line cuts the sides *BC*, *CA*, *AB*, or the sides produced, at *X*, *Y*, *Z*, show that the circles *AYZ*, *BZX*, *CXY* meet at a point *O* on the circumcircle of the triangle *ABC*.

This is the limiting case of the 'Point O' Theorem, in which one angle of the triangle XYZ tends to 180°.

19. If, in § 18, the position of the straight line *XYZ* varies in such a way that the ratio *XY* : *YZ* is constant, show that the triangles *OYZ*, *OZX*, *OXY* are of fixed species and that *O* is a fixed *O*.

[For $\angle XOY = C$, $\angle YOZ = A$ and *XY* : *YZ* is constant. See Ex. LXV, 14.]

20. Through a given point *X* in the side *BC* of the triangle *ABC*, draw a straight line to cut *CA* at *Y* and *AB* at *Z*, such that

$$XY : YZ = p : q,$$

where *p*, *q* are given numbers.

[Take points *X'*, *Y'*, *Z'* in a straight line, such that $X'Y' : Y'Z' = p : q$. On *X'Y'*, *Y'Z'* draw segments of circles containing angles equal to *C*, *A* respectively. Let the arcs meet at *O'*. The point *O* of § 18 can now be found by making $\angle CAO = \angle X'Z'O'$.]

21. Draw a straight line to cut the sides *AB*, *BC*, *CD*, *DA* of a quadrilateral *ABCD* in *X*, *Y*, *Z*, *W*, such that

$$XY : YZ : ZW = p : q : r,$$

where *p*, *q*, *r* are given numbers.

[Produce *AB*, *CD* to meet at *E*, and *BC*, *AD* to meet at *F*. Find *O* for the triangle *CEB* and the transversal *XYZ*. Find *O'* for the triangle *CDF* and the transversal YZW. Draw the circle $OO'C$, meeting BC at Y and CD at Z. Then YZ is the required straight line. Draw the figure and explain more fully.]

22. Draw a quadrilateral $XYZW$ similar to a given quadrilateral, with its vertices X, Y, Z, W on the sides AB, BC, CD, DA of a given quadrilateral $ABCD$, or on these sides produced.

[Produce AB, CD to meet at E, and BC, AD to meet at F. Find the point O for the triangle CEB and triangles of the same species as XYZ. Find O' for the triangle DFC and triangles of the same species as YZW. Draw the circle $OO'C$, cutting BC at Y and CD at Z. Then YZ is a side of the required quadrilateral. Draw the figure and explain more fully.]

23. Inscribe a square in a given quadrilateral.

LVI. MAXIMA AND MINIMA

Some of the problems in this Section have occurred previously and are repeated for convenience.

1. Through a given point O within a given circle, draw the least possible chord.

2. If one circle is within another, draw the greatest and least chords of the outer which touch the inner.

3. Given the base and the vertical angle of a triangle, prove that the area of the triangle is greatest when it is isosceles.

4. A, B are given points, and P is any point on the circumference of a given circle. Find the position of P in order that the area of the triangle PAB may have (*i*) its greatest, (*ii*) its least possible value.

5. *A, B are points on the same side of a straight line CD of indefinite length and P is any point in CD. Show that $PA+PB$ has its least value when PA, PB are equally inclined to CD.* [*Ex*, 2, *p*. 84.]

6. Given the base and the area of a triangle, show that its perimeter is least when it is isosceles.

[If PAB is a triangle of given area on the base AB, the locus of P is a straight line parallel to AB, and the result follows from § 5.]

7. If two adjacent sides AB, BC of a polygon $ABCDE$ are unequal, then it is possible to construct a polygon with the same number of sides and of equal area but with a less perimeter.

[Keep the points A, C, D, E fixed. Vary the position of B so that the area of ΔABC remains constant and apply § 6.]

8. Given the area of a triangle, prove that its perimeter is least when it is equilateral.

[For if two of its sides are unequal, we can make a triangle of equal area and less perimeter.]

9. Prove that the triangle of least perimeter which can be inscribed in a given triangle has its pairs of sides equally inclined to the sides of the given triangle. Hence show that the inscribed triangle of least perimeter is the pedal triangle of the original.

10. P is any point on a given circle, and A, B are given points outside the circle. If $PA + PB$ has its least value, show that the tangent at P is equally inclined to PA and PB. [Use § 5]

11. Find the point P within a triangle ABC at which $PA + PB + PC$ has its least possible value.

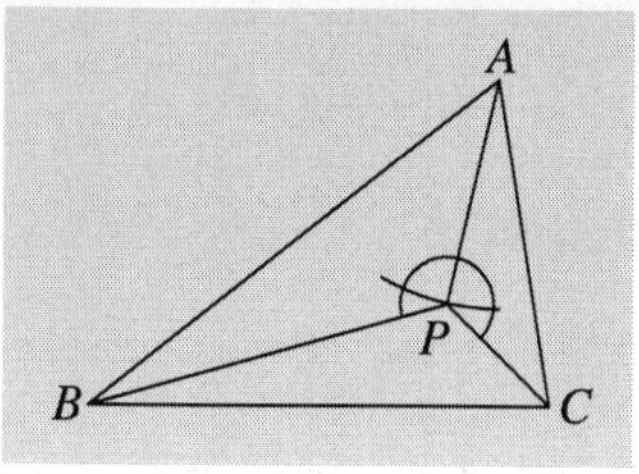

Suppose that P has the required position. Let Q be any point on the circle, with centre A and radius AP.

By hypothesis,

$$PA + PB + PC < QA + QB + QC;$$

$$\therefore \quad PB + PC < QB + QC.$$

Hence, by § 10, the tangent at P is equally inclined to PB, PC;

$$\therefore \quad \angle APB = \angle APC.$$

Similarly, it can be shown that

$$\angle APB = \angle BPC;$$

$$\therefore \quad \angle BPC = \angle CPA = \angle APB = 120°.$$

12. Find a point P in a quadrilateral $ABCD$, such that

$$PA + PB + PC + PD$$

has its least value.

13. *Given two sides of a triangle, prove that the area is greatest when these sides contain a right angle.*

14. Construct the parallelogram of greatest possible area, having given the lengths of the diagonals.

15. Given the diagonals of a quadrilateral, show that its area is greatest when they are at right angles.

16. Construct the quadrilateral of greatest possible area, having given two adjacent sides and the diagonals.

17. Construct the quadrilateral of greatest possible area, having given a pair of opposite sides and the diagonals.

18. *Given a number of straight rods AB, BC, CD, DE hinged at B, C and D, prove that the area of the figure ABCDE is greatest when B, C, D lie on a semicircle whose diameter is AE.*

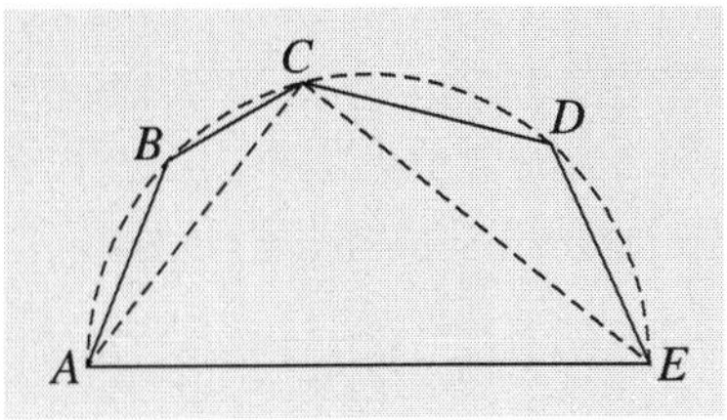

Suppose the rods to be so placed that the area *ABCDE* is greatest.

Then, however the figure is displaced by turning the rods about the hinges, the area is diminished.

Turn the figure composed of the rods *AB*, *BC* about *C*, keeping $\angle B$ constant and *CD*, *DE* fixed.

The only part of *ABCDE* whose area is altered is ΔACE;

$\therefore$ this triangle has its greatest value when *ABCDE* is greatest; and since the lengths *AC*, *CE* remain constant during the displacement,

$\therefore$ when *ABCDE* is greatest, $\angle ACE$ is a right angle. Similarly, $\angle$s *ABE*, *ADE* are right angles;

$\therefore$ the circle on *AE* as diameter passes through *B*, *C*, *D*.

This theorem may be stated as follows: *If all the sides except one of a polygon are of given length, the area is greatest when all the vertices lie on a semicircle whose diameter is the side which is not given.*

19. Take a piece of thread *AB* and throw it on the table so as to assume any shape. Prove that the area of the figure bounded by the thread and the straight line *AB*, joining its extremities *A*, *B*, is greatest when the thread lies along the arc of a semi-circle.

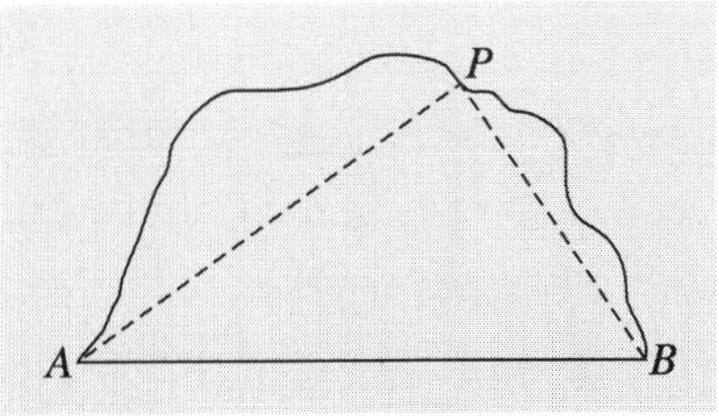

Suppose the thread to have taken such a position that the area is greatest. Let *P* be any point in the thread. Turn the figure bounded by *PA* and the part *PA* of the thread round *P*, its shape remaining unaltered. Then, by hypothesis, the area of the whole figure is decreased. But the only part of which the area varies is the triangle *APB*. Therefore the area of *APB* must be decreased by turning *PA* either way. Hence $\angle APB$ is a right angle, etc.

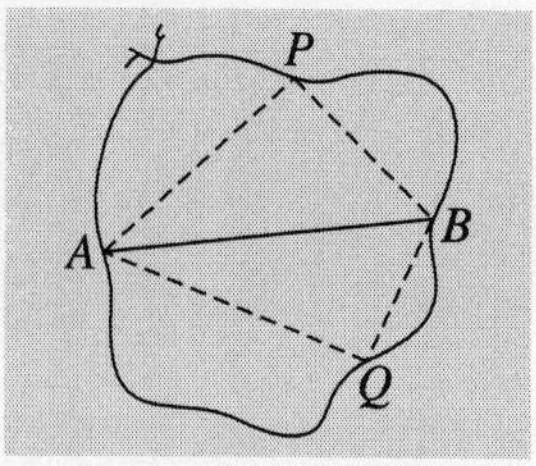

20. Take a piece of thread, knotted into a loop, and throw it on the table so as to assume any shape. Prove that the area bounded by the thread is greatest when the thread forms a circle.

[Let APB be half of the thread and AQB the remaining half. Apply (19) to the figures APB, AQB.]

21. *Of all polygons which can be constructed with sides each of given length, that which is cyclic has the greatest area.*

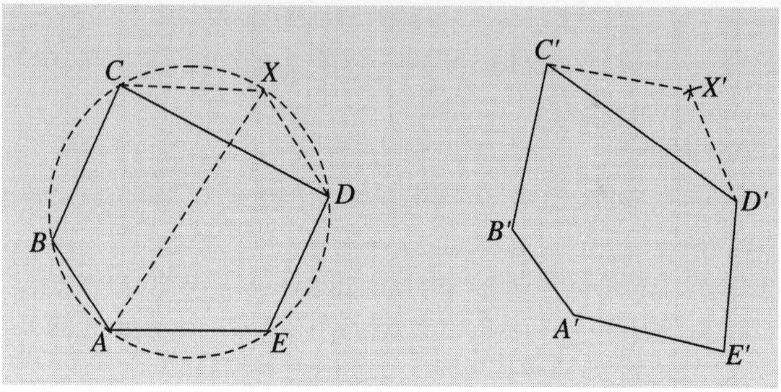

Let $ABCDE$, $A'B'C'D'E'$ be two polygons whose sides AB, $A'B'$; BC, $B'C'$, etc., are the given lengths, and let $ABCDE$ be cyclic and $A'B'C'D'E'$ not so; then the area of $ABCDE >$ that of $A'B'C'D'E'$.

Proof. Let AX be the diameter of the circumcircle of $ABCDE$ through A, and let C, D be the vertices nearest to X and on opposite sides of AX. Make $\Delta C'X'D'$ with $C'X' = CX$ and $D'X' = DX$.

Then, of the polygons $A'B'C'X'$ and $A'E'D'X'$, one at least is not inscribable in the semicircle on AX' as diameter. Let $A'B'C'X'$ be the one not so inscribable.

Since $AB = A'B'$, $BC = B'C'$, $CX = C'X'$ and $ABCX$ is a semicircle,

$\therefore$ area $ABCX >$ area $A'B'C'X'$.

Similarly, area $AEDX$ is not $<$ area $A'E'D'X'$;

$\therefore$ area $ABCXDE >$ area $A'B'C'X'D'E'$.

Now Δs CXD, $C'X'D'$ are congruent and equal in area;

$\therefore$ area $ABCDE >$ area $A'B'C'D'E'$.

22. Prove that the pentagon of greatest area, with a given perimeter p, is the regular pentagon of side $\frac{1}{5}p$. [Use §§ 7, 21.]

23. From the identities

$$(x + y)^2 - (x - y)^2 = 4xy,$$
$$(x + y)^2 + (x - y)^2 = 2(x^2 + y^2),$$

we draw the following conclusions: *Let x, y be two positive numbers, then*

(*i*) *If* $x + y$ *is constant,* xy *is greatest and* $x^2 + y^2$ *is least when* $x = y$.

(*ii*) *If* xy *is constant,* $x + y$ *is least when* $x = y$.

(*iii*) *If* $x^2 + y^2$ *is constant,* $x + y$ *is greatest when* $x = y$.

Apply these considerations in §§ 24–28: give geometrical proofs also.

24. If X is any point in a given straight line AB, then $AX \cdot XB$ is greatest when $AX = XB$.

25. Given the perimeter of a rectangle, the area is greatest when it is a square.

26. Given the area of a rectangle, the perimeter is least when it is a square.

27. The least chord of a circle which can be drawn through a given point in the circle is bisected at that point.

28. Given the base BC and the sum of the sides AB, AC of the triangle ABC, if X is the middle point of BC, then AX is least when $AB = AC$.

29. A, B are given points and CD is a given straight line. Find the point P in CD for which $PA^2 + PB^2$ has its least value.

30. A, B are given points and CD is a given straight line. Find the point P in CD for which $2PA^2 + 3PB^2$ has its least value.

31. If ABC is any triangle, find the position of P in order that $PA^2 + PB^2 + PC^2$ may have its least value.

32. A, B, C, D are four fixed points and P is a variable point. Find the position of P for which $PA^2 + PB^2 + PC^2 + PD^2$ has its least value.

33. If A, B, C are fixed points, P a variable point, and p, q, r any given positive numbers, find the position of P in order that $pPA^2 + qPB^2 + rPC^2$ may have its least value.

34. Inscribe in a given semicircle the rectangle of greatest possible area.

[If $2x$ is the length of the side along the diameter, y that of an adjacent side and r the radius, then $x^2 + y^2 = r^2$;

$\therefore\ 2xy = r^2 - (x - y)^2$, etc.]

35. ABC is a triangle and P is any point in AB. A parallelogram $PMCN$ is drawn with its vertices M, N in CA, CB respectively. Prove that the area of $PMCN$ is greatest when P is the middle point of AB. (*Euclid VI, 27*)

[Let $PM = x$, $PN = y$. Prove that $\frac{x}{a} + \frac{y}{b} = 1$, and observe that

$$\frac{4xy}{ab} = \left(\frac{x}{a} + \frac{y}{b}\right)^2 - \left(\frac{x}{a} - \frac{y}{b}\right)^2.\ \Big]$$

36. Through a point on the diagonal of a parallelogram parallels are drawn to the sides. Find the position of the point in order that the areas of the complements may be greatest.

37. In a given triangle ABC inscribe the greatest possible rectangle, with one of its sides along BC.

38. In a given triangle ABC inscribe the greatest possible parallelogram, with one vertex at a given point in BC.

39. If A is a point of intersection of two circles and a straight line PAQ is drawn through A to cut the circles again at P, Q, show that PQ is greatest when it is parallel to the straight line joining the centres.

40. Prove the identity

$$(ny - mz)^2 + (lz - nx)^2 + (mx - ly)^2$$
$$= (l^2 + m^2 + n^2)(x^2 + y^2 + z^2) - (lx + my + nz)^2.$$

Hence show that if l, m, n are given and x, y, z vary subject to the condition that $lx + my + nz$ is constant, all the numbers being positive, then $x^2 + y^2 + z^2$ has its least value when

$$\frac{x}{l} = \frac{y}{m} = \frac{z}{n}.$$

41. If P is any point within the triangle ABC and PL, PM, PN are the perpendiculars from P to BC, CA, AB, show that $PL^2 + PM^2 + PN^2$ has its least value when

$$PL : PM : PN = BC : CA : AB,$$

and give a construction for the point P.

[Let x, y, z be the lengths of PL, PM, PN. Then

$$ax + by + cz = 2\Delta,$$

and the result follows from (40) by putting $l = a$, $m = b$, $n = c$.]

42. As an example of a varying geometrical magnitude, consider the perpendicular NP drawn from any point P on the curve AB to a fixed straight line OX.

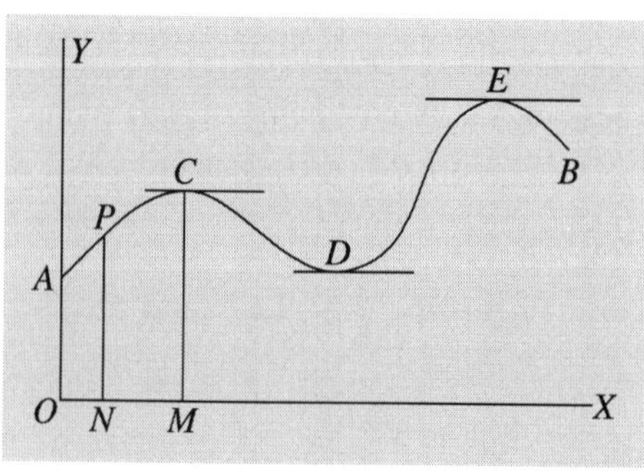

Let C, D, E be points on the curve at which the tangents are parallel to OX, and suppose that the point P moves from A to B along the curve.

As P passes through C, NP ceases to increase and begins to decrease; NP is therefore said to have a **maximum** *value at C.*

Again, *as P passes through D, NP ceases to decrease and begins to increase*; *NP is therefore said to have a* **minimum** *value at D*.

If *CM* is perpendicular to *OX*, *MC* is a maximum value of *NP*, *but this does not mean that MC is the greatest possible value of NP*; it simply means that *MC* is the greatest value of *NP* for positions of *P* near to *C*, and on either side of it.

The greatest possible value of *NP* will be either one of the maximum values of *NP*, *or a value of NP at an end-point A or B*.

In the same way, the least value of *NP* will be either a minimum value, or a value at an end-point *A* or *B*.

It is obvious that *maxima and minima values succeed one another alternately*.

43. The following considerations are of great use in questions on maxima and minima. In the figure of 41, as *P* passes through the point *C*, where *NP* is a maximum, *NP* ceases to decrease and begins to increase. Hence *there are points on the curve near to C, and on opposite sides of it, for which the values of NP are equal*.

In the same way, considering the point *D*, at which *NP* is a minimum, *there are points on the curve near to D, and on opposite sides of it, for which the values of NP are equal*.

44. As an illustration, consider the following problem: *A*, *B are points on the same side of a straight line CD, of indefinite length*.

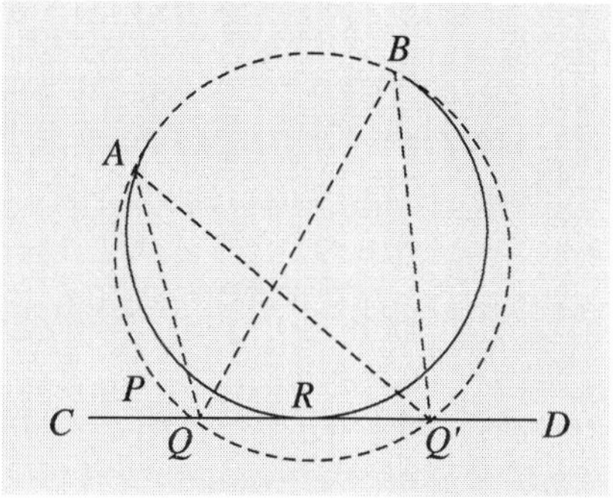

It is required to find a point P in CD at which the angle APB has a maximum value.

Draw a circle through *A*, *B* to cut *CD* in near points *Q*, *Q'*.

Since $\angle AQB = \angle AQ'B$, we infer that for some point *R* between *Q*, *Q'*, $\angle APB$ has either a maximum or a minimum value.

Now suppose the circle to vary, so that *Q*, *Q'* tend to coincidence. In the limit, the circle touches *CD*.

Hence we infer that if a circle is drawn through *A*, *B* to touch *CD* at *R*, then $\angle ARB$ is a maximum or a minimum value of $\angle APB$.

Having discovered this, we can show by simple geometrical considerations that $\angle APB$ is a maximum at *R*. This is left as an exercise.

45. A, B are points on the same side of a straight line CD of indefinite length. Suppose a point P to move along CD, starting from a remote position on the left and moving to a remote position on the right. Trace the variation of $\angle APB$, and show that, if AB is not parallel to CD, there are two positions of P for which $\angle APB$ is a maximum and one for which it is a minimum. Find these points.

Apply the method of § 43 to §§ 46-48.

46. Through a given point draw a straight line which with two given straight lines contains a minimum area.

47. Through a given point A draw a straight line cutting two given straight lines OX, OY at P, Q, such that $PA \cdot AQ$ may be a minimum.

48. A, B are points either both outside or both inside a given circle, and P is any point on the circle. Find positions of P for which $\angle APB$ is a maximum. Consider the two cases separately and distinguish between maximum and minimum values.

49. O is a given point outside a circle with centre C. Through O draw a straight line to cut the circle at P, Q, such that the triangle CPQ may be (*i*) of given area, (*ii*) of maximum area.

50.* Apply the method of § 43 to the problems in §§ 5, 34, 35 of this section.

LVII. DISPLACEMENT OF A FIGURE IN ITS OWN PLANE; ENVELOPES

1. To say that a figure is *displaced in its own plane* is to say that it is moved from one position to another, in the same plane, without altering its size or shape.

The figure is to remain in the same plane throughout the displacement: it cannot therefore be *turned over*.

2. *In general, a figure can be moved from any one position to any other, in the same plane, by rotation about a certain point.*

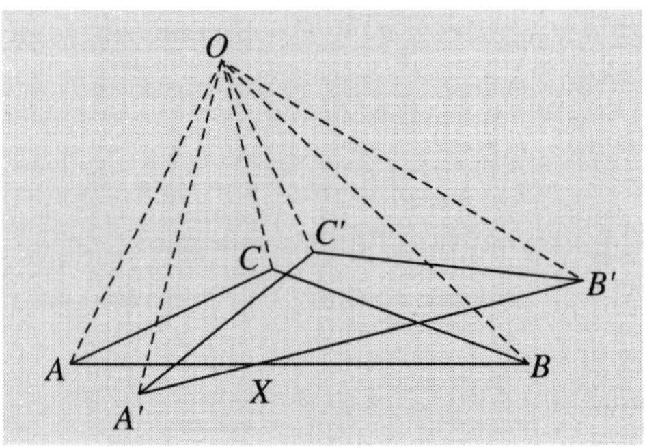

* *This question involves the ideas of the Calculus, with which the student probably has some acquaintance.*

Choose any two points A and B of the figure, and let C be any third point of it. Let A', B', C' be the new positions of A, B, C when the figure has been displaced.

(*i*) Suppose that AA' is not parallel to BB'. Draw the perpendicular bisectors of AA', BB', meeting at O.

It will be proved that the figure can be moved from the first to the second position by rotation about O.

In the triangles OAB, $OA'B'$,

$$OA = OA', OB = OB', AB = A'B';$$

$\therefore$ the triangles are congruent;

$$\therefore \quad \angle AOB = \angle A'OB';$$

$$\therefore \quad \angle AOA' = \angle BOB'.$$

Again, because the triangles OAB, $OA'B'$ are congruent,

$$\therefore \quad \angle OAB = \angle OA'B';$$

and because the triangles ABC, $A'B'C$ are congruent,

$$\therefore \quad \angle CAB = \angle C'A'B';$$

$$\therefore \quad \angle OAC = \angle OA'C'.$$

Hence, in the triangles OAC, $OA'C'$,

$$OA = OA', AC = A'C', \angle OAC = \angle OA'C';$$

$\therefore$ the triangles are congruent;

$$\therefore \quad OC = OC'$$

and

$$\angle AOC = \angle A'OC';$$

$$\therefore \quad \angle AOA' = \angle COC'.$$

Hence the point O is such that $OA = OA'$, $OB = OB'$, $OC = OC'$, and the angles AOA', BOB', COC' are equal, and they are measured in the same sense;

$\therefore$ the figure can be moved from the first position to the second by rotation about O.

(*ii*) Let AA' be parallel to BB', and suppose that AB is not parallel to $A'B'$. Let AB, $A'B'$ meet at O.

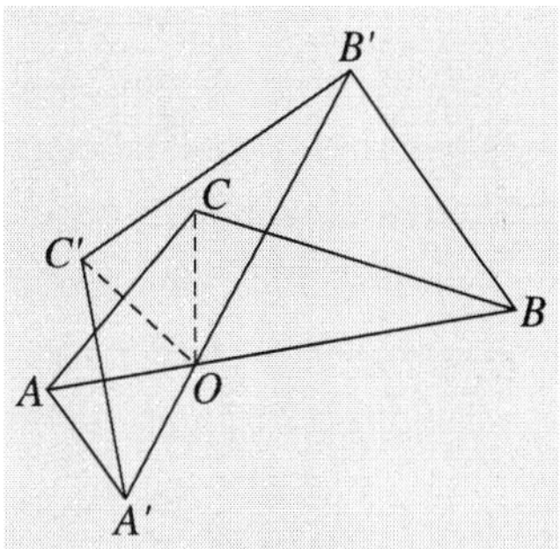

Since $AB = A'B'$ and AA' is parallel to BB',

$\therefore$ AB, $A'B'$ are equally inclined to AA' and BB';

$$\therefore \quad \angle OAA' = \angle OA'A$$

and

$$\angle OBB' = OB'B;$$

$$\therefore \quad OA = OA'$$

and

$$OB = OB'.$$

Also

$$\angle AOA' = \angle BOB',$$

and, as before, it can be shown that the figure can be moved from the first to the second position by rotation about O.

(*iii*) If AA' is parallel to BB' and AB is parallel to $A'B'$, produce AA' to D.

Then $\angle DAB = \angle DA'B'$

and $\angle CAB = \angle C'A'B'$;

$\therefore$ $\angle DAC = \angle DA'C'$;

$\therefore$ AC is parallel to $A'C'$.

Also $AC = A'C'$;

$\therefore$ CC' is equal and parallel to AA'.

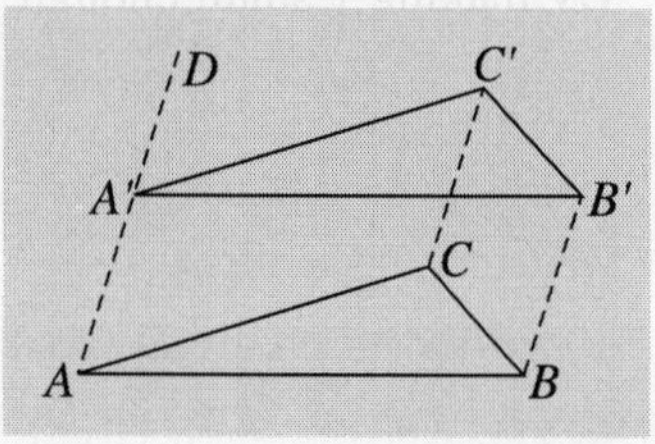

In the same way, it can be shown that the displacement of any other point in the figure is equal and parallel to that of A.

3. In §2 (*i*) prove the following construction for the point O: Let AB, $A'B'$ (produced if necessary) meet at X. *Draw the circles $AA'X$, $BB'X$; then O is the second point of intersection of these circles.*

4. If a figure is rotated about a point O through an angle α, and $A'B'$ is the new position of any straight line AB of the figure, prove that the angle between AB and $A'B'$ is equal to α.

5. If a figure moves in its own plane in such a way that its size and shape remain unaltered and two points A, B in it describe parallel straight lines in the same sense, prove that any third point C of the figure describes a parallel straight line.

Motion of this kind is called **motion of translation**.

In a translation, *the new position of any straight line of the figure is parallel to its original position.*

6. Suppose that a point is moving along a curve. Let A be its position at a certain instant, and let A' be its position after a small interval of time τ.

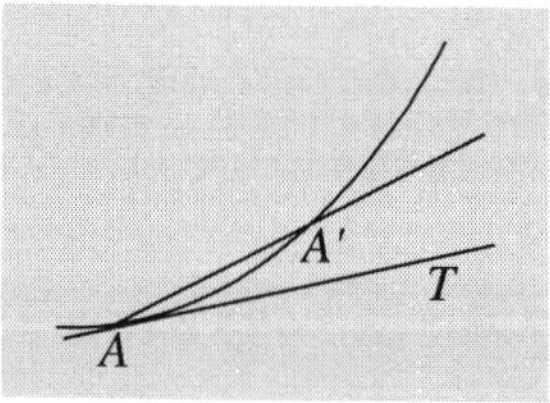

By making τ small enough, we can make AA' as small as we like, and, as τ tends to zero, the straight line through A, A' tends to a limiting position AT, which is the tangent to the curve at A.

We therefore say that *the direction of motion at A is the direction of the tangent at that point.*

7. Suppose a figure to be moving in any way, in its own plane, the motion not being one of translation.

Referring to the figure in § 2(*i*), let A, B be the positions of two points of the figure at a certain instant, and let A', B' be the positions of A, B after a small interval of time τ.

By making τ small enough, we can make AA', BB' as small as we like.

When AA', BB' are very small, their perpendicular bisectors become indistinguishable from the perpendiculars through A, B to AA', BB'.

Hence we conclude that, *as τ tends to zero, the point O tends to a limiting position, which is the point of intersection of the perpendiculars through A, B to the directions of motion of these points.*

This point is called the centre of **instantaneous rotation** of the figure.

8. The construction for the instantaneous centre O of the figure, at any instant, is as follows:

Take any two points A and B in the figure of which the directions of motion are known.

Draw perpendiculars to these directions through A and B respectively: these perpendiculars meet in the instantaneous centre O.

The motion of the figure at the instant under consideration is one of rotation about O, and the direction of motion of any point P in the figure is perpendicular to OP.

9. *A rod of given length moves so that its extremities P and Q slide along two fixed straight lines OA, OB. Explain how to find the direction of motion of any point R in the rod, at a particular instant.*

Draw PI, QI perpendicular to OA, OB, meeting at I. Then I is the instantaneous centre, and the direction of motion of R is perpendicular to IR.

Envelopes

10. If a straight line moves in such a way that it always touches a given curve, the curve is called the **envelope** of the line.

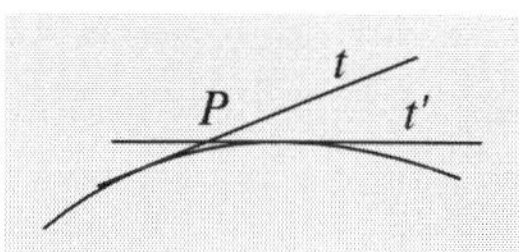

Let t be any position of a line which moves according to some fixed law, and let t' be another position of the line nearly coinciding with t. Let P be the point of intersection of t and t'.

If t' tends to coincidence with t, P tends to a limiting position which is the point of contact of t with the curve which is the envelope of the moving line.

Thus a curve may be regarded as

(*i*) the locus of a point which moves so as to satisfy a given geometrical condition ; or

(*ii*) the envelope of a line which moves so as to satisfy a given geometrical condition.

11. As an example on the plotting of the envelope of a moving line take the following:

"P is any point on a given straight line AB, and O is a fixed point. A perpendicular PQ is drawn to PO. Plot the envelope of the moving line PQ."

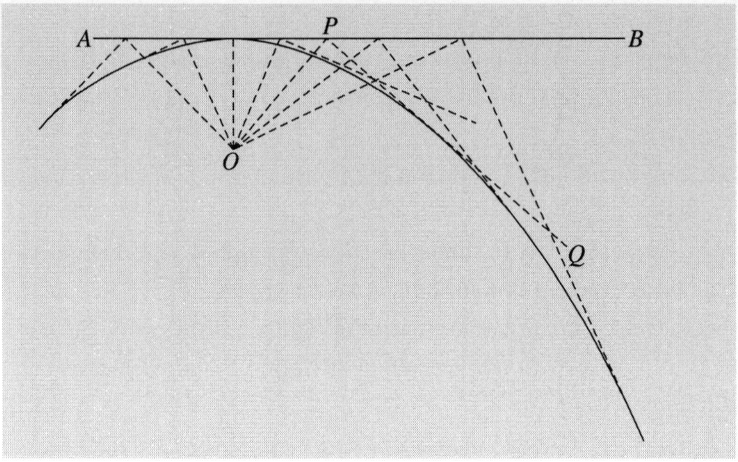

A number of different positions of PQ are constructed, and the envelope is then drawn freehand.

***Use the practical method of* § 11 *for* §§ 12–15**

12. Draw a circle of radius 1.4 in. Plot the envelope of a chord of length 1.6 in.

Prove that the envelope is a circle and find its radius.

13. Draw a circle, centre C and radius 2 in. Take a point O distant 1.2 in. from C. Take any point P on the circle, and draw the envelope of the perpendicular PQ to PO. *This curve is an* ellipse.

If the envelope cuts the perpendicular through O to CO in L, measure OL.

14. Draw two straight lines OA, OB, making an angle of 60°. Plot the envelope of a line which with OA, OB forms a triangle of area 1.5 sq. in. *This curve is called a* **hyperbola**.

Measure the distance from O of the point where the bisector of $\angle AOB$ meets the curve.

15. An equilateral triangle of 2 in. side is drawn, two of whose sides pass each through a fixed point, these points being 1.3 in. apart. Plot the envelope of the third side, which will be found to be a circle, and measure the radius.

16. In order to investigate **theoretically** the envelope of a moving line, one of two methods may be followed:

(*i*) Plot the envelope. Try to recognize the nature of the curve and to discover a geometrical construction for it, and *prove* that the moving line (in any position) touches the curve.

(*ii*) Let t, t' be near positions of the moving line, meeting at P. Find a geometrical construction for the limiting position of P as t' tends to coincide with t. Find the locus of P. This is the required envelope.

17. As an example of § 16(*ii*) take the following: *A straight line of constant length moves with its extremities A and B on a given curve. Find the point of contact of AB with its envelope.*

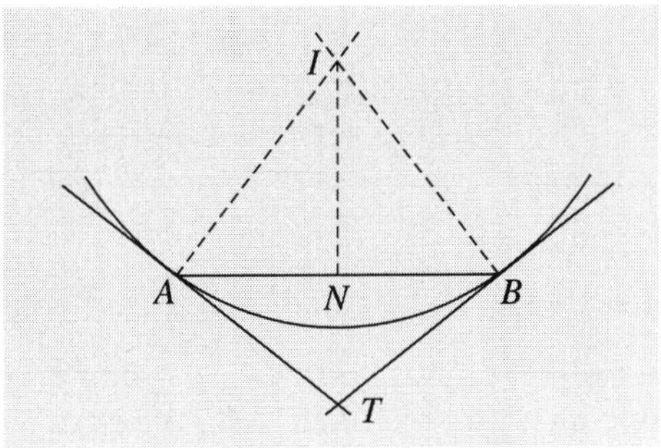

Draw the tangents AT, BT. Draw AI, BI perpendicular to AT, BT, meeting at I. Draw IN perpendicular to AB, meeting it at N.

Then N is the required point.

For I is the instantaneous centre and, referring to the figure of § 2(*i*), I, N are the limiting positions of O, X.

Now $AA'XO$ is a cyclic quadrilateral;

$$\therefore \qquad \angle OXB' = \angle OAA'.$$

But $\angle OAA'$ tends to 90°;

$\therefore \quad \angle OXB'$ tends to 90°.

Hence, in the limit, IN is perpendicular to AB.

18. *A triangle ABC of given size and shape moves so that two of its sides AB, AC pass through fixed points H, K. Prove that the envelope of the third side BC is a circle.* (Bobillier's Theorem)

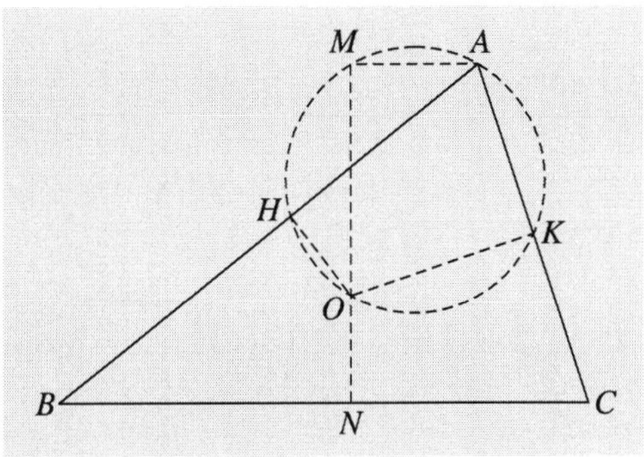

Analysis. The point of AB which is in contact with the fixed point at H starts moving along AB. Similarly, the point of AC which is in contact with K starts moving along AC. Hence, if OH, OK are perpendicular to AB, AC respectively, O is the instantaneous centre.

Through O draw ON perpendicular to BC.

Then N is the point of contact of BC with its envelope.

Hence, if the envelope of BC is a circle, ON must pass through a fixed point.

We therefore proceed to look for a fixed point on ON.

Now the circle HAK is fixed, and it passes through O. Let ON cut this circle in M;

then $$\angle MAH = \angle MOH = \angle B;$$

∴ the arc MH is constant;

∴ M is a fixed point, and MA is parallel to BC;

∴ MN is equal to the perpendicular from A to BC, and is therefore of constant length.

Hence, the envelope of BC is the circle whose centre is M and radius the length of the perpendicular from A to BC.

19. A triangle of given size and shape moves so that two of its sides always touch two fixed circles. Prove that the envelope of the third side is a circle.

MISCELLANEOUS EXERCISES

Arranged in Sets for Homework or Revision

Paper XXXVI (to Section XLVI)

1. The perpendicular from any vertex of a regular polygon, having an even number of sides, to the straight line joining any other two vertices passes through a fourth vertex of the polygon.
2. Draw a straight line such that the perpendiculars drawn to it from the vertices of a given triangle may be in a given ratio.

   Show that there are four such lines.
3. The side $AKLB$ of a rectangle $ABCD$ is three times the side AD, and $AK = KL = LB \cdot BD$ and KC intersect at R. Prove that C, R, L, B are concyclic.

   [Prove, by similar triangles, that LR passes through M, the middle point of CD, and hence, if LN, the perpendicular to DC, cuts KC in T, that Δs KLT, LNM are congruent.]
4. If the tangent at A to the circumcircle of the triangle ABC meets BC produced at D, show that

   $$BD : CD = AB^2 : AC^2.$$

   If also the tangents at B and C meet CA, AB produced at E and F, prove that D, E, F are collinear.

5. If a, b, c, d are the sides, in order, of a cyclic quadrilateral, R the radius of the circle and Q, the area of the quadrilateral, prove that

$$R^2 = (ab + cd)(bc + ad)(ac + bd)/16Q^2.$$

[If x, y are the diagonals, prove that

$$R = \frac{(ab+cd)x}{4Q} = \frac{(bc+ad)y}{4Q},$$

and then apply Ptolemy's Theorem.]

6. Describe a triangle whose vertices shall lie on three given parallel straight lines, given the area and the length of one median.

Paper XXXVII (to Section XLVII)

1. O is the centre, AC a diameter and AB a chord of a given circle; the circle on AB as diameter cuts AC in D, and the tangents to this circle at A and D intersect at E; OF is drawn, at right angles to AC, to meet AE in F. Prove that, if AF cuts the given circle in H, then $AH = 2EF$.

[If N is the middle point of AB, prove that EN is parallel to OF.]

2. Find the locus of a point such that the product of its distances from one pair of opposite vertices of a square may be equal to the product of its distances from the other pair of opposite vertices.

3. If I is the centre of the inscribed circle, and I_1, I_2, I_3 the centres of the escribed circles, of the triangle ABC, show that the sum of the squares on I_1I_3 and II_2 is equal to the square on the diameter of the circle $I_1I_2I_3$.

[Use Ex. LXXVI, 1 and 13 (*iii*).]

4. If a, b, c, d are the sides, in order, of a cyclic quadrilateral, find expressions for the diagonals.

[Ex. 5 of Paper XXXVI, gives the ratio of x to y, and Ptolemy's Theorem gives the product.]

5. The sides of a cyclic quadrilateral, taken in order, are 25, 52, 60, 39 inches. Calculate the lengths of the diagonals, the radius of the circumcircle and the area.

[Use the method of the preceding exercise to find the diagonals x and y; thence, from a triangle whose sides are a, b, x, find R, and finally Q.]

6. ABC is a triangle in which $AC = 3\frac{1}{2}$ in., $BC = 2\frac{1}{2}$ in. and AB is $1\frac{1}{2}$ in. Describe a circle intercepting chords 1 in. in length on each of the sides, and measure its radius. Verify by calculation.

[Show that the circle required is concentric with the inscribed circle.]

Paper XXXVIII (to Section XLVIII)

1. AB is a diameter of a circle and CD a parallel chord; if P is any point in AB, show that the sum of the squares on PC, PD is equal to the sum of the squares on PA, PB.

2. *A*, *B*, *C*, *D* are four points in order on a straight line. Find the locus of a point at which *AB* and *CD* subtend equal angles. [Use Ex. XLVIII, 20]
3. *It is required to divide a parallelogram ABCD, by two straight cuts, into three pieces which can be fitted together to form a square.*

   Prove the following construction: Draw a square equal in area to *ABCD*, With centre *A* and radius equal to a side of the square, draw an arc cutting *BC* at *X*. Draw *DY* perpendicular to *AX*. Cut the paper along *AX*, *DY*. The construction fails if *X* is outside *BC* or if *Y* is outside *AX*.
4. *A* line through *O*, the orthocentre of a triangle *ABC*, intersects the circles described about the triangles *BOC*, *COA*, *AOB* in points *P*, *Q*, *R*. Prove that the perpendiculars from *P*, *Q*, *R* to *BC*, *CA*, *AB* respectively intersect in a point on the circle *ABC*.
5. Straight lines are drawn from any point on a circle to the angular points of an inscribed regular pentagon. Show that the sum of two of them is equal to the sum of the other three.

   [Apply Ptolemy's Theorem to obtain various ratios equal to the ratio of a side of the pentagon to a diagonal, and use the Theorem on Equal Ratios on p. 253.]
6. Construct a triangle, having given the length of the internal bisector of one angle, the ratio of the side opposite that angle to the sum of the other two sides, and the difference of the other angles.

   [Let *ABC* be the triangle, *AX* the bisector of $\angle A$. Draw *CY* parallel to *XA* to meet *BA* produced at *Y*.]

Paper XXXIX (to Section XLIX)

1. On the sides of a triangle *ABC* as bases, equilateral triangles *XBC*, *YCA*, *ZAB* are constructed, each on the same side of its base as the triangle *ABC*. Prove that *AX*, *BY*, *CZ* are equal and pass through a point: also show that this point lies on each of the circles circumscribing the equilateral triangles.
2. *P* is a point in the diameter *AB* of the semicircle *ACDB*, *PC* is drawn at right angles to *AB*, and *PEB* is a semicircle on the same side of *AB* as *ACDB*. If *QR* is a diameter of the circle which touches *PC* in *Q* and the semicircles in *D* and *E*, show that rect. $AB \cdot QR$ = sq. on *CP*.

   [Consider the sides of the triangle whose vertices are the centres of the three circles, and work algebraically.]
3. *AB* is a fixed chord of a given circle, *P* is any point on the circumference; perpendiculars *AC* and *BD* are drawn to *BP* and *AP* respectively; find the locus of the middle point of *CD*.

   [Prove that *CD* is of constant length.]
4. A triangle *ABC* has a given base *AB* and a given vertical angle *C*. *AD*, *BE* are perpendiculars from *A*, *B* to *BC*, *CA*, meeting them in *D* and *E* respectively. Prove that the centre, of the circle *CED* is at a constant distance from *DE*.

   [Use Ex. 3]

5. L, M, N are the centres of the circles escribed to the sides BC, CA, AB of a triangle. Prove that the perpendiculars LP, MQ, NR from L, M, N to BC, CA, AB respectively are concurrent.

   [Let I be the incentre, S the circumcentre. Let LP meet IS produced at O. Prove $SO = IS$.]

6. Given one side AB of a quadrilateral in position and magnitude, the position of the middle point of CD and the area of the quadrilateral; find the loci of the vertices C and D.

   [Show that the loci of the middle points of AD and BC are straight lines.]

Paper XL (to Section L)

1. A parallelogram is such that two circles, exterior to each other, can be described in it, each touching three of its sides; a line is drawn to touch both circles, and is terminated by the longer sides of the parallelogram. Prove that the length of this line is the difference between the lengths of a long side and a short side of the parallelogram.

   [Mark the equal lengths on the figure with small letters and work algebraically.]

2. If the diagonals AC, BD of a quadrilateral $ABCD$, inscribed in a circle whose centre is O, meet at right angles in a fixed point P, prove that the feet of the perpendiculars from P and O to the sides of the quadrilateral lie on a fixed circle.

3. A point P is taken on the circle circumscribing a square $ABCD$: prove that the difference of the squares on PB, PD = 2 rect. $PA \cdot PC$.

   [Use Ptolemy's Theorem for the quadrilaterals $PBCD$, $PABD$.]

4. (*i*) The locus of the centre of a circle which passes through a fixed point and cuts a given circle orthogonally is a straight line.

   (*ii*) Draw a circle to cut a given circle orthogonally and to pass through two given points.

5. *AP, BQ, CR are the perpendiculars from the vertices of a triangle ABC to the opposite sides, and X, Y, Z are the points of intersection of BC, QR; CA, RP; AB, PQ, Prove that X, Y, Z lie on a straight line, which is the radical axis of the circumscribed and nine-point circles of the triangle.*

   [If L is the middle point of BC, prove that $XB \cdot XC = XL \cdot XP$.]

6. Prove that the square on the common tangent to two circles which intersect in P and whose centres are O and O' is equal to four times the rectangle contained by the perpendiculars from O and O' to the bisector of the angle OPO'.

   [If $OP > O'P$, produce $O'N'$, the perpendicular from O' to the bisector of the angle OPO' : let it cut OP in R and a parallel to PN through O in K : prove $4O'N' \cdot N'K = OO'^2 - OR^2$.]

PAPER XLI (to Section LI)

1. *ABC* is a triangle: D_1 is the point of contact with *BC* of the escribed circle opposite to *A*. AD_1 cuts the inscribed circle in *X* and *Y*. If *X* is the point nearer to *A*, show that the tangent at *X* is parallel to *BC*.

   [*A* is a centre of similitude of the two circles.]

2. The line joining the feet of the perpendiculars from a variable point *P* to two given straight lines is parallel to the line joining the feet of the perpendiculars from *P* to two other given straight lines. Prove that the locus of *P* is a circle.

   [The circle passes through the two points of intersection of the pairs of given lines.]

3. *ABCD* is a given quadrilateral and *X* is a variable point in *AB*. The circumcircles of the triangles *BCX*, *ADX* cut again in *Y*. Prove that (*i*) the locus of *Y* is a circle; (*ii*) *XY* passes through a fixed point for all positions of *X*.

   [Prove that $\angle DYC$ is equal or supplementary to the angle between *AD*, *BC* : *XY* cuts the circle *DYC* in a fixed point.]

4. A variable circle passes through a fixed point *O*, and the ends of a diameter *PQ*, lie on a fixed circle whose centre is *C*. Prove that the locus of *N*, the foot of the perpendicular from *O* to *PQ*, is a circle whose centre lies on *OC*, and that this circle is also the locus of the centre of the variable circle.

   [If *K* is the middle point of *OC*, and *A* the centre of the variable circle, prove that *AK* is constant and that $AK = KN$. Use Apollonius' Theorem.]

5. *APB* is an acute-angled triangle inscribed in a fixed circle. *A* and *B* being fixed points and *P* a variable point. The bisector of the angle *P* cuts the circle again in *Q*. Show that $AP + PB : PQ = AB : AQ$.

6. *It is required to divide an equilateral triangle ABC, by three straight cuts, into four pieces which can be fitted together to form a square.*

   Prove the following construction: Bisect *AB*, *AC* at *N*, *M* respectively. Construct a square equal in area to ΔABC. With centre *N* and radius equal to a side of the square, draw an arc cutting *BC* at *L*. Along *LB* set off *LX* equal to *MN*. Draw *XY*, *MZ* perpendicular to *LM*. Cut the paper along *LM*, *XY*, *MZ*.

PAPER XLII (to Section LII)

1. *ABC* is a triangle, in which *L*, *M*, *N* are the middle points of the sides *BC*, *CA*, *AB*. Through *A*, *B*, *C* are drawn three parallel straight lines cutting *MN*, *NL*, *LM* in *X*, *Y*, *Z* respectively. Prove that

   (*i*) *A*, *Y*, *Z* ; *B*, *Z*, *X* ; *C*, *X*, *Y* are collinear points.

   (*ii*) Area of $\Delta XYZ = \frac{1}{2}$ area of ΔABC.

   [If *NM* cuts *BY* in *P*, prove $PN = NX$: hence, since $BL = LC$, it follows that *Y*, *C*, *X* are collinear. For (*ii*), prove $\Delta XYZ = \Delta AYC = \Delta ANC$]

2. O is any point on the circle circumscribing the triangle ABC, and OA', OB', OC' are chords of the circle, perpendicular to the sides BC, CA, AB respectively.

Prove that the triangles ABC, $A'B'C'$ are congruent.

[Show that $A = A'$, etc.]

3. S is the circumcentre, I the incentre and G the centroid of a triangle ABC. R, r, r' are the radii of the circumcircle, incircle and the circle circumscribing the pedal triangle; and a, b, c are the lengths of the sides BC, CA, AB. Prove that

(*i*) $9SG^2 = 9R^2 - (a^2 + b^2 + c^2)$.

(*ii*) $r' = (a^2 + b^2 + c^2)/4R - 2R$.

[For (*i*), use Ex. LVI. 14, 16; for (*ii*), Ex. LXXVI, 22.]

4. A circle is circumscribed about a triangle ABC. The bisector of the angle A cuts BC in D and the circle again in X. Given that $BC = 3\frac{1}{2}$ in., $AD = 1$ in. and $DX = 1\frac{1}{2}$ in. Construct the triangle and measure BD and DC.

Between what limits must BC lie so that, with the given values of AD and DX, the construction may be possible?

5. Two circles intersect in A and B; the tangents at A to each circle cut the other circle in X and Y; a straight line through B perpendicular to AB cuts the circles in P and Q. Prove that

$$\Delta\, AXB : \Delta\, AYB = PX^2 : QY^2.$$

6. (*i*) If a circle passes through a given point and cuts a given circle orthogonally; prove that it passes through a second fixed point.

(*ii*) Draw a circle to touch a given circle, cut another given circle orthogonally and pass through a given point.

Paper XLIII (to Section LIII)

1. $DBCE$, $FCAG$, $HABK$ are three chords of a circle, such that the segments BC, CA, AB intercepted by any two of them on the third are all equal. Prove that

$$DB + FC + HA = CE + AG + BK.$$

2. Two given circles intersect in A and D; through D a variable line is drawn meeting the circles again in B and C. Find the locus of the centre of the circle inscribed in the triangle ABC.

[If I is the centre of the circle inscribed in $\Delta\, ABC$, show that BI, CI pass through fixed points and contain a constant angle.]

3. AP, BQ, CR are the perpendiculars from the vertices of a triangle ABC to the opposite sides, and O is the orthocentre. Prove that

$$AO \cdot OP = BO \cdot OQ = CO \cdot OR = 2Rr',$$

where R, r' are the radii of the circumcircle and the circle inscribed in the pedal triangle. Hence show that $r' = 2R \cos A \cos B \cos C$.

[If SL is the perpendicular from the circumcentre to BC, $AO = 2SL$]

4. (*i*) A, B are given points and Z is any point on a given circle. Join ZA, ZB, cutting the circle again at Y, Z, respectively. Draw the chord YL parallel to AB. Let LX, or LX produced, cut AB at M. Prove that M is a fixed point.

   (*ii*) *In a given circle, inscribe a triangle XYZ such that ZY, ZX pass through given points A, B, and XY is parallel to a given straight line.*

   (*iii*) *In a given circle inscribe a triangle XYZ with its sides YZ, ZX, XY passing through three given points A, B, C.*

   [Referring to (*i*), the point M can be determined.; then the triangle LXY can be drawn, as in (*ii*).]

5. ACB is a triangle such that $BC = 2\frac{1}{2}$ inches, $CA = 2$ inches, $AB = 3$ inches. A circle is drawn through B and C, cutting AC in P and AB in Q, and another circle is drawn through A, P, Q. Determine the position of P such that the radius of the first circle may equal the diameter of the second. Measure the radius of the first circle.

6. *Given an isosceles triangle ABC with the angle A a right angle. It is required to divide the triangle ABC, by three straight cuts, into four pieces which can be fitted to form an equilateral triangle.*

   Prove the following construction: Draw an equilateral triangle $A'B'C'$, equal in area to ABC. Bisect BC, CA, AB at L, M, N. With centre L and radius $\frac{1}{2}A'B'$, draw an arc to cut MC at X. On LX draw an equilateral triangle LXY. Join NY and produce it to cut AC at Z. Cut the paper along NZ, LX, LY.

PAPER XLIV (to Section LIV)

1. AB is a fixed diameter of a given circle, whose centre is C. BP and BQ are two chords, including a constant angle. Prove that $AQ \cdot AP$ increases as the angle ABP diminishes.

   [Along AQ set off $AR = AP$: prove that the circle PRQ, has a constant radius and that the angle QPR increases as the angle ABP diminishes.]

2. P is a variable point on the circumference of the circle circumscribing the triangle ABC, and PL, PM, PN are perpendicular to the sides of the triangle. Prove that the circle through the centres of the circles PMN, PNL, PLM is equal to the nine-point circle of ABC for all positions of P.

   [P is a centre of similitude of the circle ABC and the circle through the centres.]

3. Through a fixed point O any straight line OPQ is drawn cutting a fixed circle in P and Q, and on OP, OQ as chords are described circles touching the fixed circle at P and Q. Prove that the two circles so described intersect on another fixed circle.

   [If the circles cut at T, and C is the centre of the given circle, prove that T, P, C, Q are concyclic and that the angle OTC is a right angle.]

4. A, B are given points on a circle and CD is a given chord. Find a point P on the circle such that, if PA, PB meet CD at X, Y, the line XY may be of given length.

   [Draw AZ parallel to CD and equal to the given length. The position of Y can now be determined.]

5. Prove that the harmonic mean of the perpendiculars from a given point O within a circle to the tangents, drawn from any point on the polar of O is constant.

   [If P is the point on the polar of O, obtain by similar triangles values of the perpendiculars in terms of the radius and the segments into which O divides the polar of P.]

6. OX, OY are two given straight lines. A and B are fixed points on OX, C and D fixed points on OY. On OX and XO produced take OP and OQ equal to AB, and on OY and YO produced take OR and OS equal to CD.

   Prove that the locus of a point V, such that the sum or difference of the triangles VAB, VCD is constant, consists of four straight lines parallel to the sides of the parallelogram $PRQS$: distinguish between the parts corresponding to the sum and to the difference.

Paper XLV (to Section LV)

1. Two circles cut at right angles in A and B, and through B a chord is drawn, cutting the circles in C and D. AC cuts the second circle in E and AD cuts the first circle in F. Prove that E and F lie on a straight line through B, perpendicular to CD.

2. From a given point outside a circle, two straight lines, containing a given angle, are drawn to cut the circle, the straight lines being both on the same side of the line joining the point to the centre. Prove that the difference between the perpendiculars, from the centre to the lines, is least when one of the lines touches the circle.

   [Use Ex. 1 of Paper XLIV.]

3. $ABCD$ is a quadrilateral inscribed in a circle, and AXB, CXD are two circles touching one another at X. Prove that X lies on a fixed circle.

   [Make an accurate construction for one pair of circles: this should suggest the proof.]

4. BC and CE are equal chords of a given circle, and BC is produced to meet the tangent at E in D. If DL is drawn perpendicular to EC, prove that $BD = 2EL$.

   [If EC produced meets the tangent at B in F, show that the triangle DEF is isosceles.]

5. Points D, E, F are taken on the sides BC, CA, AB of a given triangle ABC, such that the angles D, E, F of the triangle DEF are respectively equal to the angles A, B, C. Prove that the orthocentre of the triangle DEF is fixed for all positions of the triangle DEF.

   [The orthocentre of DEF is the circumcentre of ABC; use Theorem 114.]

6. (*i*) The distance of the centroid of a triangle from any given straight line is equal to one-third of the sum of the distances of its vertices from the line.

   (*ii*) Points D, E, F are taken on the sides BC, CA, AB of a triangle ABC, such that $BD : DC = CE : EA = AF : FB$. Prove that the centroids of the triangles ABC, DEF coincide with one another.

   [Use (*i*) to show that the centroids of ABC, DEF are equidistant from BC, etc.]

Paper XLVI (to Section LVI)

1. AX, AY are two given straight lines; B, C are variable points in AX, AY such that $BC + CA + AB$ is constant. Prove that BC touches a fixed circle.

2. Prove that the line joining the middle points of the diagonals of a quadrilateral circumscribed to a circle passes through the centre of the circle.

   [Use Ex. 6 of Paper XLIV, showing that the sum of the triangles, standing on a pair of opposite sides, and having either one of the mid-points or the centre as a common vertex, is equal to half the quadrilateral.

3. (AB, CD), ($A'B'$, $C'D'$) *are two harmonic ranges. Two corresponding points, such as* C, C', *coincide. Prove that the straight lines joining the other pairs,* AA', BB', DD', *are concurrent.*

   [*Let* AA', BB' *meet in* O, *and let* OD *cut* $A'B'C'D'$ *in* D'': *prove that* D'' *coincides with* D'.]

4. (*i*) If a circle bisects the circumferences of two given circles, the locus of its centre is a straight line.

   (*ii*) Draw a circle to bisect the circumferences of two given circles and pass through a given point.

5. ABC, $A'B'C'$ *are two triangles so placed that* AA', BB', CC' *meet in a point* O.* *Prove that* BC, $B'C'$; CA, $C'A'$; AB, $A'B'$ *meet in three points* X, Y, Z *which are in a straight line.*

   [Apply Menelaus' Theorem to

   (*i*) $\Delta\, AOB$, transversal $A'B'Z$, to prove $\dfrac{AZ}{ZB} = \dfrac{AA'}{BB'} \cdot \dfrac{OB'}{OA'}$.

   (*ii*) Similarly prove $\dfrac{BX}{XC} = \dfrac{BB'}{CC'} \cdot \dfrac{OC'}{OB'}$ and $\dfrac{CA}{YA} = \dfrac{CC'}{AA'} \cdot \dfrac{OA'}{OB'}$.

   (*iii*) Hence prove that $\dfrac{BX}{XC} \cdot \dfrac{CY}{YA} \cdot \dfrac{AZ}{ZB} = 1$.]

6. Given the vertex A of a triangle and also the magnitude of the angle at A, and that the vertices B and C move on fixed straight lines intersecting in a point O; give a construction for drawing the triangle BAC when its area is a minimum.

---

* Two such triangles are said to be in **perspective** : O is called the centre of **perspective** and XYZ the **axis of perspective**.

[If B_1AC_1, B_2AC_2 are two positions of the triangle ABC, such that the areas of B_1AC_1, B_2AC_2 are equal, prove that $B_1A \cdot AC_1 = B_2A \cdot AC_2$, and hence that the triangles B_1AB_2, C_2AC_1 are similar. In the minimum position AB, AC make equal angles with the given straight lines respectively.]

Paper XLVII (to Section LVII)

1. Two points A and B and a circle are given, also a third point O is given within the circle. Draw a circle through A and B to cut the given circle at the ends of a chord through O.

[If X is the centre of the required circle $XA^2 \sim XO^2$ is known.]

2. ABC is a given triangle, OX and OY two given straight lines. Q is a variable point on BC and QP, QR are drawn parallel to OX, OY to cut AB, AC, produced if necessary, in P and R. The parallelogram $PQRS$ is completed. Prove that the locus of S is a straight line.

[Let S_1, S_2, be two positions of S, and let S_1S_2, cut AB in T: prove TP_2; $TP_1 = CQ_2 : CQ_1$, and hence that CT is parallel to OX.]

3. $ABCD$ is a quadrilateral in which AD, BC, when produced, meet in E, and AB, DC, when produced, meet in F.* Join EF. Bisect AC, BD, EF in P, Q, R. Prove that PQR is a straight line.

[Complete the parallelograms $AECX$, $DEBY$, and prove, by similar triangles, that E, X, Y are collinear.]

4. *ABCD is a quadrilateral in which AD, BC, when produced, meet in E, and AB, DC, when produced, meet in F. Let AC, BD meet in G. Prove that the pencils E(AB, GF), F(AD, GE), G(BC, FE) are harmonic.*

[Find points K, L, such that the ranges (DC, KF), (AB, LF) are harmonic. Use Ex. 3 of Paper XLVI to show that KL passes through both E and G.]

5. Draw any complete quadrilateral, and in it inscribe a rectangle having its sides respectively parallel to and perpendicular to the third diagonal.

[Use Ex. 2]

6. Draw a circle, centre O, having a radius of $2\frac{1}{2}$ in. Take a point P, such that $OP = 1$ in. Plot the envelope of chords of the circle which subtend a right angle at P.

Paper XLVIII

1. $ABCDEF$ is a complete quadrilateral: AD, BC meet in E, and AB, DC meet in F. P, Q, R are the middle points of AC, BD, EF. Prove that

$$\Delta PBF - \Delta PDE = \Delta QBF - \Delta QDE = \Delta RBF - \Delta RDE,$$

and give an alternative proof for the theorem of Ex. 3 of Paper XLVII.

---

* The figure made up of the four lines ADE, ECB, ABF, FCD is called a **complete quadrilateral**; the six points A, B, C, D, E, F are its **vertices**. The three lines AC, BD, EF, joining the points of intersections of the sides taken in pairs, are the **diagonals**, EF being referred to as the **third diagonal**.

2. P and Q are given points on two given straight lines OA, OB, X. and Y are variable points on these lines, such that the ratio $PX : QY$ is constant. Find the locus of the middle point of XY.

   [Let L, M be the middle points of PQ, XY : complete the parallelogram $PQXZ$ and join YZ : prove that the locus of N, the middle point of YZ, is a straight line through Q, and that $LQNM$ is a parallelogram.]

3. (*i*) Through a given point A draw a straight line, to cut two given straight lines at P and Q, such that $PA \cdot AQ$ may be a minimum.

   (*ii*) A, B are given points and P is a point on a given circle. PA, PB (produced if necessary) cut the circle again at X, Y. Find positions of P for which the chord XY is a maximum or a minimum. Consider separately the cases where A, B are both within, and both without the circle.

4. *If ABC is an obtuse-angled triangle, show that the polar circle (§12 of Section LIV), the nine-point circle and the circumcircle have the same radical axis.*

   [Let S, J, O be the centres of the circumcircle, the nine-point circle and the polar circle, and T the point of intersection of the first two circles, and let OR cut the circumcircle again at H : prove

$$OT^2 = 2SJ^2 + 2JT^2 - ST = \frac{1}{2}(OS^2 - R^2);$$

   hence $OT^2 = \frac{1}{2} OC \cdot OH = OC \cdot OR$, and T is on the polar circle.]

5. A, B, $A'B'$ are given points and PQ is a given straight line. Find points C, C' in PQ, such that the triangles ABC, $A'B'C'$ are equal in area and CC' is equal to a given length.

   [Suppose C, C' to be found; draw CP, $C'P$ parallel to AB, $A'B'$ to intersect in P: then P is a point on the locus, such that $\Delta PAB - \Delta PA'B' = O$, a straight line passing through the point of intersection of AB, $A'B'$.]

6. P is a point within the angle XOY. Draw a line through P to cut OX, OY in A and B, such that the perimeter of the triangle OAB is a minimum.

PAPER XLIX

1. $ABCD$ is a quadrilateral, such that $AD + BC = AB + CD$. Points E, F, G, H are taken on the sides AB, BC, CD, DA, so that each side is divided into segments in the ratio of the adjacent sides. I is the centre of the inscribed circle. Prove that

   (*i*) $AH = AE$ ($= r_1$ say), $BE = BF$ ($= r_2$ say), etc.

   (*ii*) The quadrilateral is divided into two sets of similar triangles by the lines IA, IE, IB, IF, etc.

   (*iii*) $IE^2 = IF^2 = IG^2 = IH^2 = AB \cdot BC \cdot CD \cdot DA/(AB + CD)(BC + AD)$.

2. $ABCD$ is a cyclic quadrilateral. AD, CB, when produced, meet in E; BC, DA, when produced, meet in F; AC, BD meet in G. Prove that the triangle EFG is self-conjugate with regard to the circle.

Let EG cut AB, CD in L, K.

[Use Ex. 4 of Paper XLVII and §16 of Section LIV to prove that K and L both lie on the polar of F.]

3. *Use the preceding exercise to construct the polar of a point (i) within or (ii) without a given circle, by means of a ruler only.*

4. A straight line DE is drawn to cut the base BC of a triangle ABC, and is terminated by the sides AB, AC, produced if necessary, in D and E, Prove that, if the quadrilateral $BDCE$ is of constant area, the middle point of DE lies on one of two fixed lines,

[If D_1E_1, D_2E_2 are two positions of DE, show that the triangles CD_1D_2, BE_1E_2 are equal in area: hence the ratio $D_1D_2 : E_1E_2$ is constant. Use Ex. 2 of Paper XLVIII. One line of the locus corresponds to D, on AB and the other to D on AB produced.]

5. *$OAPB$, $OCQD$ are two straight lines intersecting in O. AC, PQ, BD meet in another point S.* Prove that

$$OA \cdot PB : AP \cdot BO = OC \cdot QD : CQ \cdot DO.*$$

[Through Q draw a parallel to OB, cutting SC, SD, produced if necessary, in X and Y : then $AP : PB = XQ \cdot QY$, etc.]

6. ABC is a given triangle: find a point such that the feet of the perpendiculars from the point to the sides of the triangle are the vertices of an equilateral triangle.

If $BC = 2\frac{1}{2}$ in., $CA = 2$ in., $AB = 3$ in., draw the equilateral triangle and measure a side.

[Use the 'Point O' Theorem.]

Paper L

1. *$ABCD$* is a quadrilateral whose sides are of given lengths a, b, c, d, such that $a + c = b + d = s$. Prove that the radius of the inscribed circle is a maximum when the quadrilateral is also cyclic; and in that case, if Q is the area of the quadrilateral and r the radius of the inscribed circle, Q is also a maximum, and $Q = rs = \sqrt{abcd}$.

[Referring to the figure of Ex. 1 of Paper XLIX, let $IE = \rho$ and $\angle AEI = \theta$: prove $sr_1 = ad$, etc., and $\rho^2 = r_1r_3 = r_2r_4$; hence

$$sr = \sqrt{abcd}\sin\theta.]$$

2. P, Q, R are points on the sides BC, CA, AB of the triangle ABC, such that AP, BQ, CR intersect in a point U. P', Q', R' are points on the sides QR, RP, PQ, such that PP', QQ', RR' meet in a point V. Prove that AP', BQ', CR' meet in a point.

---

* The ratio $OA \cdot PB : APBO$ is called the **anharmonic ratio** or **cross ratio** of the range $\{OAPB\}$.

[Let QR, RP, PQ meet BC, CA, AB, when produced in L, M, N ; produce AP', BQ', CR' to meet BC, CA, AB, produced if necessary, in X, Y, Z. Prove that

(i) $\dfrac{BX}{XC} = \dfrac{CL}{LB} \cdot \dfrac{RL}{LQ} \cdot \dfrac{QP'}{P'R}$ and two similar results for $\dfrac{CY}{YA}$ and $\dfrac{AZ}{ZB}$, by the use of Ex. 5 of Paper XLIX.

(ii) $\dfrac{CL}{LB} \cdot \dfrac{BN}{NA} \cdot \dfrac{AM}{MC} = 1$, $\dfrac{RL}{LQ} \cdot \dfrac{QN}{NP} \cdot \dfrac{PM}{MR} = 1$, $\dfrac{QP'}{P'R} \cdot \dfrac{RQ'}{Q'P} \cdot \dfrac{PR'}{R'P} = 1$,

by the application of the theorems of Menelaus and Ceva to the triangles ABC, PQR, $P'Q'R'$.]

3. $ABCD$ is a quadrilateral in which A, B are right angles, and the diagonals AC, BD intersect at right angles in O. BA, CD, when produced, meet in P, and PO cuts AD in K and BC in L. Prove that $AB^2 = 4KO \cdot OL$.

4. A straight line cuts the sides AB, BC, CD, DA of a quadrilateral in P, Q, R, S. Prove that

$$\frac{AP}{PB} \cdot \frac{BQ}{QC} \cdot \frac{CR}{RD} \cdot \frac{DS}{SA} = 1;$$

and state whether the theorem can be generalized for a rectilinear figure of any number of sides.

5. A quadrilateral is inscribed in a circle (O, R) and circumscribed to another circle (I, r). If d is the distance between O and I, prove that

$$\frac{1}{(R+d)^2} + \frac{1}{(R-d)^2} = \frac{1}{r^2}.$$

[Let AI, CI meet the circumcircle in T, T'. Prove that TT' is a diameter, and that $R^2 - d^2 = AI \cdot IT = CI \cdot IT'$. Hence

$$2r^2(R^2 + d^2) = r^2(IT^2 + IT'^2) = (R^2 - d^2)^2.]$$

6. The radius of the Polar circle is a mean proportional to the radius of the circumcircle and the diameter of the circle inscribed in the pedal triangle. [Use Ex. 3 of Paper XLIII.]

Paper LI

1. The bisectors of the angles B, C of the triangle ABC meet the sides AC, AB in E, F; the circles ABE, ACF intersect in X; XE, XF produced meet the circles ACF, ABE in Y, Z. Prove that Z, A, Y are collinear.

2. If the bisectors of two angles of a triangle are equal, prove in a direct manner that the triangle is isosceles.

[In Ex. 1, let ZE, YF meet in K, and BE, CF in I. Prove that the circles are equal, that $FIEK$ is a parallelogram, and that the triangles ZKY, BIC are similar; hence, that

$$BI/IC = ZK/KY = (ZK + BI)/(KY + IC) = 1.]$$

3. In the triangle ABC, prove that the common tangent of the nine-point circle and the escribed circle opposite to A cuts BC in the ratio of $a + b$ to $a + c$. [See p. 351, Ex. 14]

4. A, B, C, D are four given points in a straight line; show how to find two points X, Y, which are harmonic conjugates with respect to A and B, and also with respect to C and D.

   [See § 4]

5. If a, b, c, d are the lengths of the four sides of a cyclic quadrilateral in order, prove that the square on one of the diagonals is

$$(ab + cd)(ac + bd)/(ad + bc).$$

   Hence, given the lengths of the four sides of a cyclic quadrilateral, construct it.

6. $ABCD$ is a given parallelogram, XY a given straight line, and P a given point. Show how to construct, with a ruler alone, the parallel to XY through P.

   [If AB, AD cut XY in L, M, and BM, DL cut CD, BC in R, S, respectively, prove that RS is parallel to XY; the construction follows easily.]

TYPICAL EXAMINATION PAPERS

ON UNITS 1–4

The papers below have, by kind permission of the several authorities, been compiled from questions recently set.

[LONDON MATRICULATION]

I

1. If the side BC of a triangle ABC is produced to D, prove that the angle ACD is equal to the sum of the angles ABC and BAC. (Do not assume in your proof that the sum of the three angles of a triangle is equal to two right angles.)

   The sides AB, BC, CD, DE, EA of a pentagon inscribed in a circle subtend angles 40°, 50°, 60°, 100° and 110° respectively at the centre of the circle. Find the number of degrees in each angle of the pentagon.

2. The equal angles B, C of an isosceles triangle ABC are bisected by lines BY, CX, which meet the opposite sides in Y and X. Prove that XY is parallel to BC.

   Also, by assuming the above if you cannot prove it, show that $BX = XY = YC$.

3. A trapezium $ABCD$ has its parallel sides AD, BC 11 cm. and 4.7 cm. long respectively, and its non-parallel sides AB, CD 5.2 cm; and 6.8 cm long respectively. Construct the trapezium accurately full-size, stating the steps in your construction; and calculate its area.

4. Prove the extension of Pythagoras' Theorem to an acute angled triangle; *i.e.*, that the square on the side opposite an acute angle is less than the sum of the squares on the sides that contain it by twice a certain rectangle.

In a triangle ABC, the perpendicular AD from A to BC divides BC so that $BD = 3DC$. Prove that $AB^2 - AC^2 = 8DC^2$.

5. AB is a chord of a circle ABC and SAT is the tangent to the circle at A. Mark in a carefully drawn figure any pairs of angles that you know to be equal.

ABC is a triangle; find the locus of a point P such that the angles PAB, PCA are equal, P lying inside the triangle.

Hence construct a point P such that the angle PAB = angle PCA and also the angle PBA = angle PCB.

(The figure should be constructed accurately, the sides of the triangle ABC being unequal and the angle C an acute angle.)

6. Illustrate by a diagram the truth of the formula

$$(a + b)^2 = a^2 + 2ab + b^2.$$

A, B, C, D are four points in that order on a straight line, such that $AB = CD$. Prove that $AC^2 = AB^2 + AD \cdot BC$, preferably by a diagram.

7. AD, BC are perpendicular to a straight line AB and on the same side of it. BD and AC are joined and BD is bisected at E. A point F is found in AC, produced if necessary, such that $EA = EF$. Prove that the angles AFB, ADB are equal.

8. P, Q, R, S are four points on the circumference of a circle such that another circle can be inscribed within the quadrilateral $PQRS$. Prove that the two lines which join the points of contact of opposite sides are perpendicular to one another.

II

1. ABC is an equilateral triangle. P is any point inside it. On PC an equilateral triangle PQC is drawn (so that PQ cuts AC and not BC). Prove that the triangles ACQ, BCP are congruent.

Hence, by first drawing the triangle APQ, construct an equilateral triangle ABC, being given the distances PA, PB, PC of a point P from its three vertices to be 1 inch, 1.6 inches, 2 inches respectively.

2. Prove that the sum of the angles of a triangle is equal to two right angles, without assuming that an exterior angle of the triangle is equal to the sum of the two interior opposite angles.

In a triangle ABC, BC is produced beyond C to D so that $CD = AC$; CB is produced beyond B so that $BE = AB$. Show that EAD is an obtuse angle.

If $AE = AD$, prove that ABC is an isosceles triangle.

3. Prove that the area of a triangle is one-half the area of a rectangle on the same base and with the same height.

$ABCD$ is a rectangle in which the lengths of AB and BC are 6 ins. and $3\frac{1}{2}$ ins. respectively. X and Y are points on CD between C and D such that $DX = 3$ ins. and $CY = 1$ in. Find the area of the figure $ABYX$.

4. State and prove a property of the angles of a cyclic quadrilateral.

   AOB is a diameter and *AS* a chord of a circle whose centre is *O*. The chord *AP* bisects the angle *BAS*. *AS* is produced to *Q* so that *PQ* = *PA*. Prove that (*i*) *OP* is parallel to *AQ*, (*ii*) angle *APQ* = angle *BPS*, (*iii*) angle *SPQ* = 90°, (*iv*) triangles *APB*, *QPS* are congruent.

5. *O* is a point outside a circle, centre *C*, and *OJ* is a tangent from *O* to the circle; *S* is taken on *OC* so that *OS* = *OJ*. At *O* a perpendicular is drawn to *CO* and *P* is any point on this perpendicular. *PT* is the tangent to the circle from *P*. Prove that $PT^2 = PO^2 + OC^2 - (\text{radius})^2$, and hence show that *PT* = *PS*.

6. The base *BC* of a triangle *ABC* is 6 inches long and *M* is its middle point. State without proof the value of

$$AB^2 + AC^2 - 2AM^2.$$

   If *P* is a point in the same plane as a parallelogram *ABCD*, prove that $PA^2 - PB^2 + PC^2 - PD^2$ is the same wherever *P* is, and that the expression is equal to half the difference between the squares on the diagonals of the parallelogram.

7. Prove that equal chords in a circle subtend equal or supplementary angles at any point of the circumference.

   AB, *AC* are equal chords of a circle; *P* and *Q* are any two points in *BC*, and *AP*, *AQ* meet the circle again at *R* and *S*. Prove that a circle can be described about *PQSR*.

8. State the relation which connects the angles between a tangent and a chord with the angles in the alternate segments of the circle, giving a figure.

   The tangent at *C* to the circumcircle of the triangle *ABC* cuts *AB* produced in *D*, and a point *R* is taken in *DA* such that *DR* = *DC*. Prove that *CR* bisects the angle *ACB*.

[University of London : General School Examination]

I

1. Prove that, if a straight line cutting two straight lines makes the alternate angles equal to one another, the straight lines are parallel.

   Prove that, if the lines which bisect the opposite angles *A* and *C* of a quadrilateral *ABCD* are parallel to one another, the angles *B* and *D* are equal.

2. Prove that the exterior angle of a triangle is greater than either of the interior opposite angles.

   ABC is a triangle in which the angles at *A* and *B* are acute, and *B* is greater than *A*; also, when *AC* is produced to *D*, the angle *BCD* is three times the angle *BAC*. With centre *C* and radius *CB* a circle is drawn cutting *AB* again in *E*. Prove that *E* lies between *A* and *B*, and that *AE* = *BC*.

3. *A* and *B* are two points 3.4 inches apart and *BC* is a line such that *ABC* = 55°. By geometrical constructions and measurement find the distances from *A* of (*i*) a point *P* in *BC* which is equidistant from *A* and *B*; and (*ii*) a point *Q*,

in *BC* which is 1.2 inches nearer to *B* than to *A*. The steps in the construction must be stated.

4. On the side *AB* of a parallelogram *ABCD* any point *P* is taken, on the side *CD* a point *Q*, and on the diagonal *AC* a point *R*, such that $BP = CQ = CR$. Show (*i*) that the locus of the middle point *M* of *QR* is the bisector of the angle *ACD*; and (*ii*) that the locus of the middle point *N* of *PR* is the line parallel to *CM* through the middle point of *BC*.

5. Prove that two equal chords of a circle subtend equal angles at the centre and equal angles at the circumference of the circle.

   The internal bisectors of the angle *B* and *C* of a triangle *ABC* meet the circumscribing circle of the triangle in *Q* and *R* respectively. Prove that, if *BR* is parallel to *CQ*, the angle *BAC* must be 60°.

6. If a perpendicular be drawn from the centre of a circle to a chord, prove that it bisects the chord.

   OA, *OB* are two perpendicular radii of a circle whose centre is *O*, and *C* is any other point on the circumference. *OS* is drawn perpendicular to *AC*; and *OS* and *BC*, produced if necessary, cut at *M*. Prove that *AM* is perpendicular to *BC*.

7. Give, without proof, a geometrical illustration of the algebraical identity, $a^2 - b^2 = (a + b)(a - b)$.

   AB is a straight line 4.6 inches long. Find, by a geometrical construction, a point *P* between *A* and *B* such that the square on *AP* exceeds the square on *PB* by the square on a line 2.4 inches long.

8. From a point *P* outside a circle a tangent *PT* is drawn touching the circle at *T*, and a secant cutting the circle at *Q* and *R*; on *PT* as diameter a circle is drawn cutting *RT*, produced if necessary, at *Y* and *TQ*, produced if necessary, at *X*. Prove that *XY* is perpendicular to *PR*.

II

1. *ABCD* is a parallelogram. On the opposite sides *AB*, *CD* equilateral triangles *APB*, *CSD* are drawn external to the parallelogram, and on *BC* an equilateral triangle *BQC* is drawn on the same side of *BC* as the parallelogram. Prove that *PQ* and *QS* are equal to the diagonals of the parallelogram.

2. A tree is equidistant from two straight paths *AB*, *AC* which intersect at an angle of 30°, and is 70 yards distant from a third path which crosses the other two at *B* and *C*. *AB* is 500 yards and *AC* 250 yards. Find all the possible positions of the tree and their distances from *A*.

3. *LM* and *PQ* are two equal and parallel straight lines drawn in the same sense; prove that *LP* and *MQ* are also equal and parallel.

   ABCD is a quadrilateral in which *AB* is equal to *CD*, and the angle *ABC* is equal to the angle *BCD*. Prove that *AD* is parallel to *BC*.

4. Construct a parallelogram having one side $2\frac{1}{2}$ inches long and the diagonals 3 inches and 4 inches long. State clearly (without proof) the steps of the construction.

   Prove that the parallelogram is a rhombus.

5. ABC is a triangle having the angle C obtuse. D is the foot of the perpendicular from A on BC produced. Prove that

$$AB^2 = BC^2 + CA^2 + 2BC \cdot CD.$$

   BC, AD are the parallel sides of any trapezium $ABCD$. Prove that

$$AC^2 + BD^2 = AB^2 + CD^2 + 2BC \cdot AD.$$

6. Prove that the angle which an arc of a circle subtends at the centre is double of the angle which it subtends at any point of the supplementary arc.

   A circle APB passes through the centre of a circle AQB; the circles intersecting at A and B. Prove that, if P lies within the circle AQB and APQ is a straight line, PB is equal to PQ.

7. If a pair of opposite angles of a quadrilateral are supplementary, prove that its vertices lie on a circle.

   AB is a diameter of a circle, and AC, AD are any two chords which when produced cut the tangent at B at E, F respectively. Prove that E, C, D, F lie on a circle.

8. Prove that the angles made by a tangent to a circle with a chord drawn from the point of contact are respectively equal to the angles in the alternate segments of the circle.

   Two unequal circles cut at A and B. The tangents at A to the circles cut them again at P, Q. PQ, produced if necessary, cuts the circle AQB at S, and the circle APB at T. Prove that $AS = AT$.

[Scotch Leaving Certificate : Lower Grade]

I

Section I

(All the questions in this Section should be attempted)

1. Prove that the straight line joining the middle points of two sides of a triangle is parallel to the third side and equal to half of it.
2. Prove that the sum of the squares on two sides of a triangle is twice the sum of the squares on half the third side and the median which bisects the latter.
3. Prove that the opposite angles of a cyclic quadrilateral are supplementary.
4. State and prove a construction for drawing a rectangle, with one side of given length, equal in area to a given rectangle.

SECTION II

(*Only* THREE *questions should be attempted from this Section*)

5. *ABCD* is a parallelogram. Through *C* a straight line is drawn parallel to the diagonal *BD* to meet *AD* produced in *E*. If *F* and *G* are the mid-points of *AB* and *CE*, prove that the straight line *FG* is trisected at the points in which it is cut by *BD* and *CD*. (*See* Section I, 1)

6. The base *AB* of a triangle *ABC* is given in magnitude and position, and also the sum of the squares of the sides in magnitude. Prove that the locus of the vertex *C* is a circle. (*See* Section I, 2)

7. *BE* is the bisector of the angle *ABC* of a cyclic quadrilateral *ABCD*, and it meets the circumference of the circumscribed circle in *E*. Prove that *DE* bisects the exterior angle of the quadrilateral at *D*.

8. *S*, a point in the base *QR* of an isosceles triangle *PQR*, is joined to the vertex *P*. Prove that the ratio of the sines of the angles *QPS*, *RPS* is the same as the ratio of the segments of the base.

Hence, if $\sin \alpha / \sin \beta = \sin \theta / \sin \phi$, and $\alpha + \beta = \theta + \phi$, prove that $\theta = \phi$.

9. *BCE* is a triangle inscribed in a circle, and the tangents at *B* and *C* meet in *A*. The figure is such that the straight line through *A* parallel to *CE* cuts *BE* internally in *O*. Prove that the angles *ABC*, *ACB*, *AOB*, *CEB* are all equal to one another. Also, if *CO* is drawn, show that *AO* bisects the angle *BOC*.

II

SECTION I

(*All the questions in this Section should be attempted*)

1. In a triangle *ABC*, the angle *ACB* is greater than the angle *ABC*. Prove that *AB* is greater than *AC*.

2. Show that, if the straight line joining a vertex of a triangle to the middle point of the base is half the base, then the triangle is right-angled.

3. Prove that the perpendicular bisector of a chord of a circle passes through the centre of the circle.

4. State and prove a construction for finding the mean proportional between two given straight lines.

SECTION II

(*Only* THREE *questions from this Section should be attempted*)

5. *A* is the greatest and *C* the least angle of a triangle *ABC*; and *O* is the centre of the circle inscribed in the triangle. Prove that, of the three distances *AO*, *BO*, *CO*, the least is *AO*, and the greatest is *CO*. (*See* Section I, 1)

6. *D* is the foot of the perpendicular from the vertex *C* of a triangle *ABC* on the bisector *AD* of the angle *CAB* : and *DE* is drawn parallel to *BA* to meet *AC* in *E*. Prove that *E* is the mid-point of *AC*. (*See* Section I, 2.)

7. Two circles intersect in the points *P* and *Q*. *C*, the centre of one of them, is joined to *P* and *Q* and the straight lines, *CP*, *CQ* (produced if necessary), intersect the other circle in *R*, *S*, respectively. Prove that *PR* = *QS*.

8. Prove *from a geometrical figure* that, if A is an acute angle,

$$\sin A = \frac{\tan A}{\sqrt{(1+\tan^2 A)}}.$$

Find the corresponding formula for $\tan A$ in terms of $\sin A$.

9. ABC is a triangle, $BCPR$ is a circle cutting AB, AC respectively in R and P and BP, CR intersect in O. If the angles BRC, BOC, PBC are 50°, 60°, 45° respectively, calculate the size of the angles of the triangle ABC.

TYPICAL EXAMINATION PAPERS

ON UNITS 1–5
[Oxford School Certificate]
I

1. $ABCD$ is a quadrilateral; $AB = 5.6$ cm., $BC = 4.9$ cm., $CD = 6.5$ cm., $DA = 10.4$ cm., $AC = 9.1$ cm. Construct the quadrilateral. Construct also a point P on the diagonal BD, such that the area of the triangle PAC equals half that of the quadrilateral, and measure the distance of P from the point of intersection of AC and BD.

2. O is the middle point of the hypotenuse BC of an isosceles right-angled triangle ABC. P and Q are points on AB such that $AP = BQ = AO$. Prove that

$$\frac{1}{2} POQ = QOA = 22\frac{1^\circ}{2}.$$

Hence show that if R is the point on AC such that $CR = AO$, then PQ, and QR are consecutive sides of a regular octagon whose centre is O.

3. If two opposite sides of a quadrilateral are equal and parallel, prove that it is a parallelogram.

Through A, the vertex of a triangle ABC, AM is drawn parallel to BC and equal to half of BC (M and C being on the same side of AB). If D, F, the middle points of BC and AB respectively, are joined to M, and DM, FM cut AC at E and K respectively, prove (*i*) that $AE = \frac{1}{2}AC$, and (*ii*) that $AK = \frac{1}{4}AC$.

4. Three triangles have sides, 2, 3, 4; 3, 4, 5; 4, 5, 6 inches respectively. Without drawing or measurement, prove that they are one right-angled, one obtuse-angled, and one acute angled.

5. Prove that, if a pair of opposite angles of a quadrilateral are supplementary, its vertices are concyclic.

OC is the internal bisector of an angle AOB which equals 120°. X and Y are any points on OA, OB respectively. The perpendiculars to OA, OB at X and Y intersect in P. Z is the foot of the perpendicular on OC from P. Prove that the triangle XYZ is equilateral.

6. Prove that if two chords of a circle intersect inside the circle, the rectangle contained by the parts of the one is equal to the rectangle contained by the parts of the other.

P is any point on the diagonal *BD* of a parallelogram *ABCD*. The circles through *APB*, *APD* cut *AC* again in *Q* and *R*. Prove that $OR = OQ$, where *O* is the point of intersection of *AC* and *BD*.

7. If two triangles have one angle of the one equal to one angle of the other and the sides about these equal angles proportional, prove that the triangles are similar.

A, *B*, *C*, *D* are four points taken in order on the circumference of a circle, such that $BA : AC = CD : DB$. Prove that $BA = CD$, $AC = DB$.

8. A solid consists of a cylinder, height 5 in. and diameter 4 in., and a hemisphere on each plane end. Draw on a scale of 2 cm. = 1 in. a section of the solid by a plane parallel to its axis and distant $1\frac{1}{2}$ in. from that axis. Measure the length and width of the section.

II

1. Prove that, if two sides of a triangle are unequal, the angle opposite the greater side is greater than the angle opposite the smaller side.

In the quadrilateral *ABCD*, *AB* is the greatest side and *CD* is the least. Prove $\angle C > \angle A$ and $\angle D > \angle B$.

2. If *P* is any point inside a quadrilateral *ABCD*, show that the sum of the lengths of *PA*, *PB*, *PC*, *PD* is always greater than half the sum of the sides of the quadrilateral.

Show also that the sum is least when *P* is at the point of intersection of the diagonals of the quadrilateral.

3. The diagonals *AC*, *BD* of a quadrilateral *ABCD* are inclined at equal angles of 30° to the side *AB*. The length of the diagonal *AC* is 9.2 cm., and that of the side *CD* is 5.2 cm. The perpendicular distance of *D* from *AB* is 4.4 cm. Construct the quadrilateral from these data, and measure the lengths of *AD*, *BC* and *AB*.

4. Prove that triangles on the same base which are equal in area are of the same altitude.

ABCD is a quadrilateral whose opposite sides *AB*, *DC* are parallel. Any line through the point of intersection of the diagonals *AC*, *BD* intersects the parallels to *AC* through *D* and *B* in *X* and *Y* respectively. Prove (*i*) that the areas of *XAY* and *DAB* are equal, (*ii*) that $AY \cdot XC$ are parallel.

5. If the line joining two points subtends equal angles at two other points on the same side of it, prove that the four points lie on a circle.

P, *Q*, *R* are three given points, and a variable point *X* moves so that the difference of the angles *XQP*, *XRP* is constant. Assuming that *X* and *Q* are on opposite sides of *PR*, find the locus of *X*.

6. A circle is described about a triangle XYZ, and at the vertices tangents are drawn to the circle. The tangents at Y and Z meet in A, those at Z and X in B, and those at X and Y in C. Prove that the angle XZY is half the sum of the angles A and B.

   State, without proof, the propositions which you use to prove this fact.

7. Prove that the ratio of the areas of similar triangles is equal to the ratio of the squares on corresponding sides.

   The tangent at C to the circumcircle of the triangle ABC cuts the side AB produced in P. Prove that $PA : PB = CA^2 : CB^2$.

8. AO, AB, AC are concurrent edges of a rectangular block and P, Q, R are the middle points of OB, AB, AC respectively. Prove that PQR is a right-angled triangle and that $4PR^2 = OA^2 + BC^2$.

[CAMBRIDGE SCHOOL CERTIFICATE]

I

1. Prove that the sum of the interior angles of a quadrilateral is four right angles.

   $ABCDE$ is a five-sided figure in which $AB = AE$, $BC = ED$, and the angle ABC = the angle AED. Prove that (*i*) $AC = AD$; (*ii*) the angle BCD = the angle EDC; (*iii*) BE is parallel to CD.

2. Prove that the opposite sides and angles of a parallelogram are equal.

   $ABCD$ is a parallelogram in which $AB = 2BC$, and the bisector of the angle A cuts DC in E. Prove that AEB is a right angle.

3. Prove that, if the intercepts made by three parallel straight lines on any straight line which cuts them are equal, then the intercepts on any other straight line which cuts them are equal.

   In the triangle ABC the side AC is produced to E, CE being equal to half of AC. BE is bisected at D. Prove that, if ED meets AB in F, then $DE = 2DF$.

4. P and Q are two points on the same side of the line joining two given points M and N; and the angles MPN, MQN are equal. Prove that the points M, N, P, Q are concyclic.

   A triangle ABC is inscribed in a circle whose centre is O ; E, R are the mid-points of AC, AB; and the lines EO, FO meet AB, AC respectively in X, Y. Prove that the points B, C, X, Y are concyclic.

5. Prove that, if a line touches a circle, and if from the point of contact a chord is drawn, the angles which the chord makes with the tangent are equal to the angles in the alternate segments of the circle.

   The tangents at B and C to the circle circumscribing a triangle ABC meet in O; and the circle whose centre is O, and radius OB or OC, meets AC produced in Q. QO is drawn to meet AB in R. Prove that R lies on the circle CBQ.

6. Prove that the internal bisector of an angle of a triangle divides the opposite side in the ratio of the sides containing the angle.

TA and *TC* are lines which touch a circle in *A* and *C*, and *TBD* is a line which cuts the circle at *B* and *D*. Prove that the bisectors of the angles *A* and *C* of the quadrilateral *ABCD* intersect on the diagonal *BD*.

7. Construct a triangle *ABC* in which *AD* is the perpendicular from *A* on *BC*, having given that *BC* = 10 cm., *BD* : *BA* = 3 : 5, *CD* : *CA* = 1 : 3. Calculate the length of *AD*.

8. *A*, *B*, *C*, *D* are four places on a map. *B* is due North of *A*, *C* is due East of *B*, and *D* is in the direction 30° 17′ West of North from *C*. If *AB* = 10 miles, *BC* = 18 miles, and *CD* = 15 miles, find the distance between *A* and *D*.

II

1. Prove that, if in a triangle *ABC* the side *AB* is greater than the side *AC*, then the angle *ACB* is greater than the angle *ABC*.

Show also that, if in this triangle *D* is the middle point of *BC*, then the angle *BAD* is less than the angle *CAD*.

2. Prove that parallelograms on the same base and of the same altitude are equal in area. (*N.B. The formula for the area of a parallelogram is not to be assumed.*)

The side *AB* of a parallelogram *ABCD* is produced to any point *P*. A straight line through *A*, parallel to *CP*, meets *CB* produced in *Q*, and the parallelogram *PBQR* is completed. Prove that the parallelograms *PBQR*, *ABCD* are equal in area.

3. Prove that the tangent at any point of a circle is perpendicular to the radius through the point of contact.

Two circles, external to one another, have centres *A*, *B*, and radii r_1, r_2 respectively. *N* is a point in *AB* such that

$$AN^2 - NB^2 = r_1^2 - r_2^2,$$

and *P* is any point in the straight line through *N* perpendicular to *AB*. Show that the tangents from *P* to the two circles are equal.

4. Prove that a straight line, drawn from the centre of a circle to bisect a chord which is not a diameter, is at right angles to the chord.

ABCD is a rectangle; *X* is any point in *AB*; *DY* is drawn perpendicular to *DX* cutting *BC* produced in *Y*; and *V* is the middle point of *XY*. Prove that *V* is equidistant from *B* and *D*.

What is the locus of *V* as the position of *X* in *AB* changes?

5. *ACB* is the diameter of a semicircle, whose centre is *C*; semicircles are described on *AC* and *CB*, both on the same side of *AB*. If *AB* = 12 cm., calculate the radius of a circle which shall touch all three semicircles, and construct the figure.

6. Prove that, if in the triangles *ABC*, *DEF*, the angles *A* and *D* are equal, and *AB* : *AC* = *DE* : *DF*, the triangles are similar.

A point *O* is taken within a quadrilateral *ABCD* ; and the lines *OX*, *OY*, drawn through O parallel to CB, CD meet AB, AD in X, Y respectively. Prove that, if XY is parallel to BD, then O lies on the diagonal AC.

7. A, B, C, D, are four points on a circle whose centre is O and whose radius is 10 feet, such that $AB = BC = CD = 7$ feet. Calculate the angle AOB and the length of AD.

8. A lighthouse is 8 miles due South of a port. A ship leaves the port at 10 p.m. and steams at 12 miles an hour in a direction 20° South of East. Calculate the times at which the ship is (*i*) due East of the lighthouse, (*ii*) North-East of the lighthouse.

UNIVERSITIES OF MANCHESTER, LIVERPOOL, LEEDS, SHEFFIELD AND BIRMINGHAM JOINT MATRICULATION BOARD

[SCHOOL CERTIFICATE AND MATRICULATION EXAMINATION]

Answer Section A and three questions from Section B.

I

SECTION A

1. (*a*) ABC is a triangle in which $A = 60°$, $B = 58°$, and the side BC is produced to D. The bisectors of the angles ABC and ACD meet at X. Calculate the number of degrees in the angle BXC.

(*b*) K is a point inside a square $ABCD$, such that the angle BKC is a right angle and CK is greater than BK. The perpendicular drawn from D to CK meets CK at N. Prove that DN is equal to CK.

2. Construct, in separate figures, two non-congruent triangles, ABC, DEF, with

$$\angle B = \angle E = 43°,$$
$$AB = DE = 2 \text{ inches},$$
$$AC = DF = 1.7 \text{ inches}.$$

3. Draw a triangle ABC in which $BC = 4''$, $\angle B = 40°$, $\angle A = 110°$. Construct one circle of radius $1''$ to pass through A and touch BC. (No written explanation is required, but all construction lines must be clearly shown.)

4. Prove that the straight lines joining the middle points of two parallel chords of a circle will pass through the centre of the circle.

A circular field is crossed by two parallel paths, 66 feet apart, the lengths of the paths being 144 feet and 120 feet. Find, by calculation or by an accurate drawing, the length of the diameter of the field.

5. (*a*) $PQRS$ is a quadrilateral inscribed in a circle; the angle $QPS = 110°$, the angle $PQS = 30°$, and the angle $SQR = 30°$. PR, QS intersect at T. Find (*i*) the angle QRS, (*ii*) the angle PRS, (*iii*) the angle QTR.

(*b*) The internal bisector of the angle BAC of a triangle ABC meets BC at R, and RS drawn parallel to BA meets AC at S. If $AC = 2AB$, find the ratio (*i*) of AS to SC, (*ii*) of AB to SR.

SECTION B

6. Prove that if two triangles have three sides of the one equal to three sides of the other, each to each, the triangles are congruent.

   ABCD is a quadrilateral in which *AB* = *CD*. Also there is a point *O* inside the quadrilateral such that *OA* = *OD*, and *OB* = *OC*. Prove that *BC* is parallel to *AD*.

7. *ABC* is a right-angled triangle with the right angle at *A*, and *AC* is greater than *AB*; *OBC* is an isosceles right-angled triangle on the side *BC* remote from *A*, with *OC* equal to *OB*. *M* and *N* are the feet of the perpendiculars from *O* to *AB* and *AC* respectively, produced if necessary. Prove (*i*) that the angle *ABO* is obtuse, and the angle *ACO* is acute ; (*ii*) that *AMON* is a square.

8. If the opposite angles of a quadrilateral are supplementary, prove that the vertices of the quadrilateral lie on a circle.

   A point *O* is taken on the side *BC* of a triangle *ABC*, and the tangent at *B* to the circle *AOB* meets the tangent at *C* to the circle *AOC* at the point *X*. Prove that *A*, *B*, *X*, *C* lie on a circle.

9. The internal bisector of the angle *A* of a triangle *ABC* meets the base *BC* at *X*. Show that *BX* : *XC* = *BA* : *AC*.

   The base *BC* of a triangle *ABC* is bisected at *D*. *DE*, the bisector of the angle *ADB*, meets *AB* in *E*, and *EF* is drawn parallel to *BC* meeting *AC* in *F*. Show that *DT* is the bisector of the angle *ADC*.

10. If the angles of one triangle are equal to the angles of another triangle, each to each, prove that the corresponding sides of the two triangles are proportional.

    ABC is a triangle, and *P* is a point on *AB*. *PR*, drawn parallel to *AC*, meets *BC* at *R*; *RP* is produced to *S* so that *PS* = *RP*; *AS*, *CP* are produced to meet at *X*. Prove that *XB* is parallel to *AC*.

II

SECTION A

1. (*a*) *P* and *Q* are points on the sides *AB*, *AC* of a triangle *ABC*, and *PQ* is parallel to *BC*. If *AP* = 1 inch, *PB* = 1.5 inches, *QC* = 2.7 inches, find the length of *AC*.

   (*b*) The side *BC* of a triangle *ABC* is produced to *X* so that *CX* = *BC*, and *Y* is the middle point of *AB*. Prove that the triangles *ABC*, *XPY* are equal in area.

   (*c*) Adjacent sides of a parallelogram are 2 inches and 3 inches long. If the perpendicular distance between the longer sides is 1 inch, what is the perpendicular distance between the shorter sides?

2. (*a*) Draw two straight lines *AB*, *AC* containing an angle of 45°. Find a point *P* on *AB* and a point *Q* on *AC* such that the angle *APQ* = 50° and

$PQ = 1$ in. State the steps of the construction in order, but do not give a proof.

(*b*) Draw two circles of radii 1.5 in. and 1 in., the centres being 2 in. apart; and construct a common tangent to the two circles. All construction lines must, be clearly shown.

3. Prove that if two circles touch externally, the point of contact lies on the straight line through their centres.

Three circles with centres *A*, *B*, *C* touch each other externally, each circle touching the other two circles. $BC = 4$ in., $CA = 5$ in., $AB = 7$ in. What are the radii of the circles?

4. (*a*) Find the angles subtended at the centre of a circle by chords which make angles of $23°$ and $140°$ with the tangents at their extremities.

(*b*) The inscribed circle of a triangle *ABC* touches *BC*, *CA*, *AB* at *D*, *E*, *F* respectively. $BD = 0.8$ in., $BC = 1.5$ in., $AE = 0.9$ in. Find the lengths of *AB* and *AC*.

5. *ABC* is a triangle; *X* and *Y* are points on *AB* and *AC* such that $2AX = XB$, and $AY = YC$. *BY* and *CX* intersect in *O*. Find the ratio (*i*) of *CO* to *OX*, (*ii*) of ΔBOC to ΔBAC.

SECTION B

6. Prove that, if two triangles have two angles of the one equal to two angles of the other, each to each, and also one side of the one equal to the corresponding side of the other, the triangles are congruent.

Any point *P* is taken on the side *AD* of a square *ABCD* and the straight line *BP* is drawn. *AQ* is drawn perpendicular to *BP*, and meets *CD* at *Q* and *BP* at *O*. Prove that the area of the quadrilateral *OPDQ* is equal to the area of the triangle *OAB*.

7. Prove that, if one pair of sides of a quadrilateral are equal and parallel, the other pair of sides are equal and parallel.

Points *P* and *Q* are taken within a parallelogram *ABCD*, so that *AP* and *QC* are parallel and also *BP* and *QD* are parallel. Prove that *AQ* is equal and parallel to *PC*.

8. *ABCD* is a square; two parallel lines, *APS*, *CRQ* are drawn, and *BQP*, *DSR* are drawn perpendicular to these parallels, so that *P*, *Q*, *R*, *S*, all lie within the square. Prove that *PQRS* is a square whose side is the difference between *BP* and *AP*; and deduce an *algebraical proof* of Pythagoras' Theorem.

The points *B*, *E*, *A*, *C* lie on a semicircle of which *BC* is the diameter. *AD* is drawn perpendicular to *BC* meeting it at *D*; and $BE = AD$. If *DA* is produced to *F*, so that $DF = BA$, prove that $CF = CE$.

9. Prove that, if two triangles have one angle of the one equal to one angle of the other, and the sides about these equal angles proportionals, then the triangles are similar.

The angle A of a triangle ABC is obtuse. Points P and Q are taken between B and C on the side BC such that $BP \cdot BC = BA^2$, and $CQ \cdot CB = CA^2$. Prove that $AP = AQ$.

10. The side AB of a triangle is fixed and the length of BC is given. The bisector of the angle ABC meets AC at D, and DO is drawn parallel to CB to meet AB at O. Prove (*i*) that O is a fixed point, (*ii*) that the locus of D is a circle.

[CENTRAL WELSH BOARD SCHOOL CERTIFICATE EXAMINATION]

I

1. (*i*) It is required to construct a quadrilateral $ABCD$ from the following data: $AB = 7.0$ cm., $BC = 9.4$ cm., $AC = 11.0$ cm., $\angle ACD = 47°$, $AD = 9.1$ cm. Show that there are two solutions, draw the quadrilaterals, and measure and write down the length of the diagonal BD for each.

(*ii*) Suppose the first four data are as before, but AD has now to be made as short as possible. Find by measurement this shortest length, and also the length of BD for this case.

2. O is the middle point of the hypotenuse AB of a right-angled triangle ABC, and OC is drawn. Denoting the angle BAC by A, express the other angles in the figure in terms of A.

AC is produced both ways, and points, E beyond A and F beyond C, are taken such that $AE = AB$, $CF = CO$. FO produced meets EB in K. Prove that $KE = KF$.

3. Define a *parallelogram*, and prove that its opposite angles and sides are equal.

E is any point in the side BC of a rectangle $ABCD$. EF is drawn parallel to the diagonal CA to meet AB in F, and EG is drawn parallel to BD to meet CD in G. Show that EF and EG together make up the length of a diagonal.

4. In a triangle ABC the angle C is acute, and AD is drawn perpendicular to BC. Prove $AB^2 = BC^2 + AC^2 - 2BC \cdot DC$.

Given that $\angle C = 60°$, and $BC = 4AC$, prove $AB = \sqrt{13}\, AC$.

5. Prove that the angles in the same segment of a circle are equal.

$ABCD$ is a cyclic quadrilateral; AN is parallel to BC, and meets BD, not produced, at N. Prove that the angle ADN is equal to the angle NAC. What can you say about AC and the circle circumscribing the triangle AND?

6. Show that the angle between a chord of a circle and the tangent at one end of the chord is equal to the angle in the alternate segment.

PQ is a line of length $2\frac{1}{2}$ inches. It is required to find a point distant 3 inches from P, at which PQ subtends an angle of 40°. Show that there are two such points, and find them. (Each step in your construction should be indicated.)

7. Show that a line parallel to one side of a triangle divides the other two sides proportionally.

E is the point in the side *AC* of a triangle *ABC*, such that $AE : EC = 1 : 4$. *EF* is drawn parallel to *CB*, meeting *AB* in *F*; and *ED* is drawn parallel to *AB*, meeting *BC* in *D*. Show that the area of the parallelogram *BDEF* is 8/25 of the area of the triangle *ABC*.

8. *A* and *B* are two points 4 inches apart. Find (*i*) the locus of a point *P* which moves in such a way that $AP^2 - PB^2 = 24$ sq. in.; (*ii*) the locus of a point *Q* which moves in such a way that $AQ^2 + QB^2 = 24$ sq. in.

II

1. Draw two straight lines *OA*, *OB* containing an angle of 50°, and make *OA* = 2.4 in., *OB* = 1.25 in. Describe a circle touching *OA* at *A* and passing through *B*, explaining your construction.

Produce *OB* to meet the circle again at *C*, and measure *OC*. Find two points *D*, *E* on the circle which are equidistant from *AB* (or *AB* produced) and *AC*. Measure *AD*, *AE*.

2. Given that the four sides of a quadrilateral are equal; show that its angles are bisected by its diagonals.

A line drawn parallel to the base *BC* of an isosceles triangle *ABC* meets the equal sides *AB*, *AC* in *E*, *F* respectively. Lines through *R*, *F* parallel to *AC*, *AB* meet in *G*. Show that *AG*, produced if necessary, will bisect *BC*.

3. Prove that the area of a triangle is half that of a parallelogram on the same base and between the same parallels, and deduce the formula for the area of a triangle.

AFE is a triangle with *B* the mid-point of *AF* and *C* the mid-point of *BE*; *AC* produced meets *EF* at *D*. Prove that the triangles *FBD*, *ADB*, *ADE* are equal in area.

4. (*i*) *M* is the middle point of a line *AB*, and *P* is any point outside *AB*. Prove $AP^2 + PB^2 = 2AM^2 + 2MP^2$.

(*ii*) *A*, *B*, *C*, *D*, *X* are five points on a line, such that $AB = BC = CD$. Prove that $AX^2 + 3CX^2 = 3BX^2 + DX^2$.

5. Show that two opposite angles of a cyclic quadrilateral together make up two right angles.

ABC is an acute-angled triangle. The circle described on *AB* as diameter cuts the side *BC*, *CA* in *D*, *E* respectively; *AD*, *BE* meet in *P*. Show that the angle *APB* is equal to the sum of the angles *ABC*, *BAC*.

6. Show that, when two chords intersect within a circle, the rectangle contained by the segments of one is equal to the rectangle contained by the segments of the other.

In a circle two chords *AKC*, *BKD* are at right angles. Given that $AK = 20$, $KC = 36$, $AB = 25$, calculate *KB*, *KD*, and the radius.

7. A and B are two fixed points 3 inches apart; P is a point which moves in such a way that $AP : PB$ has always the value 3 : 1. Find the points P_0, P_1, on AB and AB produced which satisfy this condition; and find at least two other points which satisfy it.

   Then *prove* that the locus of P is the circle on $P_0 P_1$ as diameter.

8. Show, with proof, how to construct a square equal to a given rectangle.

   Draw an equilateral triangle ABC of side 3 inches. From AB, AC mark off AD, AE respectively, each equal to 1 inch. Construct a square equal in area to the quadrilateral $BDEC$, explaining each step you take. Measure and write down the length of a side of your square.

   While the questions on Trigonometry and Solid Geometry in the following papers can usually be answered successfully with a good knowledge of the corresponding sections in this volume, the student is strongly advised to study these two subjects further than is here given.

TYPICAL EXAMINATION PAPERS

ON UNITS 1–6
[Scottish Leaving Certificate: Higher Grade]

I

Section I

(*All the questions in this Section should be attempted*)

1. Prove that the square on the hypotenuse of a right-angled triangle is equal to the sum of the squares on the sides containing the right angle.
2. Prove that the angles between the tangent at any point of a circle and a chord through that point are equal to the angles in the alternate segments.
3. Prove that two triangles are equiangular if they have one angle of the one equal to one angle of the other and the sides round the equal angles proportional.
4. Prove the geometrical proposition equivalent to the formula:

$$\cos A = (b^2 + c^2 - a^2)/2bc.$$

Section II

(*Only* **THREE** *questions from this Section should be attempted*)

5. $ABCD$ is a square; a quadrant of a circle whose centre is A touches BC at B and CD at D. Through P, any point in AD, a straight line is drawn parallel to AB, cutting the diagonal AC in Q, the quadrant in R, and the side BC in S. Prove that the square on PS is equal to the sum of the squares on PQ and PR. (See Section I, 1)
6. Two circles intersect at A and B. The tangent at B to one of the circles meets the other again at D; and any straight line through A meets the first circle again, in P, and the second in Q. Prove that BP is parallel to DQ. (See Section I, 2)

7. A circle is circumscribed to a triangle *ABC*, and *I* is the centre of the circle inscribed in the triangle. *AI* produced meets the circumference of the circumscribed circle at *Q*. Prove that *QI*, *QB*, and *QC* are all equal.

8. State and prove a construction for finding two points, *P*, *Q*, on the sides *AB*, *AC* of the triangle *ABC*, such that the triangle *PBQ* may be similar to the triangle *ABC*, and $BP + PQ$ may equal a given length.

9. The vertical angles at *A* of a tetrahedron *ABCD* are right angles, A plane *PQR* cuts the edges through *A* in such a way that $AP = 3''$, $AQ = AR = 5''$. Find the number of degrees in the angle *QPR*.

10. *ABC* is a triangle having the angles *A* and *B* acute. *P* is a point on *CA* produced, and *PB* cuts the circumference of the circle on *AB* as diameter in *D*. If $BD = d$, prove that θ = angle *CAD*, ϕ = angle *CPD* is a solution of the simultaneous equations

$$\theta + \phi = 180° - C, \text{ and } a \cos \theta + b \cos \phi = d.$$

What is the greatest possible value of *d* for given values of *a*, *b*, and *C*?

II

Section I

(All the questions in this Section should be attempted)

1. Prove that equal triangles on the same base and on the same side of it are between the same parallels.

2. From the vertex of a triangle a straight line is drawn perpendicular to the base. Prove that the rectangle contained by the sides of the triangle is equal to the rectangle contained by the base and the diameter of the circle circumscribed to the triangle.

3. Prove that, if a straight line is perpendicular to two straight lines at their point of intersection, it is perpendicular to the plane containing the two straight lines at that point.

4. The angles *A*, *B*, $A + B$, all being acute, prove that

$$\cos (A + B) = \cos A \cos B - \sin A \sin B.$$

Section II

(Only **THREE** *questions from this Section should be attempted)*

5. *AB*, *CD* are two given straight lines, and *BA*, *DC*, when produced, intersect in *O*. *X* and *Y* are two points on *OA* and *OC* such that $OX = AB$, and $OY = CD$. *P* is a point within the angles *XOY* such that the sum of the areas of the triangle *PAB*, *PCD* is constant. Prove that the locus of *P* is a straight line parallel to *XY*. (See Section I, 1)

6. *ABCD* is a quadrilateral inscribed in a circle, and *P* is any other point on the circumference. If p, p', q, q' are the lengths of the perpendiculars from *P* on the sides *AB*, *CD*, *BC*, *DA* respectively, the sides being produced if necessary, prove that $pp' = qq'$. (See Section I, 2)

7. *ABC* is a triangle whose angle at *B* is greater than the angle at *C*. Through *B* draw a straight line *BD* making the angle *ABD* equal to the angle at *C*, and cutting the base internally at *D*. Prove that the square on *BD* is to the square on *BC* as *AD* is to *AC*.

8. At the ends *B*, *C*, of the base *BC* of a triangle *ABC* perpendiculars *BP* and *CQ* are drawn to the plane of the triangle; *BP* is made equal to *CA* and *CQ* to *BA*. Prove that the points *P*, *Q* are equidistant from *A*.

9. From a point *P* on the top of a cliff of height *a* above the sea, the angle of depression of a boat *B* on the sea is ϕ. Prove that the tangent of the angle which a flagstaff, *PQ*, of height *b*, standing on the cliff at *P*, subtends at the boat *B* is

$$b \sin \phi \cos \phi/(a + b \sin^2 \phi).$$

(**HINT:** Draw *QN* perpendicular to *BP* produced, and find the lengths of *QN*, *NP*, *PB*.)

Answers

UNIT 1

Exercise I, *Pages 7–8*

1. (*i*) 60°; (*ii*) 150°.

2. (*i*) 120°; (*ii*) 10°.

3. (*i*) $82\frac{1}{2}^\circ$; (*ii*) $22\frac{1}{2}^\circ$.

4. $\frac{4}{3}$ rt. ∠; 140°; 60°; 81°50'.

5. $\frac{3}{4}$ rt. ∠; 50°; 54°; 49' 50".

6. ∠ *ACD* = 148°, ∠ *ECD* = 58°.

7. 120°.

8. ∠ *BOD* = 42°, ∠*AOD* = ∠ *BOC*=138°.

Exercise IV, *Page 21*

1. 40°.

2. 60°.

3. 25°.

4. 120°.

5. 60°; a parallelogram.

6. *n*, *n* – 2; 72°, 108°; 60°, 120°; 45°, 135°.

7. 80°.

8. $\frac{2}{3}$ rt. ∠.

9. $\frac{16}{9}$ rt. ∠.

10. 16.

Exercise VI, *Page 26*

1. 45°.

2. 40°.

3. 60°.

Exercise VIII, *Page 33*

2. 76; 25.

4. 2.8.

5. 0.39.

6. *AB* = 1.2 in., *BC* = 0.5 in., *CD* = 1.0 in.; *AD* = 2.8 in., all to the nearest tenth of an inch.

7. *PQ* = 26 mm., *XY* = 21 mm.
PR = 4.87 cm., *PS* = 5.03 cm., *PT* = 6.96 cm.

Exercise IX, *Pages 34–35*

1. AB = 0.59 in., BC = 0.72 in., CD = 0.92 in., DE = 0.24 in, EF = 0.90 in., AF = 3.37 in.

2. AB = 22.6 mm., BC = 12.0 mm., AC = 34.6 mm.

3. See answers to 1 and 2.

4. 2.3 in., if B and C are on opposite sides, or 0.4 in., if B and C are on the same side, of the diameter through A.

5. DE = 1.4 in., AC = 3.1 in.

6. CZ = 0.92 in.

7. XY = 1.15 in.

8. 2.31 in., 2.08 in.

9. 0.5 in.

Exercise X, *Page 36*

2. 143°.

4. 142°.

5. $A = B = 71°$.

6. $AB = AC$ = 1.55 in.

7. 360°.

8. 360°.

Exercise XI, *Pages 37–38*

1. 32.

2. $11\frac{1°}{4}$.

4. E by S, ESE, SE by E, SE, SE by S, SSE, S by E.

5. WSW, S by W, NE by N.

6. 90°, 45°, $22\frac{1°}{2}$, 135°.

7. 45°; $56\frac{1°}{4}$.

Exercise XII, *Pages 39–40*

1. 4.56 in.

2. 1.46 in.

3. 126 in.

4. 6.5 in.

5. 78 ft. nearly.

6. (*i*) AC = 16 ft.; (*ii*) BC = 40 ft.

7. 9 miles.

8. 10 miles.

9. BC = 36 yd., CD = 40 yd., DA = 75 yd.

10. 4.24 miles, NE.

11. 5.2 miles, 30° N of E.

12. 14.14 miles; 28° E of S.

13. C from A, 10° W of N ; C from B, 20° N of W.

14. C from A, 28° E of N ; C from B, 18° W of N.

15. 3.1 miles, 36° E of S.

16. $17\frac{1}{2}$ seconds.

Exercise XIII(*a*), *Pages 40–41*

1. 73 ft. nearly.

2. 84 ft. nearly.

3. 346 ft.

4. 6° nearly.

5. 4657 ft.

6. 19° nearly.

7. 7075 ft.

8. 39° nearly

Exercise XVI, *Pages 62–63*

2. $A = 90°$, $B = 53°$, $C = 37°$.
3. $A = 41°$, $B = 111°$, $C = 28°$.
4. $BC = 2.15$ in., $B = 85°$, $C = 53°$.
5. $AC = 6.4$ in., $A = 19°$, $C = 26°$.
6. $AB = 4.72$ in., $AC = 5.91$ in., $A = 37°$.
7. $AB = 2.42$ cm., $AC = 4.09$ cm., $B = 84°$.
8. $AC = 10.3$ cm., $BC = 6.2$ cm., $B = 112°$.
9. $b = 2.38$ in., $c = 1.08$ in., $B = 134°$.
10. $a = 2.43$ in., $A = 60°$, $C = 90°$.
11. $a = 8.8$ cm., $A = 117°$, $C = 18°$.
12. $a = 4.5$ cm., $A = 27°$, $C = 18°$.

13. 1.06 in. **14.** 42°. **17.** 1.39 in.

Exercise XVIII(a), *Page 67*

2. 6.6 cm.

Exercise XVIII(b), *Page 68*

In each case, if the diagonals meet in O, and the measurements are taken to the nearest 0.05 in.,

1. $AO = 0.4$ in.,	$BO = 1.25$ in.,	$CO = 1.9$ in.,	$DO = 1.15$ in.
2. $AO = 0.6$ in.,	$BO = 0.9$ in.,	$CO = 0.6$ in.,	$DO = 0.9$ in.
3. $AO = 1.1$ in.,	$BO = 0.6$ in.,	$CO = 0.6$ IN.,	$DO = 1.1$ in.
4. $AO = 0.95$ in.,	$BO = 0.8$ in.,	$CO = 0.2$ in.,	$DO = 1.25$ in.,
or $AO = 1.25$ in.,	$BO = 0.2$ in.,	$CO = 0.8$ in.,	$DO = 0.95$ in.
5. $AO = 1.85$ cm.,	$BO = 2$ cm.,	$CO = 2.35$ cm.,	$DO = 1.3$ cm.,
or $AO = 2.4$ cm.,	$BO = 1.4$ cm.,	$CO = 1.7$ cm.,	$DO = 2$ cm.
6. $AO = 0.7$ cm.,	$BO = 2.8$ cm.,	$CO = 5.1$ cm.,	$DO = 3.2$ cm.
or $AO = 3.2$ cm.,	$BO = 5.2$ cm.,	$CO = 2.6$ cm.,	$DO = 0.8$ cm.
7. $AO = 0.8$ in.,	$BO = 0.2$ in.,	$CO = 0.3$ in.,	$DO = 2.6$ in.
8. $AO = 0.25$ in.,	$BO = 0.65$ in.,	$CO = 0.85$ in.,	$DO = 2.2$ in.

Exercise XIX, *Page 73*

2. $CD = 0.96$ in. ; $EF = 2.02$ in., $GH = 1.39$ in.

Exercise XXII(a), *Page 87*

9. 10.8 ft.

Exercise XXII(b), *Page 88*

4. 1.6 in. **5.** 4.53 in. **6.** 2 in. **7.** 1.58 in.
8. 2.2 in. **9.** 2.58 in. **10.** 1.83 in. **11.** 3.06 in.

Exercise XXIII, *Pages 89–90*

1. The lengths of three edges, none of which are parallel.

2. (*i*) 6 ; (*ii*) rectangular; (*iii*) two faces *a* in. by *b* in., two faces *b* in. by *c* in., two faces *c* in. by *a* in.; (*iv*) they are in parallel planes; (*v*) 3; (*vi*) 12; (*vii*) 4; (*viii*) 3; (*ix*) 8.

3. (*i*) They are all equal; one, coupled with the fact that it is a cube; (*iii*) a square; (*iv*) 12*a* in.; (*v*) by giving the two numbers corresponding to the faces which meet in an edge, or the three numbers corresponding to the faces which meet in a point; thus, the edge 2, 3 or the vertex 4, 5, 6.

6. 8 bricks in four layers of two will form a 9 in. cube.

Exercise XXIV, *Page 91*

1. 4.
2. Tetrahedron.
3. (*i*) 4; (*ii*) 6; (*iii*) 4; (*iv*) triangular.
4. 2.5 in., 2 in., 2.3 in.

Exercise XXVI, *Page 94*

1. Rectangular.
2. Each is a parallelogram.
3. The opposite sides of parallelograms are equal.
4. Right prism with base an equilateral triangle.
7. The lines are in a plane, but do not meet if produced.

Exercise XXVII, *Page 96*

1. They are all equal.
2. Rectangular block.
3. Right regular hexagonal prism.
6. (*i*) circle; (*ii*) rectangle.
8. Rectangle.
9. 9.425 in., 4 in.
10. 39.1 ft.

Exercise XXVIII, *Page 97*

2. Triangle.
3. 216°.
4. 30°.

Unit 2

Exercise XXX(a), *Pages 108–109*

1. 12 sq. in.
2. 3 sq. in.
3. 1.5 sq. in.
4. 0.5 sq. in.
5. 3.6 sq. in.
6. 1.3 sq. in.
7. 3.03 sq. in.
8. 3.03 sq. in.
9. 0.96 sq. in.
10. 3 sq. in.
11. 2.2 sq. in.
12. 0.9 sq. in.
13. 2.55 sq. in,
14. 3 sq. in.
15. 2.94 acres.

Exercise XXX(b), *Page 109*

1. 1.2 sq. in. **2.** 2.31 sq. in.
3. 0.132 sq. in. **4.** 1.732 sq. in.
5. 2.94 sq. in., or 0.96 sq. in. **6.** 4.03 sq. in.
7. 1.17 sq. in., or 0.27 sq. in. **8.** 2.31 sq. in.
9. 1.26 sq. in. **10.** 1.004 sq. in.
11. 0.7728 sq. in. **12.** 1.847 sq. in.
13. 2.34 sq. in.

Exercise XXXI, *Pages 111–112*

1. 2.598 sq. in. **3.** (*i*) 0.123 ac. ; (*ii*) 0.2108 ac.

Exercise XXXIII(b), *Page 118*

5. 0.654 in. **6.** 0.9 in. **7.** 1.09 in. **8.** 4 in.
9. 2.5 in. **10.** 1.7 in. **11.** 4.75 in.
12. c = 1.92 in., or 3.64 in. **13.** 0.84 in.

Exercise XXXVI(a), *Pages 124–125*

2. 4, 1. **6.** 3 in. **7.** 4.8 miles. **8.** 123 ft.
9. $4\frac{1}{2}$ miles. **10.** 16 ft. **11.** $32\frac{1}{2}$ ft. **12.** 65 feet.
13. 32 in. **14.** 14 in. **15.** 2.046 ac. **16.** 7532 yd., 9700yd.

Exercise XXXVI(b), *Pages 125–126*

1. 38 ft. **2.** 1.73 in. **3.** 7 ft. **4.** 6.5 in.
5. 8.5 in. **6.** 2.4 in. **7.** 2 ft. 6 in. **8.** 85.8 sq. ft.
9. 3.9 in.

Exercise XXXVIII, *Pages 133–134*

1. 52 ft. **2.** 10° nearly. **3.** 2.83 in.
4. 0.845 in., 1.81 in. **5.** 0.964 in., 1.15 in.
6. (*i*) 7.3 ft., 41°, 49°; (*ii*) 6.3 yd., 14°, 76°; (*iii*) 2 ch. 80 lk., 58°, 32°; (*iv*) 8.28 in., 2.14 in., 75°; (*v*) 5.14 in., 6.13 in., 40°.
7. 73 ft. nearly. **8.** 35° nearly. **9.** 2830 ft. nearly.
10. 2.8 miles. **11.** 3.464 in. **12.** 135 feet.
13. 2.384 in., 0.728 m., 3.112 in. **14.** 1.324 in.
15. 1.05 in. **16.** 39°. **17.** 19.3 miles, 2.59 miles.
18. 148 ft. **19.** $8\frac{3}{4}$ sec. **20.** 8°. **21.** 2532 ft.

Exercise XXXIX, *Page 136*

1. 3.03 sq. in. **2.** 3.03 sq. in. **3.** 2.31 sq. in. **4.** 1.26 sq. in.
5. 1.73 sq. in. **6.** 1.98 sq. in. **7.** 2.3 sq. in. **8.** 4.43 sq. in.
9. 1.88 sq. in. **10.** 5.87 sq. in. **11.** 14.15 sq. in.

Miscellaneous Exercises, *Pages 136–142*

PAPER I. **6.** 2 in.

PAPER II. **5.** 2.8 in. **6.** $35\frac{1^\circ}{2}$.

PAPER III. **6.** 52 ft.

PAPER IV. **5.** 1.15 in. **6.** 1806 ft.

PAPER V. **5.** AC = 6500 yd., BC = 3520 yd.

PAPER VI. **5.** 1.53 in. **6.** 147 ft.

PAPER VII. **5.** 2 in. **6.** 2.84 miles.

PAPER VIII. **6.** 1.52 in., 1.02 in.

PAPER IX. **6.** $A = 44\frac{1^\circ}{4}$, $B = 60^\circ$, $C = 75\frac{3^\circ}{4}$.

PAPER X. **6.** 90,000 sq. yd.

PAPER XI. **5.** 2.24, 2.65. **6.** $94\frac{1}{4}$ ft.

PAPER XII. **2.** 150 sq. in., $37\frac{1}{2}$ sq. in., $37\frac{1}{2}$ sq. in., $12\frac{1}{2}$ sq. in., $62\frac{1}{2}$ sq. in.
6. 10.6 ft./sec.

UNIT 3

Exercise XLI(b), *Pages 149–150*

9. 8 in. **10.** 0.4 in. **11.** 130 yd.

Exercise XLIV(b), *Pages 165–166*

1. 0.87 in. **2.** 2.37 in. **3.** 21.9 cm.
4. 1.6 in. **5.** 1.05 in. **6.** 2.4 in.

Exercise XLV(b), *Pages 170–171*

1. 5.6 in. **2.** 60°. **3.** 2.15 in. **6.** 0.95 in.
7. 4.5 cm. **8.** 2.4 in., 1.2 in. **9.** $14\frac{1^\circ}{2}$, $44\frac{1^\circ}{2}$.

Exercise XLVI(b), *Page 176*

1. 0.8 in., 0.4 in. **2.** 0.4 in., 1.4 in. **5.** 1 in., 6 in. 3.60 in.

Exercise XLIX(a), *Pages 186–187*

9. The centre is an arc *ACB*, where it is cut by the perpendicular bisector of *AB*.

11. The centre is the diametrically opposite point to that in Ex. 9.

Exercise XLIX(b), *Page 188*

1. 2.3 in. **2.** 1 in. **3.** 3.96 in. **4.** 2.66 in.
5. 3.60 in. **6.** 3.94 in.
7. (*i*) 2.6 in., 1.6 in.; (*ii*) 3.01 in., 1.01 in.; (*iii*) 2.62 in., 2.07 in.
8. 1.02 in.

Exercise L, *Pages 190–191*

1. 1.73 in. **2.** 3.46 in. **4.** 2.82 in., 5.81 in.
5. 1.84 in., 3.75 in. **6.** 2.95 in.
7. 3.14 in. **13.** 0.7 in.

Exercise LI, *Pages 197–198*

1. 1.15 in. **6.** 1.18 in. **7.** 0.85 in. **8.** 3 in.
9. 3.63 sq. in.
11. (*i*) 1.7 in., 1.38 in.; (*ii*) 2.30 in., 2.08 in.
15. (*i*) 9.51 sq. in.; (*ii*) 13.86 sq. in.; (*iii*) 14.54 sq, in.

Miscellaneous Exercises, *Pages 200–204*

PAPER XIII.	**5.** 2.83 in.	**6.** 26 ft.
PAPER XIV.	**2.** 30°.	**6.** 1.41 in.
PAPER XV.	**6.** 2.2 in.	
PAPER XVI.	**6.** 4.2 cm., 21 cm.	
PAPER XVII.	**6.** 2.6 in., 2.8 in.	
PAPER XVIII.	**6.** 3 in.	
PAPER XIX.	**5.** 2.35 in., 1.65 in.	**6.** 1.2 in.

UNIT 4

Exercise LII, *Page 213*

12. 2 : 1.

Exercise LIII(b), *Pages 217–218*

1. 1.41 in. **2.** 1.32 in. **3.** 1.41 in. **4.** 0.92 in.
5. 0.92 in. **6.** 1.73 in. **7.** 1.45 in. **8.** 1.61 in.
9. 2.62 in. **10.** 1.56 in.
11. (*a*) 1.05 in.; (*b*) 2.12 in.; (*c*) 2.45 in.; (*d*) 2.63 in.

Exercise LIV, *Pages 223–224*

1. (*i*) Obtuse-angled; (*ii*) Acute-angled.
2. (*i*) 5, 12, 24; (*ii*) 5, 12, 126. **3.** (*i*) 36 ; (*ii*) 234
4. (*i*) 3.5 ; (*ii*) $3\frac{1}{3}$. **5.** (*i*) 40 ; (*ii*) 44
6. (*i*) $25\frac{5}{8}$; (*ii*) $18\frac{1}{8}$. **7.** 21, 26.4, 12.32.
8. (*i*) 29°, 47°, 104°; (*ii*) 49°, 59°, 72°
10. (*i*) 84; (*ii*) 660. **11.** 516.

Exercise LVI, *Pages 226–227*

1. (*i*) 7; (*ii*) 9. **2.** 22. **3.** $2\frac{1}{3}$. **4.** $82\frac{1}{2}$.

Exercise LVII, *Pages 232–233*

11. A must be greater than a right angle.

Exercise LVIII, *Pages 236–238*

1. (*i*) $\frac{1}{12}$; (*ii*) $\frac{1}{36}$; (*iii*) $\frac{1}{792}$; (*iv*) $\frac{1}{63360}$; (*v*) $\frac{1}{10560}$; (*vi*) $\frac{1}{2534.4}$
2. $4\frac{6}{11}$ in. = 4.55 in. **3.** 1.46 in., *R.F.* = $\frac{1}{288}$.
4. 7.5 in. **5.** 6.625 m. **6.** $\frac{1}{452571\frac{3}{7}}$.
7. A case of relative error. **8.** 38.6 yd.
14. 4 yd. 1 ft. 10 in.; 3 yd. 0 ft. 7 in.; 1 yd. 2 ft. 1 in.
19. The length of the full scale is 3.1 in. nearly.

Miscellaneous Exercises, *Pages 238–241*

PAPER XXI. **5.** A circle on *XY* as diameter.
PAPER XXII. **6.** 3.31 in., 1.81 in.
PAPER XXIII. **6.** 7 in., 8 in.
PAPER XXIV. **6.** 1.62 in.
PAPER XXV. **6.** About 3600 ac.

UNIT 5

Exercise LX, *Pages 258–259*

4(*v*). $CY : YA = 15 : 8$; $AZ : ZB = 4 : 5$.

Exercise LXII, *Page 268*

5. $BO : OY = \frac{m}{n} + \frac{s}{r} : 1$; $CO : OZ = \frac{p}{q} + \frac{n}{m} : 1$.

Exercise LXIII(a), *Pages 268–269*

11. 1.16 in., 3.91 in.

12. $3\frac{3}{7}$ sq. in.

Exercise LXX, *Pages 301–302*

16. The required point is the centre of similitude of the line segments *AB*, *AC*.

Exercise LXXI, *Page 304*

1. 1.58 in. **2.** 1.75 in. **3.** 0.93 in. **4.** 2.20 in.
5. 1.54 in. **6.** 1.06 in. **7.** 0.97 in. **8.** 0.93 in.

Exercise LXXII, *Pages 306–308*

9. 0.79 in. **11.** (*i*) 1.27 in.; (*ii*) 1.06 in.; (*iii*) 2.20 in.
12. 3.38 in. **15.** 1.25 in.

Miscellaneous Exercises, *Pages 321–328*

Paper XXVI.	**6.** $\frac{3}{4}$; $\frac{1}{2}$; $\frac{1}{4}$.
Paper XXVII.	**6.** 3.94 in.
Paper XXVIII.	**6.** 3.56 or 5.26.
Paper XXIX.	**6.** 2 in.
Paper XXX.	**6.** 2.76 in.
Paper XXXI.	**6.** 1.2 in.
Paper XXXII.	**6.** 1.3 in.
Paper XXXIII.	**6.** 2.58 in.
Paper XXXV.	**6.** 2.31 in., 2.86 in.

UNIT 6

Exercise LXXIX, *Pages 358–359*

1. 0.65 in. **2.** 0.59 in. **3.** 0.55 in. **4.** 0.67 in.
5. 0.62 in. **6.** (*i*) 0.5 in.; (*ii*) 1.5 in. **7.** 0.55 in.
8. 0.20 in.

Miscellaneous Exercises, *Pages 389, 392–394, 398–399*

PAPER XXXVII. **5.** 63 in., 56 in., $32\frac{1}{2}$ in., 1764 sq. in.

6. 0.66 in.

PAPER XLII. **4.** 3 in.; $\frac{1}{2}$ in.; between $\sqrt{6}$ in. and $\sqrt{10}$ in.

PAPER XLIII. **5.** 1.25 in.

PAPER XLIX. **6.** 1.15 in.

Index